T0189458

Big Data Analytics in Genomics

Ka-Chun Wong

# Big Data Analytics in Genomics

Ka-Chun Wong
Department of Computer Science
City University of Hong Kong
Kowloon Tong, Hong Kong

ISBN 978-3-319-82312-6 ISBN 978-3-319-41279-5 (eBook)
DOI 10.1007/978-3-319-41279-5


Softcover reprint of the hardcover 1st edition 2016



Printed on acid-free paper

This Springer imprint is published by Springer Nature
The registered company is Springer International Publishing AG Switzerland

# Preface

At the beginning of the 21st century, next-generation sequencing (NGS) and third-generation sequencing (TGS) technologies have enabled high-throughput sequencing data generation for genomics; international projects (e.g., the Encyclopedia of DNA Elements (ENCODE) Consortium, the 1000 Genomes Project, The Cancer Genome Atlas (TCGA), Genotype-Tissue Expression (GTEx) program, and the Functional Annotation Of Mammalian genome (FANTOM) project) have been successfully launched, leading to massive genomic data accumulation at an unprecedentedly fast pace.

To reveal novel genomic insights from those big data within a reasonable time frame, traditional data analysis methods may not be sufficient and scalable. Therefore, big data analytics have to be developed for genomics.

As an attempt to summarize the current efforts in big data analytics for genomics, an open book chapter call is made at the end of 2015, resulting in 40 book chapter submissions which have gone through rigorous single-blind review process. After the initial screening and hundreds of reviewer invitations, the authors of each eligible book chapter submission have received at least 2 anonymous expert reviews (at most, 6 reviews) for improvements, resulting in the current 13 book chapters.

Those book chapters are organized into three parts ("Statistical Analytics," "Computational Analytics," and "Cancer Analytics") in the spirit that statistics form the basis for computation which leads to cancer genome analytics. In each part, the book chapters have been arranged from general introduction to advanced topics/specific applications/specific cancer sequentially, for the interests of readership.

In the first part on statistical analytics, four book chapters (Chaps. 1–4) have been contributed. In Chap. 1, Yang et al. have compiled a statistical introduction for the integrative analysis of genomic data. After that, we go deep into the statistical methodology of expression quantitative trait loci (eQTL) mapping in Chap. 2 written by Cheng et al. Given the genomic variants mapped, Ribeiro et al. have contributed a book chapter on how to integrate and organize those genomic variants into genotype-phenotype networks using causal inference and structure learning in Chap. 3. At the end of the first part, Li and Tong have given a refreshing statistical perspective on genomic applications of the Neyman-Pearson classification paradigm in Chap. 4.

In the second part on computational analytics, four book chapters (Chaps. 5–8) have been contributed. In Chap. 5, Gupta et al. have reviewed and improved the existing computational pipelines for re-annotating eukaryotic genomes. In Chap. 6, Rucci et al. have compiled a comprehensive survey on the computational acceleration of Smith-Waterman protein sequence database search which is still central to genome research. Based on those sequence database search techniques, protein function prediction methods have been developed and demonstrated promising. Therefore, the recent algorithmic developments, remaining challenges, and prospects for future research in protein function prediction are discussed in great details by Shehu et al. in Chap. 7. At the end of the part, Nagarajan and Prabhu provided a review on the computational pipelines for epigenetics in Chap. 8.

In the third part on cancer analytics, five chapters (Chaps. 9–13) have been contributed. At the beginning, Prabahar and Swaminathan have written a reader-friendly perspective on machine learning techniques in cancer analytics in Chap. 9. To provide solid supports for the perspective, Tong and Li summarize the existing resources, tools, and algorithms for therapeutic biomarker discovery for cancer analytics in Chap.10. The NGS analysis of somatic mutations in cancer genomes are then discussed by Prieto et al. in Chap. 11. To consolidate the cancer analytics part further, two computational pipelines for cancer analytics are described in the last two chapters, demonstrating concrete examples for reader interests. In Chap. 12, Leung et al. have proposed and described a novel pipeline for statistical analysis of exonic variants in cancer genomes. In Chap. 13, Yotsukura et al. have proposed and described a unique pipeline for understanding genotype-phenotype correlation in breast cancer genomes.

Kowloon Tong, Hong Kong Ka-Chun Wong
April 2016

# Contents

## Part III Cancer Analytics

# Part I
# Statistical Analytics

# Introduction to Statistical Methods for Integrative Data Analysis in Genome-Wide Association Studies

**Can Yang, Xiang Wan, Jin Liu, and Michael Ng**

**Abstract** Scientists in the life science field have long been seeking genetic variants associated with complex phenotypes to advance our understanding of complex genetic disorders. In the past decade, genome-wide association studies (GWASs) have been used to identify many thousands of genetic variants, each associated with at least one complex phenotype. Despite these successes, there is one major challenge towards fully characterizing the biological mechanism of complex diseases. It has been long hypothesized that many complex diseases are driven by the combined effect of many genetic variants, formally known as "polygenicity," each of which may only have a small effect. To identify these genetic variants, large sample sizes are required but meeting such a requirement is usually beyond the capacity of a single GWAS. As the era of big data is coming, many genomic consortia are generating an enormous amount of data to characterize the functional roles of genetic variants and these data are widely available to the public. Integrating rich genomic data to deepen our understanding of genetic architecture calls for statistically rigorous methods in the big-genomic-data analysis. In this book chapter, we present a brief introduction to recent progresses on the development of statistical methodology for integrating genomic data. Our introduction begins with the discovery of polygenic genetic architecture, and aims at providing a unified statistical framework of integrative analysis. In particular, we highlight the

C. Yang (✉) • M. Ng
Department of Mathematics, Hong Kong Baptist University, Kowloon Tong, Hong Kong
e-mail: eeyang@hkbu.edu.hk; mng@math.hkbu.edu.hk

X. Wan
Department of Computer Science, Hong Kong Baptist University, Kowloon Tong, Hong Kong
e-mail: xwan@comp.hkbu.edu.hk

J. Liu
Center of Quantitative Medicine, Duke-NUS Graduate Medical School, Singapore, Singapore
e-mail: jin.liu@duke-nus.edu.sg



K.-C. Wong (ed.), *Big Data Analytics in Genomics*,
DOI 10.1007/978-3-319-41279-5_1

importance of integrative analysis of multiple GWAS and functional information. We believe that statistically rigorous integrative analysis can offer more biologically interpretable inference and drive new scientific insights.

**Keywords** Statistics • SNP • Population genetics • Methodology • Genomic data

# 1 Introduction

Genome-wide association studies (GWAS) aim at studying the role of genetic variations in complex human phenotypes (including quantitative traits and qualitative diseases) by genotyping a dense set of single-nucleotide polymorphisms (SNPs) across the whole genome. Compared with the candidate-gene approaches which only consider some regions chosen based on researcher's experience, GWAS are intended to provide an unbiased examination of the genetic risk variants [46]. In 2005, the identification of the complement factor H for age-related macular degeneration in a small sample set (96 cases v.s. 50 controls) was the first successful example of searching for risk genes under the GWAS paradigm [31]. It was a milestone moment in the genetics community, and this result convinced researchers that GWAS paradigm would be powerful even with such a small sample size. Since then, an increasing number of GWAS have been conducted each year and significant risk variants have been routinely reported. As of December, 2015, more than 15,000 risk genetic variants have been associated with at least one complex phenotypes at the genome-wide significance level ($p$-value$< 5 \times 10^{-8}$) [61].

Despite the accumulating discoveries from GWAS, researchers found out that the significantly associated variants only explained a small proportion of the genetic contribution to the phenotypes in 2009 [42]. This is the so-called missing heritability. For example, it is widely agreed that 70–80 % of variations in human height can be attributed to genetics based on pedigree study while the significant hits from GWAS can only explain less than 5–10 % of the height variance [1, 42]. In 2010, the seminal work of Yang et al. [66] showed that 45 % of variance in human height can be explained by 294,831 common SNPs using a linear mixed model (LMM)-based approach. This result implies that there exist a large number of SNPs jointly contributing a substantial heritability on human height but their individual effects are too small to pass the genome-wide significance level due to the limited sample size. They further provided evidence that the remaining heritability on human height (the gap between 45 % estimated from GWAS and 70–80 % estimated from pedigree studies) might be due to the incomplete linkage disequilibrium (LD) between causal variants and SNPs genotyped in GWAS. Researchers have applied this LMM approach to many other complex phenotypes, e.g., metabolic syndrome traits [56] and psychiatric disorders [11, 34]. These results suggest that complex phenotypes are often highly polygenic, i.e., they are affected by many genetic variants with small effects rather than just a few variants with large effects [57].

The polygenicity of complex phenotypes has many important implications on the development of statistical methodology for genetic data analysis. First, the methods relying on "extremely sparse and large effects" may not work well because the sum of many small effects, which is non-negligible, has not been taken into account. Second, it is often challenging to pinpoint those variants with small effects only based on information from GWAS. Fortunately, an enormous amount of data from different perspectives to characterize human genome is being generated and much richer than ever. This motivates us to search for relevant information beyond GWAS (indirect evidence) and combine it with GWAS signals (direct evidence) to make more convincing inference [15]. However, it is not an easy task to integrate indirect evidence with direct evidence. A major challenge in integrative analysis is that the direct evidence and indirect evidence are often obtained from different data sources (e.g., different sample cohorts, different experimental designs). A naive combination may potentially lead to high false positive findings and misleading interpretation. Yet, effective methods that combine indirect evidence with direct evidence are still lacking [23]. In this book chapter, we offer an introduction to the statistical methods for integrative analysis of genomic data, and highlight their importance in the big genomic data era.

To provide a bird's-eye view of integrative analysis of genomic data, we start with the introduction of heritability estimation because heritability serves as a fundamental concept which quantifies the genetic contribution to a phenotype [58]. A good understanding of heritability estimation offers valuable insights of the polygenic architecture of complex phenotypes. From a statistical point of view, it is the polygenicity that motivates integrative analysis of genomic data such that more genetic variants with small effects can be identified robustly. Our discussion of the statistical methods for integrative analysis will be divided into two sections: integrative analysis of multiple GWAS and integrative analysis of GWAS with genomic functional information. Then we demonstrate how to integrate multiple GWAS and functional information simultaneously in the case study section. At the end, we summarize this chapter with some discussions about the future directions of this area.

## 2 Heritability Estimation

The theoretical foundation of heritability estimation can be traced back to R. A. Fisher's development [20], in which the phenotypic similarity between relatives is related to the degrees of genetic resemblance. In quantitative genetics, the phenotypic value ($P$) is modeled as the sum of genetic effects ($G$) and environmental effects ($E$),

$$P = \mu + G + E, \tag{1}$$

where $\mu$ is the population mean of the phenotype. To keep our introduction simple, $G$ and $E$ are assumed to be independent, i.e., $\text{Cov}(G, E) = 0$. The genetic effect can be further decomposed into the additive effect (also known as the breeding value), the dominance effect and the interaction effect, $G = A + D + I$. Accordingly, the phenotype variance can be decomposed as

$$\sigma_P^2 = \sigma_G^2 + \sigma_E^2 = (\sigma_A^2 + \sigma_D^2 + \sigma_I^2) + \sigma_E^2, \tag{2}$$

where $\sigma_G^2$ is the variance due to genetic variations, $\sigma_A^2, \sigma_D^2, \sigma_I^2$, and $\sigma_E^2$ correspond to the variance of additive effects, dominance effects, interaction effects (also known as epistasis), and environmental effects, respectively. Based on these variance components, two types of heritability are defined. The broad-sense heritability ($H^2$) is defined as the proportion of the phenotypic variance that can be attributed to the genetic factors,

$$H^2 = \frac{\sigma_G^2}{\sigma_P^2} = \frac{\sigma_A^2 + \sigma_D^2 + \sigma_I^2}{\sigma_A^2 + \sigma_D^2 + \sigma_I^2 + \sigma_E^2}. \tag{3}$$

The narrow-sense heritability ($h^2$), however, focuses only on the contribution of the additive effects:

$$h^2 = \frac{\sigma_A^2}{\sigma_A^2 + \sigma_E^2}. \tag{4}$$

Due to the law of inheritance, individuals can only transmit one allele of each gene to their offsprings, most relatives (except full siblings and monozygotic twins) share only one allele or no allele that is identical by descent (IBD). Therefore, the dominance effects and interaction effects will not contribute to their genetic resemblance as these effects are due to the sharing two IBD alleles. Accumulating evidence suggests that non-additive genetic effects on complex phenotypes may be negligible [28, 64, 69]. For example, Yang et al. [64] reported that the additive effects of about 17 million imputed variants explained 56 % variance of human height, leaving a very small space for the non-additive effects to contribute. Zhu et al. [69] found the dominance effects on 79 quantitative traits explained little phenotypic variance. Therefore, we will ignore non-additive effects and concentrate our discussion on narrow-sense heritability in this book chapter.

### 2.1 *The Basic Idea of Heritability Estimation from Pedigree Data*

In this section, we will introduce the key idea of heritability estimation from pedigree data, which provides the basis of our discussion on integrative analysis. Interested readers are referred to [18, 27, 40, 59] for the comprehensive discussion of this issue. Assuming a number of conditions (e.g., random mating, no inbreeding, Hardy–Weinberg equilibrium, and linkage equilibrium), a simple formula for the genetic covariance between two relatives can be derived based on the additive variance component:

$$\mathrm{Cov}(G_1, G_2) = K_{1,2}\sigma_A^2, \tag{5}$$

where $K_{1,2}$ is the expected proportion of their genomes sharing one chromosome IBD. Let us take a parent–offspring pair as an example. Because the parent transmits one copy of each gene to his/her offspring, i.e., $K_{1,2} = \frac{1}{2}$, thus their genetic covariance is $\frac{1}{2}\sigma_A^2$. Let $P_1$ and $P_2$ be the phenotypic values (e.g., height) of the parent and the offspring. Based on (1), we have $\mathrm{Cov}(P_1, P_2) = \mathrm{Cov}(G_1, G_2) + \mathrm{Cov}(E_1, E_2)$. Assuming the independence of the environmental factor, $\mathrm{Cov}(E_1, E_2) = 0$, we further have

$$\mathrm{Cov}(P_1, P_2) = \frac{1}{2}\sigma_A^2. \tag{6}$$

Noticing that $\mathrm{Var}(P_1) = \mathrm{Var}(P_2) = \sigma_P^2 = \sigma_A^2 + \sigma_E^2$, the phenotypic correlation can be related to the narrow-sense heritability $h^2$:

$$\mathrm{Corr}(P_1, P_2) = \frac{\mathrm{Cov}(P_1, P_2)}{\sqrt{\mathrm{Var}(P_1)\mathrm{Var}(P_2)}} = \frac{1}{2}\frac{\sigma_A^2}{\sigma_A^2 + \sigma_E^2} = \frac{1}{2}h^2. \tag{7}$$

Suppose we have collected the phenotypic values of $n$ parent–offspring pairs. A simple way to estimate $h^2$ based on this data set is to use the linear regression:

$$P_{i2} = P_{i1}\beta + \beta_0 + \epsilon_i, \tag{8}$$

where $i = 1, \ldots, n$ is the index of samples, $\beta$ is the regression coefficient, and $\epsilon_i$ is the residual of the $i$th sample. The ordinary least square estimate of $\beta$ is

$$\hat{\beta} = \frac{\sum_i (P_{i2} - \bar{P}_2)(P_{i1} - \bar{P}_1)}{\sum_i (P_{i2} - \bar{P}_2)^2}, \ \hat{\beta}_0 = \bar{P}_1 - \hat{\beta}_1 \bar{P}_2, \tag{9}$$

where $\bar{P}_1 = \frac{1}{n}\sum_i P_{i1}$ and $\bar{P}_2 = \frac{1}{n}\sum_i P_{i2}$ are the sample means of parent phenotypic values and offspring phenotypic values. Because $\hat{\beta}$ is the sample version of the correlation given in (7), heritability estimated from parent–offspring pairs is given by twice of the regression slope, i.e., $\hat{h}^2 = 2\hat{\beta}$.

Another example of heritability estimation is based on the phenotypic values of two parents ($P_1$ and $P_2$) and one offspring ($P_3$). Let $P_M = \frac{P_1+P_2}{2}$ be the phenotypic value of the mid-parent. Similarly, we have the genetic covariance $\mathrm{Cov}(P_M, P_3) = \frac{1}{2}\mathrm{Cov}(P_1, P_3) + \frac{1}{2}\mathrm{Cov}(P_2, P_3) = \frac{1}{2}\sigma_A^2$, and correlation between the mid-parent and the offspring can be related to heritability $h^2$ as

$$\mathrm{Corr}(P_M, P_3) = \frac{\mathrm{Cov}(P_M, P_3)}{\sqrt{\mathrm{Var}(P_M)\mathrm{Var}(P_3)}} = \frac{\frac{1}{2}\sigma_A^2}{\sqrt{\frac{1}{2}}(\sigma_A^2 + \sigma_E^2)} = \sqrt{\frac{1}{2}}h^2. \tag{10}$$

Suppose we have $n$ trio samples $\{P_{i1}, P_{i2}, P_{i3}\}$, where $(P_{i1}, P_{i2}, P_{i3})$ corresponds to the phenotypic values of two parents and the offspring from the $i$th sample. Again, a convenient way to estimate $h^2$ is to still use linear regression:

$$P_{i3} = \frac{P_{i1} + P_{i2}}{2}\beta + \beta_0 + \epsilon_i. \tag{11}$$

Heritability estimated from the phenotypic values of mid-parents and offsprings can be read from the coefficient fitted in (11) as $\hat{h}^2 = \hat{\beta} = \widehat{\text{Var}(P_M)}^{-1}\widehat{\text{Cov}(P_M, P_3)}$.

It is worth pointing out that the above methods for heritability estimation only make use of covariance information. In statistics, they are referred to as the methods of moments because covariance is the second moment. In fact, we can impose normality assumptions and reformulate heritability estimation using maximum likelihood estimator. Considering the parent–offspring case, we can view all the samples independently drawn from the following distribution:

$$\begin{pmatrix} P_{i1} \\ P_{i2} \end{pmatrix} \sim \mathcal{N}\left[\begin{pmatrix} \mu \\ \mu \end{pmatrix}, \begin{pmatrix} 1 & \frac{1}{2} \\ \frac{1}{2} & 1 \end{pmatrix}\sigma_A^2 + \begin{pmatrix} 1 & 0 \\ 0 & 1 \end{pmatrix}\sigma_E^2\right], \tag{12}$$

where $P_{i1}$ and $P_{i2}$ are the phenotypic values of the parent and offspring from the $i$th family. Similarly, we can view a trio sample $P_{i1}, P_{i2}, P_{i3}$ independently drawn from the following distribution:

$$\begin{pmatrix} P_{i1} \\ P_{i2} \\ P_{i3} \end{pmatrix} \sim \mathcal{N}\left[\begin{pmatrix} \mu \\ \mu \\ \mu \end{pmatrix}, \begin{pmatrix} 1 & 0 & \frac{1}{2} \\ 0 & 1 & \frac{1}{2} \\ \frac{1}{2} & \frac{1}{2} & 1 \end{pmatrix}\sigma_A^2 + \begin{pmatrix} 1 & 0 & 0 \\ 0 & 1 & 0 \\ 0 & 0 & 1 \end{pmatrix}\sigma_E^2\right]. \tag{13}$$

The restricted maximum likelihood (REML) approach can be used to efficiently compute the estimates of model parameters $\{\mu, \sigma_A^2, \sigma_E^2\}$ in (12) and (13). Then the heritability estimation can be obtained as

$$\hat{h}^2 = \frac{\hat{\sigma}_A^2}{\hat{\sigma}_A^2 + \hat{\sigma}_E^2}. \tag{14}$$

The matrices $\begin{pmatrix} 1 & \frac{1}{2} \\ \frac{1}{2} & 1 \end{pmatrix}$ and $\begin{pmatrix} 1 & 0 & \frac{1}{2} \\ 0 & 1 & \frac{1}{2} \\ \frac{1}{2} & \frac{1}{2} & 1 \end{pmatrix}$ in (12) and (13) can be considered as expected genetic similarity (i.e., expected genome sharing) in parent–offspring samples and two-parent–offspring samples. As a result, heritability estimation based on pedigree data relates the phenotypic similarity of relatives to their *expected* genome sharing.

## 2.2 Heritability Estimation Based on GWAS

As we discussed above, the heritability estimation based on pedigree data relies on the expected genome sharing between relatives. Nowadays, genome-wide dense SNP data provides an unprecedented opportunity to accurately characterize genome sharing. However, this advantage brings new challenges. First, three billion base pairs of human genome sequences are identical at more than 99.9 % of the sites due to the inheritance from the common ancestors. SNP-based data only records genotypes at some specific genome positions with single-nucleotide mutations, and thus SNP-based measures of genetic similarity are much lower than the 99.9 % similarity based on the whole genome DNA sequence. Second, SNP-based measures depend on the subset of SNPs genotyped in GWAS and their allele frequencies. Third, SNP-based measure can be affected by the quality control procedures used in GWAS.

Our discussion assumes that the SNPs used in heritability estimation are fixed. There are many different ways to characterize genome similarity based on these fixed SNPs, as discussed in [51]. Here, we choose the GCTA approach [66, 67] as it is the most widely used one. Suppose we have collected the genotypes of $n$ subjects in matrix $\mathbf{G} = [g_{im}] \in R^{n \times M}$ and their phenotype in vector $y \in R^{n \times 1}$, where $M$ is the number of SNP markers and $g_{im} \in \{0, 1, 2\}$ is the numerical coding of the genotypes at the $m$th SNP of the $i$th individual. Yang et al. [66, 67] proposed to standardize the genotype matrix $\mathbf{G}$ as follows:

$$w_{im} = \frac{(g_{im} - f_m)}{\sqrt{2f_m(1 - f_m)M}}, \tag{15}$$

where $f_m$ is the frequency of the reference allele. An underlying assumption in this standardization is that lower frequency variants tend to have larger effects. Speed et al. [52] examined this assumption and concluded that it would be robust in both simulation studies and real data analysis. After standardization, an LMM is used to model the relationship between the phenotypic value and the genotypes:

$$\begin{aligned} \mathbf{y} &= \mathbf{X}\boldsymbol{\beta} + \mathbf{W}\mathbf{u} + \mathbf{e}, \\ \mathbf{u} &\sim \mathcal{N}(\mathbf{0}, \sigma_u^2\mathbf{I}), \\ \mathbf{e} &\sim \mathcal{N}(\mathbf{0}, \sigma_e^2\mathbf{I}), \end{aligned} \tag{16}$$

where $\mathbf{X} \in R^{n \times c}$ is the fixed-effect design matrix collecting the intercept of the regression model and all covariates, such as age, sex, and a few principal components (PC) of the genotype data (PCs are used for adjustment of the population structure [45]); $\boldsymbol{\beta}$ is the vector of fixed effects; $\mathbf{u}$ collects all the individual SNP effects which are considered as random, and $\mathbf{e}$ collects the random errors due to the environmental factors. Since both $\mathbf{u}$ and $\mathbf{e}$ are Gaussian, they can be integrated out analytically, which yields the marginal distribution of $\mathbf{y}$:

$$\mathbf{y} \sim \mathcal{N}(\mathbf{X}\beta, \mathbf{W}\mathbf{W}^T\sigma_u^2 + \sigma_e^2 I), \tag{17}$$

Efficient algorithms, such as AI-REML[25] and expectation-maximization (EM) algorithms [43], are available for estimating model parameters. Let $\{\hat{\beta}, \hat{\sigma}_u^2, \hat{\sigma}_e^2\}$ be the REML estimates. Then heritability can be estimated as

$$\hat{h}_g^2 = \frac{\hat{\sigma}_u^2}{\hat{\sigma}_u^2 + \hat{\sigma}_e^2}, \tag{18}$$

where $\hat{h}_g^2$ is called chip heritability because it depends on the SNPs genotyped from chip. Since the genotyped SNPs only form a subset of all SNPs in the human genome, the chip heritability should be smaller than the narrow-sense heritability, i.e., $h_g^2 \leq h^2$. One can compare (17) with (12) and (13) to get some intuitive understandings. The matrix $\mathbf{WW}^T$ can be regarded as the genetic similarity measured by the SNP data, which is the so-called genetic relatedness matrix (GRM). In this sense, heritability estimation based on GWAS data makes use of the realized genome similarity rather than the expected genome sharing in pedigree data analysis.

Although the idea of heritability estimation based on pedigree data and GWAS data looks similar, there is an important difference. The chip heritability can be largely inflated in presence of cryptical relatedness. Let us briefly discuss this issue so that readers can gain more insights on chip heritability estimation. Notice that chip heritability relies on GRM calculated using genotyped SNPs. However, this does not mean that GRM only captures information from genotyped SNPs because there exists linkage disequilibrium (LD, i.e., correlation) among genotyped SNPs and un-genotyped SNPs. In this situation, GRM indeed "sees" the un-genotyped SNPs partially due to the imperfect LD. Suppose a GWAS data set is comprised of many unrelated samples and a few relatives, which is ready for the chip heritability estimation. Consider an extreme case that there is a pair of identical twins whose genomes will be the same ideally. Thus, their genotyped SNPs can capture more information from their un-genotyped SNPs because their chromosomes are highly correlated. For unrelated individuals, however, their chromosomes can be expected to be nearly uncorrelated such that their genotyped SNPs capture less information from the un-genotyped SNPs. As a result, the chip heritability estimation will be inflated even though a few relatives are included. To avoid the inflation due to the cryptical relatedness, Yang et al. [66, 67] advocated to use samples that are less related than the second degree relative.

The GCTA approach has been widely used to explore the genetic architecture of complex phenotypes besides human height. For example, SNPs at the genome-wide significant level can explain little heritability of psychiatric disorders (e.g., schizophrenia and bipolar disorders (BPD)) but all genotyped SNPs can explain a substantial proportion [11, 34], which implies the polygenicity of these psychiatric disorders. Polygenic architectures have been reported for some other complex phenotypes [57], such as metabolic syndrome traits [56] and alcohol dependence [62].

From the statistical point of view, a remaining issue is whether the statistical estimate can be done efficiently using unrelated samples, where sample size $n$ is much smaller than the number of SNPs $M$. This is about whether variance component estimation can be done in the high dimensional setting. The problem is challenging because all the SNPs are included for heritability estimation but most of them are believed to be irrelevant to the phenotype of interest. In other words, the GCTA approach assumed the nonzero effects of all genotyped SNPs in LMM, leading to misspecified LMM when most of the included SNPs have no effects. Recently, a theoretical study [30] has showed that the REML estimator in the misspecified LMM is still consistent under some regularity conditions, which provides a justification of the GCTA approach. Heritability estimation is still a hot research topic. For more detailed discussion, interested readers are referred to [13, 26, 32, 68].

## 3 Integrative Analysis of Multiple GWAS

In this section, we will introduce the statistical methods for integrative analysis of multiple GWAS of different phenotypes, which is motivated from both biological and statistical perspectives. The biological basis to perform integrative analysis is the fact that a single locus can affect multiple seemly unrelated phenotypes, which is known as "pleiotropy" [53]. Recently, an increasing number of reports have indicated abundant pleiotropy among complex phenotypes [49, 50]. Examples include *TERT-CLPTM1L* associated with both bladder and lung cancers [21] and *PTPN22* associated with multiple auto-immune disorders [10]. On the other hand, polygenicity imposes great statistical challenges in identification of weak genetic effects. The existence of pleiotropy allows us to combine information from multiple seemingly unrelated phenotypes. Indeed, recent discoveries along this line are fruitful [63], e.g., the discovery of pleiotropic loci affecting multiple psychiatric disorders [12] and the identification of pleiotropy between schizophrenia and immune disorders [48, 60].

Before we proceed, we first introduce a concept closely related to pleiotropy—genetic correlation (denoted as $\rho$; also known as co-heritability) [11]. Let us consider GWAS of two distinct phenotypes without overlapped samples. Denote the phenotypes and standardized genotype matrices as $\mathbf{y}^{(k)} \in R^{n_k \times 1}$ and $\mathbf{W}^{(k)} \in R^{n_k \times M}$, respectively, where $M$ is the total number of genotyped SNPs and $n_k$ is the sample size of the $k$th GWAS, $k = 1, 2$. Bivariate LMM can be written as follows:

$$\mathbf{y}^{(1)} = \mathbf{X}^{(1)}\boldsymbol{\beta}^{(1)} + \mathbf{W}^{(1)}\mathbf{u}^{(1)} + \mathbf{e}^{(1)}, \tag{19}$$

$$\mathbf{y}^{(2)} = \mathbf{X}^{(2)}\boldsymbol{\beta}^{(2)} + \mathbf{W}^{(2)}\mathbf{u}^{(2)} + \mathbf{e}^{(2)}, \tag{20}$$

where $\mathbf{X}^{(k)}$ collects all the covariates of the $k$th GWAS and $\boldsymbol{\beta}^{(k)}$ is the corresponding fixed effects, $\mathbf{u}^{(k)}$ is the vector of random effects for genotyped SNPs in $\mathbf{W}^{(k)}$ and

$\mathbf{e}^{(k)}$ is the independent noise due to environment. Denote the $m$th element of $\mathbf{u}^{(1)}$ and $\mathbf{u}^{(2)}$ as $u_m^{(1)}$ and $u_m^{(2)}$, respectively. In bivariate LMM, $[u_m^{(1)}, u_m^{(2)}]^T$ $(m = 1, \ldots, M)$ are assumed to be independently drawn from the bivariate normal distribution:

$$\begin{bmatrix} u_m^{(1)} \\ u_m^{(2)} \end{bmatrix} \sim \mathcal{N}(\begin{bmatrix} 0 \\ 0 \end{bmatrix}, \begin{bmatrix} \sigma_1^2 & \rho\sigma_1\sigma_2 \\ \rho\sigma_1\sigma_2 & \sigma_2^2 \end{bmatrix}),$$

where $\rho$ is defined to be the co-heritability of the two phenotypes. In this regard, co-heritability is a global measure of the genetic relationship between two phenotypes while detection of loci with pleiotropy is a local characterization.

In the past decades, accumulating GWAS data allows us to investigate co-heritability and pleiotropy in a comprehensive manner. First, European Genome-phenome Archive (EGA) and The database of Genotypes and Phenotypes (dbGap) have collected an enormous amount of genotype and phenotype data at the individual level. Second, the summary statistics from many GWAS are directly downloadable through public gateways, such as the websites of the GIANT consortium and the Psychiatric Genomics Consortium (PGC). Third, databases have been built up to collect the output of published GWAS. For example, the Genome-Wide Repository of Associations between SNPs and Phenotypes (GRASP) database has been developed for such a purpose [36]. Very recently, GRASP has been updated [17] to provide latest summary of GWAS output—about 8.87 million SNP-phenotype associations in 2082 studies with $p$-values $\leq 0.05$.

Various statistical methods have been developed to explore co-heritability and pleiotropy. First, a straightforward extension of univariate LMM to multivariate LMM can be used for co-heritability estimation [35]. Second, co-heritability can be explored to improve risk prediction, as demonstrated in [37, 41]. The idea is that the random vectors $\mathbf{u}^{(1)}$ and $\mathbf{u}^{(2)}$ of effect sizes can be predicted more accurately when $\rho \neq 0$, because more information can be combined in bivariate LMM by introducing one more parameter, i.e., co-heritability $\rho$. An extreme case is $\rho = 1$, which means the sample size in bivariate LMM is doubled compared with univariate LMM. In the absence of co-heritability, i.e., $\rho = 0$, bivariate LMM will have one redundant parameter compared to univariate LMM, resulting in a slightly less efficiency. But the inefficiency caused by one redundant parameter can be neglected as there are hundreds or thousands of samples in GWAS. In other words, compared to univariate LMM, bivariate LMM has a flexible model structure to combine relevant information and does not sacrifice too much efficiency in absence of such information. Third, pleiotropy can be used for co-localization of risk variants in multiple GWAS [8, 22, 24, 38]. We will use a real data example to illustrate the impact of pleiotropy in our case study.

# 4 Integrative Analysis of GWAS with Functional Information

Besides integrating multiple GWAS, integrative analysis of GWAS with functional information is also a very promising strategy to explore the genetic architectures of complex phenotypes. Accumulating evidence suggests that this strategy can effectively boost the statistical power of GWAS data analysis [5]. The reason for such an improvement is that SNPs do not make equal contributions to a phenotype and a group of functionally related SNPs can contribute much more than the average, which is known as "functional enrichment" [19, 54]. For example, an SNP that plays a role in the central nervous system (CNS) is more likely to be involved in psychiatric disorders than a randomly selected SNP [11]. As a matter of fact, not only can functional information help to improve the statistical power, but also offer deeper understanding on biological mechanisms of complex phenotypes. For instance, the integration of functional information into GWAS analysis suggests a possible connection between the immune system and schizophrenia [48, 60]. However, the fine-grained characterization of the functional role of genetic variations was not widely available until recent years.

In 2012, the Encyclopedia of DNA Elements (ENCODE) project [9] reported a high-quality functional characterization of the human genome. This report highlighted the regulatory role of non-coding variants, which helped to explain the fact that about 85 % of the GWAS hits are in the non-coding region of human genome [29]. More specifically, the analysis results from the ENCODE project showed that 31 % of the GWAS hits overlap with transcription factor binding sites and 71 % overlap with DNase I hypersensitive sites, indicating the functional roles of GWAS hits. Afterwards, large genomic consortia started generating an enormous amount of data to provide functional annotation of the human genome. The Roadmap Epigenomics project [33] aims at providing the epigenome reference of more than one hundred tissues and cell types to tackle human diseases. Besides the epigenome reference, the Genotype-Tissue Expression project (GTEx) [39] has been initiated to collect about 20,000 tissues from 900 donors, serving as a comprehensive atlas of gene expression and regulation. Based on the data collected from 175 individuals across 43 tissues, GTEx [2] has reported a pilot analysis result of the gene expression patterns across tissues, including identification of thousands of shared and tissue-specific eQTL. Clearly, the integration of GWAS and functional information is calling effective methods that hardness such a rich data resources [47].

To introduce the key idea of integrative analysis of GWAS with functional information, we briefly discuss a Bayesian method [6] to see the advantages of statistically rigorous methods. Suppose we have collected $n$ samples with their phenotypic values $\mathbf{y} \in R^n$ and genotypes in $\mathbf{X} \in R^{n \times M}$. Following the typical practice, we assume the linear relationship between $\mathbf{y}$ and $\mathbf{X}$:

$$y_i = \beta_0 + \sum_{j=1}^{M} x_{ij}\beta_j + e_i, \tag{21}$$

where $\beta_j, j = 1, \ldots, M$ are the coefficients and $e_i$ is the independent noise $e_i \sim \mathcal{N}(0, \sigma_e^2)$. Identification of risk variants can be viewed as determination of the nonzero coefficients in $\boldsymbol{\beta} = [\beta_1, \ldots, \beta_M]^T$. Next, we use a binary variable $\boldsymbol{\gamma} = [\gamma_1, \ldots, \gamma_M]$ to indicate whether the corresponding $\beta_j$ is zero or not: $\beta_j = 0$ if and only if $\gamma_j = 0$. The spike and slab prior [44] is assigned for $\beta_j$:

$$\begin{aligned} \beta_j &\sim \mathcal{N}(0, \sigma_\beta^2), && \text{if } \gamma_j = 1, \\ \beta_j &= 0, && \text{if } \gamma_j = 0, \end{aligned} \tag{22}$$

where $\Pr(\gamma_j = 1) = \pi$ and $\Pr(\gamma_j = 0) = 1 - \pi$. Following the standard procedure in Bayesian inference, the remaining is to calculate the posterior $\Pr(\gamma|\mathbf{y}, \mathbf{X})$ based on Markov chain Monte Carlo (MCMC) method. Although the computational cost of MCMC can be expensive, efficient variational approximation can be used [3, 7].

Suppose we have extracted functional information from the reference data of high quality, such as Roadmap [33] and GTEx [39] and collected them in an $M \times D$ matrix, denoted as $\mathbf{A}$. Each row of $\mathbf{A}$ corresponds to an SNP and each column corresponds to a functional category. For example, if the $i$th SNP is known to play a role in the $d$-functional category from the reference data, then we put $A_{jd} = 1$ and $A_{jd} = 0$ otherwise. To keep our notation simple, we use $\mathbf{A}_j \in R^{1 \times D}$ to index the $j$th row of $\mathbf{A}$. Note that functional information in $\mathbf{A}$ may come from different studies. It is inappropriate to conclude that SNPs being annotated in $\mathbf{A}$ are more useful because the relevance of such functional information has not been examined yet.

To determine the relevance of functional information, statistical modeling plays a critical role. Indeed, functional information $\mathbf{A}_j$ of the $j$th SNP can be naturally related to its association status $\gamma_j$ is using a logistic model [6]:

$$\log \frac{\Pr(\gamma_j = 1|\mathbf{A}_j)}{\Pr(\gamma_j = 0|\mathbf{A}_j)} = \mathbf{A}_j \boldsymbol{\theta} + \boldsymbol{\theta}_0, \tag{23}$$

where $\boldsymbol{\theta} \in R^D$ and $\theta_0 \in R$ are the logistic regression coefficients to be estimated. Clearly, when there are nonzero entries in $\boldsymbol{\theta}$, the prior of the association status $\gamma_j$ will be modulated by its functional annotation $\mathbf{a}_j$, indicating the relevance of functional annotation. More rigorously, a Bayes factor of $\boldsymbol{\theta}$ can be computed to determine the relevance of function information. In summary, statistical methods allow a flexible way to incorporate functional information into the model and adaptively determine the relevance of such kind of information.

## 5 Case Study

So far, we have discussed the integrative analysis of multiple GWAS and the integrative analysis of a single GWAS with functional information. Taking one step forward, we can integrate multiple GWAS and functional information simultaneously. To be more specific, we consider our GPA (Genetic analysis incorporating Pleiotropy and Annotation) approach [8] as a case study.

In contrast to the method discussed in the previous sections, GPA takes summary statistics and functional annotations as its input. Let us begin with the simplest case where we have only $p$-values from one GWAS data set, denoted as $\{p_1, p_2, \ldots, p_j, \ldots, p_M\}$, where $M$ is the number of SNPs. Following the "two-groups model" [16], we assume the observed $p$-values from a mixture of null and non-null distributions, with probability $\pi_0$ and $\pi_1 = 1 - \pi_0$, respectively. Here we choose the null distribution to be the Uniform distribution on [0,1], denoted as $\mathscr{U}[0,1]$, and the non-null distribution to be the Beta distribution with parameters $(\alpha, 1)$, denoted as $\mathscr{B}(\alpha, 1)$, respectively. Again, we introduce a binary variable $Z_j \in \{0, 1\}$ to indicate the association status of the $j$th SNP: $Z_j = 0$ means null and $Z_j = 1$ means non-null. Then the two-groups model can be written as

$$\begin{aligned}\pi_0 &= \Pr(Z_j = 0) : p_j \sim \mathscr{U}[0,1], \text{if } Z_j = 0,\\ \pi_1 &= \Pr(Z_j = 1) : p_j \sim \mathscr{B}(\alpha,1), \text{if } Z_j = 1,\end{aligned} \tag{24}$$

where $\pi_0 + \pi_1 = 1$ and $0 < \alpha < 1$. An efficient EM algorithm can be easily derived if the independence among the SNP markers is assumed, as detailed in the GPA paper. Let $\hat{\boldsymbol{\Theta}} = \{\hat{\pi}_0, \hat{\pi}_1, \hat{\alpha}\}$ be the estimated model parameters, then the posterior is given as

$$\widehat{\Pr}(Z_j = 0|p_j; \hat{\boldsymbol{\Theta}}) = \frac{\hat{\pi}_0}{\hat{\pi}_0 + \hat{\pi}_1 f_B(p_j; \hat{\alpha})}, \tag{25}$$

where $f_B(p;\alpha) = \alpha p^{\alpha-1}$ is the density function of $\mathscr{B}(\alpha, 1)$. Indeed, this posterior is known as the local false discovery rate [14], which is widely used in the type I error control.

To explore pleiotropy between two GWAS, the above two-groups model can be extended to a four-groups model. Suppose we have collected $p$-values from two GWAS and denote the $p$-value of the $j$th SNP as $\{p_{j1}, p_{j2}\}, j = 1, \ldots, M$. Let $Z_{j1} \in \{0, 1\}$ and $Z_{j2} \in \{0, 1\}$ be the indicator of association status of the $j$th SNP in two GWAS. Then the four-groups model can be written as

$$\begin{aligned}\pi_{00} &= \Pr(Z_{j1} = 0, Z_{j2} = 0) : p_{j1} \sim \mathscr{U}[0,1], p_{j2} \sim \mathscr{U}[0,1], \text{if } Z_{j1} = 0, Z_{j2} = 0,\\ \pi_{10} &= \Pr(Z_{j1} = 1, Z_{j2} = 0) : p_{j1} \sim \mathscr{B}(\alpha_1,1), p_{j2} \sim \mathscr{U}[0,1], \text{if } Z_{j1} = 1, Z_{j2} = 0,\\ \pi_{01} &= \Pr(Z_{j1} = 0, Z_{j2} = 1) : p_{j1} \sim \mathscr{U}[0,1], p_{j2} \sim \mathscr{B}(\alpha_2,1), \text{if } Z_{j1} = 0, Z_{j2} = 1,\\ \pi_{11} &= \Pr(Z_{j1} = 1, Z_{j2} = 1) : p_{j1} \sim \mathscr{B}(\alpha_1,1), p_{j2} \sim \mathscr{B}(\alpha_2,1), \text{if } Z_{j1} = 1, Z_{j2} = 1,\end{aligned}$$

where $0 < \alpha_1 < 1, 0 < \alpha_2 < 1$ and $\pi_{00} + \pi_{10} + \pi_{01} + \pi_{11} = 1$. The four-groups model takes pleiotropy into account by allowing the correlation between $Z_{j1}$ and $Z_{j2}$. It is easy to see that the correlation $\text{Corr}(Z_{j1}, Z_{j2}) \neq 0$ if $\pi_{11} \neq (\pi_{10} + \pi_{11})(\pi_{01} + \pi_{11})$. In this regard, a hypothesis test $(H_0 : \pi_{11} = (\pi_{10} + \pi_{11})\ (\pi_{01} + \pi_{11}))$ can be designed to examine whether the overlapping of risk variants between two GWAS is different from the overlapping just by chance. The testing result can be viewed as an indicator of pleiotropy.

To incorporate functional annotations, GPA assumes that all the functional annotations are independent after conditioning on the association status. Again, let $\mathbf{A} \in R^{M \times D}$ be the annotation matrix, where $A_{jd} = 1$ corresponds to the $j$th SNP being annotated in the $d$th functional category, and $A_{jd} = 0$ otherwise. Therefore, in the two-groups model (24), the conditional probability of the $d$th annotation can be written as

$$q_{0d} = \Pr(A_{jd} = 1|Z_j = 0), \quad q_{1d} = \Pr(A_{jd} = 1|Z_j = 1), \tag{26}$$

where $q_{0d}$ and $q_{1d}$ are GPA model parameters which can be estimated by the EM algorithm. Readers who are familiar with classification can easily recognize that (26) is the Naive Bayes formulation with latent class label, while (23) is a logistic regression with latent class label. Latent space plays a very important role in integrative analysis, in which indirect information (annotation data) can be combined with direct information ($p$-values). Under a coherent statistical framework, we are able to employ statistically efficient methods for parameter estimation rather than relying on ad-hoc rules. Let $\hat{\mathbf{\Theta}} = \{\hat{\pi}_0, \hat{\pi}_1, \hat{\alpha}, (\hat{q}_{1d}, \hat{q}_{0d})_{d=1,\ldots,D}\}$ be the estimated parameters. Then the posterior $\Pr(Z_j = 0|p_j, \mathbf{A}_j; \hat{\mathbf{\Theta}})$ can be written as

$$\Pr(Z_j = 0|p_j, \mathbf{A}_j; \hat{\mathbf{\Theta}}) = \frac{\hat{\pi}_0 \prod_{d=1}^{D} \hat{q}_{0d}^{1-A_{jd}} \hat{q}_{1d}^{A_{jd}}}{\hat{\pi}_0 \prod_{d=1}^{D} \hat{q}_{0d}^{1-A_{jd}} \hat{q}_{1d}^{A_{jd}} + \hat{\pi}_1 \prod_{d=1}^{D} \hat{q}_{0d}^{1-A_{jd}} \hat{q}_{1d}^{A_{jd}} f_B(p_j; \hat{\alpha})} \tag{27}$$

Compared with (25), when $q_{0d} \neq q_{1d}$, posterior (27) will be updated according to functional enrichment in the $d$th annotation. Hypothesis testing $H_0 : q_{0d} = q_{1d}$ can be used to declare the significance of the enrichment. Similarly, functional annotations can be incorporated into the four-groups model as follows:

$$\begin{aligned}
q_{00d} &= \Pr(A_{jd} = 1|Z_{j1} = 0, Z_{j2} = 0), \\
q_{10d} &= \Pr(A_{jd} = 1|Z_{j1} = 1, Z_{j2} = 0), \\
q_{01d} &= \Pr(A_{jd} = 1|Z_{j1} = 0, Z_{j2} = 1), \\
q_{11d} &= \Pr(A_{jd} = 1|Z_{j1} = 1, Z_{j2} = 1).
\end{aligned} \tag{28}$$

As a demonstration, we apply the GPA approach to the GWAS of schizophrenia (SCZ) and BPD with the CNS genes as the functional annotation. The detailed description of the dataset can be found in the GPA paper. To make our demonstration easily reproducible, the R package of GPA and the demonstration dataset have been made freely accessible at https://sites.google.com/site/eeyangc/software/gpa. The analysis results are summarized in Tables 1 and 2 and Fig. 1. Here we give some brief discussions. First, more significant GWAS hits with controlled false discovery rates can be identified by integrative analysis of GWAS and functional information, as shown in Tables 1 and 2. Second, we can see the pleiotropic effects exist between SCZ and BPD (the estimated shared proportion $\hat{\pi}_{11} \approx 0.15$). Indeed, such pleiotropy information boosts the statistical power a lot. Third, functional information (the CNS annotation) further helps improve the statistical power, although its contribution is less than pleiotropy in this real data analysis. This suggests that pleiotropy and functional information are complementary to each other and both of them are necessary.

**Table 1** Single-GWAS analysis of SCZ and BPD (with or without the CNS annotation)

| | $\hat{\pi}_1$ | $\hat{\alpha}$ | $\hat{q}_0$ | $\hat{q}_1$ | No. hits ($fdr \leq 0.05$) | No. hits ($fdr \leq 0.1$) |
|---|---|---|---|---|---|---|
| SCZ (without annotation) | 0.195 (0.004) | 0.596 (0.004) | – | – | 391 | 875 |
| BPD (without annotation) | 0.181 (0.007) | 0.700 (0.007) | – | – | 13 | 23 |
| SCZ (with annotation) | 0.196 (0.004) | 0.596 (0.004) | 0.203 (0.001) | 0.283 (0.003) | 409 | 902 |
| BPD (with annotation) | 0.179 (0.004) | 0.697 (0.004) | 0.202 (0.001) | 0.297 (0.004) | 14 | 43 |

The values in the brackets are standard errors of the corresponding estimates

**Table 2** Integrative analysis of SCZ and BPD (with or without the CNS annotation)

| | $\hat{\pi}_{00}$ | $\hat{\pi}_{10}$ | $\hat{\pi}_{01}$ | $\hat{\pi}_{11}$ | $\hat{\alpha}_1$ | $\hat{\alpha}_2$ |
|---|---|---|---|---|---|---|
| Without annotation | 0.816 (0.004) | 0.006 (0.005) | 0.027 (0.006) | 0.152 (0.006) | 0.579 (0.004) | 0.671 (0.007) |
| With annotation | 0.815 (0.004) | 0.007 (0.005) | 0.029 (0.007) | 0.149 (0.006) | 0.577 (0.003) | 0.670 (0.007) |
| | $\hat{q}_{00}$ | $\hat{q}_{10}$ | $\hat{q}_{01}$ | $\hat{q}_{11}$ | No. hits ($fdr \leq 0.05$) | No. hits ($fdr \leq 0.1$) |
| Without annotation | – | – | – | – | 801 (SCZ); 157 (BPD) | 1442 (SCZ); 645 (BPD) |
| With annotation | 0.207 (0.001) | 0.014 (0.243) | 0.103 (0.088) | 0.318 (0.006) | 818 (SCZ); 237 (BPD) | 1492 (SCZ); 706 (BPD) |

The values in the brackets are standard errors of the corresponding estimates. There are some very minor differences between the values reported here and those in the original paper. This is because all the results are reported with the maximum number of EM iterations at 2000 (the default setting of the R package) while those reported in the original paper are based on the maximum number of EM iterations at 10,000

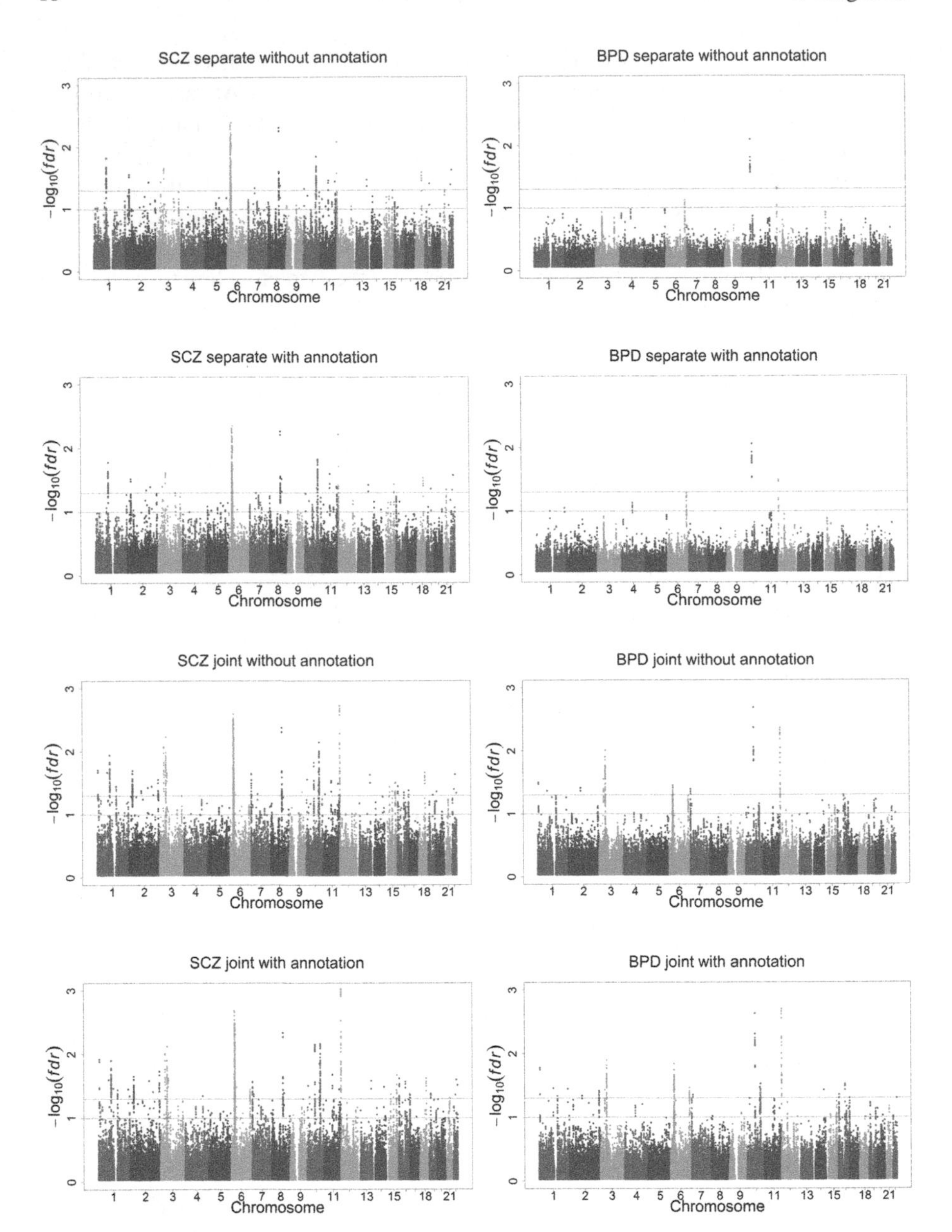


**Fig. 1** Manhattan plots of GPA analysis result for SCZ and BPD. From *top* to *bottom* panels: separate analysis of SCZ (*left*) and BPD (*right*) without annotation, separate analysis of SCZ (*left*) and BPD (*right*) with the CNS annotation, joint analysis of SCZ (*left*) and BPD (*right*) without annotation and joint analysis of SCZ (*left*) and BPD (*right*) with the CNS annotation. The *horizontal red and blue lines* indicate local false discovery rate at 0.05 and 0.1, respectively. The numbers of significant GWAS hits at $fdr \leq 0.05$ and $fdr \leq 0.1$ are given in Tables 1 and 2

## 6 Future Directions and Conclusion

Although the analysis result from the GPA approach looks promising, there are some limitations. First, the GPA approach assumed the independence among the SNP markers, implying that the linkage disequilibrium (LD) among SNP markers was not taken into account. Second, the GPA approach assumed the conditional independence among functional annotations, which may not be true in presence of multiple annotations. All these limitations should be addressed in the future. Recently, a closely related approach, the LD-score method [4], has been proposed to analyze GWAS data based on summary statistics, in which LD has been explicitly taken into account. This method can be used for heritability (and co-heritability) estimation, as well as the detection of functional enrichment [19]. However, some empirical studies have shown that the standard error of the LD-score method is nearly twice of that of the REML estimate [65], indicating that this method is far less efficient than REML and thus the large sample size is required to ensure its effectiveness. More statistically efficient methods are still in high demand to address this issue.

In summary, we have provided a brief introduction to integrative analysis of GWAS and functional information, including heritability estimation and risk variant identification. Facing the challenges raised by the polygenicity, it is highly demanded to perform integrative analysis from both biological and statistical perspectives. Novel approaches which take LD into account when integrating summary statistics with functional information will be greatly needed in the future. There are also many issues remaining in the study of functional enrichment. Recently, more and more functional enrichments have been observed in a variety of studies [19, 55]. However, most of the enrichment is often too general to provide phenotype-specific information. For example, coding regions and transcription factor binding sites are generally enriched in various types of phenotypes. We are drowning in cross-phenotype functional enrichment but starving for phenotype-specific knowledge—how does a functional unit of human genome affect a phenotype of interest. Adjusting for the common enrichment (viewed as confounding factors here), rigorous methods for detecting phenotype-specific patterns will be highly appreciated.

**Acknowledgements** This work was supported in part by grant NO. 61501389 from National Natural Science Foundation of China (NSFC), grants HKBU_22302815 and HKBU_12202114 from Hong Kong Research Grant Council, and grants FRG2/14-15/069, FRG2/15-16/011, and FRG2/14-15/077 from Hong Kong Baptist University, and Duke-NUS Medical School WBS: R-913-200-098-263.

## References

1. Hana Lango Allen, Karol Estrada, Guillaume Lettre, Sonja I Berndt, Michael N Weedon, Fernando Rivadeneira, and et al. Hundreds of variants clustered in genomic loci and biological pathways affect human height. *Nature*, 467(7317):832–838, 2010.
2. Kristin G Ardlie, David S Deluca, Ayellet V Segrè, Timothy J Sullivan, Taylor R Young, Ellen T Gelfand, Casandra A Trowbridge, Julian B Maller, Taru Tukiainen, Monkol Lek, et al. The Genotype-Tissue Expression (GTEx) pilot analysis: Multitissue gene regulation in humans. *Science*, 348(6235):648–660, 2015.
3. Christopher M Bishop and Nasser M Nasrabadi. *Pattern recognition and machine learning*, volume 1. Springer New York, 2006.
4. Brendan K Bulik-Sullivan, Po-Ru Loh, Hilary K Finucane, Stephan Ripke, Jian Yang, Nick Patterson, Mark J Daly, Alkes L Price, Benjamin M Neale, Schizophrenia Working Group of the Psychiatric Genomics Consortium, et al. LD score regression distinguishes confounding from polygenicity in genome-wide association studies. *Nature genetics*, 47(3):291–295, 2015.
5. Rita M Cantor, Kenneth Lange, and Janet S Sinsheimer. Prioritizing GWAS results: a review of statistical methods and recommendations for their application. *The American Journal of Human Genetics*, 86(1):6–22, 2010.
6. Peter Carbonetto and Matthew Stephens. Integrated enrichment analysis of variants and pathways in genome-wide association studies indicates central role for il-2 signaling genes in type 1 diabetes, and cytokine signaling genes in Crohn's disease. *PLoS Genet*, 9(10):1003770, 2013.
7. Peter Carbonetto, Matthew Stephens, et al. Scalable variational inference for Bayesian variable selection in regression, and its accuracy in genetic association studies. *Bayesian Analysis*, 7(1):73–108, 2012.
8. Dongjun Chung, Can Yang, Cong Li, Joel Gelernter, and Hongyu Zhao. GPA: A Statistical Approach to Prioritizing GWAS Results by Integrating Pleiotropy and Annotation. *PLoS genetics*, 10(11):e1004787, 2014.
9. ENCODE Project Consortium et al. An integrated encyclopedia of DNA elements in the human genome. *Nature*, 489(7414):57–74, 2012.
10. Chris Cotsapas, Benjamin F Voight, Elizabeth Rossin, Kasper Lage, Benjamin M Neale, Chris Wallace, Gonçalo R Abecasis, Jeffrey C Barrett, Timothy Behrens, Judy Cho, et al. Pervasive sharing of genetic effects in autoimmune disease. *PLoS genetics*, 7(8):e1002254, 2011.
11. Cross-Disorder Group of the Psychiatric Genomics Consortium. Genetic relationship between five psychiatric disorders estimated from genome-wide SNPs. *Nature genetics*, 45(9):984–994, 2013.
12. Cross-Disorder Group of the Psychiatric Genomics Consortium. Identification of risk loci with shared effects on five major psychiatric disorders: a genome-wide analysis. *Lancet*, 2013.
13. Gustavo de los Campos, Daniel Sorensen, and Daniel Gianola. Genomic heritability: what is it? *PLoS Genetics*, 10(5):e1005048, 2015.
14. B. Efron. *Large-Scale Inference: Empirical Bayes Methods for Estimation, Testing, and Prediction*. Cambridge University Press, 2010.
15. Bradley Efron. The future of indirect evidence. *Statistical science: a review journal of the Institute of Mathematical Statistics*, 25(2):145, 2010.
16. Bradley Efron et al. Microarrays, empirical Bayes and the two-groups model. *STAT SCI*, 23(1):1–22, 2008.
17. John D Eicher, Christa Landowski, Brian Stackhouse, Arielle Sloan, Wenjie Chen, Nicole Jensen, Ju-Ping Lien, Richard Leslie, and Andrew D Johnson. GRASP v2. 0: an update on the Genome-Wide Repository of Associations between SNPs and phenotypes. *Nucleic acids research*, 43(D1):D799–D804, 2015.
18. Douglas S Falconer, Trudy FC Mackay, and Richard Frankham. Introduction to quantitative genetics (4th edn). *Trends in Genetics*, 12(7):280, 1996.

19. Hilary K Finucane, Brendan Bulik-Sullivan, Alexander Gusev, Gosia Trynka, Yakir Reshef, Po-Ru Loh, Verneri Anttila, Han Xu, Chongzhi Zang, Kyle Farh, et al. Partitioning heritability by functional annotation using genome-wide association summary statistics. *Nature genetics*, 47(11):1228–1235, 2015.
20. R. A. Fisher. The correlations between relatives on the supposition of Mendelian inheritance. *Philosophical Transactions of the Royal Society of Edinburgh*, 52:399–433, 1918.
21. Olivia Fletcher and Richard S Houlston. Architecture of inherited susceptibility to common cancer. *Nature Reviews Cancer*, 10(5):353–361, 2010.
22. Mary D Fortune, Hui Guo, Oliver Burren, Ellen Schofield, Neil M Walker, Maria Ban, Stephen J Sawcer, John Bowes, Jane Worthington, Anne Barton, et al. Statistical colocalization of genetic risk variants for related autoimmune diseases in the context of common controls. *Nature genetics*, 47(7):839–846, 2015.
23. Eric R Gamazon, Heather E Wheeler, Kaanan P Shah, Sahar V Mozaffari, Keston Aquino-Michaels, Robert J Carroll, Anne E Eyler, Joshua C Denny, Dan L Nicolae, Nancy J Cox, et al. A gene-based association method for mapping traits using reference transcriptome data. *Nature genetics*, 47(9):1091–1098, 2015.
24. Claudia Giambartolomei, Damjan Vukcevic, Eric E Schadt, Lude Franke, Aroon D Hingorani, Chris Wallace, and Vincent Plagnol. Bayesian test for colocalisation between pairs of genetic association studies using summary statistics. *PLoS Genetics*, 10(5):e1004383, 2014.
25. Arthur R Gilmour, Robin Thompson, and Brian R Cullis. Average information REML: an efficient algorithm for variance parameter estimation in linear mixed models. *Biometrics*, pages 1440–1450, 1995.
26. David Golan, Eric S Lander, and Saharon Rosset. Measuring missing heritability: Inferring the contribution of common variants. *Proceedings of the National Academy of Sciences*, 111(49):E5272–E5281, 2014.
27. Anthony J.F. Griffiths, Susan R. Wessler, Sean B. Carroll, and John Doebley. *An introduction to genetic analysis, 11 edition*. W. H. Freeman, 2015.
28. William G Hill, Michael E Goddard, and Peter M Visscher. Data and theory point to mainly additive genetic variance for complex traits. *PLoS Genet*, 4(2):e1000008, 2008.
29. L.A. Hindorff, P. Sethupathy, H.A. Junkins, E.M. Ramos, J.P. Mehta, F.S. Collins, and T.A. Manolio. Potential etiologic and functional implications of genome-wide association loci for human diseases and traits. *Proceedings of the National Academy of Sciences*, 106(23):9362, 2009.
30. Jiming Jiang, Cong Li, Debashis Paul, Can Yang, and Hongyu Zhao. High-dimensional genome-wide association study and misspecified mixed model analysis. *arXiv preprint arXiv:1404.2355, to appear in Annals of statistics*, 2014.
31. Robert J Klein, Caroline Zeiss, Emily Y Chew, Jen-Yue Tsai, Richard S Sackler, Chad Haynes, Alice K Henning, John Paul SanGiovanni, Shrikant M Mane, Susan T Mayne, et al. Complement factor h polymorphism in age-related macular degeneration. *Science*, 308(5720):385–389, 2005.
32. Siddharth Krishna Kumar, Marcus W Feldman, David H Rehkopf, and Shripad Tuljapurkar. Limitations of GCTA as a solution to the missing heritability problem. *Proceedings of the National Academy of Sciences*, 113(1):E61–E70, 2016.
33. Anshul Kundaje, Wouter Meuleman, Jason Ernst, Misha Bilenky, Angela Yen, Alireza Heravi-Moussavi, Pouya Kheradpour, Zhizhuo Zhang, Jianrong Wang, Michael J Ziller, et al. Integrative analysis of 111 reference human epigenomes. *Nature*, 518(7539):317–330, 2015.
34. S Hong Lee, Teresa R DeCandia, Stephan Ripke, Jian Yang, Patrick F Sullivan, Michael E Goddard, and et al. Estimating the proportion of variation in susceptibility to schizophrenia captured by common SNPs. *Nature genetics*, 44(3):247–250, 2012.
35. SH Lee, J Yang, ME Goddard, PM Visscher, and NR Wray. Estimation of pleiotropy between complex diseases using SNP-derived genomic relationships and restricted maximum likelihood. *Bioinformatics*, page bts474, 2012.
36. Richard Leslie, Christopher J O'Donnell, and Andrew D Johnson. GRASP: analysis of genotype–phenotype results from 1390 genome-wide association studies and corresponding open access database. *Bioinformatics*, 30(12):i185–i194, 2014.

37. Cong Li, Can Yang, Joel Gelernter, and Hongyu Zhao. Improving genetic risk prediction by leveraging pleiotropy. *Human genetics*, 133(5):639–650, 2014.
38. James Liley and Chris Wallace. A pleiotropy-informed Bayesian false discovery rate adapted to a shared control design finds new disease associations from GWAS summary statistics. *PLoS genetics*, 11(2):e1004926, 2015.
39. John Lonsdale, Jeffrey Thomas, Mike Salvatore, Rebecca Phillips, Edmund Lo, Saboor Shad, Richard Hasz, Gary Walters, Fernando Garcia, Nancy Young, et al. The genotype-tissue expression (GTEx) project. *Nature genetics*, 45(6):580–585, 2013.
40. Michael Lynch, Bruce Walsh, et al. *Genetics and analysis of quantitative traits*, volume 1. Sinauer Sunderland, MA, 1998.
41. Robert Maier, Gerhard Moser, Guo-Bo Chen, Stephan Ripke, William Coryell, James B Potash, William A Scheftner, Jianxin Shi, Myrna M Weissman, Christina M Hultman, et al. Joint analysis of psychiatric disorders increases accuracy of risk prediction for schizophrenia, bipolar disorder, and major depressive disorder. *The American Journal of Human Genetics*, 96(2):283–294, 2015.
42. Teri A Manolio, Francis S Collins, Nancy J Cox, David B Goldstein, Lucia A Hindorff, David J Hunter, Mark I McCarthy, Erin M Ramos, Lon R Cardon, Aravinda Chakravarti, et al. Finding the missing heritability of complex diseases. *Nature*, 461(7265):747–753, 2009.
43. Geoffrey McLachlan and Thriyambakam Krishnan. *The EM algorithm and extensions*, volume 382. John Wiley & Sons, 2008.
44. Toby J Mitchell and John J Beauchamp. Bayesian variable selection in linear regression. *Journal of the American Statistical Association*, 83(404):1023–1032, 1988.
45. Alkes L Price, Nick J Patterson, Robert M Plenge, Michael E Weinblatt, Nancy A Shadick, and David Reich. Principal components analysis corrects for stratification in genome-wide association studies. *Nature genetics*, 38(8):904–909, 2006.
46. Neil Risch, Kathleen Merikangas, et al. The future of genetic studies of complex human diseases. *Science*, 273(5281):1516–1517, 1996.
47. Marylyn D Ritchie, Emily R Holzinger, Ruowang Li, Sarah A Pendergrass, and Dokyoon Kim. Methods of integrating data to uncover genotype-phenotype interactions. *Nature Reviews Genetics*, 16(2):85–97, 2015.
48. Schizophrenia Working Group of the Psychiatric Genomics Consortium. Biological insights from 108 schizophrenia-associated genetic loci. *Nature*, 511(7510):421–427, 2014.
49. Shanya Sivakumaran, Felix Agakov, Evropi Theodoratou, et al. Abundant pleiotropy in human complex diseases and traits. *AM J HUM GENET*, 89(5):607–618, 2011.
50. Nadia Solovieff, Chris Cotsapas, Phil H Lee, Shaun M Purcell, and Jordan W Smoller. Pleiotropy in complex traits: challenges and strategies. *Nature Reviews Genetics*, 14(7): 483–495, 2013.
51. Doug Speed and David J Balding. Relatedness in the post-genomic era: is it still useful? *Nature Reviews Genetics*, 16(1):33–44, 2015.
52. Doug Speed, Gibran Hemani, Michael R Johnson, and David J Balding. Improved heritability estimation from genome-wide SNPs. *The American Journal of Human Genetics*, 91(6):1011–1021, 2012.
53. Frank W Stearns. One hundred years of pleiotropy: a retrospective. *Genetics*, 186(3):767–773, 2010.
54. Aravind Subramanian, Pablo Tamayo, Vamsi K Mootha, Sayan Mukherjee, Benjamin L Ebert, Michael A Gillette, Amanda Paulovich, Scott L Pomeroy, Todd R Golub, Eric S Lander, et al. Gene set enrichment analysis: a knowledge-based approach for interpreting genome-wide expression profiles. *Proceedings of the National Academy of Sciences of the United States of America*, 102(43):15545–15550, 2005.
55. Jason M Torres, Eric R Gamazon, Esteban J Parra, Jennifer E Below, Adan Valladares-Salgado, Niels Wacher, Miguel Cruz, Craig L Hanis, and Nancy J Cox. Cross-tissue and tissue-specific eQTLs: partitioning the heritability of a complex trait. *The American Journal of Human Genetics*, 95(5):521–534, 2014.

56. Shashaank Vattikuti, Juen Guo, and Carson C Chow. Heritability and genetic correlations explained by common SNPs for metabolic syndrome traits. *PLoS genetics*, 8(3):e1002637, 2012.
57. Peter M Visscher, Matthew A Brown, Mark I McCarthy, and Jian Yang. Five years of GWAS discovery. *The American Journal of Human Genetics*, 90(1):7–24, 2012.
58. Peter M Visscher, William G Hill, and Naomi R Wray. Heritability in the genomics era-concepts and misconceptions. *Nature Reviews Genetics*, 9(4):255–266, 2008.
59. Peter M Visscher, Sarah E Medland, MA Ferreira, Katherine I Morley, Gu Zhu, Belinda K Cornes, Grant W Montgomery, and Nicholas G Martin. Assumption-free estimation of heritability from genome-wide identity-by-descent sharing between full siblings. *PLoS Genet*, 2(3):e41, 2006.
60. Qian Wang, Can Yang, Joel Gelernter, and Hongyu Zhao. Pervasive pleiotropy between psychiatric disorders and immune disorders revealed by integrative analysis of multiple GWAS. *Human genetics*, 134(11–12):1195–1209, 2015.
61. Danielle Welter, Jacqueline MacArthur, Joannella Morales, Tony Burdett, Peggy Hall, Heather Junkins, Alan Klemm, Paul Flicek, Teri Manolio, Lucia Hindorff, et al. The NHGRI GWAS Catalog, a curated resource of SNP-trait associations. *Nucleic acids research*, 42(D1):D1001–D1006, 2014.
62. Can Yang, Cong Li, Henry R Kranzler, Lindsay A Farrer, Hongyu Zhao, and Joel Gelernter. Exploring the genetic architecture of alcohol dependence in African-Americans via analysis of a genomewide set of common variants. *Human Genetics*, 133(5):617–624, 2014.
63. Can Yang, Cong Li, Qian Wang, Dongjun Chung, and Hongyu Zhao. Implications of pleiotropy: challenges and opportunities for mining big data in biomedicine. *Frontiers in genetics*, 6, 2015.
64. Jian Yang, Andrew Bakshi, Zhihong Zhu, Gibran Hemani, Anna AE Vinkhuyzen, Sang Hong Lee, Matthew R Robinson, John RB Perry, Ilja M Nolte, Jana V van Vliet-Ostaptchouk, et al. Genetic variance estimation with imputed variants finds negligible missing heritability for human height and body mass index. *Nature genetics*, 2015.
65. Jian Yang, Andrew Bakshi, Zhihong Zhu, Gibran Hemani, Anna AE Vinkhuyzen, Ilja M Nolte, Jana V van Vliet-Ostaptchouk, Harold Snieder, Tonu Esko, Lili Milani, et al. Genome-wide genetic homogeneity between sexes and populations for human height and body mass index. *Human molecular genetics*, 24(25):7445–7449, 2015.
66. Jian Yang, Beben Benyamin, Brian P McEvoy, Scott Gordon, Anjali K Henders, Dale R Nyholt, Pamela A Madden, Andrew C Heath, Nicholas G Martin, Grant W Montgomery, et al. Common SNPs explain a large proportion of the heritability for human height. *Nature genetics*, 42(7):565–569, 2010.
67. Jian Yang, S Hong Lee, Michael E Goddard, and Peter M Visscher. GCTA: a tool for genome-wide complex trait analysis. *The American Journal of Human Genetics*, 88(1):76–82, 2011.
68. Jian Yang, Sang Hong Lee, Naomi R Wray, Michael E Goddard, and Peter M Visscher. Commentary on "Limitations of GCTA as a solution to the missing heritability problem". *bioRxiv*, page 036574, 2016.
69. Zhihong Zhu, Andrew Bakshi, Anna AE Vinkhuyzen, Gibran Hemani, Sang Hong Lee, Ilja M Nolte, Jana V van Vliet-Ostaptchouk, Harold Snieder, Tonu Esko, Lili Milani, et al. Dominance genetic variation contributes little to the missing heritability for human complex traits. *The American Journal of Human Genetics*, 96(3):377–385, 2015.

# Robust Methods for Expression Quantitative Trait Loci Mapping

**Wei Cheng, Xiang Zhang, and Wei Wang**

**Abstract** As a promising tool for dissecting the genetic basis of common diseases, expression quantitative trait loci (eQTL) study has attracted increasing research interest. The traditional eQTL methods focus on testing the associations between individual single-nucleotide polymorphisms (SNPs) and gene expression traits. A major drawback of this approach is that it cannot model the joint effect of a set of SNPs on a set of genes, which may correspond to biological pathways. In this chapter, we study the problem of identifying group-wise associations in eQTL mapping. Based on the intuition of group-wise association, we examine how the integration of heterogeneous prior knowledge on the correlation structures between SNPs, and between genes can improve the robustness and the interpretability of eQTL mapping.

**Keywords** Robust methods • eQTL • Gene expression • Parameter analysis • Biostatistics

## 1 Introduction

The most abundant sources of genetic variations in modern organisms are single-nucleotide polymorphisms (SNPs). An SNP is a DNA sequence variation occurring when a single nucleotide (A, T, G, or C) in the genome differs between individuals of a species. For inbred diploid organisms, such as inbred mice, an SNP usually shows variation between only two of the four possible nucleotide types [26], which

W. Cheng (✉)
NEC Laboratories America, Inc., Princeton, NJ, USA
e-mail: weicheng@nec-labs.com; chengw02@gmail.com

X. Zhang
Department of Electrical Engineering and Computer Science, Case Western Reserve University, Cleveland, OH, USA
e-mail: xiang.zhang@case.edu

W. Wang
Department of Computer Science, University of California, Los Angeles, CA, USA
e-mail: weiwang@cs.ucla.edu



K.-C. Wong (ed.), *Big Data Analytics in Genomics*,
DOI 10.1007/978-3-319-41279-5_2

allows us to represent it by a binary variable. The binary representation of an SNP is also referred to as the *genotype* of the SNP. The genotype of an organism is the genetic code in its cells. This genetic constitution of an individual influences, but is not solely responsible for, many of its traits. A *phenotype* is an observable trait or characteristic of an individual. The phenotype is the visible, or expressed trait, such as hair color. The phenotype depends upon the genotype but can also be influenced by environmental factors. Phenotypes can be either quantitative or binary.

Driven by the advancement of cost-effective and high-throughput genotyping technologies, genome-wide association studies (GWAS) have revolutionized the field of genetics by providing new ways to identify genetic factors that influence phenotypic traits. Typically, GWAS focus on associations between SNPs and traits like major diseases. As an important subsequent analysis, quantitative trait locus (QTL) analysis is aiming at to detect the associations between two types of information—quantitative phenotypic data (trait measurements) and genotypic data (usually SNPs)—in an attempt to explain the genetic basis of variation in complex traits. QTL analysis allows researchers in fields as diverse as agriculture, evolution, and medicine to link certain complex phenotypes to specific regions of chromosomes.

Gene expression is the process by which information from a gene is used in the synthesis of a functional gene product, such as proteins. It is the most fundamental level at which the genotype gives rise to the phenotype. Gene expression profile is the quantitative measurement of the activity of thousands of genes at once. The gene expression levels can be represented by continuous variables. Figure 1 shows an example dataset consisting of 1000 SNPs $\{x_1, x_2, \cdots, x_{1000}\}$ and a gene expression level $z_1$ for 12 individuals.

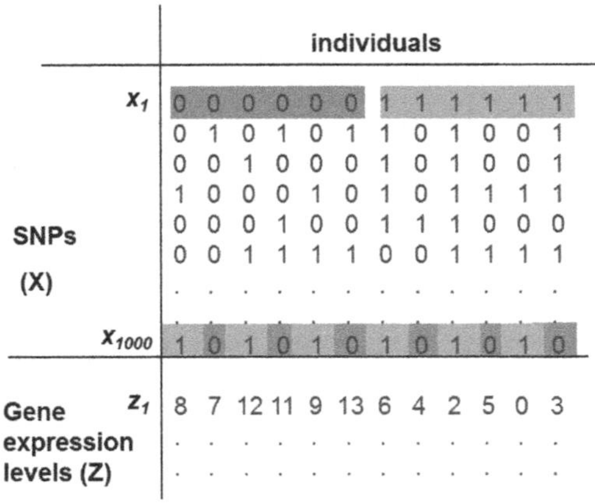

**Fig. 1** An example dataset in eQTL mapping

| | | individuals | | | | | | | | | | | |
|---|---|---|---|---|---|---|---|---|---|---|---|---|---|
| SNPs (X) | $x_1$ | 0 | 0 | 0 | 0 | 0 | 0 | 1 | 1 | 1 | 1 | 1 | 1 |
| | | 0 | 1 | 0 | 1 | 0 | 1 | 1 | 0 | 1 | 0 | 0 | 1 |
| | | 0 | 0 | 1 | 0 | 0 | 0 | 1 | 0 | 1 | 0 | 0 | 1 |
| | | 1 | 0 | 0 | 0 | 1 | 0 | 1 | 0 | 1 | 1 | 1 | 1 |
| | | 0 | 0 | 0 | 1 | 0 | 0 | 1 | 1 | 1 | 0 | 0 | 0 |
| | | 0 | 0 | 1 | 1 | 1 | 1 | 0 | 0 | 1 | 1 | 1 | 1 |
| | | . | . | . | . | . | . | . | . | . | . | . | . |
| | $x_{1000}$ | 1 | 0 | 1 | 0 | 1 | 0 | 1 | 0 | 1 | 0 | 1 | 0 |
| Gene expression levels (Z) | $z_1$ | 8 | 7 | 12 | 11 | 9 | 13 | 6 | 4 | 2 | 5 | 0 | 3 |
| | | . | . | . | . | . | . | . | . | . | . | . | . |
| | | . | . | . | . | . | . | . | . | . | . | . | . |

## 2 eQTL Mapping

For a QTL analysis, if the phenotype to be analyzed is the gene expression level data, then the analysis is referred to as the expression quantitative trait loci (eQTL) mapping. It aims to identify SNPs that influence the expression level of genes. It has been widely applied to dissect the genetic basis of gene expression and molecular mechanisms underlying complex traits [5, 45, 58]. More formally, let $\mathbf{X} = \{\mathbf{x}_d | 1 \leq d \leq D\} \in \mathbb{R}^{K \times D}$ be the SNP matrix denoting genotypes of $K$ SNPs of $D$ individuals and $\mathbf{Z} = \{\mathbf{z}_d | 1 \leq d \leq D\} \in \mathbb{R}^{N \times D}$ be the gene expression matrix denoting phenotypes of $N$ gene expression levels of the same set of $D$ individuals. Each column of $\mathbf{X}$ and $\mathbf{Z}$ stands for one individual. The goal of eQTL mapping is to find SNPs in $\mathbf{X}$, that are highly associated with genes in $\mathbf{Z}$.

Various statistics, such as the ANOVA (analysis of variance) test and the chi-square test, can be applied to measure the association between SNPs and the gene expression level of interest. Sparse feature selection methods, e.g., Lasso [63], are also widely used for eQTL mapping problems. Here, we take Lasso as an example. Lasso is a method for estimating the regression coefficients $\mathbf{W}$ using $\ell_1$ penalty. The objective function of Lasso is

$$\min_{\mathbf{W}} \frac{1}{2} ||\mathbf{Z} - \mathbf{W}\mathbf{X}||_F^2 + \eta ||\mathbf{W}||_1 \tag{1}$$

where $|| \cdot ||_F$ denotes the Frobenius norm, $|| \cdot ||_1$ is the $\ell_1$-norm. $\eta$ is the empirical parameter for the $\ell_1$ penalty. $\mathbf{W}$ is the parameter (also called weight) matrix setting the limits for the space of linear functions mapping from $\mathbf{X}$ to $\mathbf{Z}$. Each element of $\mathbf{W}$ is the effect size of corresponding SNP and expression level. Lasso uses the least squares method with $\ell_1$ penalty. $\ell_1$-norm sets many non-significant elements of $\mathbf{W}$ to be exactly zero, since many SNPs have no associations to a given gene. Lasso works even when the number of SNPs is significantly larger than the sample size ($K \gg D$) under the sparsity assumption.

Using the dataset shown in Fig. 1, Fig. 2a shows an example of strong association between gene expression $z_1$ and SNP $x_1$. 0 and 1 on the y-axis represent the binary SNP genotype and the x-axis represents the gene expression level. Each point in the figure represents an individual. It is clear from the figure that the gene expression

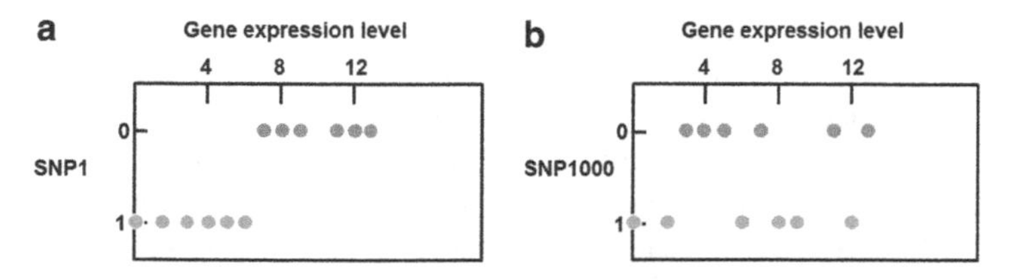


**Fig. 2** Examples of associations between a gene expression level and two different SNPs. (**a**) Strong association. (**b**) No association

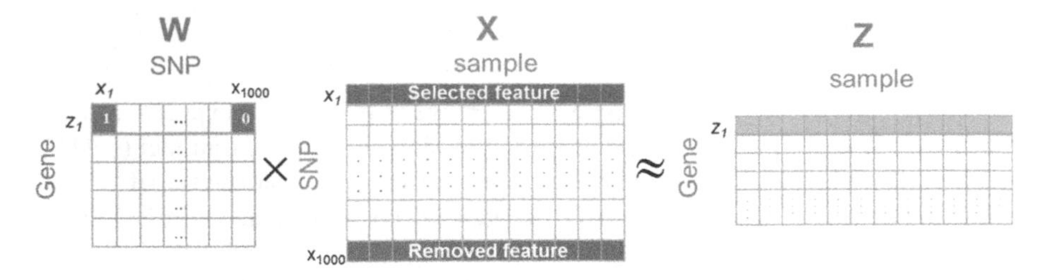


**Fig. 3** Association weights estimated by Lasso on the example data

level values are partitioned into two groups with distinct means, hence indicating a strong association between the gene expression and the SNP. On the other hand, if the genotype of an SNP partitions the gene expression level values into groups as shown in Fig. 2b, the gene expression and the SNP are not associated with each other. An illustration result of Lasso is shown in Fig. 3. $\mathbf{W}_{ij} = 0$ means no association between *j*th SNP and *i*th gene expression. $\mathbf{W}_{ij} \neq 0$ means there exists an association between the *j*th SNP and the *i*th gene expression.

## 2.1 Group-Wise eQTL Mapping and Challenges

In a typical eQTL study, the association between each expression trait and each SNP is assessed separately [11, 63, 72]. This approach does not consider the interactions among SNPs and among genes. However, multiple SNPs may jointly influence the phenotypes [33], and genes in the same biological pathway are often co-regulated and may share a common genetic basis [48, 55].

To better elucidate the genetic basis of gene expression, it is highly desirable to develop efficient methods that can automatically infer associations between a group of SNPs and a group of genes. We refer to the process of identifying such associations as *group-wise* eQTL mapping. In contrast, we refer to those associations between individual SNPs and individual genes as *individual* eQTL mapping. An example is shown in Fig. 4. Note that an ideal model should allow overlaps between SNP sets and between gene sets; that is, an SNP or gene may participate in multiple individual and group-wise associations. This is because genes and the SNPs influencing them may play different roles in multiple biological pathways [33].

Besides, advanced bio-techniques are generating a large volume of heterogeneous datasets, such as protein–protein interaction (PPI) networks [2] and genetic interaction networks [13]. These datasets describe the partial relationships between SNPs and relationships between genes. Because SNPs and genes are not independent of each other, and there exist group-wise associations, the integration of these

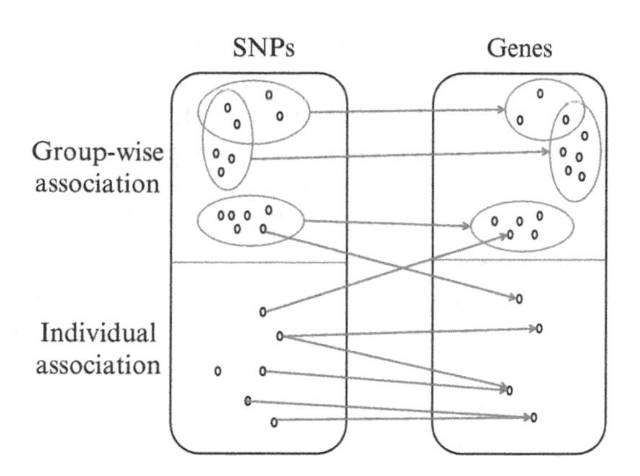


**Fig. 4** An illustration of individual and group-wise associations

multi-domain heterogeneous data sets is able to improve the accuracy of eQTL mapping since more domain knowledge can be integrated. In literature, several methods based on Lasso have been proposed [4, 32, 35, 36] to leverage the network prior knowledge [28, 32, 35, 36]. However, these methods suffer from poor quality or incompleteness of this prior knowledge.

In summary, there are several issues that greatly limit the applicability of current eQTL mapping approaches.

1. It is a crucial challenge to understand *how multiple, modestly associated SNPs interact to influence the phenotypes* [33]. However, little prior work has studied the group-wise eQTL mapping problem.
2. The prior knowledge about the relationships between SNPs and between genes is often partial and usually includes noise.
3. Confounding factors such as expression heterogeneity may result in spurious associations and mask real signals [20, 46, 60].

## 2.2 *Overview of the Developed Algorithms*

This book chapter proposes and studies the problem of group-wise eQTL mapping. We can decouple the problem into the following sub-problems:

- How can we detect group-wise eQTL associations with eQTL data only, i.e., with SNPs and gene expression profile data?
- How can we incorporate the prior interaction structures between SNPs and between genes into eQTL mapping to improve the robustness of the model and the interpretability of the results?

To address the first sub-problem, the book chapter proposes three approaches based on sparse linear-Gaussian graphical models to infer novel associations between SNP sets and gene sets. In literature, many efforts have focused on single-locus eQTL mapping. However, a multi-locus study dramatically increases the computation burden. The existing algorithms cannot be applied on a genome-wide scale. In order to accurately capture possible interactions between multiple genetic factors and their joint contribution to a group of phenotypic variations, we propose three algorithms. The first algorithm, SET-eQTL, makes use of a three-layer sparse linear-Gaussian model. The upper layer nodes correspond to the set of SNPs in the study. The middle layer consists of a set of hidden variables. The hidden variables are used to model both the joint effect of a set of SNPs and the effect of confounding factors. The lower layer nodes correspond to the genes in the study. The nodes in different layers are connected via arcs. SET-eQTL can help unravel true functional components in existing pathways. The results could provide new insights on how genes act and coordinate with each other to achieve certain biological functions. We further extend the approach to be able to consider confounding factors and decouple *individual* associations and *group-wise* associations for eQTL mapping.

To address the second sub-problem, this chapter presents an algorithm, Graph-regularized Dual Lasso (GDL), to simultaneously learn the association between SNPs and genes and refine the prior networks. Traditional sparse regression problems in data mining and machine learning consider both predictor variables and response variables individually, such as sparse feature selection using Lasso. In the eQTL mapping application, both predictor variables and response variables are not independent of each other, and we may be interested in the joint effects of multiple predictors to a group of response variables. In some cases, we may have partial prior knowledge, such as the correlation structures between predictors, and correlation structures between response variables. This chapter shows how prior graph information would help improve eQTL mapping accuracy and how refinement of prior knowledge would further improve the mapping accuracy. In addition, other different types of prior knowledge, e.g., location information of SNPs and genes, as well as pathway information, can also be integrated for the graph refinement.

## 2.3 Chapter Outline

The book chapter is organized as follows:

- The algorithms to detect group-wise eQTL associations with eQTL data only (SET-eQTL, etc.) are presented in Sect. 3.
- The algorithm (GDL) to incorporate the prior interaction structures or grouping information of SNPs or genes into eQTL mapping is presented in Sect. 4.
- Section 5 concludes the chapter work.

# 3 Group-Wise eQTL Mapping

## 3.1 Introduction

To better elucidate the genetic basis of gene expression and understand the underlying biology pathways, it is desirable to develop methods that can automatically infer associations between a group of SNPs and a group of genes. We refer to the process of identifying such associations as *group-wise* eQTL mapping. In contrast, we refer to the process of identifying associations between individual SNPs and genes as *individual* eQTL mapping. In this chapter, we propose several algorithms to detect group-wise associations. The first algorithm, SET-eQTL, makes use of a three-layer sparse linear-Gaussian model. It is able to identify novel associations between sets of SNPs and sets of genes. The results could provide new insights on how genes act and coordinate with each other to achieve certain biological functions. We further propose a fast and robust approach that is able to consider confounding factors and decouple *individual* associations and *group-wise* associations for eQTL mapping. The model is a multi-layer linear-Gaussian model and uses two different types of hidden variables: one capturing group-wise associations and the other capturing confounding factors [8, 18, 19, 29, 38, 42]. We apply an $\ell_1$-norm on the parameters [37, 63], which yields a sparse network with a large number of association weights being zero [50]. We develop an efficient optimization procedure that makes this approach suitable for large scale studies.

## 3.2 Related Work

Recently, various analytic methods have been developed to address the limitations of the traditional single-locus approach. Epistasis detection methods aim to find the interaction between SNP-pairs [3, 21, 22, 47]. The computational burden of epistasis detection is usually very high due to the large number of interactions that need to be examined [49, 57]. Filtering-based approaches [17, 23, 69], which reduce the search space by selecting a small subset of SNPs for interaction study, may miss important interactions in the SNPs that have been filtered out.

Statistical graphical models and Lasso-based methods [63] have been applied to eQTL study. A tree-guided group lasso has been proposed in [32]. This method directly combines statistical strength across multiple related genes in gene expression data to identify SNPs with pleiotropic effects by leveraging the hierarchical clustering tree over genes. Bayesian methods have also been developed [39, 61]. Confounding factors may greatly affect the results of the eQTL study. To model confounders, a two-step approach can be applied [27, 61]. These methods first learn the confounders that may exhibit broad effects to the gene expression traits. The learned confounders are then used as covariates in the subsequent analysis.

Statistical models that incorporate confounders have been proposed [51]. However, none of these methods are specifically designed to find novel associations between SNP sets and gene sets.

Pathway analysis methods have been developed to aggregate the association signals by considering a set of SNPs together [7, 16, 54, 64]. A pathway consists of a set of genes that coordinate to achieve a specific cell function. This approach studies a set of known pathways to find the ones that are highly associated with the phenotype [67]. Although appealing, this approach is limited to the a priori knowledge on the predefined gene sets/pathways. On the other hand, the current knowledgebase on the biological pathways is still far from being complete.

A method is proposed to identify eQTL association cliques that expose the hidden structure of genotype and expression data [25]. By using the cliques identified, this method can filter out SNP-gene pairs that are unlikely to have significant associations. It models the SNP, progeny, and gene expression data as an eQTL association graph, and thus depends on the availability of the progeny strain data as a bridge for modeling the eQTL association graph.

## 3.3 The Problem

Important notations used in this section are listed in Table 1. Throughout the section, we assume that, for each sample, the SNPs and genes are represented by column vectors. Let $\mathbf{x} = [x_1, x_2, \ldots, x_K]^{\mathrm{T}}$ represent the $K$ SNPs in the study, where $x_i \in \{0, 1, 2\}$ is a random variable corresponding to the $i$th SNP. For example, 0, 1, 2

**Table 1** Summary of notations

| Symbols | Description |
|---|---|
| $K$ | Number of SNPs |
| $N$ | Number of genes |
| $D$ | Number of samples |
| $M$ | Number of group-wise associations |
| $H$ | Number of confounding factors |
| $\mathbf{x}$ | Random variables of $K$ SNPs |
| $\mathbf{z}$ | Random variables of $N$ genes |
| $\mathbf{y}$ | Latent variables to model group-wise association |
| $\mathbf{X} \in \mathbb{R}^{K \times H}$ | SNP matrix data |
| $\mathbf{Z} \in \mathbb{R}^{N \times H}$ | Gene expression matrix data |
| $\mathbf{A} \in \mathbb{R}^{M \times K}$ | Group-wise association coefficient matrix between $\mathbf{x}$ and $\mathbf{y}$ |
| $\mathbf{B} \in \mathbb{R}^{N \times M}$ | Group-wise association coefficient matrix between $\mathbf{y}$ and $\mathbf{z}$ |
| $\mathbf{C} \in \mathbb{R}^{N \times K}$ | Individual association coefficient matrix between $\mathbf{x}$ and $\mathbf{y}$ |
| $\mathbf{P} \in \mathbb{R}^{N \times H}$ | Coefficient matrix of confounding factors |
| $\lambda, \gamma$ | Regularization parameters |

may encode the homozygous major allele, heterozygous allele, and homozygous minor allele, respectively. Let $\mathbf{z} = [z_1, z_2, \ldots, z_N]^T$ represent the $N$ genes in the study, where $z_j$ is a continuous random variable corresponding to the $j$th gene.

The traditional linear regression model for association mapping between $\mathbf{x}$ and $\mathbf{z}$ is

$$\mathbf{z} = \mathbf{W}\mathbf{x} + \boldsymbol{\mu} + \boldsymbol{\epsilon}, \tag{2}$$

where $\mathbf{z}$ is a linear function of $\mathbf{x}$ with coefficient matrix $\mathbf{W}$. $\boldsymbol{\mu}$ is an $N \times 1$ translation factor vector. $\boldsymbol{\epsilon}$ is the additive noise of Gaussian distribution with zero-mean and variance $\psi\mathbf{I}$, where $\psi$ is a scalar. That is, $\boldsymbol{\epsilon} \sim N(\mathbf{0}, \psi\mathbf{I})$.

The question now is how to define an appropriate objective function to decompose $\mathbf{W}$ which (1) can effectively detect both individual and group-wise eQTL associations, and (2) is efficient to compute so that it is suitable for large scale studies. In the next, we will propose a group-wise eQTL detection method first, and then improve it to capture both individual and group-wise associations. Finally, we will discuss how to boost the computational efficiency.

## 3.4 *Detecting Group-Wise Associations*

### 3.4.1 SET-eQTL Model

To infer associations between SNP sets and gene sets, we propose a graphical model as shown in Fig. 5, which is able to capture any potential confounding factors in a natural way. This model is a two-layer linear-Gaussian model. The hidden variables in the middle layer are used to capture the group-wise association between SNP sets and gene sets. These latent variables are presented as $\mathbf{y} = [y_1, y_2, \ldots, y_M]^T$, where $M$ is the total number of latent variables bridging SNP sets and gene sets. Each hidden variable may represent a latent factor regulating a set of genes, and its associated genes may correspond to a set of genes in the same pathway or participating in certain biological function. Note that this model allows an SNP or gene to participate in multiple (SNP set, gene set) pairs. This is reasonable because SNPs and genes may play different roles in multiple biology pathways. Since the model bridges SNP sets and gene sets, we refer this method as SET-eQTL.

The exact role of these latent factors can be inferred from the network topology of the resulting sparse graphical model learned from the data (by imposing $\ell_1$-norm on the likelihood function, which will be discussed later in this section). Figure 6 shows an example of the resulting graphical model. There are two types of hidden variables. One type consists of hidden variables with zero in-degree (i.e., no connections with the SNPs). These hidden variables correspond to the confounding factors. Other types of hidden variables serve as bridges connecting SNP sets and gene sets. In Fig. 6, $y_k$ is a hidden variable modeling confounding effects. $y_i$ and $y_j$ are bridge nodes connecting the SNPs and genes associated with them. Note that this

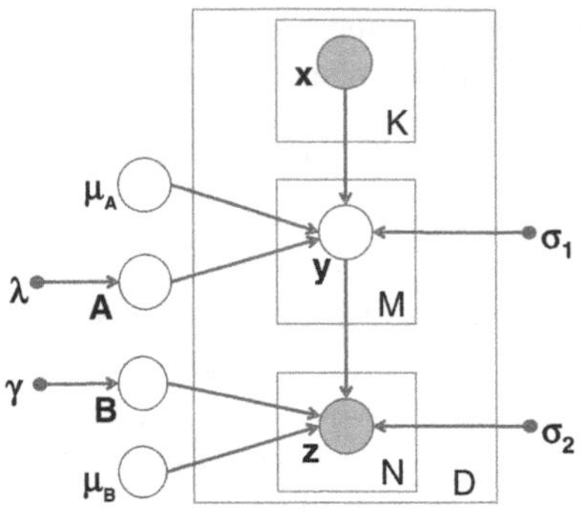

**Fig. 5** The proposed graphical model with hidden variables

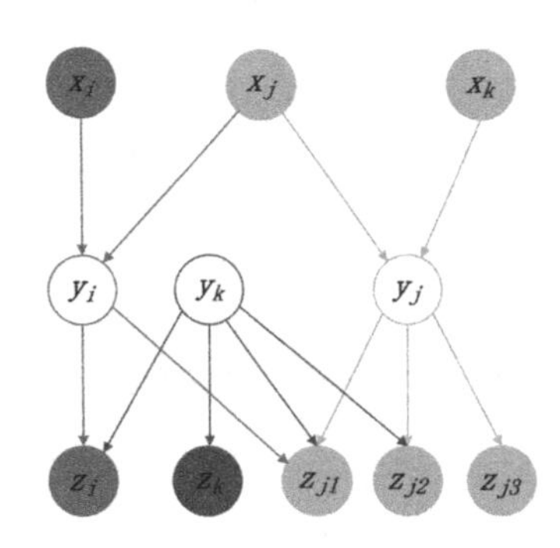


**Fig. 6** An example of the inferred sparse graphical model

model allows overlaps between different (SNP set, gene set) pairs. It is reasonable because SNPs and genes may play multiple roles in different biology pathways.

### 3.4.2 Objective Function

From the probability theory, we have that the joint probability of $\mathbf{x}$ and $\mathbf{z}$ is

$$p(\mathbf{x}, \mathbf{z}) = \int_{\mathbf{y}} p(\mathbf{x}, \mathbf{y}, \mathbf{z}) d\mathbf{y}. \tag{3}$$

From the factorization properties of the joint distribution for a directed graphical model, we have

$$p(\mathbf{x}, \mathbf{y}, \mathbf{z}) = p(\mathbf{y}|\mathbf{x}) p(\mathbf{z}|\mathbf{y}) p(\mathbf{x}). \tag{4}$$

Thus, we have

$$p(\mathbf{z}|\mathbf{x}) = \frac{p(\mathbf{x}, \mathbf{z})}{p(\mathbf{x})} = \int_{\mathbf{y}} p(\mathbf{y}|\mathbf{x}) p(\mathbf{z}|\mathbf{y}) d\mathbf{y}. \tag{5}$$

We assume that the two conditional probabilities follow normal distributions:

$$\mathbf{y}|\mathbf{x} \sim \mathcal{N}(\mathbf{y}|\mathbf{A}\mathbf{x} + \boldsymbol{\mu}_{\mathbf{A}}, \sigma_1^2\mathbf{I_M}),$$

and

$$\mathbf{z}|\mathbf{y} \sim \mathcal{N}(\mathbf{z}|\mathbf{B}\mathbf{y} + \boldsymbol{\mu}_{\mathbf{B}}, \sigma_2^2\mathbf{I_N}),$$

where $\mathbf{A} \in \mathbb{R}^{M\times K}$ is the coefficient matrix between $\mathbf{x}$ and $\mathbf{y}$, $\mathbf{B} \in \mathbb{R}^{N\times M}$ is the coefficient matrix between $\mathbf{y}$ and $\mathbf{z}$. $\boldsymbol{\mu}_{\mathbf{A}} \in \mathbb{R}^{M\times 1}$ and $\boldsymbol{\mu}_{\mathbf{B}} \in \mathbb{R}^{N\times 1}$ are the translation factor vectors, of which $\sigma_1^2\mathbf{I}_M$ and $\sigma_2^2\mathbf{I}_N$ are their variances, respectively, ($\sigma_1$ and $\sigma_2$ are constant scalars and $\mathbf{I}_M$ and $\mathbf{I}_N$ are identity matrices).

To impose sparsity, we assume that entries of $\mathbf{A}$ and $\mathbf{B}$ follow Laplace distributions:

$$\mathbf{A} \sim \mathbf{Laplace}(\mathbf{0}, 1/\lambda),$$

and

$$\mathbf{B} \sim \mathbf{Laplace}(\mathbf{0}, 1/\gamma).$$

$\lambda$ and $\gamma$ are parameters of the $\ell_1$-regularization penalty on the objective function. This model is a two-layer linear model and $p(\mathbf{y}|\mathbf{x})$ serves as the conjugate prior of $p(\mathbf{z}|\mathbf{y})$. Thus we have

$$\boldsymbol{\beta} \cdot \mathcal{N}(\mathbf{y}|\boldsymbol{\mu}_{\mathbf{y}}, \boldsymbol{\Sigma}_{\mathbf{y}}) = \mathcal{N}(\mathbf{y}|\mathbf{A}\mathbf{x} + \boldsymbol{\mu}_{\mathbf{A}}, \sigma_1^2\mathbf{I_M}) \cdot \mathcal{N}(\mathbf{z}|\mathbf{B}\mathbf{y} + \boldsymbol{\mu}_{\mathbf{B}}, \sigma_2^2\mathbf{I_N}) \tag{6}$$

where $\boldsymbol{\beta}$ is a scalar, $\boldsymbol{\mu}_{\mathbf{y}}$ and $\boldsymbol{\Sigma}_{\mathbf{y}}$ are the mean and variance of a new normal distribution, respectively.

From Eqs. (5) and (6), we have that

$$p(\mathbf{z}|\mathbf{x}) = \int_{\mathbf{y}} \boldsymbol{\beta} \cdot \mathcal{N}(\mathbf{y}|\boldsymbol{\mu}_{\mathbf{y}}, \boldsymbol{\Sigma}_{\mathbf{y}})\mathrm{d}\mathbf{y} = \boldsymbol{\beta} \tag{7}$$

Thus, maximizing $p(\mathbf{z}|\mathbf{x})$ is equivalent to maximizing $\boldsymbol{\beta}$. Next, we show the derivation of $\boldsymbol{\beta}$. We first derive the value of $\boldsymbol{\mu}_{\mathbf{y}}$ and $\boldsymbol{\Sigma}_{\mathbf{y}}^{-1}$ by comparing the exponential terms on both sides of Eq. (6).

$$\begin{aligned} &\mathcal{N}(\mathbf{y}|\mathbf{A}\mathbf{x} + \boldsymbol{\mu}_{\mathbf{A}}, \sigma_1^2\mathbf{I_M}) \cdot \mathcal{N}(\mathbf{z}|\mathbf{B}\mathbf{y} + \boldsymbol{\mu}_{\mathbf{B}}, \sigma_2^2\mathbf{I_N}) \\ &= \frac{1}{(2\pi)^{\frac{M+N}{2}}\sigma_1^M\sigma_2^N} \exp\{-\tfrac{1}{2}[\tfrac{1}{\sigma_1^2}(\mathbf{y} - \mathbf{A}\mathbf{x} - \boldsymbol{\mu}_{\mathbf{A}})^{\mathrm{T}}(\mathbf{y} - \mathbf{A}\mathbf{x} - \boldsymbol{\mu}_{\mathbf{A}}) \\ &\qquad + \tfrac{1}{\sigma_2^2}(\mathbf{z} - \mathbf{B}\mathbf{y} - \boldsymbol{\mu}_{\mathbf{B}})^{\mathrm{T}}(\mathbf{z} - \mathbf{B}\mathbf{y} - \boldsymbol{\mu}_{\mathbf{B}})]\} \end{aligned} \tag{8}$$

The exponential term in Eq. (8) can be expanded as

$$\begin{aligned}
\boldsymbol{\Psi} &= -\tfrac{1}{2}[\tfrac{1}{\sigma_1^2}(\mathbf{y}-\mathbf{A}\mathbf{x}-\boldsymbol{\mu}_{\mathbf{A}})^{\mathrm{T}}(\mathbf{y}-\mathbf{A}\mathbf{x}) \\
&\quad +\sigma_2^2(\mathbf{z}-\mathbf{B}\mathbf{y}-\boldsymbol{\mu}_{\mathbf{B}})^{\mathrm{T}}(\mathbf{z}-\mathbf{B}\mathbf{y})] \\
&= -\tfrac{1}{2}[\tfrac{1}{\sigma_1^2}(\mathbf{y}^{\mathrm{T}}\mathbf{y}-\mathbf{y}^{\mathrm{T}}\mathbf{A}\mathbf{x}-\mathbf{y}^{\mathrm{T}}\boldsymbol{\mu}_{\mathbf{A}}-\mathbf{x}^{\mathrm{T}}\mathbf{A}^{\mathrm{T}}\mathbf{y}+\mathbf{x}^{\mathrm{T}}\mathbf{A}^{\mathrm{T}}\mathbf{A}\mathbf{x} \\
&\quad +\mathbf{x}^{\mathrm{T}}\mathbf{A}^{\mathrm{T}}\boldsymbol{\mu}_{\mathbf{A}}-\boldsymbol{\mu}_{\mathbf{A}}^{\mathrm{T}}\mathbf{y}+\boldsymbol{\mu}_{\mathbf{A}}^{\mathrm{T}}\mathbf{A}X+\boldsymbol{\mu}_{\mathbf{A}}^{\mathrm{T}}\boldsymbol{\mu}_{\mathbf{A}})+\tfrac{1}{\sigma_2^2}(\mathbf{z}^{\mathrm{T}}\mathbf{z}-\mathbf{z}^{\mathrm{T}}\mathbf{B}\mathbf{y} \\
&\quad -\mathbf{z}^{\mathrm{T}}\boldsymbol{\mu}_{\mathbf{B}}-\mathbf{y}^{\mathrm{T}}\mathbf{B}^{\mathrm{T}}\mathbf{z}+\mathbf{y}^{\mathrm{T}}\mathbf{B}^{\mathrm{T}}\mathbf{B}\mathbf{y}+\mathbf{y}^{\mathrm{T}}\mathbf{B}^{\mathrm{T}}\boldsymbol{\mu}_{\mathbf{B}}-\boldsymbol{\mu}_{\mathbf{B}}^{\mathrm{T}}\mathbf{z}+\boldsymbol{\mu}_{\mathbf{B}}^{\mathrm{T}}\mathbf{B}\mathbf{y} \\
&\quad +\boldsymbol{\mu}_{\mathbf{B}}^{\mathrm{T}}\boldsymbol{\mu}_{\mathbf{B}})] \\
&= -\tfrac{1}{2}[\mathbf{y}^{\mathrm{T}}(\tfrac{1}{\sigma_1^2}\mathbf{I}_M+\tfrac{1}{\sigma_2^2}\mathbf{B}^{\mathrm{T}}\mathbf{B})\mathbf{y}-\tfrac{2}{\sigma_1^2}(\mathbf{x}^{\mathrm{T}}\mathbf{A}^{\mathrm{T}}\mathbf{y}+\boldsymbol{\mu}_{\mathbf{A}}^{\mathrm{T}}\mathbf{y}) \\
&\quad -\tfrac{2}{\sigma_2^2}(\mathbf{z}^{\mathrm{T}}\mathbf{B}\mathbf{y}-\boldsymbol{\mu}_{\mathbf{B}}^{\mathrm{T}}\mathbf{B}\mathbf{y})+\tfrac{1}{\sigma_1^2}(\mathbf{x}^{\mathrm{T}}\mathbf{A}^{\mathrm{T}}\mathbf{A}\mathbf{x}+2\boldsymbol{\mu}_{\mathbf{A}}^{\mathrm{T}}\mathbf{A}\mathbf{x}+\boldsymbol{\mu}_{\mathbf{A}}^{\mathrm{T}}\boldsymbol{\mu}_{\mathbf{A}}) \\
&\quad +\tfrac{1}{\sigma_2^2}(\mathbf{z}^{\mathrm{T}}\mathbf{z}-2\boldsymbol{\mu}_{\mathbf{B}}^{\mathrm{T}}\mathbf{z}+\boldsymbol{\mu}_{\mathbf{B}}^{\mathrm{T}}\boldsymbol{\mu}_{\mathbf{B}})]
\end{aligned} \tag{9}$$

Thus, by comparing the exponential terms on both sides of Eq. (6), we get

$$\boldsymbol{\Sigma}_{\mathbf{y}}^{-1} = \frac{1}{\sigma_1^2}\mathbf{I}_M + \frac{1}{\sigma_2^2}\mathbf{B}^{\mathrm{T}}\mathbf{B}, \tag{10}$$

$$\boldsymbol{\mu}_{\mathbf{y}}^{\mathrm{T}}\boldsymbol{\Sigma}_{\mathbf{y}}^{-1} = \frac{1}{\sigma_1^2}(\mathbf{x}^{\mathrm{T}}\mathbf{A}^{\mathrm{T}}+\boldsymbol{\mu}_{\mathbf{A}}^{\mathrm{T}}) + \frac{1}{\sigma_2^2}(\mathbf{z}^{\mathrm{T}}\mathbf{B}-\boldsymbol{\mu}_{\mathbf{B}}^{\mathrm{T}}\mathbf{B}). \tag{11}$$

Further, we have

$$\boldsymbol{\mu}_{\mathbf{y}} = \boldsymbol{\Sigma}_{\mathbf{y}}[\frac{1}{\sigma_1^2}(\mathbf{A}\mathbf{x}+\boldsymbol{\mu}_{\mathbf{A}}) + \frac{1}{\sigma_2^2}(\mathbf{B}^{\mathrm{T}}\mathbf{z}-\mathbf{B}^{\mathrm{T}}\boldsymbol{\mu}_{\mathbf{B}})]. \tag{12}$$

With $\boldsymbol{\Sigma}_{\mathbf{y}}^{-1}$ and $\boldsymbol{\mu}_{\mathbf{y}}$, we can derive the explicit form of $\boldsymbol{\beta}$ easily by setting $\mathbf{y} = \mathbf{0}$, which leads to the equation below:

$$\begin{aligned}
&\boldsymbol{\beta}\cdot\frac{1}{(2\pi)^{\frac{M}{2}}|\boldsymbol{\Sigma}_{\mathbf{y}}|^{\frac{1}{2}}}\exp\{-\tfrac{1}{2}\boldsymbol{\mu}_{\mathbf{y}}^{\mathrm{T}}\boldsymbol{\Sigma}_{\mathbf{y}}^{-1}\boldsymbol{\mu}_{\mathbf{y}}\} \\
&= \frac{1}{(2\pi)^{\frac{M+N}{2}}\sigma_1^M\sigma_2^N}\exp\{\boldsymbol{\Psi}_{\mathbf{y}=\mathbf{0}}\},
\end{aligned} \tag{13}$$

where $\boldsymbol{\Psi}_{\mathbf{y}=\mathbf{0}}$ is the value of $\boldsymbol{\Psi}$ when $\mathbf{y} = \mathbf{0}$, and thereby

$$\begin{aligned}
\boldsymbol{\Psi}_{\mathbf{y}=\mathbf{0}} &= -\tfrac{1}{2}[\tfrac{1}{\sigma_1^2}(\mathbf{x}^{\mathrm{T}}\mathbf{A}^{\mathrm{T}}\mathbf{A}\mathbf{x}+2\boldsymbol{\mu}_{\mathbf{A}}^{\mathrm{T}}\mathbf{A}\mathbf{x}+\boldsymbol{\mu}_{\mathbf{A}}^{\mathrm{T}}\boldsymbol{\mu}_{\mathbf{A}}) \\
&\quad +\tfrac{1}{\sigma_2^2}(\mathbf{z}^{\mathrm{T}}\mathbf{z}-2\boldsymbol{\mu}_{\mathbf{B}}^{\mathrm{T}}\mathbf{z}+\boldsymbol{\mu}_{\mathbf{B}}^{\mathrm{T}}\boldsymbol{\mu}_{\mathbf{B}})]
\end{aligned} \tag{14}$$

Thus, we get the explicit form of $\boldsymbol{\beta}$ as

$$\boldsymbol{\beta} = \frac{|\boldsymbol{\Sigma}_{\mathbf{y}}|^{\frac{1}{2}}}{(2\pi)^{\frac{N}{2}}\sigma_1^M\sigma_2^N}\exp\{\boldsymbol{\Psi}_{\mathbf{y}=\mathbf{0}}+\tfrac{1}{2}(\boldsymbol{\mu}_{\mathbf{y}}^{\mathrm{T}}\boldsymbol{\Sigma}_{\mathbf{y}}^{-1}\boldsymbol{\mu}_{\mathbf{y}})\}. \tag{15}$$

Here, $\boldsymbol{\beta} = p(\mathbf{z}|\mathbf{x}, \mathbf{A}, \mathbf{B}, \boldsymbol{\mu}_{\mathbf{A}}, \boldsymbol{\mu}_{\mathbf{B}}, \sigma_1, \sigma_2)$ is the likelihood function for one data point $\mathbf{x}$. Let $\mathbf{X} = \{\mathbf{x}_d\}$ and $\mathbf{Z} = \{\mathbf{z}_d\}$ be the sets of $D$ observed data points (genotype and the gene expression profiles for the samples in the study). To maximize $\boldsymbol{\beta}_d$, we can minimize the negative log-likelihood of $\boldsymbol{\beta}_d$. Thus, our loss function is

$$\begin{aligned} \mathcal{J} &= -\log \textstyle\prod_{d=1}^{D} p(\mathbf{z}_d|\mathbf{x}_d) \\ &= -\textstyle\sum_{d=1}^{D} \log p(\mathbf{z}_d|\mathbf{x}_d) \\ &= -\textstyle\sum_{d=1}^{D} \log \boldsymbol{\beta}_d \end{aligned} \tag{16}$$

Substituting Eq. (15) into Eq. (16), the expanded form of the loss function is

$$\begin{aligned} &\mathcal{J}(\mathbf{A}, \mathbf{B}, \boldsymbol{\mu}_{\mathbf{A}}, \boldsymbol{\mu}_{\mathbf{B}}, \sigma_1, \sigma_2) \\ &= \tfrac{D \cdot N}{2} \ln(2\pi) + D \cdot M \ln(\sigma_1) + D \cdot N \ln(\sigma_2) + \tfrac{D}{2} \ln |\boldsymbol{\Sigma}_{\mathbf{y}}^{-1}| \\ &+ \tfrac{1}{2} \textstyle\sum_{d=1}^{D} \{ \tfrac{1}{\sigma_1^2} (\mathbf{x}_d^{\mathrm{T}} \mathbf{A}^{\mathrm{T}} \mathbf{A} \mathbf{x}_d + 2\boldsymbol{\mu}_{\mathbf{A}}^{\mathrm{T}} \mathbf{A} \mathbf{x}_d + \boldsymbol{\mu}_{\mathbf{A}}^{\mathrm{T}} \boldsymbol{\mu}_{\mathbf{A}}) \\ &+ \tfrac{1}{\sigma_2^2} (\mathbf{z}_d^{\mathrm{T}} \mathbf{z}_d - 2\boldsymbol{\mu}_{\mathbf{B}}^{\mathrm{T}} \mathbf{z}_d + \boldsymbol{\mu}_{\mathbf{B}}^{\mathrm{T}} \boldsymbol{\mu}_{\mathbf{B}}) - [\tfrac{1}{\sigma_1^2} (\mathbf{x}_d^{\mathrm{T}} \mathbf{A}^{\mathrm{T}} + \boldsymbol{\mu}_A^{\mathrm{T}}) \\ &+ \tfrac{1}{\sigma_2^2} (\mathbf{z}_d^{\mathrm{T}} \mathbf{B} - \boldsymbol{\mu}_{\mathbf{B}}^{\mathrm{T}} \mathbf{B})] \boldsymbol{\Sigma}_{\mathbf{y}} [\tfrac{1}{\sigma_1^2} (\mathbf{A} \mathbf{x}_d + \boldsymbol{\mu}_A) + \tfrac{1}{\sigma_2^2} (\mathbf{B}^{\mathrm{T}} \mathbf{z}_d - \mathbf{B}^{\mathrm{T}} \boldsymbol{\mu}_{\mathbf{B}})] \} \end{aligned} \tag{17}$$

Taking into account the prior distributions of $\mathbf{A}$ and $\mathbf{B}$, we have that

$$\begin{aligned} &p(\mathbf{z}, \mathbf{A}, \mathbf{B}|\mathbf{x}, \boldsymbol{\mu}_{\mathbf{A}}, \boldsymbol{\mu}_{\mathbf{B}}, \sigma_1, \sigma_2) \\ &= \boldsymbol{\beta} \cdot \textbf{Laplace}(\mathbf{A}|\mathbf{0}, 1/\lambda) \cdot \textbf{Laplace}(\mathbf{B}|\mathbf{0}, 1/\gamma) \end{aligned} \tag{18}$$

Thus, we can have the $\ell_1$-regularized objective function

$$\max_{\mathbf{A}, \mathbf{B}, \boldsymbol{\mu}_{\mathbf{A}}, \boldsymbol{\mu}_{\mathbf{B}}, \sigma_1, \sigma_2} \log \prod_{d=1}^{D} p(\mathbf{z}_d, \mathbf{A}, \mathbf{B}|\mathbf{x}_d, \boldsymbol{\mu}_{\mathbf{A}}, \boldsymbol{\mu}_{\mathbf{B}}, \sigma_1, \sigma_2),$$

which is identical to

$$\min_{\mathbf{A}, \mathbf{B}, \boldsymbol{\mu}_{\mathbf{A}}, \boldsymbol{\mu}_{\mathbf{B}}, \sigma_1, \sigma_2} [\mathcal{J} + D \cdot (\lambda ||\mathbf{A}||_1 + \gamma ||\mathbf{B}||_1)], \tag{19}$$

where $|| \cdot ||_1$ is the $\ell_1$-norm. $\lambda$ and $\gamma$ are the *precision* of the prior Laplace distributions of $\mathbf{A}$ and $\mathbf{B}$, respectively, serving as the regularization parameters which can be determined by cross or holdout validation.

The gradient of the loss function $\mathcal{J}$ with respect to $\mathbf{A}$, $\mathbf{B}$, $\boldsymbol{\mu}_{\mathbf{A}}$, $\boldsymbol{\mu}_{\mathbf{B}}$, $\sigma_1$, and $\sigma_2$ are

$$\begin{aligned} \nabla_{\mathbf{A}} \mathcal{J} &= \textstyle\sum_{d=1}^{D} (\tfrac{1}{\sigma_1^2} \mathbf{A} \mathbf{x}_d \mathbf{x}_d^{\mathrm{T}} - \tfrac{1}{\sigma_1^4} \boldsymbol{\Sigma}_{\mathbf{y}} \mathbf{A} \mathbf{x}_d \mathbf{x}_d^{\mathrm{T}} - \tfrac{1}{\sigma_1^2 \sigma_2^2} \boldsymbol{\Sigma}_{\mathbf{y}} \mathbf{B}^{\mathrm{T}} \mathbf{z}_d \mathbf{x}_d^{\mathrm{T}} \\ &+ \tfrac{1}{\sigma_1^2} \boldsymbol{\mu}_{\mathbf{A}} \mathbf{x}_d^{\mathrm{T}} - \tfrac{1}{\sigma_1^4} \boldsymbol{\Sigma}_{\mathbf{y}} \boldsymbol{\mu}_{\mathbf{A}} \mathbf{x}_d^{\mathrm{T}} + \tfrac{1}{\sigma_1^2 \sigma_2^2} \boldsymbol{\Sigma}_{\mathbf{y}} \mathbf{B}^{\mathrm{T}} \boldsymbol{\mu}_{\mathbf{B}} \mathbf{x}_d^{\mathrm{T}}) \end{aligned} \tag{20}$$

$$\begin{aligned}\nabla_{\mathbf{B}}\mathcal{J} &= \tfrac{D}{\sigma_2^2}\mathbf{B}\boldsymbol{\Sigma}_{\mathbf{y}} + \tfrac{1}{\sigma_2^4}(\tfrac{1}{\sigma_2^2}\mathbf{B}\boldsymbol{\Sigma}_{\mathbf{y}}\mathbf{B}^{\mathrm{T}} - \mathbf{I}_N)\textstyle\sum_{d=1}^{D}[(\mathbf{z}_d - \boldsymbol{\mu}_{\mathbf{B}})\\ &\cdot(\mathbf{z}_d - \boldsymbol{\mu}_{\mathbf{B}})^{\mathrm{T}}]\mathbf{B}\boldsymbol{\Sigma}_{\mathbf{y}} + \tfrac{1}{\sigma_1^2\sigma_2^4}\textstyle\sum_{d=1}^{D}\{\mathbf{B}\boldsymbol{\Sigma}_{\mathbf{y}}[(\mathbf{A}\mathbf{x}_d + \boldsymbol{\mu}_{\mathbf{A}})(\mathbf{z}_d - \boldsymbol{\mu}_{\mathbf{B}})^{\mathrm{T}}\mathbf{B}\\ &+\mathbf{B}^{\mathrm{T}}(\mathbf{z}_d - \boldsymbol{\mu}_{\mathbf{B}})(\mathbf{A}\mathbf{x}_d + \boldsymbol{\mu}_{\mathbf{A}})^{\mathrm{T}}]\boldsymbol{\Sigma}_{\mathbf{y}} - \sigma_2^2(\mathbf{z}_d - \boldsymbol{\mu}_{\mathbf{B}})(\mathbf{A}\mathbf{x}_d + \boldsymbol{\mu}_{\mathbf{A}})^{\mathrm{T}}\boldsymbol{\Sigma}_{\mathbf{y}}\}\\ &+\tfrac{1}{\sigma_1^4\sigma_2^2}B\boldsymbol{\Sigma}_{\mathbf{y}}\textstyle\sum_{d=1}^{D}[(\mathbf{A}\mathbf{x}_d + \boldsymbol{\mu}_{\mathbf{A}})(\mathbf{x}_d^{\mathrm{T}}\mathbf{A}^{\mathrm{T}} + \boldsymbol{\mu}_{\mathbf{A}}^{\mathrm{T}})]\boldsymbol{\Sigma}_{\mathbf{y}}\end{aligned} \tag{21}$$

$$\nabla_{\boldsymbol{\mu}_{\mathbf{A}}}\mathcal{J} = \tfrac{1}{2}\textstyle\sum_{d=1}^{D}[\tfrac{2}{\sigma_1^2}(\mathbf{A}\mathbf{x}_d + \boldsymbol{\mu}_{\mathbf{A}}) - \tfrac{2}{\sigma_1^4}\boldsymbol{\Sigma}_{\mathbf{y}}(\boldsymbol{\mu}_{\mathbf{A}} + \mathbf{A}\mathbf{x}_d) - \tfrac{2}{\sigma_1^2\sigma_2^2}\boldsymbol{\Sigma}_{\mathbf{y}}(\mathbf{B}^{\mathrm{T}}\mathbf{z}_d - \mathbf{B}^{\mathrm{T}}\boldsymbol{\mu}_{\mathbf{B}})] \tag{22}$$

$$\nabla_{\boldsymbol{\mu}_{\mathbf{B}}}\mathcal{J} = \tfrac{1}{2}\textstyle\sum_{d=1}^{D}[\tfrac{2}{\sigma_1^2}(-\mathbf{z}_d + \boldsymbol{\mu}_{\mathbf{B}}) + \tfrac{2}{\sigma_1^4}\mathbf{B}\boldsymbol{\Sigma}_{\mathbf{y}}\mathbf{B}^{\mathrm{T}}(\mathbf{z}_d - \boldsymbol{\mu}_{\mathbf{B}}) + \tfrac{2}{\sigma_1^2\sigma_2^2}\mathbf{B}\boldsymbol{\Sigma}_{\mathbf{y}}(\mathbf{A}\mathbf{x}_d + \boldsymbol{\mu}_{\mathbf{A}})] \tag{23}$$

$$\begin{aligned}\nabla_{\sigma_1}\mathcal{J} &= \tfrac{D\cdot M}{\sigma_1} - \tfrac{D\cdot\mathrm{tr}(\boldsymbol{\Sigma}_{\mathbf{y}})}{\sigma_1^3} + \textstyle\sum_{d=1}^{D}[-\tfrac{\mathbf{x}_d^{\mathrm{T}}\mathbf{A}^{\mathrm{T}}\mathbf{A}\mathbf{x}_d + 2\boldsymbol{\mu}_{\mathbf{A}}^{\mathrm{T}}\mathbf{A}\mathbf{x}_d + \boldsymbol{\mu}_{\mathbf{A}}^{\mathrm{T}}\boldsymbol{\mu}_{\mathbf{A}}}{\sigma_1^3}\\ &+\tfrac{2(\mathbf{x}_d^{\mathrm{T}}\mathbf{A}^{\mathrm{T}} + \boldsymbol{\mu}_{\mathbf{A}}^{\mathrm{T}})\boldsymbol{\Sigma}_{\mathbf{y}}(\mathbf{A}\mathbf{x}_d + \boldsymbol{\mu}_{\mathbf{A}})}{\sigma_1^5} - \tfrac{(\mathbf{x}_d^{\mathrm{T}}\mathbf{A}^{\mathrm{T}} + \boldsymbol{\mu}_{\mathbf{A}}^{\mathrm{T}})\boldsymbol{\Sigma}_{\mathbf{y}}^2(\mathbf{A}\mathbf{x}_d + \boldsymbol{\mu}_{\mathbf{A}})}{\sigma_1^7}\\ &+\tfrac{2(\mathbf{x}_d^{\mathrm{T}}\mathbf{A}^{\mathrm{T}} + \boldsymbol{\mu}_{\mathbf{A}}^{\mathrm{T}})\boldsymbol{\Sigma}_{\mathbf{y}}(\mathbf{B}^{\mathrm{T}}\mathbf{z}_d - \mathbf{B}^{\mathrm{T}}\boldsymbol{\mu}_{\mathbf{B}})}{\sigma_1^3\sigma_2^2} - \tfrac{2(\mathbf{x}_d^{\mathrm{T}}\mathbf{A}^{\mathrm{T}} + \boldsymbol{\mu}_{\mathbf{A}}^{\mathrm{T}})\boldsymbol{\Sigma}_{\mathbf{y}}^2(\mathbf{B}^{\mathrm{T}}\mathbf{z}_d - \mathbf{B}^{\mathrm{T}}\boldsymbol{\mu}_{\mathbf{B}})}{\sigma_1^5\sigma_2^2}\\ &-\tfrac{(\mathbf{z}_d^{\mathrm{T}}\mathbf{B} - \boldsymbol{\mu}_{\mathbf{B}}^{\mathrm{T}}\mathbf{B})\boldsymbol{\Sigma}_{\mathbf{y}}^2(\mathbf{B}^{\mathrm{T}}\mathbf{z}_d - \mathbf{B}^{\mathrm{T}}\boldsymbol{\mu}_{\mathbf{B}})}{\sigma_1^3\sigma_2^4}]\end{aligned} \tag{24}$$

$$\begin{aligned}\nabla_{\sigma_2}\mathcal{J} &= \tfrac{D\cdot N}{\sigma_2} - \tfrac{D\cdot\mathrm{tr}(\boldsymbol{\Sigma}_{\mathbf{y}}\mathbf{B}^{\mathrm{T}}\mathbf{B})}{\sigma_2^3} + \textstyle\sum_{d=1}^{D}[-\tfrac{\mathbf{z}_d^{\mathrm{T}}\mathbf{z}_d - 2\boldsymbol{\mu}_{\mathbf{B}}^{\mathrm{T}}\mathbf{z}_d + \boldsymbol{\mu}_{\mathbf{B}}^{\mathrm{T}}\boldsymbol{\mu}_{\mathbf{B}}}{\sigma_2^3}\\ &+\tfrac{2(\mathbf{z}_d^{\mathrm{T}}\mathbf{B} - \boldsymbol{\mu}_{\mathbf{B}}^{\mathrm{T}}\mathbf{B})\boldsymbol{\Sigma}_{\mathbf{y}}(\mathbf{B}^{\mathrm{T}}\mathbf{z}_d - \mathbf{B}^{\mathrm{T}}\boldsymbol{\mu}_{\mathbf{B}})}{\sigma_2^5} - \tfrac{(\mathbf{z}_d^{\mathrm{T}}\mathbf{B} - \boldsymbol{\mu}_{\mathbf{B}}^{\mathrm{T}}\mathbf{B})\boldsymbol{\Sigma}_{\mathbf{y}}\mathbf{B}^{\mathrm{T}}\mathbf{B}\boldsymbol{\Sigma}_{\mathbf{y}}(\mathbf{B}^{\mathrm{T}}\mathbf{z}_d - \mathbf{B}^{\mathrm{T}}\boldsymbol{\mu}_{\mathbf{B}})}{\sigma_2^7}\\ &+\tfrac{2(\mathbf{z}_d^{\mathrm{T}}\mathbf{B} - \boldsymbol{\mu}_{\mathbf{B}}^{\mathrm{T}}\mathbf{B})\boldsymbol{\Sigma}_{\mathbf{y}}(\mathbf{A}\mathbf{x}_d + \boldsymbol{\mu}_{\mathbf{A}})}{\sigma_1^2\sigma_2^3} - \tfrac{2(\mathbf{z}_d^{\mathrm{T}}\mathbf{B} - \boldsymbol{\mu}_{\mathbf{B}}^{\mathrm{T}}\mathbf{B})\boldsymbol{\Sigma}_{\mathbf{y}}\mathbf{B}^{\mathrm{T}}\mathbf{B}\boldsymbol{\Sigma}_{\mathbf{y}}(\mathbf{A}\mathbf{x}_d + \boldsymbol{\mu}_{\mathbf{A}})}{\sigma_1^2\sigma_2^5}\\ &-\tfrac{(\mathbf{x}_d^{\mathrm{T}}\mathbf{A}^{\mathrm{T}} + \boldsymbol{\mu}_{\mathbf{A}}^{\mathrm{T}})\boldsymbol{\Sigma}_{\mathbf{y}}\mathbf{B}^{\mathrm{T}}\mathbf{B}\boldsymbol{\Sigma}_{\mathbf{y}}(\mathbf{A}\mathbf{x}_d + \boldsymbol{\mu}_{\mathbf{A}})}{\sigma_1^4\sigma_2^3}]\end{aligned} \tag{25}$$

### *3.5 Considering Confounding Factors*

To infer associations between SNP sets and gene sets while taking into consideration confounding factors, we further propose a graphical model as shown in Fig. 7. Different from the previous model, a new type of hidden variable, $\mathbf{s} = [s_1, s_2, \ldots, s_H]^{\mathrm{T}}$, is used to model confounding factors. For simplicity, we refer to this model as *Model 1*. The objective function of this model can be derivated using similar strategy as SET-eQTL.

### *3.6 Incorporating Individual Effect*

In the graphical model shown in Fig. 7, we use a hidden variable $y$ as a bridge between an SNP set and a gene set to capture the group-wise effect. In addition, individual effects may exist as well [42]. An example is shown in Fig. 4. Note that an ideal model should allow overlaps between SNP sets and between gene sets; that is,

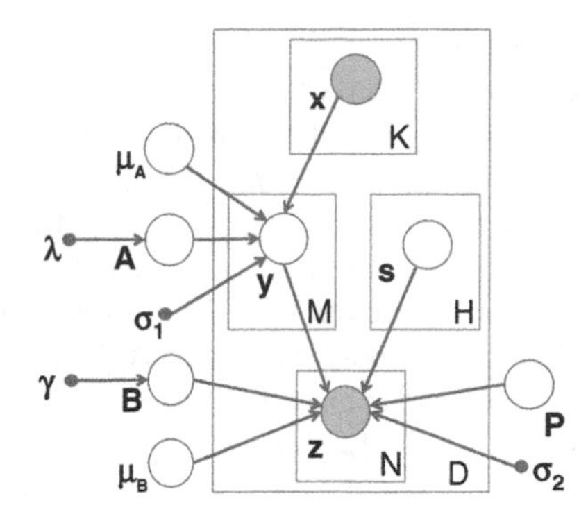


**Fig. 7** Graphical model with two types of hidden variables

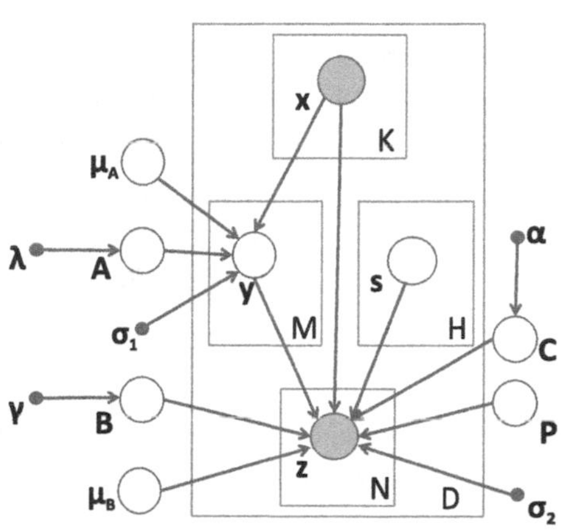


**Fig. 8** Refined graphical model to capture both individual and group-wise associations

an SNP or gene may participate in multiple individual and group-wise associations. To incorporate both individual and group-wise effects, we extend the model in Fig. 7 and add one edge between $\mathbf{x}$ and $\mathbf{z}$ to capture individual associations as shown in Fig. 8. We will show that this refinement will significantly improve the accuracy of model and enhance its computational efficiency. For simplicity, we refer to the new model that considers both individual and group-wise associations as *Model 2*.

### 3.6.1 Objective Function

Next, we give the derivation of the objective function for the model in Fig. 8. We assume that the two conditional probabilities follow normal distributions:

$$\mathbf{y}|\mathbf{x} \sim N(\mathbf{y}|\mathbf{A}\mathbf{x} + \boldsymbol{\mu}_{\mathbf{A}}, \sigma_1^2\mathbf{I}_{\mathbf{M}}), \tag{26}$$

and

$$\mathbf{z}|\mathbf{y}, \mathbf{x} \sim N(\mathbf{z}|\mathbf{B}\mathbf{y} + \mathbf{C}\mathbf{x} + \mathbf{P}\mathbf{s} + \boldsymbol{\mu}_{\mathbf{B}}, \sigma_2^2\mathbf{I}_{\mathbf{N}}), \tag{27}$$

where $\mathbf{A} \in \mathbb{R}^{M\times K}$ is the coefficient matrix between $\mathbf{x}$ and $\mathbf{y}$, $\mathbf{B} \in \mathbb{R}^{N\times M}$ is the coefficient matrix between $\mathbf{y}$ and $\mathbf{z}$, $\mathbf{C} \in \mathbb{R}^{N\times K}$ is the coefficient matrix between $\mathbf{x}$ and $\mathbf{z}$ to capture the individual associations, $\mathbf{P} \in \mathbb{R}^{N\times H}$ is the coefficient matrix of confounding factors. $\boldsymbol{\mu}_{\mathbf{A}} \in \mathbb{R}^{M\times 1}$ and $\boldsymbol{\mu}_{\mathbf{B}} \in \mathbb{R}^{N\times 1}$ are the translation factor vectors, $\sigma_1^2\mathbf{I}_M$ and $\sigma_2^2\mathbf{I}_N$ are the variances of the two conditional probabilities, respectively ($\sigma_1$ and $\sigma_2$ are constant scalars and $\mathbf{I}_M$ and $\mathbf{I}_N$ are identity matrices).

Since the expression level of a gene is usually affected by a small fraction of SNPs, we impose sparsity on **A**, **B**, and **C**. We assume that the entries of these matrices follow Laplace distributions: $\mathbf{A}_{i,j} \sim \mathbf{Laplace}(0, 1/\lambda)$, $\mathbf{B}_{i,j} \sim \mathbf{Laplace}(0, 1/\gamma)$, and $\mathbf{C}_{i,j} \sim \mathbf{Laplace}(0, 1/\alpha)$. $\lambda$, $\gamma$, and $\alpha$ will be used as parameters in the objective function. The probability density function of $\mathbf{Laplace}(\mu, b)$ distribution is $f(x|\mu, b) = \frac{1}{2b}\exp(-\frac{|x-\mu|}{b})$.

Thus, we have

$$\mathbf{y} = \mathbf{A}\mathbf{x} + \boldsymbol{\mu}_{\mathbf{A}} + \boldsymbol{\epsilon}_1, \tag{28}$$

$$\mathbf{z} = \mathbf{B}\mathbf{y} + \mathbf{C}\mathbf{x} + \mathbf{P}\mathbf{s} + \boldsymbol{\mu}_{\mathbf{B}} + \boldsymbol{\epsilon}_2, \tag{29}$$

where $\boldsymbol{\epsilon}_1 \sim N(\mathbf{0}, \sigma_1^2\mathbf{I}_M)$,$\boldsymbol{\epsilon}_2 \sim N(\mathbf{0}, \sigma_2^2\mathbf{I}_N)$. From Eq. (26) we have

$$\mathbf{B}\mathbf{y}|\mathbf{x} \sim N(\mathbf{B}\mathbf{A}\mathbf{x} + \mathbf{B}\boldsymbol{\mu}_{\mathbf{A}}, \sigma_1^2\mathbf{B}\mathbf{B}^{\mathrm{T}}), \tag{30}$$

Assuming that the confounding factors follow normal distribution [42], $\mathbf{s} \sim N(\mathbf{0}, \mathbf{I}_H)$, then we have

$$\mathbf{P}\mathbf{s} \sim N(\mathbf{0}, \mathbf{P}\mathbf{P}^{\mathrm{T}}). \tag{31}$$

We substitute Eqs. (30), (31) into Eq. (29), and get

$$\mathbf{z}|\mathbf{x} \sim N(\mathbf{B}\mathbf{A}\mathbf{x} + \mathbf{B}\boldsymbol{\mu}_{\mathbf{A}} + \mathbf{C}\mathbf{x} + \boldsymbol{\mu}_{\mathbf{B}}, \sigma_1^2\mathbf{B}\mathbf{B}^{\mathrm{T}} + \mathbf{P}\mathbf{P}^{\mathrm{T}} + \sigma_2^2\mathbf{I}_N).$$

From the formula above, we observe that the summand $\mathbf{B}\boldsymbol{\mu}_{\mathbf{A}}$ can also be integrated in $\boldsymbol{\mu}_{\mathbf{B}}$. Thus to simplify the model, we set $\boldsymbol{\mu}_{\mathbf{A}} = \mathbf{0}$ and obtain

$$\mathbf{z}|\mathbf{x} \sim N(\mathbf{B}\mathbf{A}\mathbf{x} + \mathbf{C}\mathbf{x} + \boldsymbol{\mu}_{\mathbf{B}}, \sigma_1^2\mathbf{B}\mathbf{B}^{\mathrm{T}} + \mathbf{P}\mathbf{P}^{\mathrm{T}} + \sigma_2^2\mathbf{I}_N).$$

To learn the parameters, we can use maximize likelihood estimation or maximum a posteriori. Then, we get the likelihood function as $p(\mathbf{z}|\mathbf{x}) = \prod_{d=1}^{D} p(\mathbf{z}_d|\mathbf{x}_d)$. Maximizing the likelihood function is identical to minimizing the negative log-likelihood. Here, the negative log-likelihood (loss function) is

$$\begin{aligned}\mathscr{J} &= \sum_{d=1}^{D} \mathscr{J}_d \\ &= -1 \cdot \log \prod_{d=1}^{D} p(\mathbf{z}_d|\mathbf{x}_d) \\ &= \sum_{d=1}^{D} (-1) \cdot \log p(\mathbf{z}_d|\mathbf{x}_d) \\ &= \frac{D \cdot N}{2} \log(2\pi) + \frac{D}{2} \log |\boldsymbol{\Sigma}| + \frac{1}{2} \sum_{d=1}^{D} [(\mathbf{z}_d - \boldsymbol{\mu}_d)^{\mathbf{T}} \boldsymbol{\Sigma}^{-1} (\mathbf{z}_d - \boldsymbol{\mu}_d)],\end{aligned} \tag{32}$$

where

$$\begin{aligned}\boldsymbol{\mu}_d &= \mathbf{BAx}_d + \mathbf{Cx}_d + \boldsymbol{\mu}_{\mathbf{B}}, \\ \boldsymbol{\Sigma} &= \sigma_1^2 \mathbf{BB}^{\mathrm{T}} + \mathbf{WW}^{\mathrm{T}} + \sigma_2^2 \mathbf{I}_N.\end{aligned}$$

Moreover, taking into account the prior distributions of **A**, **B**, and **C**, we have

$$\begin{aligned}&p(\mathbf{z}_d, \mathbf{A}, \mathbf{B}, \mathbf{C}|\mathbf{x}_d, \mathbf{P}, \sigma_1, \sigma_2) = \\ &\exp(-\mathscr{J}_d) \cdot \tfrac{\lambda}{2} \textstyle\prod_{i,j} \exp(-\lambda|\mathbf{A}_{i,j}|) \cdot \tfrac{\gamma}{2} \prod_{i,j} \exp(-\gamma|\mathbf{B}_{i,j}|) \cdot \tfrac{\alpha}{2} \prod_{i,j} \exp(-\alpha|\mathbf{C}_{i,j}|).\end{aligned} \tag{33}$$

Thus, we have the $\ell_1$-regularized objective function

$$\max_{\mathbf{A},\mathbf{B},\mathbf{C},\mathbf{P},\sigma_1,\sigma_2} \log \prod_{d=1}^{D} p(\mathbf{z}_d, \mathbf{A}, \mathbf{B}, \mathbf{C}|\mathbf{x}_d, \mathbf{P}, \sigma_1, \sigma_2),$$

which is identical to

$$\min_{\mathbf{A},\mathbf{B},\mathbf{C},\mathbf{P},\sigma_1,\sigma_2} [\mathscr{J} + D \cdot (\lambda||\mathbf{A}||_1 + \gamma||\mathbf{B}||_1 + \alpha||\mathbf{C}||_1)], \tag{34}$$

where $||\cdot||_1$ is the $\ell_1$-norm. $\lambda$, $\gamma$, and $\alpha$ are the *precision* of the prior Laplace distributions of **A**, **B**, and **C**, respectively. They serve as the regularization parameters and can be determined by cross or holdout validation.

The explicit expression of $\boldsymbol{\mu}_{\mathbf{B}}$ can be derived as follows. When **A**, **B**, and **C** are fixed, we have $\mathscr{J} = \frac{D \cdot N}{2} \log(2\pi) + \frac{D}{2} \log |\boldsymbol{\Sigma}| + \frac{1}{2} \sum_{d=1}^{D} [(\mathbf{z}_d - \mathbf{BAx}_d - \mathbf{Cx}_d - \boldsymbol{\mu}_{\mathbf{B}})^{\mathrm{T}} \boldsymbol{\Sigma}^{-1} (\mathbf{z}_d - \mathbf{BAx}_d - \mathbf{Cx}_d - \boldsymbol{\mu}_{\mathbf{B}})]$. When $D = 1$, this is a classic maximum likelihood estimation problem, and we have $\boldsymbol{\mu}_{\mathbf{B}} = \mathbf{z}_d - \mathbf{BAx}_d - \mathbf{Cx}_d$. When $D > 1$, leveraging the fact that $\boldsymbol{\Sigma}^{-1}$ is symmetric, we convert the problem into a least square problem, which leads to

$$\boldsymbol{\mu}_{\mathbf{B}} = \frac{1}{D} \sum_{d=1}^{D} (\mathbf{z}_d - \mathbf{BAx}_d - \mathbf{Cx}_d).$$

Substituting it into Eq. (32), we have

$$\begin{aligned}\mathscr{J} &= \tfrac{D\cdot N}{2}\log(2\pi) + \tfrac{D}{2}\log|\boldsymbol{\Sigma}| + \tfrac{1}{2}\textstyle\sum_{d=1}^{D}\{[(\mathbf{z}_d - \bar{\mathbf{z}}) \\ &-(\mathbf{BA}+\mathbf{C})(\mathbf{x}_d - \bar{\mathbf{x}})]^{\mathrm{T}}\boldsymbol{\Sigma}^{-1}[(\mathbf{z}_d - \bar{\mathbf{z}}) - (\mathbf{BA}+\mathbf{C})(\mathbf{x}_d - \bar{\mathbf{x}})]\},\end{aligned} \tag{35}$$

where

$$\bar{\mathbf{x}} = \frac{1}{D}\sum_{d=1}^{D}\mathbf{x}_d, \qquad \bar{\mathbf{z}} = \frac{1}{D}\sum_{d=1}^{D}\mathbf{z}_d.$$

The gradient of the loss function, which (without detailed derivation) is given in the below. For notational simplicity, we denote

$$\mathbf{t}_d = (\mathbf{z}_d - \bar{\mathbf{z}}) - (\mathbf{BA}+\mathbf{C})(\mathbf{x}_d - \bar{\mathbf{x}}),$$

$$\boldsymbol{\Psi}_d = \frac{1}{2}(\boldsymbol{\Sigma}^{-1} - \boldsymbol{\Sigma}^{-1}\mathbf{t}_d\mathbf{t}_d{}^{\mathrm{T}}\boldsymbol{\Sigma}^{-1}).$$

1. Derivative with respect to $\sigma_1$

$$\nabla_{\sigma_1}\mathscr{O} = 2\sigma_1\sum_{d=1}^{D}\{\mathrm{tr}[\boldsymbol{\Psi}_d]\mathbf{BB}^{\mathrm{T}}\}. \tag{36}$$

2. Derivative with respect to $\sigma_2$

$$\nabla_{\sigma_2}\mathscr{O} = 2\sigma_2\sum_{d=1}^{D}\{\mathrm{tr}[\boldsymbol{\Psi}_d]\}. \tag{37}$$

3. Derivative with respect to $\mathbf{A}$

$$\nabla_{\mathbf{A}}\mathscr{O} = -\sum_{d=1}^{D}[\mathbf{B}^{\mathrm{T}}\boldsymbol{\Sigma}^{-1}\mathbf{t}_d(\mathbf{x}_d - \bar{\mathbf{x}})^{\mathrm{T}}]. \tag{38}$$

4. Derivative with respect to $\mathbf{B}$

$$\nabla_{\mathbf{B}}\mathscr{O} = \boldsymbol{\Xi}_1 + \boldsymbol{\Xi}_2, \tag{39}$$

where

$$\boldsymbol{\Xi}_1 = -\sum_{d=1}^{D}[\boldsymbol{\Sigma}^{-1}\mathbf{t}_d(\mathbf{x}_d - \bar{\mathbf{x}})^{\mathrm{T}}\mathbf{A}^{\mathrm{T}}], \tag{40}$$

$$(\boldsymbol{\Xi}_2)_{ij} = \sigma_1^2\sum_{d=1}^{D}\{\mathrm{tr}[\boldsymbol{\Psi}_d(\mathbf{E}_{ij}\mathbf{B}^{\mathrm{T}} + \mathbf{BE}_{ji})]\}. \tag{41}$$

(tr[·] stands for trace; $\mathbf{E}_{ij}$ is the single-entry matrix: 1 at $(i, j)$ and 0 elsewhere.)

We speed up this calculation by exploiting sparsity of $\boldsymbol{E}_{ij}$ and tr[·]. (The following equation uses *Einstein summation convention* to better illustrate the idea.)

$$
\begin{aligned}
(\boldsymbol{\Xi}_2)_{ij} &= \sigma_1^2 \sum_{d=1}^{D} \{\mathrm{tr}[\boldsymbol{\Psi}_d(\boldsymbol{E}_{ij}\boldsymbol{B}^T + \boldsymbol{B}\boldsymbol{E}_{ji})]\} \\
&= \sigma_1^2 \sum_{d=1}^{D} \{\mathrm{tr}[\boldsymbol{\Psi}_d\boldsymbol{E}_{ij}\boldsymbol{B}^T + \boldsymbol{\Psi}_d\boldsymbol{B}\boldsymbol{E}_{ji}]\} \\
&= \sigma_1^2 \sum_{d=1}^{D} \{\sum_{k=1}^{N}[(\boldsymbol{B}^T)_{j,k}(\boldsymbol{\Psi}_d)_{k,i}] + \sum_{l=1}^{N}[(\boldsymbol{\Psi}_d)_{i,l}(\boldsymbol{B})_{l,j}]\}.
\end{aligned}
\tag{42}
$$

Therefore,

$$
\begin{aligned}
\boldsymbol{\Xi}_2 &= \sigma_1^2 \sum_{d=1}^{D} [(\boldsymbol{B}^T\boldsymbol{\Psi}_d)^T + \boldsymbol{\Psi}_d\boldsymbol{B}] \\
&= \sigma_1^2 \sum_{d=1}^{D} [\boldsymbol{\Psi}_d^T\boldsymbol{B} + \boldsymbol{\Psi}_d\boldsymbol{B}] \\
&= 2\sigma_1^2 \sum_{d=1}^{D} \boldsymbol{\Psi}_d\boldsymbol{B}.
\end{aligned}
\tag{43}
$$

5. Derivative with respect to $\mathbf{C}$

$$
\nabla_{\mathbf{C}}\mathscr{O} = -\sum_{d=1}^{D}[\boldsymbol{\Sigma}^{-1}\mathbf{t}_d(\mathbf{x}_d - \bar{\mathbf{x}})^{\mathrm{T}}]. \tag{44}
$$

6. Derivative with respect to $\mathbf{P}$

$$
\nabla_{\mathbf{P}}\mathscr{O} = \sum_{d=1}^{D}\{\mathrm{tr}[\boldsymbol{\Psi}_d(\mathbf{E}_{ij}\mathbf{P}^{\mathrm{T}} + \mathbf{P}\mathbf{E}_{ji})]\} = 2\sum_{d=1}^{D}\boldsymbol{\Psi}_d\mathbf{P}. \tag{45}
$$

### 3.6.2 Increasing Computational Speed

In this section, we discuss how to increase the speed of the optimization process for the proposed model. In the previous section, we have shown that **A**, **B**, **C**, **P**, $\sigma_1$, and $\sigma_2$ are the parameters to be solved. Here, we first derive an updating scheme for $\sigma_2$ when other parameters are fixed by following a similar technique as discussed in [30]. For other parameters, we develop an efficient method for calculating the inverse of the covariance matrix which is the main bottleneck of the optimization process.

**Updating $\sigma_2$** When all other parameters are fixed, using spectral decomposition on $(\sigma_1^2\mathbf{B}\mathbf{B}^T + \mathbf{W}\mathbf{W}^T)$, we have

$$\begin{aligned}\boldsymbol{\Sigma} &= (\sigma_1^2\mathbf{B}\mathbf{B}^T + \mathbf{W}\mathbf{W}^T) + \sigma_2^2\mathbf{I}_N \\ &= [\mathbf{U},\mathbf{V}]\,\mathrm{diag}(\lambda_1+\sigma_2^2,\ldots,\lambda_{N-q}+\sigma_2^2,0,\ldots,0)[\mathbf{U},\mathbf{V}]^T \\ &= \mathbf{U}\,\mathrm{diag}(\lambda_1+\sigma_2^2,\ldots,\lambda_{N-q}+\sigma_2^2)\mathbf{U}^T,\end{aligned} \tag{46}$$

where $\mathbf{U}$ is an $N \times (N-q)$ eigenvector matrix corresponding to the nonzero eigenvalues; $\mathbf{V}$ is an $N \times q$ eigenvector matrix corresponding to the zero eigenvalues. A reasonable solution should have no zero eigenvalues in $\boldsymbol{\Sigma}$, otherwise the loss function would be infinitely big. Therefore, $q = 0$.

Thus

$$\boldsymbol{\Sigma}^{-1} = \mathbf{U}\,\mathrm{diag}(\frac{1}{\lambda_1+\sigma_2^2},\ldots,\frac{1}{\lambda_N+\sigma_2^2})\mathbf{U}^T.$$

Let $\mathbf{U}^T(\mathbf{z}_d - \mathbf{B}\mathbf{A}\mathbf{x}_d - \mathbf{C}\mathbf{x}_d - \boldsymbol{\mu}_\mathbf{B}) =: [\eta_{d,1},\eta_{d,2},\ldots,\eta_{d,N}]^T$. Then solving $\sigma_2$ is equivalent to minimizing

$$l(\sigma_2^2) = \frac{D\cdot N}{2}\log(2\pi) + \frac{D}{2}\sum_{s=1}^{N}\log(\lambda_s+\sigma_2^2) + \frac{1}{2}\sum_{d=1}^{D}\sum_{s=1}^{N}\frac{\eta_{d,s}^2}{\lambda_s+\sigma_2^2}, \tag{47}$$

whose derivative is

$$l'(\sigma_2^2) = \frac{D}{2}\sum_{s=1}^{N}\frac{1}{\lambda_s+\sigma_2^2} - \frac{1}{2}\sum_{d=1}^{D}\sum_{s=1}^{N}\frac{\eta_{d,s}^2}{(\lambda_s+\sigma_2^2)^2}.$$

This is a 1-dimensional optimization problem that can be solved very efficiently.

**Efficiently Inverting the Covariance Matrix** From objective function Eq. (35) and the gradient of the parameters, the time complexity of each iteration in the optimization procedure is $\mathcal{O}(DN^2M + DN^2H + DN^3 + DNMK)$. Since $M \ll N$ and $H \ll N$, the third term of the time complexity ($\mathcal{O}(DN^3)$) is the bottleneck of the overall performance. This is for computing the inverse of the covariance matrix

$$\boldsymbol{\Sigma} = \sigma_1^2\boldsymbol{B}\boldsymbol{B}^T + \boldsymbol{P}\boldsymbol{P}^T + \sigma_2^2\boldsymbol{I}_N,$$

which is much more time-consuming than other matrix multiplication operations.

We devise an acceleration strategy that calculates $\boldsymbol{\Sigma}^{-1}$ using formula (48) in the following theorem. The complexity of computing the inverse reduces to $\mathcal{O}(M^3 + H^3)$.

**Theorem 1.** *Given* $\boldsymbol{B} \in \mathbb{R}^{N\times M}$, $\boldsymbol{P} \in \mathbb{R}^{N\times H}$, *and*

$$\boldsymbol{\Sigma} = \sigma_2^2 \boldsymbol{I}_N + \sigma_1^2 \boldsymbol{B}\boldsymbol{B}^{\mathrm{T}} + \boldsymbol{P}\boldsymbol{P}^{\mathrm{T}}.$$

*Then*

$$\boldsymbol{\Sigma}^{-1} = \boldsymbol{T} - \boldsymbol{T}\boldsymbol{P}\boldsymbol{S}^{-1}\boldsymbol{P}^{\mathrm{T}}\boldsymbol{T}, \tag{48}$$

*where*

$$\boldsymbol{S} = \boldsymbol{I}_H + \boldsymbol{P}^{\mathrm{T}}\boldsymbol{T}\boldsymbol{P}, \tag{49}$$

$$\boldsymbol{T} = \sigma_2^{-2}(\boldsymbol{I}_N - \sigma_1^2 \boldsymbol{B}(\sigma_2^2 \boldsymbol{I}_M + \sigma_1^2 \boldsymbol{B}^{\mathrm{T}}\boldsymbol{B})^{-1}\boldsymbol{B}^{\mathrm{T}}). \tag{50}$$

The proof of Theorem 1 is provided in the following.

*Proof of Theorem 1.* Before giving the formal proof for Theorem 1, we first introduce Lemma 1, which follows from the definition of matrix inverse.

**Lemma 1.** *For all* $\boldsymbol{U} \in \mathbb{R}^{N\times M}$, *if* $\boldsymbol{I}_M + \boldsymbol{U}^{\mathrm{T}}\boldsymbol{U}$ *is invertible, then*

$$(\boldsymbol{I}_N + \boldsymbol{U}\boldsymbol{U}^{\mathrm{T}})^{-1} = \boldsymbol{I}_N - \boldsymbol{U}(\boldsymbol{I}_M + \boldsymbol{U}^{\mathrm{T}}\boldsymbol{U})^{-1}\boldsymbol{U}^{\mathrm{T}}.$$

Here we provide a more general proof, which can be modified to derive more involved cases.

*Proof of Theorem 1.* We denote

$$\boldsymbol{Q} = \sigma_2^2 \boldsymbol{I}_N + \sigma_1^2 \boldsymbol{B}\boldsymbol{B}^{\mathrm{T}}, \tag{51}$$

that is,

$$\boldsymbol{\Sigma} = \sigma_2^2 \boldsymbol{I}_N + \sigma_1^2 \boldsymbol{B}\boldsymbol{B}^{\mathrm{T}} + \boldsymbol{P}\boldsymbol{P}^{\mathrm{T}} = \boldsymbol{Q} + \boldsymbol{P}\boldsymbol{P}^{\mathrm{T}}. \tag{52}$$

By Lemma 1, we have

$$\boldsymbol{Q}^{-1} = \boldsymbol{T} = \sigma_2^{-2}(\boldsymbol{I}_N - \sigma_1^2 \boldsymbol{B}(\sigma_2^2 \boldsymbol{I}_M + \sigma_1^2 \boldsymbol{B}^{\mathrm{T}}\boldsymbol{B})^{-1}\boldsymbol{B}^{\mathrm{T}}).$$

$\boldsymbol{Q}$ is symmetric positive definite, hence its inverse, $\boldsymbol{T}$, is symmetric positive definite. Since every symmetric positive definite matrix has exactly one symmetric positive definite square root, we can write

$$\boldsymbol{T} = \boldsymbol{R}\boldsymbol{R},$$

where $\boldsymbol{R}$ is an $N \times N$ symmetric positive definite matrix.

It is clear that $\boldsymbol{Q} = \boldsymbol{T}^{-1} = (\boldsymbol{R}\boldsymbol{R})^{-1} = \boldsymbol{R}^{-1}\boldsymbol{R}^{-1}$, which leads to $\boldsymbol{R}\boldsymbol{Q}\boldsymbol{R} = \boldsymbol{R}\boldsymbol{R}^{-1}\boldsymbol{R}^{-1}\boldsymbol{R} = \boldsymbol{I}_N$, and therefore

$$\boldsymbol{R}\boldsymbol{\Sigma}\boldsymbol{R} = \boldsymbol{I}_N + \boldsymbol{R}\boldsymbol{P}\boldsymbol{P}^{\mathrm{T}}\boldsymbol{R} = \boldsymbol{I}_N + \boldsymbol{R}\boldsymbol{P}\boldsymbol{P}^{\mathrm{T}}\boldsymbol{R}^{\mathrm{T}}.$$

Note that the above and the following formulas follow the fact that $\boldsymbol{R}$ is symmetric.

Once again, by Lemma 1, we have

$$(\boldsymbol{R}\boldsymbol{\Sigma}\boldsymbol{R})^{-1} = \boldsymbol{I}_N - \boldsymbol{R}\boldsymbol{P}\boldsymbol{S}^{-1}\boldsymbol{P}^{\mathrm{T}}\boldsymbol{R}^{\mathrm{T}},$$

where

$$\boldsymbol{S} = \boldsymbol{I}_H + \boldsymbol{P}^{\mathrm{T}}\boldsymbol{R}^{\mathrm{T}}\boldsymbol{R}\boldsymbol{P} = \boldsymbol{I}_H + \boldsymbol{P}^{\mathrm{T}}\boldsymbol{T}\boldsymbol{P}.$$

Therefore,

$$\boldsymbol{\Sigma}^{-1} = \boldsymbol{R}(\boldsymbol{R}\boldsymbol{\Sigma}\boldsymbol{R})^{-1}\boldsymbol{R} = \boldsymbol{R}\boldsymbol{R} - \boldsymbol{R}\boldsymbol{R}\boldsymbol{P}\boldsymbol{S}^{-1}\boldsymbol{P}^{\mathrm{T}}\boldsymbol{R}^{\mathrm{T}}\boldsymbol{R},$$

and thus

$$\boldsymbol{\Sigma}^{-1} = \boldsymbol{T} - \boldsymbol{T}\boldsymbol{P}\boldsymbol{S}^{-1}\boldsymbol{P}^{\mathrm{T}}\boldsymbol{T}$$

### 3.7 Optimization

To optimize the objective function, there are many off-the-shelf $\ell_1$-penalized optimization tools. We use the Orthant-Wise Limited-memory Quasi-Newton (OWL-QN) algorithm described in [1]. The OWL-QN algorithm minimizes functions of the form

$$f(w) = loss(w) + c||w||_1,$$

where $loss(\cdot)$ is an arbitrary differentiable loss function, and $||w||_1$ is the $\ell_1$-norm of the parameter vector. It is based on the L-BFGS Quasi-Newton algorithm, with modifications to deal with the fact that the $\ell_1$-norm is not differentiable [52]. The algorithm is proven to converge to a local optimum of the parameter vector. The algorithm is very fast, and capable of scaling efficiently to problems with millions of parameters. Thus it is a good option for our problem where the parameter space is large when dealing with large scale eQTL data.

## 3.8 Experimental Results

We apply our methods (SET-eQTL, *Model*1, and *Model*2) to both simulation datasets and yeast eQTL datasets [56] to evaluate its performance. For comparison, we select several recent eQTL methods, including LORS [70], MTLasso2G [10], FaST-LMM [42], and Lasso [63]. The tuning parameters in the selected methods are learned using cross-validation. All experiments are performed on a PC with 2.20 GHz Intel i7 eight-core CPU and 8 GB memory.

### 3.8.1 Simulation Study

We first evaluate whether Model 2 can identify both individual and group-wise associations. We adopt a similar setup for simulation study to that in [35, 70] and generate synthetic datasets as follows. 100 SNPs are randomly selected from the yeast eQTL dataset [56]. $N$ gene expression profiles are generated by $\mathbf{Z}_{j*} = \boldsymbol{\beta}_{j*}\mathbf{X} + \Xi_{j*} + \mathbf{E}_{j*}$ $(1 \leq j \leq N)$, where $\mathbf{E}_{j*} \sim \mathcal{N}(0, \eta I)$ $(\eta = 0.1)$ denotes Gaussian noise. $\Xi_{j*}$ is used to model non-genetic effects, which is drawn from $N(\mathbf{0}, \rho\Lambda)$, where $\rho = 0.1$. $\Lambda$ is generated by $\mathbf{F}\mathbf{F}^{\mathrm{T}}$, where $\mathbf{F} \in \mathbb{R}^{D\times U}$ and $\mathbf{F}_{ij} \sim \mathcal{N}(0, 1)$. $U$ is the number of hidden factors and is set to 10 by default. The association matrix $\boldsymbol{\beta}$ is shown in the top-left plot in Fig. 9. The association strength is 1 for all selected SNPs. There are four group-wise associations of different scales in total. The associations on the diagonal are used to represent individual association signals in *cis*-regulation.

The remaining three plots in Fig. 9 show associations estimated by *Model*2. From the figure, we can see that *Model*2 well captures both individual and group-wise signals. For comparison, Fig. 10 visualizes the association weights estimated by *Model*1 and *Model*2 when varying the number of hidden variables ($M$). We observe that for *Model*1, when $M = 20$, most of the individual association signals on the diagonal are not captured. As $M$ increases, more individual association signals are detected by *Model*1. In contrast, *Model*2 recovers both individual and group-wise linkage signals with small $M$.

Next, we generate 50 simulated datasets with different signal-to-noise ratios (defined as $SNR = \sqrt{\frac{Var(\boldsymbol{\beta}\mathbf{X})}{Var(\Xi+\mathbf{E})}}$) in the eQTL datasets [70] to compare the performance of the selected methods. Here, we fix $H = 10, \rho = 0.1$, and use different $\eta$'s to control *SNR*. For each setting, we report the average result from the 50 datasets. For the proposed methods, we use $\mathbf{BA} + \mathbf{C}$ as the overall associations. Since FaST-LMM needs extra information (e.g., the genetic similarities between individuals) and uses PLINK format, we do not list it here and will compare it on the real data set.

Figure 11 shows the ROC curves of TPR-FPR for performance comparison. The corresponding areas under the TPR-FPR curve and the areas under the precision-recall curve (AUCs) [10] are shown in Fig. 12. It can be seen that *Model*2 outperforms all alternative methods by a large margin. *Model*2 outperforms *Model*1

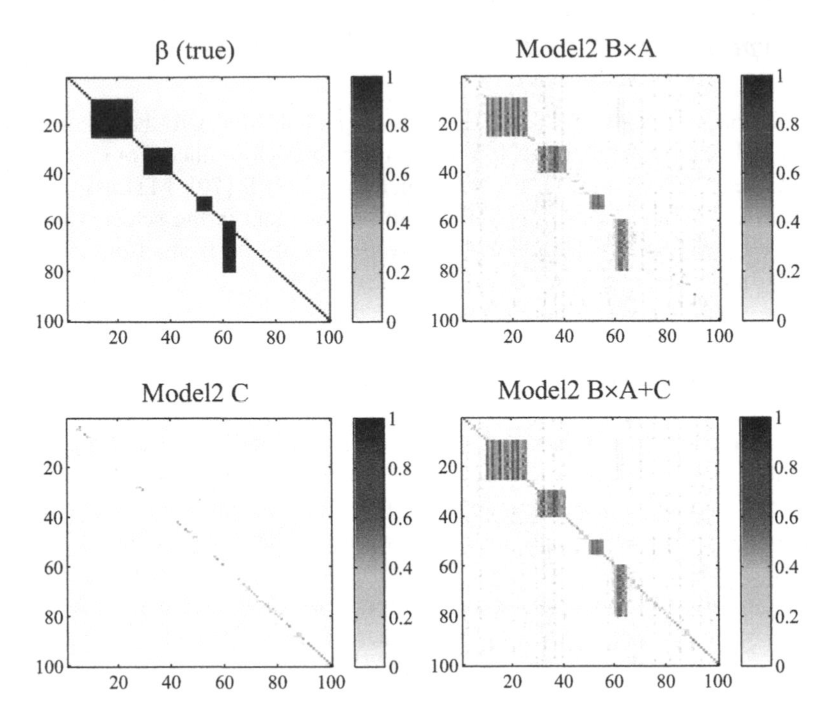


**Fig. 9** Ground truth of $\boldsymbol{\beta}$ and linkage weights estimated by *Model2* on simulated data

because it considers both group-wise and individual associations. *Model1* outperforms SET-eQTL because it considers confounding factors that is not considered by SET-eQTL. SET-eQTL considers all associations as group-wise, thus it may miss some individual associations. MTLasso2G is comparable to LORS because MTLasso2G considers the group-wise associations while neglecting confounding factors. LORS considers the confounding factors, but does not distinguish individual and group-wise associations. LORS outperforms Lasso since confounding factors are not considered in Lasso.

## Shrinkage of $\mathbf{C}$ and $\mathbf{B} \times \mathbf{A}$

As discussed in the previous section, the group-wise associations are encoded in $\mathbf{B}\times\mathbf{A}$ and individual associations are encoded in $\mathbf{C}$. To enforce sparsity on $\mathbf{A}$, $\mathbf{B}$, and $\mathbf{C}$, we use Laplace prior on the elements of these matrices. Thus, it is interesting to study the overall shrinkage of $\mathbf{B} \times \mathbf{A}$ and $\mathbf{C}$. We randomly generate seven predictors $(\{\mathbf{x}_1, \mathbf{x}_2, \ldots, \mathbf{x}_7\})$ and one response $(\mathbf{z})$ with sample size 100. $\mathbf{x_i} \sim \mathcal{N}(\mathbf{0}, 0.6\cdot\mathbf{I})(i \in [1, 7])$. The response vector was generated with the formula: $\mathbf{z} = 5\cdot(\mathbf{x}_1 + \mathbf{x}_2) - 3\cdot(\mathbf{x}_3 + \mathbf{x}_4) + 2\cdot\mathbf{x}_5 + \tilde{\boldsymbol{\epsilon}}$ and $\tilde{\boldsymbol{\epsilon}} \in \mathcal{N}(\mathbf{0}, \mathbf{I})$. Thus, there are two groups of predictors $(\{\mathbf{x}_1, \mathbf{x}_2\}$ and $\{\mathbf{x}_3, \mathbf{x}_4\})$ and one individual predictor $\mathbf{x}_5$. Figure 13 shows the Model

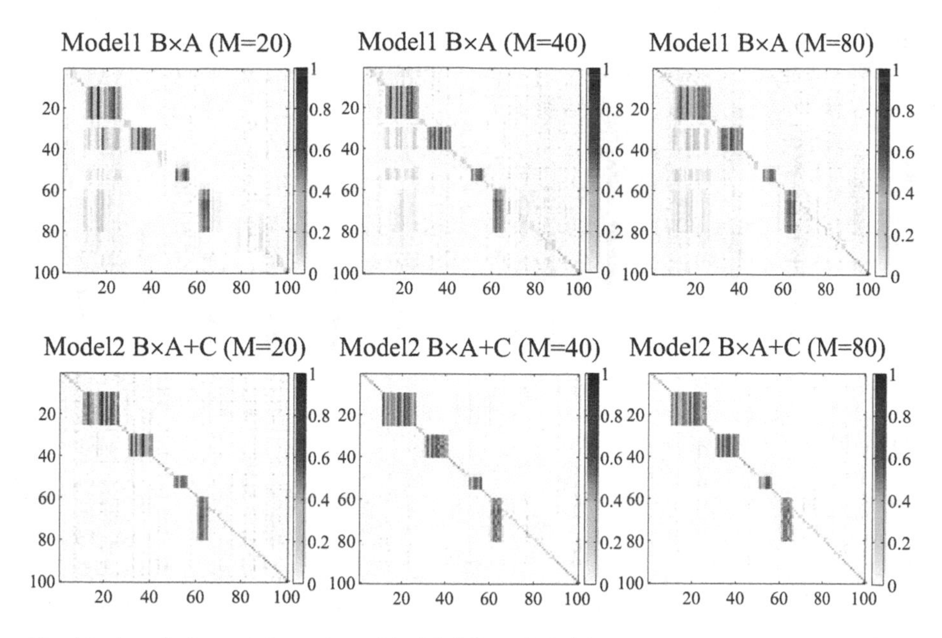


**Fig. 10** Association weights estimated by *Model*1 and *Model*2

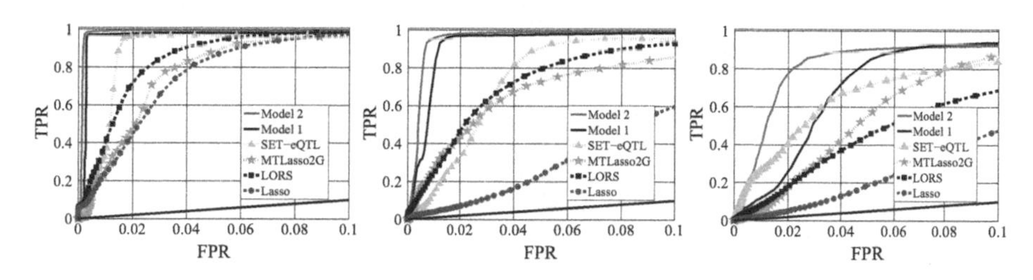


**Fig. 11** The ROC curve of FPR-TPR on simulated data. (**a**) ROC curve ($SNR = 1.16$). (**b**) ROC curve ($SNR = 0.13$). (**c**) ROC curve ($SNR = 0.08$)

2 shrinkage of coefficients for $\mathbf{B} \times \mathbf{A}$ and $\mathbf{C}$, respectively. Each curve represents a coefficient as a function of the scaled parameter $s = \frac{|\mathbf{B}\times\mathbf{A}|}{\max|\mathbf{B}\times\mathbf{A}|}$ or $s = \frac{|\mathbf{C}|}{\max|\mathbf{C}|}$. We can see that the two groups of predictors can be identified by $\mathbf{B} \times \mathbf{A}$ as the most important variables, and the individual predictor can be identified by $\mathbf{C}$.

## Computational Efficiency Evaluation

Scalability is an important issue for eQTL study. To evaluate the techniques for speeding up the computational efficiency, we compare the running time with/without these techniques. Figure 14 shows the running time when varying the number of hidden variables ($M$) and number of traits ($N$). The results are consistent with the theoretical analysis in previous part that the time complexity

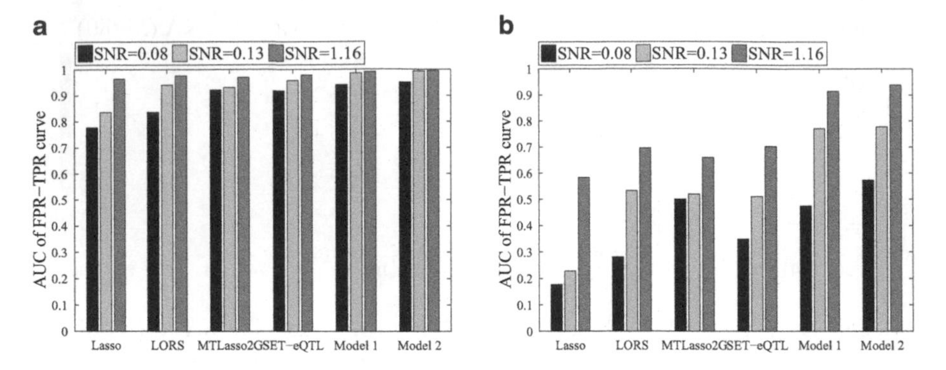


**Fig. 12** The areas under the precision-recall/FPR-TPR curve (AUCs). (**a**) AUC of FPR-TPR curve. (**b**) AUC of precision-recall curve

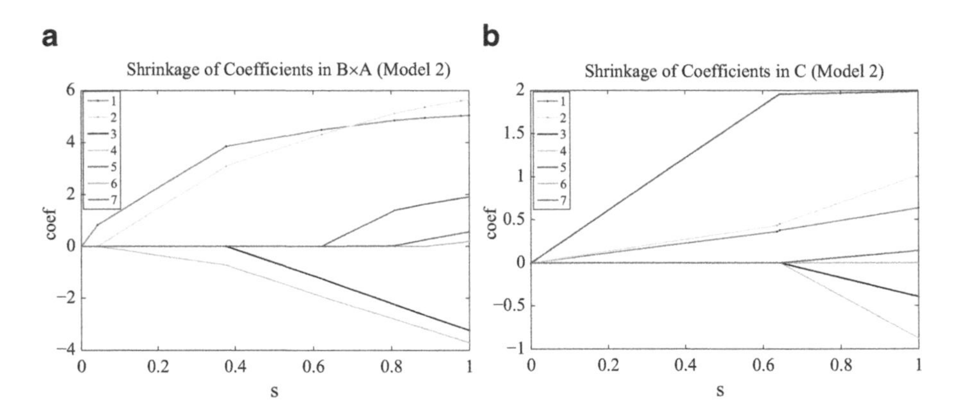


**Fig. 13** Model 2 shrinkage of coefficients for $\mathbf{B} \times \mathbf{A}$ and $\mathbf{C}$, respectively. (**a**) $\mathbf{B} \times \mathbf{A}$. (**b**) $\mathbf{C}$

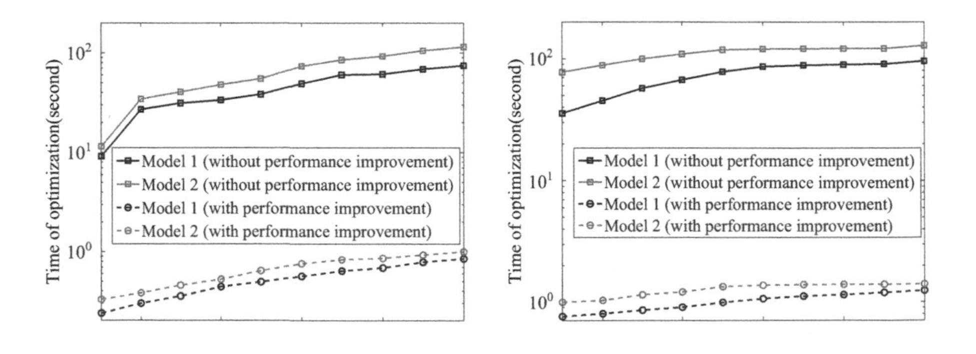


**Fig. 14** Running time performance on simulated data when varying $N$ and $M$

is reduced to $\mathcal{O}(M^3 + H^3)$ from $\mathcal{O}(N^3)$ when using the improved method for inverting the covariance matrix. We also observe that *Model2* uses slightly more time than *Model1*, since it has more parameters to optimize. However, to get similar performance, *Model1* needs a significantly larger number of hidden variables $M$. As shown in Fig. 14b, a larger $M$ results in a longer running time. In some cases, *Model2* is actually faster than *Model1*. As an example, to obtain the same performance (i.e., AUC), *Model1* needs 60 hidden variables ($M$), while *Model2* only needs $M = 20$. In this case, from Fig. 14a, we can observe that *Model2* needs less time than *Model1* to obtain the same results.

### 3.8.2 Yeast eQTL Study

We apply the proposed methods to a yeast (Saccharomyces cerevisiae) eQTL dataset of 112 yeast segregants [56] generated from a cross of two inbred strains. The dataset originally includes expression profiles of 6229 gene expression traits and genotype profiles of 2956 SNP markers. After removing SNPs with more than 10 % missing values and merging consecutive SNPS with high linkage disequilibrium, we obtain 1017 SNPs with distinct genotypes [24]. In total, 4474 expression profiles are selected after removing the ones with missing values. It takes about 5 h for *Model1*, and 3 h for *Model2* to run to completion. The regularization parameters are set by grid search in $\{0.1, 1, 10, 50, 100, 500, 1000, 2000\}$. Specifically, grid search trains the model with each combination of three regularization parameters in the grid and evaluates their performance (by measuring out-of-sample loss function value) for a two-fold cross validation. Finally, the grid search algorithm outputs the settings that achieved the smallest loss in the validation procedure.

We use hold-out validation to find the optimal number of hidden variables $M$ and $H$ for each model. Specifically, we partition the samples into two subsets of equal size. We use one subset as training data and test the learned model using the other subset of samples. By measuring out-of-sample predictions, we can find optimal combination of $M$ and $H$ that avoids over-fitting. For each combination, optimal values for regularization parameters were determined with two-fold cross validation. The loss function values for different $\{M, H\}$ combinations of *Model2* are shown in Fig. 15. We find that $M = 30$ and $H = 10$ for *Model2* deliver the best overall performance. Similarly, we find that the optimal $M$ and $H$ values for *Model1* are 150 and 10, respectively. The significant associations given by *Model1*, *Model2*, LORS, MTLasso2G, and Lasso are shown in Fig. 16. For *Model2*, we can clearly see that the estimated matrices $\mathbf{C}$ and $\mathbf{B} \times \mathbf{A}$ well capture the non-group-wise and group-wise signals, respectively. $\mathbf{C} + \mathbf{B} \times \mathbf{A}$ and $\mathbf{C}$ of *Model2* have stronger *cis*-regulatory signals and weaker *trans*-regulatory bands than that of *Model1*, LORS, and Lasso. $\mathbf{C}$ of *Model2* has the weakest *trans*-regulatory bands. LORS has weaker *trans*-regulatory bands than Lasso since it considers confounding factors. With more hidden variables (larger $M$), *Model1* obtains stronger *cis*-regulatory signals.

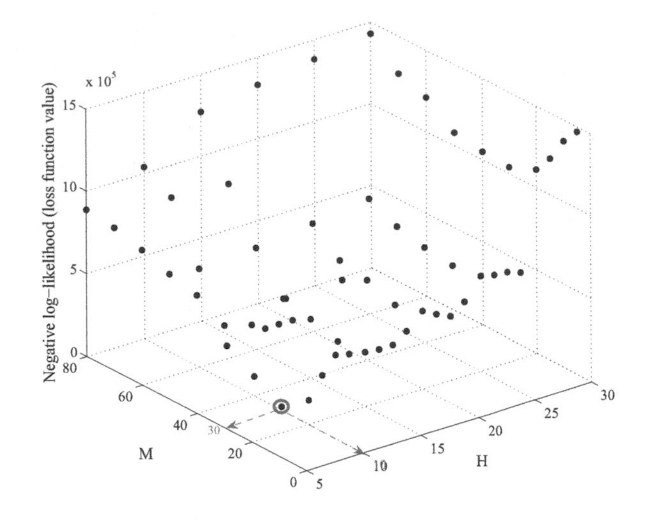


**Fig. 15** Parameter tuning for *M* and *H* (*Model*2)

### cis- and trans-Enrichment Analysis

In total, the proposed two methods detect about 6000 associations with non-zero weight values ($\mathbf{B} \times \mathbf{A}$ for *Model*1 and $\mathbf{C} + \mathbf{B} \times \mathbf{A}$ for *Model*2). We estimate their FDR values by following the method proposed in [70]. With FDR $\leq 0.01$, both models obtain about 4500 associations. The visualization of significant associations detected by different methods is provided in Fig. 16.

We apply *cis*- and *trans*-enrichment analysis on the discovered associations. In particular, we follow the standard *cis*-enrichment analysis [41, 44] to compare the performance of two competing models. The intuition behind *cis*-enrichment analysis is that more *cis*-acting SNPs are expected than *trans*-acting SNPs. A two-step procedure is used in the *cis*-enrichment analysis [41]: (1) for each model, we apply a one-tailed Mann-Whitney test on each SNP to test the null hypothesis that the model ranks its *cis* hypotheses (we use <500 bp for yeast) no better than its *trans* hypotheses, (2) for each pair of models compared, we perform a two-tailed paired Wilcoxon sign-rank test on the *p*-values obtained from the previous step. The null hypothesis is that the median difference of the *p*-values in the Mann-Whitney test for each SNP is zero. The *trans*-enrichment is implemented using a similar strategy as in [71], in which genes regulated by transcription factors are used as *trans*-acting signals.

The results of pairwise comparison of selected models are shown in Table 2. A *p*-value shows how significant a method on the left column outperforms a method in the top row in terms of *cis*-enrichment or *trans*-enrichment. We observe that the proposed *Model*2 has significantly better *cis*-enrichment scores than other methods. For *trans*-enrichment, *Model*2 is the best, and FaST-LMM comes in second. This is because both *Model*2 and FaST-LMM consider confounding factors (FaST-LMM considers confounders from population structure) and joint effects of SNPs, but only

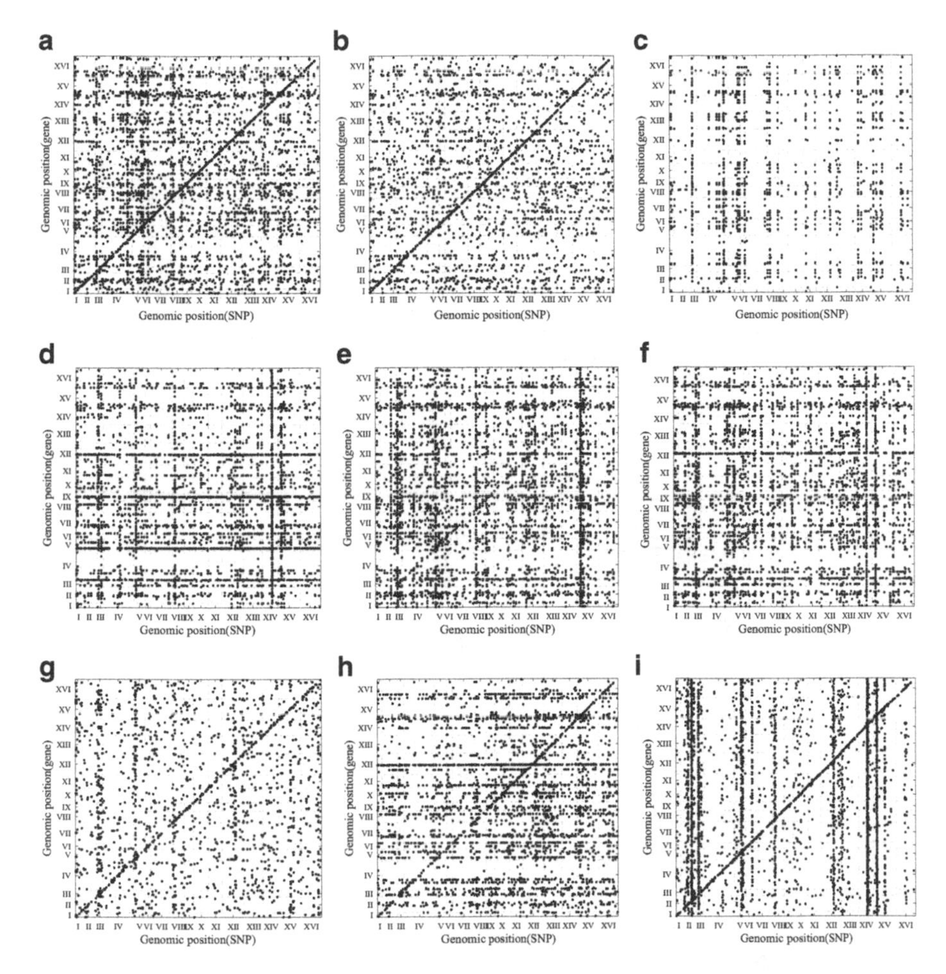


**Fig. 16** Significant associations discovered by different methods in yeast. (**a**) Model 2 **C** + **B** × **A**($M$ = 30, top 4500). (**b**) Model 2 **C**($M$ = 30, top 3000). (**c**) Model 2 **B** × **A**($M$ = 30, top 1500). (**d**) Model 1 **B** × **A**($M$ = 120). (**e**) Model 1 **B** × **A**($M$ = 150). (**f**) Model 1 **B** × **A**($M$ = 200). (**g**) MTLasso2G. (**h**) LORS. (**i**) Lasso

*Model*2 considers grouping of genes. *Model*1 has poor performance because a larger $M$ may be needed for *Model*1 to capture those individual associations.

## Reproducibility of trans Regulatory Hotspots Between Studies

We also evaluate the consistency of calling eQTL hotspots between two independent glucose yeast datasets [59]. The glucose environment from Smith et al. [59] shares a common set of segregants. It includes 5493 probes measured in 109 segregates.

**Table 2** Pairwise comparison of different models using *cis*- and *trans*- enrichment

| | | FaST-LMM | **C** of *Model*2 | SET-eQTL | MTLasso2G | **B** × **A** of *Model*1 | LORS | Lasso |
|---|---|---|---|---|---|---|---|---|
| *cis*-enrichment | **C** + **B** × **A** of *Model*2 | 0.4351 | <0.0001 | <0.0001 | <0.0001 | <0.0001 | <0.0001 | <0.0001 |
| | FaST-LMM | – | 0.2351 | <0.0001 | <0.0001 | <0.0001 | <0.0001 | <0.0001 |
| | **C** of *Model*2 | – | – | 0.0253 | 0.0221 | <0.0001 | <0.0001 | <0.0001 |
| | SET-eQTL | – | – | – | 0.0117 | <0.0001 | <0.0001 | <0.0001 |
| | MTLasso2G | – | – | – | – | <0.0001 | <0.0001 | <0.0001 |
| | **B** × **A** of *Model*1 | – | – | – | – | – | <0.0001 | <0.0001 |
| | LORS | – | – | – | – | – | – | 0.0052 |
| | | **B** × **A** of *Model*2 | FaST-LMM | MTLasso2G | LORS | **B** × **A** of *Model*1 | SET-eQTL | Lasso |
| *trans*-enrichment | **C** + **B** × **A** of *Model*2 | 0.4245 | 0.3123 | 0.0034 | 0.0029 | 0.0027 | 0.0025 | 0.0023 |
| | **B** × **A** of *Model*2 | – | 0.3213 | 0.0132 | 0.0031 | 0.0028 | 0.0027 | 0.0026 |
| | FaST-LMM | – | – | 0.0148 | 0.0033 | 0.0031 | 0.003 | 0.0029 |
| | MTLasso2G | – | – | – | 0.0038 | 0.0037 | 0.0036 | 0.0032 |
| | LORS | – | – | – | – | 0.0974 | 0.0387 | 0.0151 |
| | **B** × **A** of *Model*1 | – | – | – | – | – | 0.0411 | 0.0563 |
| | SET-eQTL | – | – | – | – | – | – | 0.0578 |

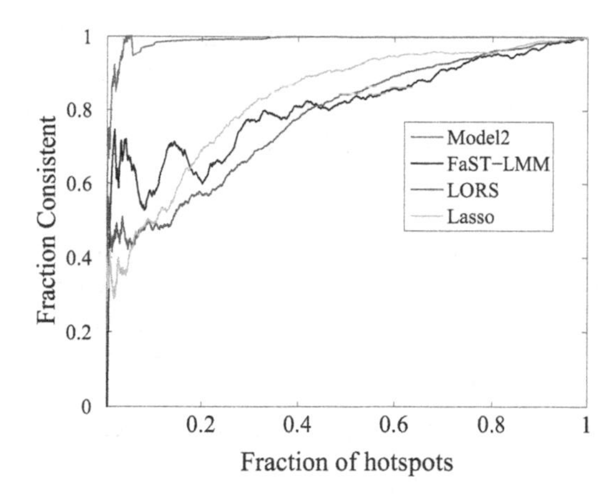


**Fig. 17** Consistency of detected eQTL hotspots

Since our algorithm aims at finding group-wise associations, we focus on the consistency of regulatory hotspots.

We examine the reproducibility of *trans* regulatory hotspots based on the following criteria [18, 29, 70]. For each SNP, we count the number of associated genes from the detected SNP-gene associations. We use this number as the regulatory degree of each SNP. For Model2, LORS, and Lasso, all SNP-Gene pairs with non-zero association weights are defined as associations. Note that Model2 uses $\mathbf{BA} + \mathbf{C}$ as the overall associations. For FaST-LMM, SNP-Gene pairs with a $q$-value $< 0.001$ are defined as associations. Note that we also tried different cutoffs for FaST-LMM (from 0.01 to 0.001), the results are similar. SNPs with large regulatory degrees are often referred to as hotspots. We sort SNPs by the extent of *trans* regulation (regulatory degrees) in a descending order. We denote the sorted SNPs lists as $S_1$ and $S_2$ for the two yeast datasets. Let $S_1^T$ and $S_2^T$ be the top $T$ SNPs in the sorted SNP lists. The trans calling consistency of detected hotspots is defined as $\frac{|S_1^T \cap S_2^T|}{T}$. Figure 17 compares the reproducibility of *trans* regulatory hotspots given by different studies. It can be seen that the proposed Model2 gives much higher consistency than any other competitors do. In particular, the consistency of *trans* hotspots suggests the superiority of Model2 in identifying hotspots that are likely to have a true genetic underpinning.

#### Gene Ontology Enrichment Analysis

As discussed in previous section, hidden variables $\mathbf{y}$ in the middle layer may model the joint effect of SNPs that have influence on a group of genes. To better understand the learned model, we look for correlations between a set of genes associated with a hidden variable and GO categories (Biological Process Ontology) [62]. In particular, for each gene set $G$, we identify the GO category whose set of genes is most correlated with $G$. We measure the correlation by a $p$-value determined by the Fisher's exact test. Since multiple gene sets $G$ need to be examined, the raw $p$-values need to be calibrated because of the multiple testing problem [68]. To compute the calibrated $p$-values for each gene set $G$, we perform a randomization test, wherein we apply the same test to randomly created gene sets that have the same number of genes as $G$. Specifically, the enrichment test is performed using DAVID [24]. And gene sets with calibrated $p$-values less than 0.01 are considered as significantly enriched. The results from *Model2* are reported in Table 3. Each row of Table 3 represents the gene set associated with a hidden variable. All of these detected gene sets are significantly enriched in certain GO categories. In total, 77 out of 90 gene sets detected by SET-eQTL are significant. For SET-eQTL, Fig. 18 shows the number of genes and SNPs within each group-wise association and the corresponding calibrated $p$-value (Fisher's exact test) of each discovered gene set. The hidden variable IDs are used as the cluster IDs. We can observe that for SET-eQTL, the gene sets with large calibrated $p$-values tend to have a very small SNP set associated with them. Those clusters are labeled in both figures. This is a strong indicator that these hidden variables may correspond to confounding factors.

For comparison, we visualize the number of SNPs and genes in each group-wise association in Fig. 19. We observe that 90 out of 150 gene sets reported by *Model1* are significantly enriched, and all 30 gene sets reported by *Model2* are significantly enriched. This indicates that *Model2* is able to detect group-wise linkages more precisely than *Model1*. We also study the hotspots detected by LORS, which affect $> 10$ gene traits [35]. Specifically, we delve into the top 15 hotspots detected by LORS (ranking by number of associated genes for each SNP). We can see that only 9 out of 15 top ranked hotspots are significantly enriched.

## 3.9 Conclusion

A crucial challenge in eQTL study is to understand how multiple SNPs interact with each other to jointly affect the expression level of genes. In this section, we propose three sparse graphical model-based approaches to identify novel group-wise eQTL associations. $\ell_1$-regularization is applied to learn the sparse structure of the graphical model. The three models incrementally take into consideration more aspects, such as group-wise association, potential confounding factors, and the existence of individual associations. We illustrate how each aspect would benefit the eQTL mapping. We also introduce computational techniques to make this approach suitable for large scale studies. Extensive experimental evaluations using both simulated and real datasets demonstrate that the proposed methods can effectively capture both individual and group-wise signals and significantly outperform the state-of-the-art eQTL mapping methods.

**Table 3** Summary of all detected groups of genes from *Model2* on yeast data

| [a]Group ID | [b]SNPs set size | [c]gene set size | [d]GO category |
|---|---|---|---|
| 1 | 63 | 294 | Oxidation-reduction process* |
| 2 | 78 | 153 | Thiamine biosynthetic process* |
| 3 | 94 | 871 | rRNA processing*** |
| 4 | 64 | 204 | Nucleosome assembly** |
| 5 | 70 | 288 | ATP synthesis coupled proton transport*** |
| 6 | 43 | 151 | Branched chain family amino acid biosynthetic...** |
| 7 | 76 | 479 | Mitochondrial translation*** |
| 8 | 47 | 349 | Transmembrane transport** |
| 9 | 64 | 253 | Cytoplasmic translation*** |
| 10 | 72 | 415 | Response to stress** |
| 11 | 64 | 225 | Mitochondrial translation* |
| 12 | 62 | 301 | Oxidation-reduction process** |
| 13 | 83 | 661 | Oxidation-reduction process* |
| 14 | 69 | 326 | Cytoplasmic translation* |
| 15 | 71 | 216 | Oxidation-reduction process* |
| 16 | 66 | 364 | Methionine metabolic process* |
| 17 | 74 | 243 | Cellular amino acid biosynthetic process*** |
| 18 | 63 | 224 | Transmembrane transport** |
| 19 | 23 | 50 | De novo' pyrimidine base biosynthetic process* |
| 20 | 66 | 205 | Cellular amino acid biosynthetic process*** |
| 21 | 81 | 372 | Oxidation-reduction process** |
| 22 | 33 | 126 | Oxidation-reduction process*** |
| 23 | 81 | 288 | Pheromone-dependent signal transduction...** |
| 24 | 53 | 190 | Pheromone-dependent signal transduction...** |
| 25 | 91 | 572 | Oxidation-reduction process*** |
| 26 | 66 | 46 | Cellular cell wall organization* |
| 27 | 111 | 1091 | Translation*** |
| 28 | 89 | 362 | Cellular amino acid biosynthetic process** |
| 29 | 62 | 217 | Transmembrane transport** |
| 30 | 71 | 151 | Cellular aldehyde metabolic process** |

[a] Group ID corresponding to Fig. 19
[b] Number of SNPs in the group
[c] Number of genes in the group
[d] The most significant GO category enriched in the associated gene set. The enrichment test was performed using DAVID [29]. The gene function is defined by GO category. Adjusted $p$-values are reported by using permutation test. Adjusted $p$-values are indicated by $*$, where $*10^{(-2)} \sim 10^{(-3)}$, $**10^{(-3)} \sim 10^{(-5)}$, $***10^{(-5)} \sim 10^{(-10)}$

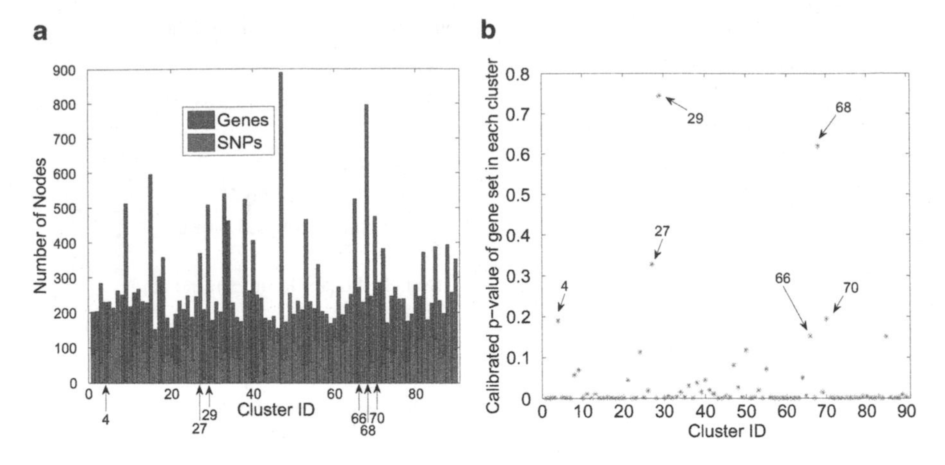


**Fig. 18** Number of nodes and calibrated $p$-values in each group-wise association. (**a**) #nodes in each set (SNP/gene). (**b**) Calibrated $p$-values of gene sets

# 4 Incorporating Prior Knowledge for Robust eQTL Mapping

## *4.1 Introduction*

Several important issues need to be considered in eQTL mapping. First, the number of SNPs is usually much larger than the number of samples [63]. Second, the existence of confounding factors, such as expression heterogeneity, may result in spurious associations [41]. Third, SNPs (and genes) usually work together to cause variation in complex traits [45]. The interplay among SNPs and the interplay among genes can be represented as networks and used as prior knowledge [48, 55]. However, such prior knowledge is far from being complete and may contain a lot of noise. Developing effective models to address these issues in eQTL studies has recently attracted increasing research interests [4, 32, 35, 36].

In eQTL studies, two types of networks can be utilized. One is the genetic interaction network [9]. Modeling genetic interaction (e.g., epistatic effect between SNPs) is essential to understanding the genetic basis of common diseases, since many diseases are complex traits [33]. Another type of network is the network among traits, such as the PPI network or the gene co-expression network. Interacting proteins or genes in a PPI network are likely to be functionally related, i.e., part of a protein complex or in the same biological pathway [65]. Effectively utilizing such prior network information can significantly improve the performance of eQTL mapping [35, 36].

Figure 20 shows an example of eQTL mapping with prior network knowledge. The interactions among SNPs and genes are represented by matrices **S** and **G**,

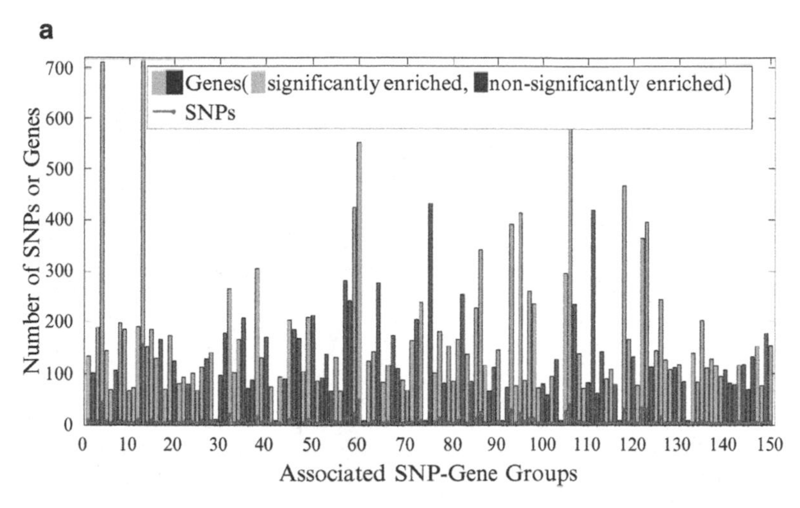


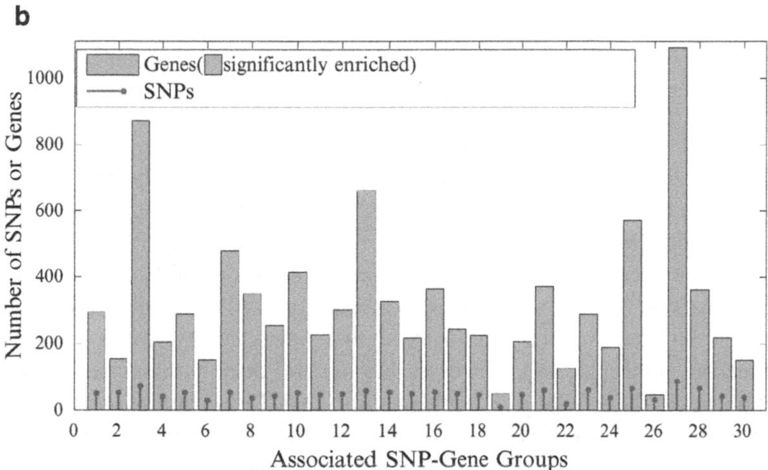


**Fig. 19** Number of SNPs and genes in each group-wise association. (**a**) *Model*1 (150 groups). (**b**) *Model*2 (30 groups)

respectively. The goal of eQTL mapping is to infer associations between SNPs and genes represented by the coefficient matrix **W**. Suppose that SNP ② is strongly associated with gene Ⓒ. Using the network prior, the moderate association between SNP ① and gene Ⓐ may be identified since ① and ②, Ⓐ and Ⓒ have interactions.

To leverage the network prior knowledge, several methods based on Lasso have been proposed [32, 35, 36]. The group-lasso penalty is applied to model the genetic interaction network. Xing et al. consider groupings of genes and apply a multi-task lasso penalty [32, 36]. They further extend the model to consider grouping information of both SNPs and genes [35]. These methods apply a "hard" clustering of SNPs (genes) so that an SNP (gene) cannot belong to multiple groups. However,

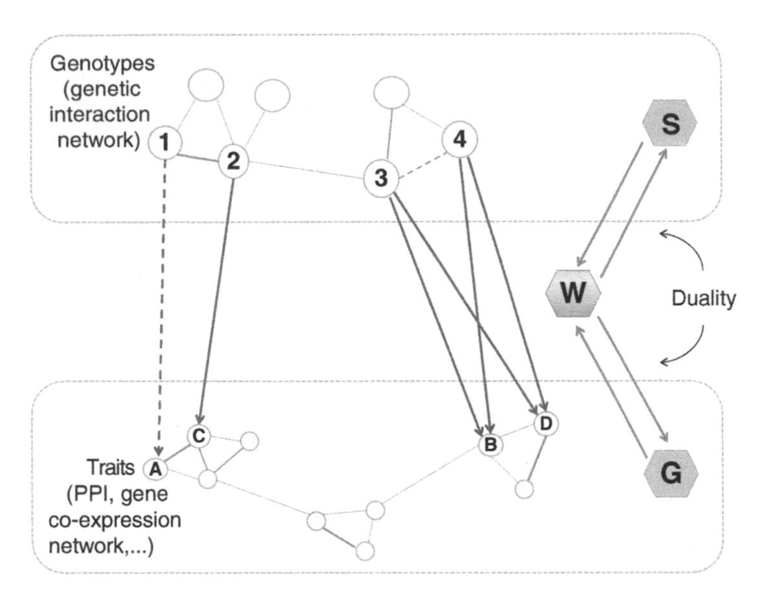


**Fig. 20** Examples of prior knowledge on **S** and **G**

an SNP may affect multiple genes and a gene may function in multiple pathways. To address this limitation, Jenatton et al. develop a model allowing overlap between different groups [28].

Despite their success, there are three common limitations of these group penalty-based approaches. First, a clustering step is usually needed to obtain the grouping information. To address this limitation, Xing et al. introduce a network-based fusion penalty on the genes [31, 40]. However, this method does not consider the genetic interaction network. A two-graph-guided multi-task Lasso approach is developed by Chen et al. [10] to make use of gene co-expression network and SNP correlation network. However, this method does not consider the network prior knowledge. The second limitation of the existing methods is that they do not take into consideration the incompleteness of the networks and the noise in them [65]. For example, PPI networks may contain false interactions and miss true interactions [65]. Directly using the grouping penalty inferred from the noisy and partial prior networks may introduce new bias and thus impair the performance. Third, in addition to the network information, other prior knowledge, such as location of genetic markers and gene pathway information, is also available. The existing methods cannot incorporate such information.

To address the limitations of the existing methods, this section proposes a novel approach, GDL, which simultaneously learns the association between SNPs and genes and refines the prior networks. To support "soft" clustering (allowing genes and SNPs to be members of multiple clusters), we adopt the graph regularizer to encode structured penalties from the prior networks. The penalties encourage the connected nodes (SNPs/genes) to have similar coefficients. This enables us to find multiple-correlated genetic markers with pleiotropic effects that affect multiple-correlated genes jointly. To tackle the problem of noisy and incomplete prior networks, we exploit the *duality* between learning the associations and refining the prior networks to achieve smoother regularization. That is, learning regression coefficients can help to refine the prior networks, and vice versa. For example, in Fig. 20, if SNPs ③ and ④ have strong associations with the same group of genes, they are likely to have interaction, which is not captured in the prior network. An ideal model should allow an update to the prior network according to the learned regression coefficients. GDL can also incorporate other available prior knowledge such as the physical location of SNPs and biology pathways to which the genes belong. The resultant optimization problem is convex and can be efficiently solved by using an alternating minimization procedure. We perform extensive empirical evaluation of the proposed method using both simulated and real eQTL datasets. The results demonstrate that GDL is robust to the incomplete and noisy prior knowledge and can significantly improve the accuracy of eQTL mapping compared to the state-of-the-art methods.

## 4.2 Background: Linear Regression with Graph Regularizer

Throughout the section, we assume that, for each sample, the SNPs and genes are represented by column vectors. Important notations are listed in Table 4. Let $\mathbf{x} = [x_1, x_2, \ldots, x_K]^T$ represent the $K$ SNPs in the study, where $x_i \in \{0, 1, 2\}$ is a random variable corresponding to the $i$th SNP. For example, 0, 1, 2 may encode the homozygous major allele, heterozygous allele, and homozygous minor allele, respectively. Let $\mathbf{z} = [z_1, z_2, \ldots, z_N]^T$ represent expression levels of the $N$ genes in the study, where $z_j$ is a continuous random variable corresponding to the $j$th gene. The traditional linear regression model for association mapping between $\mathbf{x}$ and $\mathbf{z}$ is

$$\mathbf{z} = \mathbf{W}\mathbf{x} + \boldsymbol{\mu} + \boldsymbol{\epsilon}, \tag{53}$$

where $\mathbf{z}$ is a linear function of $\mathbf{x}$ with coefficient matrix $\mathbf{W}$. $\boldsymbol{\mu}$ is an $N \times 1$ translation factor vector. $\boldsymbol{\epsilon}$ is the additive noise of Gaussian distribution with zero-mean and variance $\gamma\mathbf{I}$, where $\gamma$ is a scalar. That is, $\boldsymbol{\epsilon} \sim \mathcal{N}(\mathbf{0}, \gamma\mathbf{I})$.

The question now is how to define an appropriate objective function over $\mathbf{W}$ that (1) can effectively incorporate the prior network knowledge, and (2) is robust to the noise and incompleteness in the prior knowledge. Next, we first briefly review Lasso and its variations and then introduce the proposed GD-Lasso method.

**Table 4** Summary of notations

| Symbols | Description |
|---|---|
| $K$ | Number of SNPs |
| $N$ | Number of genes |
| $D$ | Number of samples |
| $\mathbf{X} \in \mathbb{R}^{K \times D}$ | The SNP matrix data |
| $\mathbf{Z} \in \mathbb{R}^{N \times D}$ | The gene matrix data |
| $\mathbf{L} \in \mathbb{R}^{N \times D}$ | A low-rank matrix |
| $\mathbf{S}_0 \in \mathbb{R}^{K \times K}$ | The input affinity matrices of the genetic interaction network |
| $\mathbf{G}_0 \in \mathbb{R}^{N \times N}$ | The input affinity matrices of the network of traits |
| $\mathbf{S} \in \mathbb{R}^{K \times K}$ | The refined affinity matrices of the genetic interaction network |
| $\mathbf{G} \in \mathbb{R}^{N \times N}$ | The refined affinity matrices of the network of traits |
| $\mathbf{W} \in \mathbb{R}^{N \times K}$ | The coefficient matrix to be inferred |
| $\mathcal{R}^{(S)}$ | The graph regularizer from the genetic interaction network |
| $\mathcal{R}^{(G)}$ | The graph regularizer from the PPI network |
| $\mathcal{D}(\cdot, \cdot)$ | A nonnegative distance measure |

### 4.2.1 Lasso and LORS

Lasso [63] is a method for estimating the regression coefficients $\mathbf{W}$ using $\ell_1$ penalty for sparsity. It has been widely used for association mapping problems. Let $\mathbf{X} = \{\mathbf{x}_d | 1 \leq d \leq D\} \in \mathbb{R}^{K \times D}$ be the SNP matrix and $\mathbf{Z} = \{\mathbf{z}_d | 1 \leq d \leq D\} \in \mathbb{R}^{N \times D}$ be the gene expression matrix. Each column of $\mathbf{X}$ and $\mathbf{Z}$ stands for one sample. The objective function of Lasso is

$$\min_{\mathbf{W}} \frac{1}{2} ||\mathbf{Z} - \mathbf{W}\mathbf{X} - \boldsymbol{\mu}\mathbf{1}||_F^2 + \eta ||\mathbf{W}||_1 \tag{54}$$

where $||\cdot||_F$ denotes the Frobenius norm, $||\cdot||_1$ is the $\ell_1$-norm. $\mathbf{1}$ is an $1 \times D$ vector of all 1's. $\eta$ is the empirical parameter for the $\ell_1$ penalty. $\mathbf{W}$ is the parameter (also called weight) matrix parameterizing the space of linear functions mapping from $\mathbf{X}$ to $\mathbf{Z}$.

Confounding factors, such as unobserved covariates, experimental artifacts, and unknown environmental perturbations, may mask real signals and lead to spurious findings. LORS [70] uses a low-rank matrix $\mathbf{L} \in \mathbb{R}^{N \times D}$ to account for the variations caused by hidden factors. The objective function of LORS is

$$\min_{\mathbf{W},\boldsymbol{\mu},\mathbf{L}} \frac{1}{2}||\mathbf{Z} - \mathbf{W}\mathbf{X} - \boldsymbol{\mu}\mathbf{1} - \mathbf{L}||_F^2 + \eta||\mathbf{W}||_1 + \lambda||\mathbf{L}||_* \tag{55}$$

where $||\cdot||_*$ is the nuclear norm. $\eta$ is the empirical parameter for the $\ell_1$ penalty to control the sparsity of $\mathbf{W}$, and $\lambda$ is the regularization parameter to control the rank of $\mathbf{L}$. $\mathbf{L}$ is a low-rank matrix assuming that there are only a small number of hidden factors influencing the gene expression levels.

### 4.2.2 Graph-Regularized Lasso

To incorporate the network prior knowledge, group sparse Lasso [4], multi-task Lasso [53], and SIOL [35] have been proposed. Group sparse Lasso makes use of grouping information of SNPs; multi-task Lasso makes use of grouping information of genes, while SIOL uses information from both networks. A common drawback of these methods is that the number of groups (SNP and gene clusters) has to be predetermined. To overcome this drawback, we propose to use two graph regularizers to encode the prior network information. Compared with the previous group penalty-based methods, our method does not need to pre-cluster the networks and thus may obtain smoother regularization. Moreover, these methods do not consider confounding factors that may mask real signals and lead to spurious findings. In this section, we further incorporate the idea in LORS [70] to tackle the confounding factors simultaneously.

Let $\mathbf{S}_0 \in \mathbb{R}^{K\times K}$ and $\mathbf{G}_0 \in \mathbb{R}^{N\times N}$ be the affinity matrices of the genetic interaction network (e.g., epistatic effect between SNPs) and network of traits (e.g., PPI network or gene co-expression network), and $\mathbf{D}_{S_0}$ and $\mathbf{D}_{G_0}$ be their degree matrices. Given the two networks, we can employ a pairwise comparison between $\mathbf{w}_{*i}$ and $\mathbf{w}_{*j}$ $(1 \leq i < j \leq K)$: if SNPs $i$ and $j$ are closely related, $||\mathbf{w}_{*i} - \mathbf{w}_{*j}||_2^2$ is small. The pairwise comparison can be naturally encoded in the *weighted fusion penalty* $\sum_{ij} ||\mathbf{w}_{*i} - \mathbf{w}_{*j}||_2^2(\mathbf{S}_0)_{i,j}$. This penalty will enforce $||\mathbf{w}_{*i} - \mathbf{w}_{*j}||_2^2 = 0$ for closely related SNP pairs (with large $(\mathbf{S}_0)_{i,j}$ value). Then, the graph regularizer from the genetic interaction network takes the following form:

$$\begin{aligned} \mathscr{R}^{(S)} &= \frac{1}{2}\sum_{ij} ||\mathbf{w}_{*i} - \mathbf{w}_{*j}||_2^2(\mathbf{S}_0)_{i,j} \\ &= \mathrm{tr}(\mathbf{W}(\mathbf{D}_{S_0} - \mathbf{S}_0)\mathbf{W}^{\mathrm{T}}) \end{aligned} \tag{56}$$

Similarly, the graph regularizer for the network of traits is

$$\mathscr{R}^{(G)} = \mathrm{tr}(\mathbf{W}^{\mathrm{T}}(\mathbf{D}_{G_0} - \mathbf{G}_0)\mathbf{W}) \tag{57}$$

These two regularizers encourage the connected nodes in a graph to have similar coefficients. A heavy penalty occurs if the learned regression coefficients for neighboring SNPs (genes) are disparate. $(\mathbf{D}_{S_0} - \mathbf{S}_0)$ and $(\mathbf{D}_{G_0} - \mathbf{G}_0)$ are known as the combinatorial graph Laplacian, which are positive semi-definite [12]. Graph-regularized Lasso (G-Lasso) solves the following optimization problem:

$$\begin{aligned} \min_{\mathbf{W},\boldsymbol{\mu},\mathbf{L}} & \frac{1}{2}||\mathbf{Z} - \mathbf{W}\mathbf{X} - \boldsymbol{\mu}\mathbf{1} - \mathbf{L}||_F^2 \\ & + \eta||\mathbf{W}||_1 + \lambda||\mathbf{L}||_* + \alpha\mathscr{R}^{(S)} + \beta\mathscr{R}^{(G)} \end{aligned} \tag{58}$$

where $\alpha, \beta > 0$ are regularization parameters.

## 4.3 *Graph-Regularized Dual Lasso*

In eQTL studies, the prior knowledge is usually incomplete and contains noise. It is desirable to refine the prior networks according to the learned regression coefficients. There is a *duality* between the prior networks and the regression coefficients: learning coefficients can help to refine the prior networks, and vice versa. This leads to mutual reinforcement when learning the two parts simultaneously.

Next, we introduce the GDL. We further relax the constraints from the prior networks (two graph regularizers) introduced in Sect. 4.2.2, and integrate the graph-regularized Lasso and the dual refinement of graphs into a unified objective function

$$\begin{aligned} \min_{\mathbf{W},\boldsymbol{\mu},\mathbf{L},\mathbf{S}\geq 0,\mathbf{G}\geq 0} & \frac{1}{2}||\mathbf{Z} - \mathbf{W}\mathbf{X} - \boldsymbol{\mu}\mathbf{1} - \mathbf{L}||_F^2 + \eta||\mathbf{W}||_1 + \lambda||\mathbf{L}||_* \\ & + \alpha \text{tr}(\mathbf{W}(\mathbf{D}_S - \mathbf{S})\mathbf{W}^{\mathrm{T}}) + \beta \text{tr}(\mathbf{W}^{\mathrm{T}}(\mathbf{D}_G - \mathbf{G})\mathbf{W}) \\ & + \gamma||\mathbf{S} - \mathbf{S}_0||_F^2 + \rho||\mathbf{G} - \mathbf{G}_0||_F^2 \end{aligned} \tag{59}$$

where $\gamma, \rho > 0$ are positive parameters controlling the extent to which the refined networks should be consistent with the original prior networks. $\mathbf{D}_S$ and $\mathbf{D}_G$ are the degree matrices of $\mathbf{S}$ and $\mathbf{G}$. Note that the objective function considers the non-negativity of $\mathbf{S}$ and $\mathbf{G}$. As an extension, the model can be extended easily to incorporate prior knowledge from multiple sources. We only need to revise the last two terms in Eq. (59) to $\gamma \sum_{i=1}^{f} ||\mathbf{S} - \mathbf{S}_i||_F^2 + \rho \sum_{i=1}^{e} ||\mathbf{G} - \mathbf{G}_i||_F^2$, where $f$ and $e$ are the number of sources for genetic interaction networks and gene trait networks, respectively.

---

**Algorithm 1:** Graph-regularized Dual Lasso (GD-Lasso)

---

**Input**: $\mathbf{X} = \{\mathbf{x}_d\} \in \mathbb{R}^{K\times D}$, $\mathbf{Z} = \{\mathbf{z}_d\} \in \mathbb{R}^{N\times D}$, $\mathbf{S}_0 \in \mathbb{R}^{K\times K}$, $\mathbf{G}_0 \in \mathbb{R}^{N\times N}$, $\eta,\alpha,\beta,\gamma,\rho$
**Output**: $\mathbf{W},\mu,\mathbf{S},\mathbf{G},\mathbf{L}$

---

### 4.3.1 Optimization: An Alternating Minimization Approach

In this section, we present an alternating scheme to optimize the objective function in Eq. (59) based on block coordinate techniques. We divide the variables into three sets: $\{\mathbf{L}\}$,$\{\mathbf{S},\mathbf{G}\}$, and $\{\mathbf{W},\mu\}$. We iteratively update one set of variables while fixing the other two sets. This procedure continues until convergence. Since the objective function is convex, the algorithm will converge to a global optima. The optimization process is as follows. The detailed algorithm is included in Algorithm 1.

1. While fixing $\{\mathbf{W},\mu\}$, $\{\mathbf{S},\mathbf{G}\}$, optimize $\{\mathbf{L}\}$ using singular value decomposition (SVD).

   **Lemma 2.** *[43] Suppose that matrix* $\mathbf{A}$ *has rank r. The solution to the optimization problem*

$$\min_{\mathbf{B}} \frac{1}{2}||\mathbf{A}-\mathbf{B}||_F^2 + \lambda||\mathbf{B}||_* \tag{60}$$

   *is given by* $\widehat{B} = \mathbf{H}_\lambda(\mathbf{A})$, *where* $\mathbf{H}_\lambda(\mathbf{A}) = \mathbf{U}\mathbf{D}_\lambda\mathbf{V}^T$ *with* $\mathbf{D}_\lambda = \text{diag}[(d_1-\lambda)_+,\ldots,(d_r-\lambda)_+]$, $\mathbf{U}\mathbf{D}\mathbf{V}^T$*is the Singular Value Decomposition (SVD) of* $\mathbf{A}$, $\mathbf{D} = \text{diag}[d_1,\ldots,d_r]$, *and* $(d_i-\lambda)_+ = \max((d_i-\lambda),0)$, $(1 \le i \le r)$.

   Thus, for fixed $\mathbf{W},\mu,\mathbf{S},\mathbf{G}$, the formula for updating $\mathbf{L}$ is

$$\mathbf{L} \leftarrow \mathbf{H}_\lambda(\mathbf{Z}-\mathbf{W}\mathbf{X}-\mu\mathbf{1}) \tag{61}$$

2. While fixing $\{\mathbf{W},\mu\}$, $\{\mathbf{L}\}$, optimize $\{\mathbf{S},\mathbf{G}\}$ using semi-nonnegative matrix factorization (semi-NMF) multiplicative updating on $\mathbf{S}$ and $\mathbf{G}$ iteratively. For the optimization with non-negative constraints, our updating rule is based on the following two theorems. The proofs of the theorems are given in Sect. 4.3.2.

**Theorem 1.** *For fixed* $\mathbf{L}$, $\mu$, $\mathbf{W}$, *and* $\mathbf{G}$, *updating* $\mathbf{S}$ *according to Eq.* (62) *monotonically decreases the value of the objective function in Eq.* (59) *until convergence.*

$$\mathbf{S} \leftarrow \mathbf{S} \circ \frac{\alpha(\mathbf{W}^T\mathbf{W})^+ + 2\gamma\mathbf{S}_0}{2\gamma\mathbf{S} + \alpha(\mathbf{W}^T\mathbf{W})^- + \alpha\,\text{diag}(\mathbf{W}^T\mathbf{W})\mathbf{J}_K} \tag{62}$$

*where* $\mathbf{J}_K$ *is a* $K\times K$ *matrix of all 1's.* $\circ$, $\frac{[\cdot]}{[\cdot]}$ *are element-wise operators. Since* $\mathbf{W}^T\mathbf{W}$ *may take mixed signs, we denote* $\mathbf{W}^T\mathbf{W} = (\mathbf{W}^T\mathbf{W})^+ - (\mathbf{W}^T\mathbf{W})^-$, *where* $(\mathbf{W}^T\mathbf{W})^+_{i,j} = (|(\mathbf{W}^T\mathbf{W})_{i,j}| + (\mathbf{W}^T\mathbf{W})_{i,j})/2$ *and* $(\mathbf{W}^T\mathbf{W})^-_{i,j} = (|(\mathbf{W}^T\mathbf{W})_{i,j}| - (\mathbf{W}^T\mathbf{W})_{i,j})/2$.

**Theorem 2.** *For fixed* $\mathbf{L}$, $\mu$, $\mathbf{W}$, *and* $\mathbf{S}$, *updating* $\mathbf{G}$ *according to Eq.* (63) *monotonically decreases the value of the objective function in Eq.* (59) *until convergence.*

$$\mathbf{G} \leftarrow \mathbf{G} \circ \frac{\beta(\mathbf{W}\mathbf{W}^{\mathrm{T}})^{+} + 2\rho\mathbf{G}_0}{2\rho\mathbf{G} + \beta(\mathbf{W}\mathbf{W}^{\mathrm{T}})^{-} + \beta\,\mathrm{diag}(\mathbf{W}\mathbf{W}^{\mathrm{T}})\mathbf{J}_N} \tag{63}$$

*where* $\mathbf{J}_N$ *is an* $N \times N$ *matrix of all 1's.*

The above two theorems are derived from the KKT complementarity condition [6]. We show the updating rule for **S** below. The analysis for **G** is similar and omitted. We first formulate the Lagrange function of **S** for optimization

$$L(\mathbf{S}) = \alpha \mathrm{tr}(\mathbf{W}(\mathbf{D}_S - \mathbf{S})\mathbf{W}^{\mathrm{T}}) + \gamma||\mathbf{S} - \mathbf{S}_0||_F^2 \tag{64}$$

The partial derivative of the Lagrange function with respect to **S** is

$$\nabla_{\mathbf{S}} L = -\alpha \mathbf{W}^{\mathrm{T}}\mathbf{W} - 2\gamma \mathbf{S}_0 + 2\gamma \mathbf{S} + \alpha\,\mathrm{diag}(\mathbf{W}^{\mathrm{T}}\mathbf{W})\mathbf{J}_K \tag{65}$$

Using the KKT complementarity condition for the non-negative constraint on **S**, we have

$$\nabla_{\mathbf{S}} L \circ \mathbf{S} = \mathbf{0} \tag{66}$$

The above formula leads to the updating rule for **S** in Eq. (62). It has been shown that the multiplicative updating algorithm has first order convergence rate [15].

3. While fixing $\{\mathbf{L}\}$, $\{\mathbf{S}, \mathbf{G}\}$, optimize $\{\mathbf{W}, \boldsymbol{\mu}\}$ using the coordinate descent algorithm.

   Because we use the $\ell_1$ penalty on **W**, we can use the coordinate descent algorithm for the optimization of **W**, which gives the following updating formula:

$$\mathbf{W}_{i,j} = \frac{F(m(i,j), \eta)}{(\mathbf{X}\mathbf{X}^{\mathrm{T}})_{j,j} + 2\alpha(\mathbf{D}_{\mathbf{S}} - \mathbf{S})_{j,j} + 2\beta(\mathbf{D}_{\mathbf{G}} - \mathbf{G})_{i,i}} \tag{67}$$

   where $F(m(i,j), \eta) = sign(m(i,j)) \max(|m(i,j)| - \eta, 0)$, and

$$\begin{aligned} m(i,j) = {} & (\mathbf{Z}\mathbf{X}^{\mathrm{T}})_{i,j} - \sum_{\substack{k=1 \\ k \neq j}}^{K} \mathbf{W}_{i,k}(\mathbf{X}\mathbf{X}^{\mathrm{T}})_{k,j} \\ & - 2\alpha \sum_{\substack{k=1 \\ k \neq j}}^{K} \mathbf{W}_{i,k}(\mathbf{D}_{\mathbf{S}} - \mathbf{S})_{k,j} - 2\beta \sum_{\substack{k=1 \\ k \neq i}}^{N} (\mathbf{D}_{\mathbf{G}} - \mathbf{G})_{i,k}\mathbf{W}_{k,j} \end{aligned} \tag{68}$$

   The solution of updating $\boldsymbol{\mu}$ can be derived by setting $\nabla_{\boldsymbol{\mu}} L(\boldsymbol{\mu}) = 0$, which gives

$$\boldsymbol{\mu} = \frac{(\mathbf{Z} - \mathbf{W}\mathbf{X})\mathbf{1}^{\mathrm{T}}}{D} \tag{69}$$

### 4.3.2 Convergence Analysis

In the following, we investigate the convergence of the algorithm. First, we study the convergence for the second step. We use the auxiliary function approach [34] to analyze the convergence of the multiplicative updating formulas. Here we first introduce the definition of auxiliary function.

**Definition 1.** Given a function $L(h)$ of any parameter $h$, a function $Z(h, \tilde{h})$ is an auxiliary function for $L(h)$ if the conditions

$$Z(h, \tilde{h}) \geq L(h) \qquad and \qquad Z(h, h) = L(h), \tag{70}$$

are satisfied for any given $h, \tilde{h}$ [34].

**Lemma 2.** *If $Z$ is an auxiliary function for function $L(h)$, then $L(h)$ is non-increasing under the update [34].*

$$h^{(t+1)} = \arg\min_h Z(h, h^{(t)}) \tag{71}$$

**Theorem 3.** *Let $L(\mathbf{S})$ denote the Lagrange function of $\mathbf{S}$ for optimization. The following function:*

$$\begin{aligned} Z(\mathbf{S}, \tilde{\mathbf{S}}) = {} & \alpha \sum_{ijk} \mathbf{W}_{i,j}^2 \frac{\mathbf{S}_{j,k}^2 + \tilde{\mathbf{S}}_{j,k}^2}{2\tilde{\mathbf{S}}_{j,k}} + \alpha \sum_{ijk} (\mathbf{W}_{i,j}\mathbf{W}_{i,k})^- \frac{\mathbf{S}_{j,k}^2 + \tilde{\mathbf{S}}_{j,k}^2}{2\tilde{\mathbf{S}}_{j,k}} \\ & - \alpha \sum_{ijk} (\mathbf{W}_{i,j}\mathbf{W}_{i,k})^+ \tilde{\mathbf{S}}_{j,k} (1 + \log \frac{\mathbf{S}_{j,k}}{\tilde{\mathbf{S}}_{j,k}}) + \gamma \sum_{jk} \mathbf{S}_{j,k}^2 \\ & - 2\gamma \sum_{jk} (\mathbf{S}_0)_{j,k} \tilde{\mathbf{S}}_{j,k} (1 + \log \frac{\mathbf{S}_{j,k}}{\tilde{\mathbf{S}}_{j,k}}) + \gamma \sum_{jk} (\mathbf{S}_0)_{j,k}^2. \end{aligned} \tag{72}$$

*is an auxiliary function for $L(\mathbf{S})$. Furthermore, it is a convex function in $\mathbf{S}$ and its global minimum is*

$$\mathbf{S} = \tilde{\mathbf{S}} \circ \frac{\alpha(\mathbf{W}^T\mathbf{W})^+ + 2\gamma \mathbf{S}_0}{2\gamma\tilde{\mathbf{S}} + \alpha(\mathbf{W}^T\mathbf{W})^- + \alpha \operatorname{diag}(\mathbf{W}^T\mathbf{W})\mathbf{J}_K}. \tag{73}$$

Theorem 3 can be proved using a similar idea to that in [14] by validating three **Properties**: (1) $L(\mathbf{S}) \leq Z(\mathbf{S}, \tilde{\mathbf{S}})$; (2) $L(\mathbf{S}) = Z(\mathbf{S}, \mathbf{S})$; (3) $Z(\mathbf{S}, \tilde{\mathbf{S}})$ is convex with respect to $\mathbf{S}$. The formal proof is provided below.

*Proof.* We will prove the three properties, respectively. The Lagrange function of $\mathbf{S}$ for optimization is

$$L(\mathbf{S}) = \alpha \mathrm{tr}(\mathbf{W}(\mathbf{D_S} - \mathbf{S})\mathbf{W}^T) + \gamma ||\mathbf{S} - \mathbf{S}_0||_F^2. \tag{74}$$

To prove **Properties** 1 and 2, we first deduce the following identities:

$$\mathrm{tr}(\mathbf{W}\mathbf{D}_{\mathbf{S}}\mathbf{W}^T) \quad = \sum_{ijk} \mathbf{W}_{i,j}^2 \mathbf{S}_{j,k}. \tag{75}$$

Similarly,

$$\mathrm{tr}(\mathbf{W}\mathbf{S}\mathbf{W}^{\mathrm{T}}) = \sum_{ijk} \mathbf{W}_{i,j}\mathbf{W}_{i,k}\mathbf{S}_{j,k}. \tag{76}$$

And,

$$\begin{aligned} ||\mathbf{S}-\mathbf{S}_0||_F^2 =& \mathrm{tr}(\mathbf{S}\mathbf{S}^{\mathrm{T}}) - 2\mathrm{tr}(\mathbf{S}_0\mathbf{S}^{\mathrm{T}}) + \mathrm{tr}(\mathbf{S}_0\mathbf{S}_0^{\mathrm{T}}) \\ =& \sum_{jk} \mathbf{S}_{j,k}^2 - 2\sum_{jk}(\mathbf{S}_0)_{j,k}\mathbf{S}_{j,k} + \sum_{jk}(\mathbf{S}_0)_{j,k}^2. \end{aligned} \tag{77}$$

Using identities (75), (76), and (77), and substituting $\tilde{\mathbf{S}}$ with $\mathbf{S}$ in function (72), we get the identity for **Property 2**.

Further, note that $a \leq \frac{a^2+b^2}{2b}$ and $a \geq b(1 + \log \frac{a}{b})$ for all positive $a$ and $b$, and we have

- for (75),

$$\sum_{ijk} \mathbf{W}_{i,j}^2 \mathbf{S}_{j,k} \leq \sum_{ijk} \mathbf{W}_{i,j}^2 \frac{\mathbf{S}_{j,k}^2 + \tilde{\mathbf{S}}_{j,k}^2}{2\tilde{\mathbf{S}}_{j,k}};$$

- for (76),

$$\begin{aligned} &\sum_{ijk} \mathbf{W}_{i,j}\mathbf{W}_{i,k}\mathbf{S}_{j,k} \\ =& \sum_{ijk}(\mathbf{W}_{i,j}\mathbf{W}_{i,k})^+\mathbf{S}_{j,k} - \sum_{ijk}(\mathbf{W}_{i,j}\mathbf{W}_{i,k})^-\mathbf{S}_{j,k} \\ \geq& \sum_{ijk}(\mathbf{W}_{i,j}\mathbf{W}_{i,k})^+\tilde{\mathbf{S}}_{j,k}(1 + \log \frac{\mathbf{S}_{j,k}}{\tilde{\mathbf{S}}_{j,k}}) \\ &- \sum_{ijk}(\mathbf{W}_{i,j}\mathbf{W}_{i,k})^- \frac{\mathbf{S}_{j,k}^2 + \tilde{\mathbf{S}}_{j,k}^2}{2\tilde{\mathbf{S}}_{j,k}}; \end{aligned} \tag{78}$$

- for the second term in (77),

$$\sum_{jk}(\mathbf{S}_0)_{j,k}\mathbf{S}_{j,k} \geq 2\sum_{jk}(\mathbf{S}_0)_{j,k}\tilde{\mathbf{S}}_{j,k}(1 + \log \frac{\mathbf{S}_{j,k}}{\tilde{\mathbf{S}}_{j,k}})$$

These inequalities together prove **Property 1**.

For **Property 3**, we instead prove the Hessian matrix $\nabla\nabla_{\mathbf{S}} Z(\mathbf{S}, \tilde{\mathbf{S}}) \succeq \mathbf{0}$

$$\begin{aligned} &\frac{\partial Z(\mathbf{S}, \tilde{\mathbf{S}})}{\partial \mathbf{S}_{m,n}} \\ =& \alpha \sum_i \mathbf{W}_{i,m}^2 \frac{\mathbf{S}_{m,n}}{\tilde{\mathbf{S}}_{m,n}} + \alpha \sum_i (\mathbf{W}_{i,m}\mathbf{W}_{i,n})^- \frac{\mathbf{S}_{m,n}}{\tilde{\mathbf{S}}_{m,n}} \\ &- \alpha \sum_i (\mathbf{W}_{i,m}\mathbf{W}_{i,n})^+ \frac{\tilde{\mathbf{S}}_{m,n}}{\mathbf{S}_{m,n}} + 2\gamma\mathbf{S}_{m,n} - 2\gamma(\mathbf{S}_0)_{m,n}\frac{\tilde{\mathbf{S}}_{m,n}}{\mathbf{S}_{m,n}}. \end{aligned} \tag{79}$$

Hence,

$$
\begin{aligned}
&\frac{\partial^2 Z(\mathbf{S},\tilde{\mathbf{S}})}{\partial \mathbf{S}_{s,t}\partial \mathbf{S}_{m,n}}\\
=&\alpha\sum_i \delta_{ms}\delta_{nt}\mathbf{W}_{i,m}^2\frac{1}{\tilde{\mathbf{S}}_{m,n}} + \alpha\sum_i \delta_{ms}\delta_{nt}(\mathbf{W}_{i,m}\mathbf{W}_{i,n})^-\frac{1}{\tilde{\mathbf{S}}_{m,n}}\\
&+\alpha\sum_i \delta_{ms}\delta_{nt}(\mathbf{W}_{i,m}\mathbf{W}_{i,n})^+\frac{\tilde{\mathbf{S}}_{m,n}}{\mathbf{S}_{m,n}^2}\\
&+2\gamma\delta_{ms}\delta_{nt} + 2\gamma\delta_{ms}\delta_{nt}(\mathbf{S}_0)_{m,n}\frac{\tilde{\mathbf{S}}_{m,n}}{\mathbf{S}_{m,n}^2}\\
\geq& 0.
\end{aligned}
\tag{80}
$$

Therefore, $\nabla_{\mathbf{S}}^2 Z(\mathbf{S},\tilde{\mathbf{S}})$ is diagonal with positive entries. Thus $\nabla_{\mathbf{S}}^2 Z(\mathbf{S},\tilde{\mathbf{S}})$ is positively defined, namely $Z(\mathbf{S},\tilde{\mathbf{S}})$ is convex, which concludes **Property 3**.

To solve for $\mathbf{S}$, we set $\nabla_{\mathbf{S}} Z(\mathbf{S},\tilde{\mathbf{S}}) = \mathbf{0}$, and get the following formula for all $m$ and $n$:

$$
\begin{aligned}
&\frac{\partial}{\partial \mathbf{S}_{m,n}} Z(\mathbf{S},\tilde{\mathbf{S}})\\
=&\alpha\sum_i \mathbf{W}_{i,m}^2\frac{\mathbf{S}_{m,n}}{\tilde{\mathbf{S}}_{m,n}} + \alpha\sum_i (\mathbf{W}_{i,m}\mathbf{W}_{i,n})^-\frac{\mathbf{S}_{m,n}}{\tilde{\mathbf{S}}_{m,n}}\\
&-\alpha\sum_i (\mathbf{W}_{i,m}\mathbf{W}_{i,n})^+\frac{\tilde{\mathbf{S}}_{m,n}}{\mathbf{S}_{m,n}} + 2\gamma\mathbf{S}_{m,n} - 2\gamma(\mathbf{S}_0)_{m,n}\frac{\tilde{\mathbf{S}}_{m,n}}{\mathbf{S}_{m,n}}\\
=&0.
\end{aligned}
\tag{81}
$$

After sorting the equation, we have

$$
\mathbf{S}_{m,n} = \tilde{\mathbf{S}}_{m,n}\cdot\frac{\alpha\sum_i(\mathbf{W}_{i,m}\mathbf{W}_{i,n})^+ + 2\gamma(\mathbf{S}_0)_{m,n}}{2\gamma\tilde{\mathbf{S}}_{m,n} + \alpha\sum_i(\mathbf{W}_{i,m}\mathbf{W}_{i,n})^- + \alpha\sum_i\mathbf{W}_{i,m}^2}.
\tag{82}
$$

That is equivalent to the formula (73), which is consistent with the updating formula derived from the KKT condition aforementioned. □

**Theorem 4.** *Updating* $\mathbf{S}$ *using Eq.* (62) *will monotonically decrease the value of the objective in Eq.* (59)*, the objective is invariant if and only if* $\mathbf{S}$ *is at a stationary point.*

*Proof.* By Lemma 2 and Theorem 3, for each subsequent iteration of updating $\mathbf{S}$, we have $L((\mathbf{S})^0) = Z((\mathbf{S})^0,(\mathbf{S})^0) \geq Z((\mathbf{S})^1,(\mathbf{S})^0) \geq Z((\mathbf{S})^1,(\mathbf{S})^1) = L((\mathbf{S})^1) \geq \ldots \geq L((\mathbf{S})^{Iter})$. Thus $L(\mathbf{S})$ monotonically decreases. Since the objective function Eq. (59) is obviously bounded below, the correctness of Theorem 1 is proved. Theorem 2 can be proved similarly. □

In addition to Theorem 4, since the computation of $\mathbf{L}$ in the first step decreases the value of the objective in Eq. (59), and the coordinate descent algorithm for updating $\mathbf{W}$ in the third step also monotonically decreases the value of the objective, the algorithm is guaranteed to converge.

### 4.4 Generalized Graph-Regularized Dual Lasso

In this section, we extend our model to incorporate additional prior knowledge such as SNP locations and biological pathways. If the physical locations of two SNPs are close or two genes belong to the same pathway, they are likely to have interactions. Such information can be integrated to help refine the prior networks.

Continue with our example in Fig. 20. If SNPs ③ and ④ affect the same set of genes (Ⓑ and Ⓓ), and at the same time, they are close to each other, then it is likely there exists interaction between ③ and ④.

Formally, we would like to solve the following optimization problem:

$$\begin{aligned} \min_{\mathbf{W},\mu,\mathbf{L},\mathbf{S}\geq 0,\mathbf{G}\geq 0} & \frac{1}{2}||\mathbf{WX}-\mathbf{Z}-\mu\mathbf{1}-\mathbf{L}||_F^2+\eta||\mathbf{W}||_1+\lambda||\mathbf{L}||_* \\ & +\alpha\sum_{i,j}\mathscr{D}(\mathbf{w}_{*i},\mathbf{w}_{*j})\mathbf{S}_{i,j}+\beta\sum_{i,j}\mathscr{D}(\mathbf{w}_{i*},\mathbf{w}_{j*})\mathbf{G}_{i,j} \end{aligned} \tag{83}$$

Here $\mathscr{D}(\cdot,\cdot)$ is a non-negative distance measure. Note that the Euclidean distance is used in previous sections. $\mathbf{S}$ and $\mathbf{G}$ are initially given by inputs $\mathbf{S}_0$ and $\mathbf{G}_0$. We refer to this generalized model as the Generalized Graph-regularized Dual Lasso (GGD-Lasso). GGD-Lasso executes the following two steps iteratively until the termination condition is met: (1) update $\mathbf{W}$ while fixing $\mathbf{S}$ and $\mathbf{G}$; (2) update $\mathbf{S}$ and $\mathbf{G}$ according to $\mathbf{W}$, while guarantee that both $\sum_{i,j}\mathscr{D}(\mathbf{w}_{*i},\mathbf{w}_{*j})\mathbf{S}_{i,j}$ and $\sum_{i,j}\mathscr{D}(\mathbf{w}_{i*},\mathbf{w}_{j*})\mathbf{G}_{i,j}$ decrease.

These two steps are based on the aforementioned duality between learning $\mathbf{W}$ and refining $\mathbf{S}$ and $\mathbf{G}$. The detailed algorithm is provided in Algorithm 2. Next, we illustrate the updating process assuming that $\mathbf{S}$ and $\mathbf{G}$ are unweighted graphs. It can be easily extended to weighted graphs.

Step 1 can be done by using the coordinate descent algorithm. In Step 2, to guarantee that both $\sum_{i,j}\mathscr{D}(\mathbf{w}_{*i},\mathbf{w}_{*j})\mathbf{S}_{i,j}$ and $\sum_{i,j}\mathscr{D}(\mathbf{w}_{i*},\mathbf{w}_{j*})\mathbf{G}_{i,j}$ decrease, we can maintain a fixed number of 1's in $\mathbf{S}$ and $\mathbf{G}$. Taking $\mathbf{G}$ as an example, once $\mathbf{G}_{i,j}$ is selected to change from 0 to 1, another element $\mathbf{G}_{i',j'}$ with $\mathscr{D}(\mathbf{w}_{i*},\mathbf{w}_{j*}) < \mathscr{D}(\mathbf{w}_{i'*},\mathbf{w}_{j'*})$ should be changed from 1 to 0.

The selection of $(i,j)$ and $(i',j')$ is based on the ranking of $\mathscr{D}(\mathbf{w}_{i*},\mathbf{w}_{j*})$ $(1 \leq i < j \leq N)$. Specifically, we examine $\kappa$ pairs with the smallest distances. Among them, we pick those having no edges in $\mathbf{G}$. Let $\mathscr{P}_0$ be this set of pairs. Accordingly, we examine $\kappa$ pairs with the largest distances. Among these pairs, we pick up only those having an edge in $\mathbf{G}$. Let $\mathscr{P}_1$ be this set of pairs. The elements of $\mathbf{G}$ corresponding to pairs in $\mathscr{P}_0$ are candidates for updating from 0 to 1, since these pairs of genes

---

**Algorithm 2:** Generalized Graph-regularized Dual Lasso (GGD-Lasso)

---

**Input**: $\mathbf{X}=\{\mathbf{x}_d\}\in\mathbb{R}^{K\times D}$, $\mathbf{Z}=\{\mathbf{z}_d\}\in\mathbb{R}^{N\times D}$, $\mathbf{S}_0\in\mathbb{R}^{K\times K}$, $\mathbf{G}_0\in\mathbb{R}^{N\times N}$, Pathway information, SNPs location information, $\eta$,$\alpha$,$\beta$,$\kappa_1$,$\kappa_2$
**Output**: $\mathbf{W}$,$\mu$,$\mathbf{S}$,$\mathbf{G}$,$\mathbf{L}$

---

are associated with similar SNPs. Similarly, elements of $\mathbf{G}$ corresponding to pairs in $\mathscr{P}_1$ are candidates for updating from 1 to 0.

In this process, the prior knowledge of gene pathways can be easily incorporated to better refine $\mathbf{G}$. For instance, we can further require that only the gene pairs in $\mathscr{P}_0$ belonging to the same pathway are eligible for updating, and only the gene pairs in $\mathscr{P}_1$ belonging to different pathways are eligible for updating. We denote the set of gene pairs eligible for updating by $\mathscr{P}'_0$ and $\mathscr{P}'_1$, respectively. Then, we choose $\min(|\mathscr{P}'_0|, |\mathscr{P}'_1|)$ pairs in set $\mathscr{P}'_0$ with smallest $\mathscr{D}(\mathbf{w}_{i*}, \mathbf{w}_{j*})$ $((i,j) \in \mathscr{P}'_0)$ and update $\mathbf{G}_{i,j}$ from 0 to 1. Similarly, we choose $\min(|\mathscr{P}'_0|, |\mathscr{P}'_1|)$ pairs in set $\mathscr{P}'_1$ with largest $\mathscr{D}(\mathbf{w}_{i'*}, \mathbf{w}_{j'*})$ $((i',j') \in \mathscr{P}'_1)$ and update $\mathbf{G}_{i',j'}$ from 1 to 0.

Obviously, all $\mathscr{D}(\mathbf{w}_{i*}, \mathbf{w}_{j*})$'s are smaller than $\mathscr{D}(\mathbf{w}_{i'*}, \mathbf{w}_{j'*})$ if $\kappa < \frac{N(N-1)}{4}$. Therefore, $\sum_{i,j} \mathscr{D}(\mathbf{w}_{i*}, \mathbf{w}_{j*})\mathbf{G}_{i,j}$ is guaranteed to decrease. The updating process for $\mathbf{S}$ is similar except that we compare columns rather than rows of $\mathbf{W}$ and use SNP locations rather than pathway information for evaluating the eligibility for updating. The updating process ends when no such pairs can be found so that switching their values will result in a decrease of the objective function.

The convergence of GGD-Lasso can be observed as follows. The decrease of the objective function value in the first step is straightforward since we minimize it using coordinate descent. In the second step, the change of the objective function value is given by

$$-\alpha\mathscr{D}(\mathbf{w}_{*i_S}, \mathbf{w}_{*j_S}) + \alpha\mathscr{D}(\mathbf{w}_{*i'_S}, \mathbf{w}_{*j'_S}) - \beta\mathscr{D}(\mathbf{w}_{i_G*}, \mathbf{w}_{j_G*}) + \beta\mathscr{D}(\mathbf{w}_{i'_G*}, \mathbf{w}_{j'_G*}) \tag{84}$$

which is always negative. Thus, in each iteration, the objective function value decreases. Since the objective function is non-negative, the process eventually converges.

**Theorem 5.** *GGD-Lasso converges to the global optimum if both* $\sum_{i,j} \mathscr{D}(\mathbf{w}_{i*}, \mathbf{w}_{j*})$ *and* $\sum_{i,j} \mathscr{D}(\mathbf{w}_{*i}, \mathbf{w}_{*j})$ *are convex to* $\mathbf{W}$.

*Proof.* The last two terms in Eq. (83) are linear with respect to $\mathbf{S}$ and $\mathbf{G}$, and convex to $\mathbf{W}$ according to the conditions listed. Thus the objective function is convex over all variables. A convergent result to the global optimum can be guaranteed. □

## 4.5 Experimental Results

In this section, we perform extensive experiments to evaluate the performance of the proposed methods. We use both simulated datasets and real yeast eQTL dataset [56]. For comparison, we select several state-of-the-art methods, including SIOL [35], two graph guided multi-task lasso (mtlasso2G) [10], sparse group Lasso [4], sparse multi-task Lasso [4], LORS [70], and Lasso [63]. For all the methods, the tuning parameters were learned using cross validation.

### 4.5.1 Simulation Study

We first evaluate the performance of the selected methods using simulation study. Note that GGD-Lasso requires additional prior knowledge and will be evaluated using real dataset.

We adopt the same setup for the simulation study as that in [35, 70] and generate synthetic datasets as follows. 100 SNPs are randomly selected from the yeast eQTL dataset [56] (112 samples). Ten gene expression profiles are generated by $\mathbf{Z}_{j*} = \mathbf{W}_{j*}\mathbf{X} + \Xi_{j*} + \mathbf{E}_{j*}$ $(1 \le j \le 10)$, where $\mathbf{E}_{j*} \sim \mathcal{N}(0, \sigma^2 I)$ $(\sigma = 1)$ denotes Gaussian noise. $\Xi_{j*}$ is used to model non-genetic effects, which are drawn from $\mathcal{N}(\mathbf{0}, \tau\Sigma)$, where $\tau = 0.1$. $\Sigma$ is generated by $\mathbf{M}\mathbf{M}^{\mathrm{T}}$, where $\mathbf{M} \in \mathbb{R}^{D \times C}$ and $\mathbf{M}_{ij} \sim \mathcal{N}(0, 1)$. $C$ is the number of hidden factors and is set to 10 by default. The association matrix $\mathbf{W}$ is generated as follows. Three sets of randomly selected four SNPs are associated with three gene clusters (1–3), (4–6), (7–10), respectively. In addition, one SNP is associated with two gene clusters (1–3) and (4–6), and one SNP is associated with all genes. The association strength is set to 1 for all selected SNPs. The clustering structures among SNPs and genes serve as the *ground truth* of the prior network knowledge. Only two of the three SNP (gene) clusters are used in $\mathbf{W}$ to simulate incomplete prior knowledge.

Figure 21 shows the estimated $\mathbf{W}$ matrix by various methods. The x-axis represents traits (1–10) and y-axis represents SNPs (1–100). From the figure, we can see that GD-Lasso is more effective than G-Lasso. This is because the dual refinement enables a more robust model. G-Lasso outperforms SIOL and mtlasso2G, indicating that the graph regularizer provides a smoother regularization than the hard clustering-based penalty. In addition, SIOL and mtlasso2G do not consider confounding factors. SIOL and mtlasso2G outperform multi-task Lasso and sparse group Lasso since it uses both SNP and gene grouping information, while multi-task Lasso and sparse group Lasso only use one of them. We also observe that all methods utilizing prior grouping knowledge outperform LORS and Lasso which

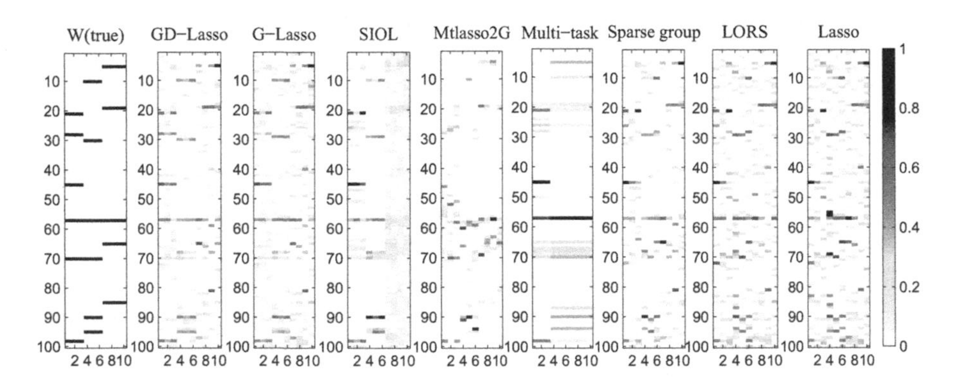


**Fig. 21** Ground truth of **W** and that estimated by different methods

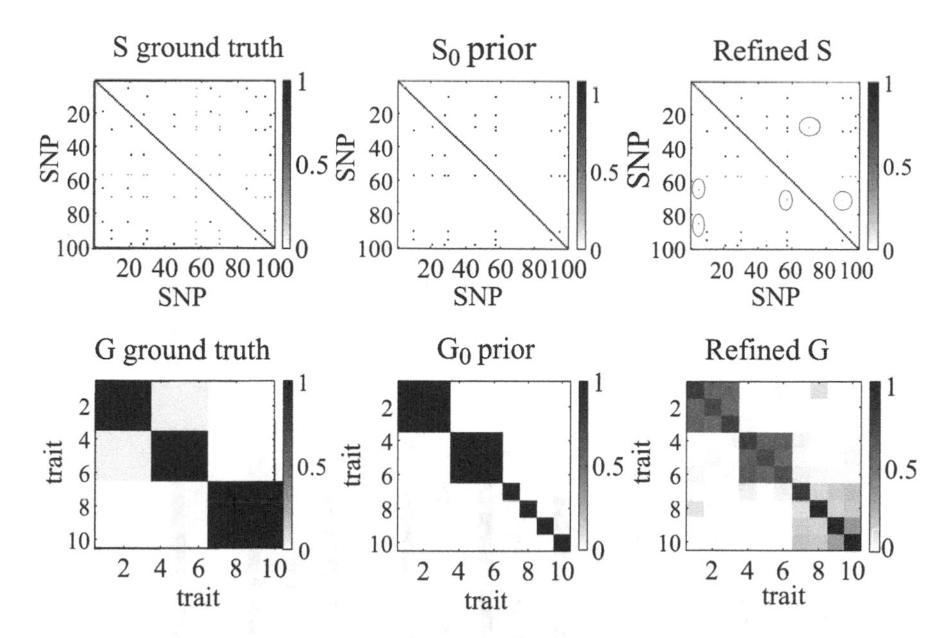


**Fig. 22** The ground truth networks, prior partial networks, and the refined networks

cannot incorporate prior knowledge. LORS outperforms Lasso since it considers the confounding factors.

The ground truth networks, prior networks, and GD-Lasso refined networks are shown in Fig. 22. Note that only a portion of the ground truth networks are used as prior networks. In particular, the information related to gene cluster (7–10) is missing in the prior networks. We observe that the refined matrix **G** well captures the missing grouping information of gene cluster (7–10). Similarly, many missing pairwise relationships in **S** are recovered in the refined matrix (points in red ellipses).

Using 50 simulated datasets with different Gaussian noise ($\sigma^2 = 1$ and $\sigma^2 = 5$), we compare the proposed methods with alternative state-of-the-art approaches. For each setting, we use 30 samples for test and 82 samples for training. We report the average result from 50 realizations. Figure 23 shows the ROC curves of TPR-FPR for performance comparison, together with the areas under the precision-recall curve (AUCs) [10]. The association strengths between SNPs and genes are set to be 0.1, 1, and 3, respectively. It is clear that GD-Lasso outperforms all alternative methods by effectively using and refining the prior network knowledge. We also computed test errors. On average, GD-Lasso achieved the best test error rate of 0.9122, and the order of the other methods in terms of the test errors is: G-Lasso (0.9276), SIOL (0.9485), Mtlasso2G (0.9521), Multi-task Lasso (0.9723), Sparse group Lasso (0.9814), LORS (1.0429), and Lasso (1.2153).

To evaluate the effectiveness of dual refinement, we compare GD-Lasso and G-Lasso since the only difference between these two methods is whether the prior networks are refined during the optimization process. We add noises to the prior

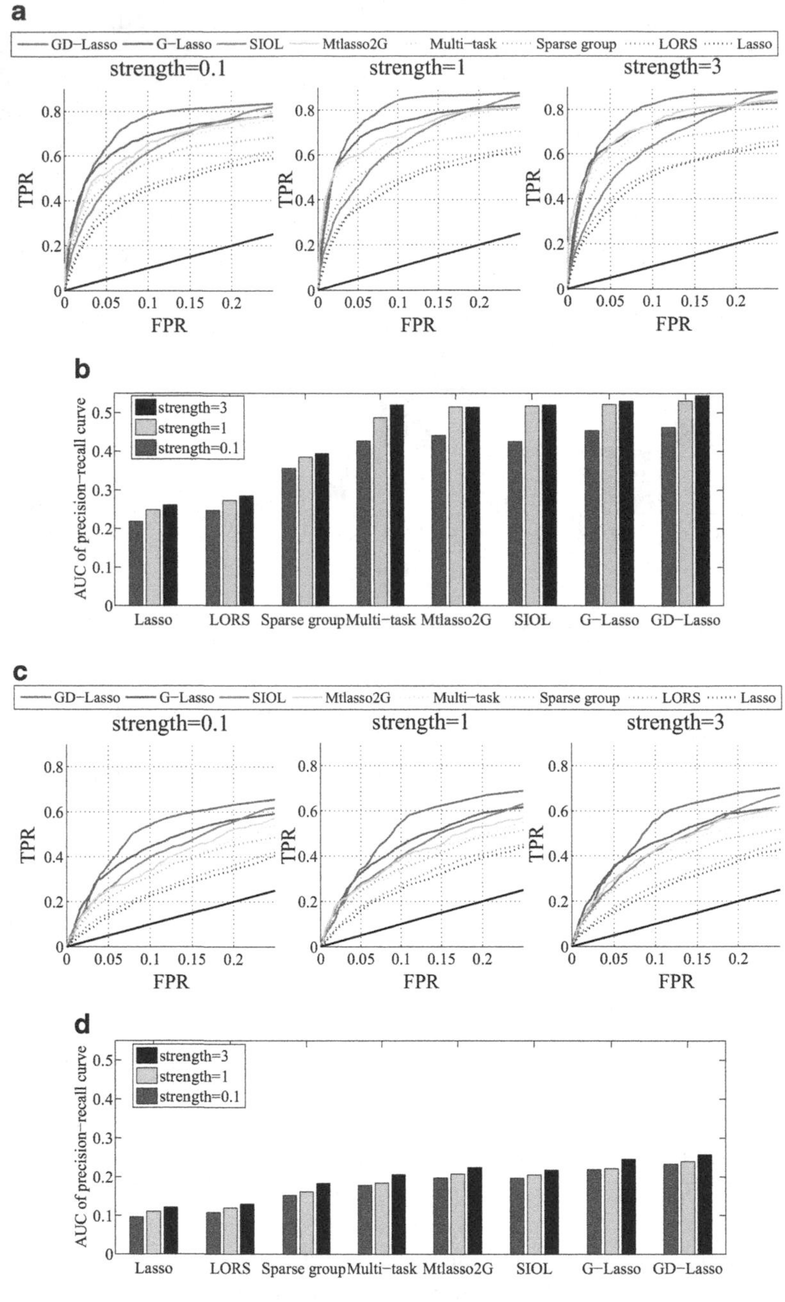


**Fig. 23** The ROC curve and AUCs of different methods. (**a**) variance of errors ($\sigma^2 = 1$). (**b**) AUC of precision-recall curve ($\sigma^2 = 1$). (**c**) variance of errors ($\sigma^2 = 5$). (**d**) AUC of precision-recall curve ($\sigma^2 = 5$)

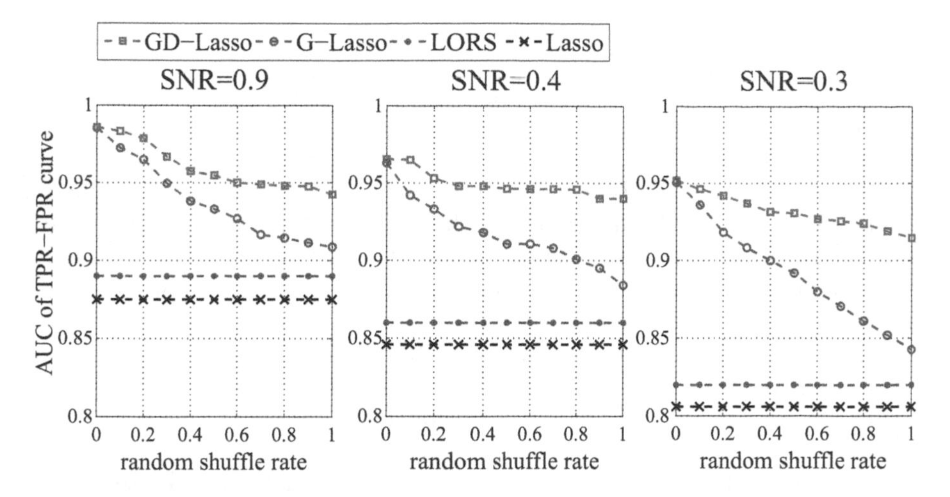


**Fig. 24** The AUCs of the TPR-FPR curve of different methods

networks by randomly shuffling the elements in them. Furthermore, we use the signal-to-noise ratio defined as $SNR = \sqrt{\frac{Var(\mathbf{WX})}{Var(\Xi+\mathbf{E})}}$ [70] to measure the noise ratio in the eQTL datasets. Here, we fix $C = 10, \tau = 0.1$, and use different $\sigma$'s to control SNR.

Figure 24 shows the results for different SNRs. For a fixed SNR, we vary the percentage of noises in the prior networks and compare the performance of selected methods. From the results, we can see that G-Lasso is more sensitive to noises in the prior networks than GD-Lasso is. Moreover, when the SNR is low, the advantage of GD-Lasso is more prominent. These results indicate using dual refinement can dramatically improve the accuracy of the identified associations.

### 4.5.2 Yeast eQTL Study

We apply the proposed methods to a yeast (Saccharomyces cerevisiae) eQTL dataset of 112 yeast segregants [56] generated from a cross of two inbred strains. The dataset originally includes expression profiles of 6229 gene expression traits and genotype profiles of 2956 SNPs. After removing SNPs with more than 10 % missing values and merging consecutive SNPs high linkage disequilibrium, we get 1017 SNPs with unique genotypes [24]. 4474 expression profiles are selected after removing the ones with missing values. The genetic interaction network is generated as in [35]. We use the PPI network downloaded from BioGRID (http://thebiogrid.org/) to represent the prior network among genes. It takes around 1 day for GGD-Lasso, and around 10 h for GD-Lasso to run into completion.

### 4.5.3 cis and trans Enrichment Analysis

We follow the standard *cis*-enrichment analysis [41] to compare the performance of two competing models. The intuition behind *cis*-enrichment analysis is that more *cis*-acting SNPs are expected than *trans*-acting SNPs. A two-step procedure is used in the *cis*-enrichment analysis [41]: (1) for each model, we apply a one-tailed Mann-Whitney test on each SNP to test the null hypothesis that the model ranks its *cis* hypotheses no better than its *trans* hypotheses, (2) for each pair of models compared, we perform a two-tailed paired Wilcoxon sign-rank test on the $p$-values obtained from the previous step. The null hypothesis is that the median difference of the $p$-values in the Mann-Whitney test for each SNP is zero. The *trans*-enrichment is implemented using a similar strategy [71], in which genes regulated by transcription factors (obtained from http://www.yeastract.com/download.php) are used as *trans*-acting signals.

In addition to the methods evaluated in the simulation study, GGD-Lasso is also evaluated here (with $\kappa = 100,000$,$\eta = 5, \lambda = 8, \alpha = 15, \beta = 1$). For GD-Lasso, $\eta = 5, \lambda = 8, \alpha = 15, \beta = 1, \gamma = 15, \rho = 1$. The Euclidean distance is used as the distance metric. We rank pairs of SNPs and genes according to the learned $\mathbf{W}$. $\mathbf{S}$ is refined if the locations of the two SNPs are less than 500 bp. $\mathbf{G}$ is refined if the two genes are in the same pathway. The pathway information is downloaded from Saccharomyces Genome Database [SGD (http://www.yeastgenome.org/)].

The results of pairwise comparison of selected models are shown in Table 5. In this table, a $p$-value shows how significant a method on the left column outperforms a method in the top row in terms of *cis* and *trans* enrichments. We observe that the proposed GGD-Lasso and GD-Lasso have significantly better enrichment scores than the other models. By incorporating genomic location and pathway information, GGD-Lasso performs better than GD-Lasso with $p$-value less than 0.0001. The effectiveness of the dual refinement on prior graphs is demonstrated by GD-Lasso's better performance over G-Lasso. Note that the performance ranking of these models is consistent with that in the simulation study.

The top-1000 significant associations given by GGD-Lasso, GD-Lasso, and G-Lasso are shown in Fig. 25. We can see that GGD-Lasso and GD-Lasso have stronger cis-regulatory signals than G-Lasso does. In total, these methods each detected about 6000 associations according to non-zero $\mathbf{W}$ values. We estimate FDR using 50 permutations as proposed in [70]. With FDR $\leq 0.01$, GGD-Lasso obtains about 4500 significant associations. The plots of all identified significant associations for different methods are given in Fig. 26.

### 4.5.4 Refinement of the Prior Networks

To investigate to what extent GGD-Lasso is able to refine the prior networks and study the effect of different parameter settings on $\kappa$, we intentionally change 75 % of the elements in the original prior PPI network and genetic interaction network to random noises. We feed the new networks to GGD-Lasso and evaluate the

**Table 5** Pairwise comparison of different models using *cis*- and *trans*- enrichment

| | | GD-Lasso | G-Lasso | SIOL | Mtlasso2G | Multi-task | Sparse group | LORS | Lasso |
|---|---|---|---|---|---|---|---|---|---|
| *cis*-enrichment | GGD-Lasso | 0.0003 | <0.0001 | <0.0001 | <0.0001 | <0.0001 | <0.0001 | <0.0001 | <0.0001 |
| | GD-Lasso | – | 0.0009 | <0.0001 | <0.0001 | <0.0001 | <0.0001 | <0.0001 | <0.0001 |
| | G-Lasso | – | – | <0.0001 | <0.0001 | <0.0001 | <0.0001 | <0.0001 | <0.0001 |
| | SIOL | – | – | – | 0.1213 | 0.0331 | 0.0173 | <0.0001 | <0.0001 |
| | Mtlasso2G | – | – | – | – | 0.0487 | 0.0132 | <0.0001 | <0.0001 |
| | Multi-task | – | – | – | – | – | 0.4563 | 0.4132 | <0.0001 |
| | Sparse group | – | – | – | – | – | – | 0.4375 | <0.0001 |
| | LORS | – | – | – | – | – | – | – | <0.0001 |
| *trans*-enrichment | GGD-Lasso | 0.0881 | 0.0119 | 0.0102 | 0.0063 | 0.0006 | 0.0003 | <0.0001 | <0.0001 |
| | GD-Lasso | – | 0.0481 | 0.0253 | 0.0211 | 0.0176 | 0.0004 | <0.0001 | <0.0001 |
| | G-Lasso | – | – | 0.0312 | 0.0253 | 0.0183 | 0.0007 | <0.0001 | <0.0001 |
| | SIOL | – | – | – | 0.1976 | 0.1053 | 0.0044 | 0.0005 | <0.0001 |
| | Mtlasso2G | – | – | – | – | 0.1785 | 0.0061 | 0.0009 | <0.0001 |
| | Multi-task | – | – | – | – | – | 0.0235 | 0.0042 | 0.0011 |
| | Sparse group | – | – | – | – | – | – | 0.0075 | 0.0041 |
| | LORS | – | – | – | – | – | – | – | 0.2059 |

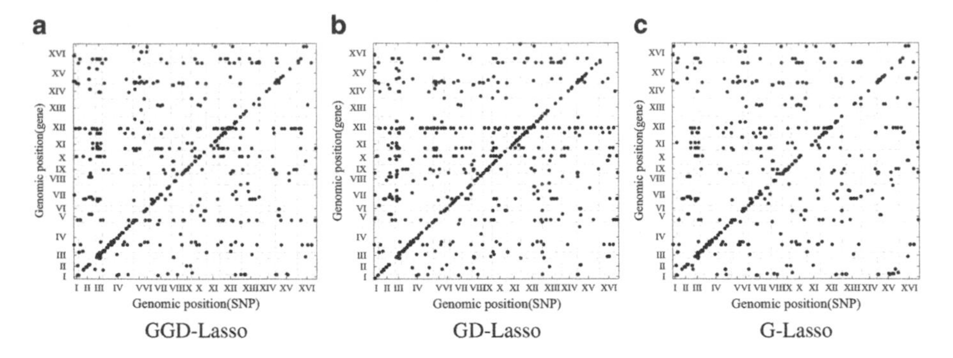


**Fig. 25** The top-1000 significant associations identified by different methods. (**a**) GGD-Lasso. (**b**) GD-Lasso. (**c**) G-Lasso

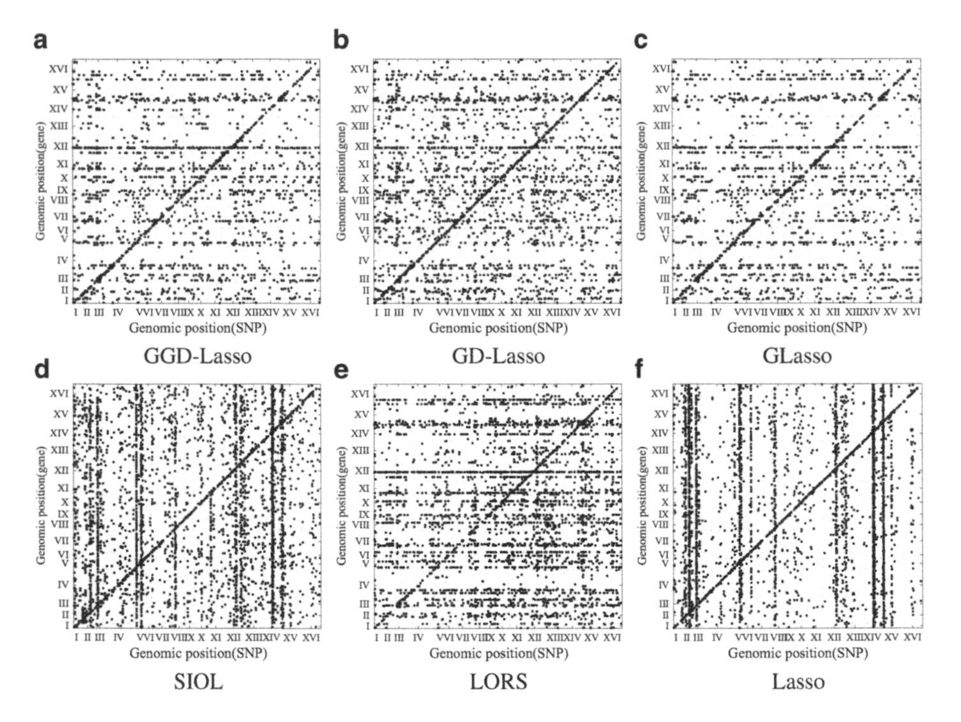


**Fig. 26** The plot of linkage peaks in the study by different methods. (**a**) GGD-Lasso. (**b**) GD-Lasso. (**c**) GLasso. (**d**) SIOL. (**e**) LORS. (**f**) Lasso

refined networks. The results are shown in Fig. 27. We can see that for both PPI and genetic interaction networks, many elements are recovered by GGD-Lasso. This demonstrates the effectiveness of GGD-Lasso. Moreover, when the number of SNP (gene) pairs ($\kappa$) examined for updating reaches 100,000, both PPI and genetic iteration networks are well refined.

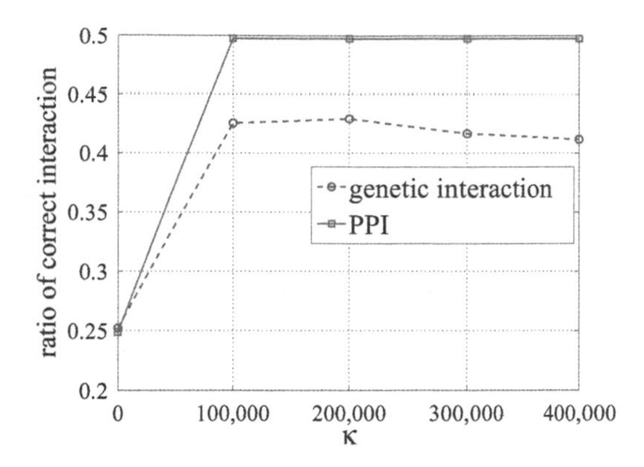


**Fig. 27** Ratio of correct interactions refined when varying $\kappa$

Hotspots Analysis

In this subsection, we study whether GGD-Lasso can help detect more biologically relevant associations than the alternatives. Specifically, we examine the hotspots which affect more than ten gene traits [35]. The top 15 hotspots detected by GGD-Lasso are listed in Table 6. The top 15 hotspots detected by other methods are included in Tables 7, 8, and 9. From Table 6, we observe that for all hotspots, the associated genes are enriched with at least one GO category. Note that GGD-Lasso and GD-Lasso detect one hotspot (12), which cannot be detected by G-Lasso. They also detect one hotspot (6), which cannot be detected by SIOL. The number of hotspots that are significant enriched is listed in Table 10. From the table, we can see that GGD-Lasso slightly outperforms GD-Lasso since it incorporates the location of SNPs and gene pathway information.

## *4.6 Conclusion*

As a promising tool for dissecting the genetic basis of common diseases, eQTL study has attracted increasing research interest. The traditional eQTL methods focus on testing the associations between individual SNPs and gene expression traits. A major drawback of this approach is that it cannot model the joint effect of a set of SNPs on a set of genes, which may correspond to biological pathways.

Recent advancement in high-throughput biology has made a variety of biological interaction networks available. Effectively integrating such prior knowledge is essential for accurate and robust eQTL mapping. However, the prior networks are often noisy and incomplete. In this section, we propose novel graph regularized regression models to take into account the prior networks of SNPs and genes simultaneously. Exploiting the duality between the learned coefficients and incomplete prior networks enables more robust model. We also generalize our model to integrate other types of information, such as SNP locations and gene pathways. The

**Table 6** Summary of the top 15 hotspots detected by GGD-Lasso

| ID | size[a] | Loci[b] | GO[c] | Hits[d] | GD-Lasso (all)[e] | GD-Lasso (hits)[f] | G-Lasso (all)[g] | G-Lasso (hits)[h] | SIOL (all)[i] | SIOL (hits)[j] | LORS (all)[k] | LORS (hits)[l] |
|---|---|---|---|---|---|---|---|---|---|---|---|---|
| 1 | 31 | XII:1056097 | (1)*** | 7 | 31 | 7 | 32 | 7 | 8 | 6 | 31 | 7 |
| 2 | 28 | III:81832..92391 | (2)** | 5 | 29 | 5 | 28 | 5 | 58 | 5 | 22 | 4 |
| 3 | 28 | `XII:1056103` | `(1)`*** | `7` | `29` | `6` | `28` | `6` | `1` | `1` | `2` | `0` |
| 4 | 27 | III:79091 | (2)*** | 6 | 29 | 6 | 28 | 6 | 28 | 7 | 10 | 2 |
| 5 | 27 | III:175799..177850 | (3)* | 3 | 26 | 3 | 23 | 3 | 9 | 2 | 18 | 4 |
| *6* | *27* | *XII:1059925..1059930* | *(1)**** | *7* | *27* | *7* | *27* | *7* | *0* | *0* | *5* | *1* |
| 7 | 25 | III:105042 | (2)*** | 6 | 23 | 6 | 25 | 6 | 5 | 3 | 19 | 4 |
| 8 | 23 | III:201166..201167 | (3)*** | 3 | 23 | 3 | 22 | 3 | 13 | 2 | 23 | 3 |
| 9 | 22 | XII:1054278..1054302 | (1)*** | 7 | 26 | 7 | 24 | 7 | 24 | 5 | 12 | 4 |
| 10 | 21 | III:100213 | (2)** | 5 | 23 | 5 | 23 | 5 | 5 | 3 | 5 | 1 |
| 11 | 20 | III:209932 | (3)* | 3 | 21 | 3 | 19 | 3 | 16 | 4 | 15 | 4 |
| **12** | **20** | **XII:659357..662627** | **(4)*** | **4** | **19** | **4** | **3** | **0** | **37** | **9** | **36** | **6** |
| 13 | 19 | III:210748..210748 | (5)* | 4 | 24 | 4 | 18 | 4 | 2 | 3 | 11 | 4 |
| 14 | 19 | VIII:111679..111680 | (6)* | 3 | 20 | 3 | 19 | 3 | 3 | 3 | 12 | 2 |
| 15 | 19 | VIII:111682..111690 | (7)** | 5 | 21 | 5 | 20 | 5 | 57 | 6 | 22 | 3 |
| Total hits | | | | 75 | | 74 | | 70 | | 59 | | 49 |

[a] Number of genes associated with the hotspot
[b] The chromosome position of the hotspot
[c] The most significant GO category enriched with the associated gene set. The enrichment test was performed using DAVID (Huang et al., 2009). The gene function is defined by GO category. The involved GO categories are: (1) telomere maintenance via recombination; (2) branched chain family amino acid biosynthetic process; (3) regulation of mating-type specific transcription, DNA-dependent; (4) sterol biosynthetic process; (5) pheromone-dependent signal transduction involved in conjugation with cellular fusion; (6) cytogamy; (7) response to pheromone
[d] Number of genes that have enriched GO categories
[e,g,I,k] Number of associated genes that can also be identified using GDL, G-Lasso, SIOL and LORS, respectively.
[f,h,j,l] Number of genes that have enriched GO categories and can also be identified by GDL, G-Lasso, SIOL and LORS, respectively. Among these hotspots, hotspot (12) in bold cannot be detected by GLasso. Hotspot (6) in italic cannot be detected by SIOL. Hotspot (3) in teletype cannot be detected by LORS. Adjusted P-values using permutation tests. $*10^{(-2)} \sim 10^{(-3)}$, $**10^{(-3)} \sim 10^{(-5)}$, $***10^{(-5)} \sim 10^{(-10)}$

Bold groups are not significantly enriched

**Table 7** Summary of the top 15 detected hotspots by GD-Lasso

| Chr | Start | End | Size | GO category | Adjusted $p$-value |
|---|---|---|---|---|---|
| XII | 1,056,097 | 1,056,097 | 31 | Telomere maintenance via recombination | 4.72498E-9 |
| III | 79,091 | 79,091 | 29 | Branched chain family amino acid biosynthetic process | 1.59139E-8 |
| III | 81,832 | 92,391 | 29 | Branched chain family amino acid biosynthetic process | 2.62475E-05 |
| XII | 1,056,103 | 1,056,103 | 29 | Telomere maintenance via recombination | 1.90447E-4 |
| XII | 1,059,925 | 1,059,930 | 27 | Telomere maintenance via recombination | 2.6379E-8 |
| III | 175,799 | 177,850 | 26 | Regulation of mating-type specific transcription, DNA-dependent | 2.07885E-03 |
| XII | 1,054,278 | 1,054,302 | 26 | Telomere maintenance via recombination | 2.30417E-9 |
| III | 210,748 | 210,748 | 24 | Regulation of mating-type specific transcription, DNA-dependent | 1.61983E-04 |
| III | 100,213 | 100,213 | 23 | Branched chain family amino acid biosynthetic process | 7.4936E-3 |
| III | 105,042 | 105,042 | 23 | Branched chain family amino acid biosynthetic process | 3.8412E-8 |
| III | 201,166 | 201,167 | 23 | Regulation of mating-type specific transcription, DNA-dependent | 0.001998002 |
| III | 209,932 | 209,932 | 21 | Regulation of mating-type specific transcription, DNA-dependent | 1.06592E-03 |
| VIII | 111,682 | 111,690 | 21 | Response to pheromone | 7.04262E-04 |
| **V** | **395,442** | **395,442** | **20** | **SRP-dependent cotranslational protein targeting to membrane...** | **0.100899101** |
| VIII | 111,679 | 111,680 | 20 | Cytogamy | 0.001998002 |

Bold groups are not significantly enriched

experimental results on both simulated and real eQTL datasets demonstrate that our models outperform alternative methods. In particular, the proposed dual refinement regularization can significantly improve the performance of eQTL mapping.

**Table 8** Summary of the top 15 detected hotspots by G-Lasso

| Chr | Start | End | Size | GO category | Adjusted $p$-value |
|---|---|---|---|---|---|
| XII | 1,056,097 | 1,056,097 | 32 | Telomere maintenance via recombination | 5.52E-08 |
| III | 79,091 | 79,091 | 28 | Branched chain family amino acid biosynthetic process | 1.28E-07 |
| III | 81,832 | 92,391 | 28 | Branched chain family amino acid biosynthetic process | 2.17E-05 |
| XII | 1,056,103 | 1,056,103 | 28 | Telomere maintenance via recombination | 1.52E-06 |
| XII | 1,059,925 | 1,059,930 | 27 | Telomere maintenance via recombination | 2.64E-08 |
| III | 105,042 | 105,042 | 25 | Branched chain family amino acid biosynthetic process | 6.35E-08 |
| XII | 1,054,278 | 1,054,302 | 24 | Telomere maintenance via recombination | 1.78E-08 |
| III | 100,213 | 100,213 | 23 | Branched chain family amino acid biosynthetic process | 7.49E-06 |
| III | 175,799 | 177,850 | 23 | Regulation of mating-type specific transcription, DNA-dependent | 0.001998002 |
| XII | 674,651 | 674,651 | 23 | Sterol biosynthetic process | 3.56E-04 |
| III | 201,166 | 201,167 | 22 | Regulation of mating-type specific transcription, DNA-dependent | 1.23E-03 |
| **V** | **395,442** | **395,442** | **21** | **SRP-dependent cotranslational protein targeting to membrane...** | **0.086913087** |
| **I** | **51,324** | **52,943** | **20** | **Fatty acid metabolic process** | **0.281718282** |
| VIII | 111,682 | 111,690 | 20 | Response to pheromone | 5.39E-04 |
| III | 209,932 | 209,932 | 19 | Regulation of mating-type specific transcription, DNA-dependent | 7.77E-03 |

Bold groups are not significantly enriched

# 5 Discussion

Driven by the advancement of cost-effective and high-throughput genotyping technologies, eQTL mapping has revolutionized the field of genetics by providing new ways to identify genetic factors that influence gene expression. Traditional eQTL mapping approaches consider both SNPs and genes individually, such as sparse feature selection using Lasso and single-locus statistical tests using $t$-test or

**Table 9** Summary of the top 15 detected hotspots by SIOL

| Chr | Start | End | Size | GO category | Adjust $p$-value |
|---|---|---|---|---|---|
| XIV | 449,639 | 449,639 | 339 | Mitochondrial translation | 2.92E-07 |
| V | 109,310 | 117,705 | 240 | Translation | 2.39E-08 |
| V | 350,744 | 350,744 | 183 | Translation | 1.32E-07 |
| **XV** | **154,177** | **154,309** | **94** | **Replicative cell aging** | **0.264735265** |
| XII | 899,898 | 927,421 | 81 | Translation | 1.45E-06 |
| XIV | 486,861 | 486,861 | 81 | Mitochondrial translation | 1.49E-06 |
| **II** | **548,401** | **548,401** | **78** | **Endonucleolytic cleavage in ITS1 to separate SSU-rRNA from 5.8S...** | **0.030969031** |
| III | 75,021 | 75,021 | 78 | Cellular amino acid biosynthetic process | 1.35E-06 |
| **XIV** | **502,316** | **502,496** | **76** | **Mitochondrial genome maintenance** | **0.824175824** |
| XII | 674,651 | 674,651 | 73 | Electron transport chain | 8.52E-04 |
| III | 81,832 | 92,391 | 58 | Branched chain family amino acid biosynthetic process | 9.78E-05 |
| VIII | 111,682 | 111,690 | 57 | Response to pheromone | 5.15E-03 |
| **XV** | **202,370** | **210,839** | **49** | **Vesicle-mediated transport** | **0.592407592** |
| XIII | 27,644 | 28,334 | 45 | Dephosphorylation | 0.007992008 |
| **XV** | **170,945** | **180,961** | **44** | **(1->6)-beta-D-glucan biosynthetic process** | **0.132867133** |

Bold groups are not significantly enriched

**Table 10** Hotspots detected by different methods

| | GGD-Lasso | GD-Lasso | G-Lasso | SIOL | LORS |
|---|---|---|---|---|---|
| **#hotspots significantly enriched (top 15 hotposts)** | **15** | **14** | **13** | **10** | **9** |
| #total reported hotspots (size > 10) | 65 | 82 | 96 | 89 | 64 |
| #hotspots significantly enriched | 45 | 56 | 61 | 53 | 41 |
| Ratio of significantly enriched hotspots | 70 % | 68 % | 64 % | 60 % | 56 % |

Bold groups are not significantly enriched

ANOVA test. However, it is commonly believed that many complex traits are caused by the joint effect of multiple genetic factors, and genes in the same biological pathway are often co-regulated and may share a common genetic basis. Thus, it is a crucial challenge to understand *how multiple, modestly-associated SNPs interact to influence the phenotypes*. However, little prior work has studied the grow-wise eQTL mapping problem. Moreover, many prior correlation structures in the form of either physical or inferred molecular networks in the genome and phenome are available in many knowledge bases, such as PPI network and genetic interaction network. Developing effective models to incorporate prior knowledge on the relationships between SNPs and relationships between genes for more robust eQTL mapping has recently attracted increasing research interests. However, the structures of prior networks are often highly noisy and far from complete. More robust models that are less sensitive to noise and incompleteness of prior knowledge are required to integrate these prior networks for eQTL mapping.

This book chapter presents a series of algorithms that take advantage of multiple domain knowledge to help with the eQTL mapping and systematically study the problem of group-wise eQTL mapping. In this section, we come to the conclusions of this book chapter and discuss the future directions of inferring group-wise associations for eQTL mapping.

## 5.1 Summary

In this book chapter, we presented our solutions for group-wise eQTL mapping. In general, we made the following contributions:

- **Algorithm to Detect Group-wise eQTL Associations with eQTL Data Only**
  Three algorithms (Sect. 3) are proposed to address this problem. The three approaches incrementally take into consideration more aspects, such as group-wise association, potential confounding factors, and the existence of individual associations. Besides, we illustrate how each aspect could benefit the eQTL mapping. Specifically, in order to accurately capture possible interactions between multiple genetic factors and their joint contribution to a group of phenotypic variations, a sparse linear-Gaussian model (SET-eQTL) is proposed to infer novel associations between multiple SNPs and genes. The proposed method can help unravel true functional components in existing pathways. The results could provide new insights on how genes act and coordinate with each other to achieve certain biological functions. The book chapter further extends the approach to consider the confounding factors and also be able to decouple *individual* associations and *group-wise* associations. The results show the superiority over those eQTL mapping algorithms that do not consider the group-wise associations.
- **Robust Algorithm to Incorporate Prior Interaction Structures into eQTL Mapping**
  To incorporate the prior SNP-SNP interaction structure and grouping information of genes into eQTL mapping, the proposed algorithm, GDL (Sect. 4), significantly improves the robustness and the interpretability of eQTL mapping. We study how prior graph information would help improve eQTL mapping accuracy and how refinement of prior knowledge would further improve the mapping accuracy. In addition, other different types of prior knowledge, e.g., location information of SNPs and genes, and pathway information, can also be integrated for the graph refinement.

## 5.2 *Future Directions*

We envision that the integration of multi-domain knowledge will be the center of interests for eQTL mapping in the future. In the past decade, many efforts have been devoted to developing methods for eQTL mapping. In this book chapter, we present approaches that address the group-wise eQTL mapping problem. To further advance the field, there are several important research issues that should be explored.

1. **Large Scale Data Sets**
   Scalability is another important issue in eQTL mapping. Especially, for human genetics, the whole genome eQTL mapping includes analysis of millions of SNPs and tens of thousands of genes. Traditional eQTL mapping approaches detect associated SNPs for each gene separately. Thus, mapping algorithm can be deployed in parallel for each gene expression. For each run, many approaches were proposed to speed up the mapping, such as screening method [66]. However, these approaches do not work for the group-wise eQTL mapping since the SNPs and genes need to be considered jointly. In our algorithm (Sect. 3), we have developed an effective approach to speed up the computing. However, it is still not able to tackle the whole genome eQTL mapping for human data set. Thus, it is desirable to design new algorithms that are capable of scaling genetic association studies across the whole-genome and support identification of multi-way interactions.
2. **Mining Biological and Medical Data Using Heterogeneous Models**
   Biological and medical research have been facing big data challenges for a long time. With the burst of many new technologies, the data are becoming larger and more complex. Our ability to identify and characterize the effects of genetic factors that contribute to complex traits depends crucially on the development of new computational approaches to integrate, analyze, and interpret these data. It is desirable to develop integrative and scalable methods to study how genetic factors interact with each other to cause common diseases. The developed techniques will dissect the relationships among different components and automatically discover most relevant patterns from the data.

# References


1. Andrew, G. and Gao, J. (2007). Scalable training of l1-regularized log-linear models. *International Conference on Machine Learning*.
2. Asur, S., Ucar, D., and Parthasarathy, S. (2007). An ensemble framework for clustering protein-protein interaction networks. In *Bioinformatics*, pages 29–40.
3. Balding, D. J. (2006). A tutorial on statistical methods for population association studies. *Nature Reviews Genetics*, 7(10):781–791.
4. Biganzoli, E. M., Boracchi, P., Ambrogi, F., and Marubini, E. (2006). Artificial neural network for the joint modelling of discrete cause-specific hazards. *Artif Intell Med*, 37(2):119–130.

5. Bochner, B. R. (2003). New technologies to assess genotype phenotype relationships. *Nature Reviews Genetics*, 4:309–314.
6. Boyd, S. and Vandenberghe, L. (2004). *Convex Optimization*. Cambridge University Press.
7. Cantor, R. M., Lange, K., and Sinsheimer, J. S. (2010). Prioritizing GWAS results: A review of statistical methods and recommendations for their application. *American journal of human genetics*, 86(1):6–22.
8. Carlos M. Carvalhoa, Jeffrey Changa, J. E. L. J. R. N. Q. W. and West, M. (2008). High-Dimensional Sparse Factor Modeling: Applications in Gene Expression Genomics. *Journal of the American Statistical Association*, pages 1438–1456.
9. Charles Boone, H. B. and Andrews, B. J. (2007). Exploring genetic interactions and networks with yeast. *Nature Reviews Genetic*, 8:437–449.
10. Chen, X., Shi, X., Xu, X., Wang, Z., Mills, R., Lee, C., and Xu, J. (2012). A two-graph guided multi-task lasso approach for eQTL mapping. In *AISTATS'12*, pages 208–217.
11. Cheung, V. G., Spielman, R. S., Ewens, K. G., Weber, T. M., Morley, M., and Burdick, J. T. (2005). Mapping determinants of human gene expression by regional and genome-wide association. *Nature*, pages 1365–1369.
12. Chung (1997). Spectral graph theory (reprinted with corrections). In *CBMS: Conference Board of the Mathematical Sciences, Regional Conference Series*.
13. Cordell, H. J. (2009). Detecting gene-gene interactions that underlie human diseases. *Nat. Rev. Genet.*, 10:392–404.
14. Ding, C., Li, T., Peng, W., and Park, H. (2006). Orthogonal nonnegative matrix t-factorizations for clustering. In *KDD*, pages 126–135.
15. Ding, C. H. Q., Li, T., and Jordan, M. I. (2010). Convex and semi-nonnegative matrix factorizations. *IEEE Trans. Pattern Anal. Mach. Intell*, 32(1):45–55.
16. Elbers, C. C., Eijk, K. R. v., Franke, L., Mulder, F., Schouw, Y. T. v. d., Wijmenga, C., and Onland-Moret, N. C. (2009). Using genome-wide pathway analysis to unravel the etiology of complex diseases. *Genetic epidemiology*, 33(5):419–31.
17. Evans, D. M., Marchini, J., Morris, A. P., and Cardon, L. R. (2006). Two-stage two-locus models in genome-wide association. *PLoS Genetics*, 2: e157.
18. Fusi, N., Stegle, O., and Lawrence, N. D. (2012). Joint modelling of confounding factors and prominent genetic regulators provides increased accuracy in genetical genomics studies. *PLoS Comput. Biol.*, 8(1):e1002330.
19. Gao, C., Brown, C. D., and Engelhardt, B. E. (2013). A latent factor model with a mixture of sparse and dense factors to model gene expression data with confounding effects. *ArXiv e-prints*.
20. Gilad, Y., Rifkin, S. A., and Pritchard, J. K. (2008). Revealing the architecture of gene regulation: the promise of eQTL studies. *Trends Genet.*, 24:408–415.
21. Hirschhorn, J. N. and Daly, M. J. (2005). Genome-wide association studies for common diseases and complex traits. *Nature Reviews Genetics*, 6:95–108.
22. Hoh, J. and Ott, J. (2003). Mathematical multi-locus approaches to localizing complex human trait genes. *Nature Reviews Genetics*, 4:701–709.
23. Hoh, J., Wille, A., Zee, R., Cheng, S., Reynolds, R., Lindpaintner, K., and Ott, J. (2000). Selecting SNPs in two-stage analysis of disease association data: a model-free approach. *Annals of Human Genetics*, 64:413–417.
24. Huang, d. a. W., Sherman, B. T., and Lempicki, R. A. (2009a). Systematic and integrative analysis of large gene lists using DAVID bioinformatics resources. *Nat Protoc*, 4(1):44–57.
25. Huang, Y., Wuchty, S., Ferdig, M. T., and Przytycka, T. M. (2009b). Graph theoretical approach to study eQTL: a case study of Plasmodium falciparum. *ISMB*, pages i15–i20.
26. Ideraabdullah, F., Casa-Esper, E., and et al. (2004). Genetic and haplotype diversity among wild-derived mouse inbred strains. *Genome Research*, 14(10a):1880–1887.
27. Jeffrey T. Leek, J. D. S. (2007). Capturing heterogeneity in gene expression studies by surrogate variable analysis. *PLoS Genet*, pages 1724–35.
28. Jenatton, R., Audibert, J.-Y., and Bach, F. (2011). Structured variable selection with sparsity-inducing norms. *JMLR*, 12:2777–2824.

29. Joo, J. W., Sul, J. H., Han, B., Ye, C., and Eskin, E. (2014). Effectively identifying regulatory hotspots while capturing expression heterogeneity in gene expression studies. *Genome Biol.*, 15(4):r61.
30. Kang, H. M., Zaitlen, N. A., Wade, C. M., Kirby, A., Heckerman, D., Daly, M. J., and Eskin, E. (2008). Efficient control of population structure in model organism association mapping. *Genetics*, 178(3):1709–1723.
31. Kim, S. and Xing, E. P. (2009). Statistical estimation of correlated genome associations to a quantitative trait network. *PLoS Genet.*, 5(8):e1000587.
32. Kim, S. and Xing, E. P. (2012). Tree-guided group lasso for multi-response regression with structured sparsity, with applications to eQTL mapping. In *ICML*.
33. Lander, E. S. (2011). Initial impact of the sequencing of the human genome. *Nature*, 470(7333):187–197.
34. Lee, D. D. and Seung, H. S. (2000). Algorithms for non-negative matrix factorization. In *NIPS*, pages 556–562.
35. Lee, S. and Xing, E. P. (2012). Leveraging input and output structures for joint mapping of epistatic and marginal eQTLs. *Bioinformatics*, 28(12):i137–146.
36. Lee, S., Zhu, J., and Xing, E. P. (2010). Adaptive multi-task lasso: with application to eQTL detection. In *NIPS*.
37. Lee, S.-I., Dudley, A. M., Drubin, D., Silver, P. A., Krogan, N. J., Pe'er, D., and Koller, D. (2009). Learning a prior on regulatory potential from eQTL data. *PLoS Genet*, page e1000358.
38. Leek, J. T. and Storey, J. D. (2007). Capturing heterogeneity in gene expression studies by surrogate variable analysis. *PLoS Genet.*, 3(9):1724–1735.
39. Leopold Parts1, Oliver Stegle, J. W. R. D. (2011). *Joint Genetic Analysis of Gene Expression Data with Inferred Cellular Phenotypes*. PLos Genetics.
40. Li, C. and Li, H. (2008). Network-constrained regularization and variable selection for analysis of genomic data. *Bioinformatics*, 24(9):1175–1182.
41. Listgarten, J., Kadie, C., Schadt, E. E., and Heckerman, D. (2010). Correction for hidden confounders in the genetic analysis of gene expression. *Proc. Natl. Acad. Sci. U.S.A.*, 107(38):16465–16470.
42. Listgarten, J., Lippert, C., Kang, E. Y., Xiang, J., Kadie, C. M., and Heckerman, D. (2013). A powerful and efficient set test for genetic markers that handles confounders. *Bioinformatics*, 29(12):1526–1533.
43. Mazumder, R., Hastie, T., and Tibshirani, R. (2010). Spectral regularization algorithms for learning large incomplete matrices. *JMLR*, 11:2287–2322.
44. McClurg, P., Janes, J., Wu, C., Delano, D. L., Walker, J. R., Batalov, S., Takahashi, J. S., Shimomura, K., Kohsaka, A., Bass, J., Wiltshire, T., and Su, A. I. (2007). Genomewide association analysis in diverse inbred mice: power and population structure. *Genetics*, 176(1):675–683.
45. Michaelson, J., Loguercio, S., and Beyer, A. (2009a). Detection and interpretation of expression quantitative trait loci (eQTL). *Methods*, 48(3):265–276.
46. Michaelson, J. J., Loguercio, S., and Beyer, A. (2009b). Detection and interpretation of expression quantitative trait loci (eQTL). *Methods*, 48:265–276.
47. Musani, S., Shriner, D., Liu, N., Feng, R., Coffey, C., Yi, N., Tiwari, H., and Allison, D. (2007a). Detection of gene - gene interactions in genome-wide association studies of human population data. *Human Heredity*, 63(2):67–84.
48. Musani, S. K., Shriner, D., Liu, N., Feng, R., Coffey, C. S., Yi, N., Tiwari, H. K., and Allison, D. B. (2007b). Detection of gene x gene interactions in genome-wide association studies of human population data. *Human Heredity*, pages 67–84.
49. Nelson, M. R., Kardia, S. L., Ferrell, R. E., and Sing, C. F. (2001). A combinatorial partitioning method to identify multilocus genotypic partitions that predict quantitative trait variation. *Genome Research*, 11:458–470.
50. Ng, A. (2004). Feature selection, l1 vs. l2 regularization, and rotational invariance. *International Conference on Machine Learning*.
51. Nicolo Fusi, O. S. and Lawrence, N. D. (2012). Joint modelling of confounding factors and prominent genetic regulators provides increased accuracy in genetical genomics studies. *PLoS Computational Biology*, page e1002330.

52. Nocedal, J. and Wright, S. J. (2006). *Numerical optimization*. Springer.
53. Obozinski, G. and Taskar, B. (2006). Multi-task feature selection. Technical report.
54. Perry, J. R. B., Mccarthy, M. I., Hattersley, A. T., Zeggini, E., Case, T., Consortium, C., Weedon, M. N., and Frayling, T. M. (2009). Interrogating type 2 diabetes genome-wide association data using a biological pathway-based approach. *Diabetes*, 58(June).
55. Pujana, M. A., Han, J.-D. J., Starita, L. M., Stevens, K. N., and Muneesh Tewari, e. a. (2007). Network modeling links breast cancer susceptibility and centrosome dysfunction. *Nature Genetics*, pages 1338–1349.
56. Rachel B. Brem, John D. Storey, J. W. and Kruglyak, L. (2005). Genetic interactions between polymorphisms that affect gene expression in yeast. *Nature*, pages 701–03.
57. Ritchie, M. D., Hahn, L. W., Roodi, N., Bailey, L. R., Dupont, W. D., Parl, F. F., and Moore, J. H. (2001). Multifactor-dimensionality reduction reveals high-order interactions among estrogen-metabolism genes in sporadic breast cancer. *American Journal of Human Genetics*, 69:138–147.
58. Rockman, M. V. and Kruglyak, L. (2006). Genetics of global gene expression. *Nature Reviews Genetics*, 7:862–872.
59. Smith, E. N. and Kruglyak, L. (2008). Gene-environment interaction in yeast gene expression. *PLoS Biol*, page e83.
60. Stegle, O., Kannan, A., Durbin, R., and Winn, J. (2008). Accounting for non-genetic factors improves the power of eQTL studies. In *RECOMB*, pages 411–422.
61. Stegle, O., Parts, L., Durbin, R., and Winn, J. (2010). A Bayesian framework to account for complex non-genetic factors in gene expression levels greatly increases power in eQTL studies. *PLoS Computational Biology*, page e1000770.
62. The Gene Ontology Consortium (2000). Gene ontology: tool for the unification of biology. *Nature Genetics*, 25(1):25–29.
63. Tibshirani, R. (1996). Regression shrinkage and selection via the lasso. *J. Royal. Statist. Soc B.*, 58(1):267–288.
64. Torkamani, A., Topol, E. J., and Schork, N. J. (2008). Pathway analysis of seven common diseases assessed by genome-wide association. *Genomics*, 92(5):265–72.
65. von Mering, C., Krause, R., Snel, B., Cornell, M., Oliver, S. G., Fields, S., and Bork, P. (2002). Comparative assessment of large-scale data sets of protein-protein interactions. *Nature*, 417:399–403.
66. Wang, J., Zhou, J., Wonka, P., and Ye, J. (2013). Lasso screening rules via dual polytope projection. In *NIPS*, pages 1070–1078.
67. Wang, K., Li, M., and Hakonarson, H. (2010). Analysing biological pathways in genome-wide association studies. *Nature Reviews Genetics*, 11(12):843–854.
68. Westfall, P. H. and Young, S. S. (1993). *Resampling-based Multiple Testing*. Wiley, New York.
69. Yang, C., He, Z., Wan, X., Yang, Q., Xue, H., and Yu, W. (2009). SNPHarvester: a filtering-based approach for detecting epistatic interactions in genomewide association studies. *Bioinformatics*, 25(4):504–511.
70. Yang, C., Wang, L., Zhang, S., and Zhao, H. (2013). Accounting for non-genetic factors by low-rank representation and sparse regression for eQTL mapping. *Bioinformatics*, pages 1026–1034.
71. Yvert, G., Brem, R. B., Whittle, J., Akey, J. M., Foss, E., Smith, E. N., Mackelprang, R., and Kruglyak, L. (2003). Transacting regulatory variation in Saccharomyces cerevisiae and the role of transcription factors. *Nat. Genet.*, 35(1):57–64.
72. Zhu, J., Zhang, B., Smith, E. N., Drees, B., Brem, R. B., Kruglyak, L., Bumgarner, R. E., and Schadt, E. E. (2008). Integrating large-scale functional genomic data to dissect the complexity of yeast regulatory networks. *Nature Genetics*, pages 854–61.

# Causal Inference and Structure Learning of Genotype–Phenotype Networks Using Genetic Variation

**Adèle H. Ribeiro, Júlia M. P. Soler, Elias Chaibub Neto, and André Fujita**

**Abstract** A major challenge in biomedical research is to identify causal relationships among genotypes, phenotypes, and clinical outcomes from high-dimensional measurements. Causal networks have been widely used in systems genetics for modeling gene regulatory systems and for identifying causes and risk factors of diseases. In this chapter, we describe fundamental concepts and algorithms for constructing causal networks from observational data. In biological context, causal inferences can be drawn from the natural experimental setting provided by Mendelian randomization, a term that refers to the random assignment of genotypes at meiosis. We show that genetic variants may serve as instrumental variables, improving estimation accuracy of the causal effects. In addition, identifiability issues that commonly arise when learning network structures may be overcome by using prior information on genotype–phenotype relations. We discuss four recent algorithms for genotype–phenotype network structure learning, namely (1) QTL-directed dependency graph, (2) QTL+Phenotype supervised orientation, (3) QTL-driven phenotype network, and (4) sparsity-aware maximum likelihood (SML).

**Keywords** Structural learning • Network analysis • Genotypes • Phenotypes • Genetic variations

A.H. Ribeiro • A. Fujita (✉)
Department of Computer Science, Institute of Mathematics and Statistics, University of São Paulo, São Paulo, Brazil
e-mail: adele@ime.usp.br; fujita@ime.usp.br; andrefujita@gmail.com

J.M.P. Soler
Department of Statistics, Institute of Mathematics and Statistics, University of São Paulo, São Paulo, Brazil
e-mail: pavan@ime.usp.br

E.C. Neto
Department of Computational Biology, Sage Bionetworks, Seattle, WA, USA
e-mail: elias.chaibub.neto@sagebase.org



K.-C. Wong (ed.), *Big Data Analytics in Genomics*,
DOI 10.1007/978-3-319-41279-5_3

# 1 Introduction

The development of high-throughput technologies, such as DNA microarrays and next- generation sequencing, has allowed the study of complex biological systems. However, the vast quantities of data from such large scale studies have been challenging researchers aiming to discover the complex network describing causal associations among genotypes, phenotypes, and other clinical outcomes.

In general, the associations found in observational studies cannot be interpreted as causal. However, genetic variants information has proven useful to determine causal effects from observational data. In several recent studies, genomic data and information on quantitative variation in phenotypes have been combined in order to discover causal relationships among phenotypes. Some of the most promising phenome projects, including the Consortium for Neuropsychiatric Phenomics at UCLA (www.phenomics.ucla.edu) and the National BioResource Project—Rat (http://www.anim.med.kyoto-u.ac.jp/NBR/), are listed in David Houle et al.'s paper [53].

Causal discovery methods from observational data are of great interest to researchers in many fields, such as functional genomics and proteomics, molecular biology, and epidemiology [39, 49, 76, 106, 107, 134].

In observational epidemiological studies, genetic variants that mimic the influence of modifiable environmental exposures have a key role in causal inference. If the link between a genetic variant and environmental exposure can indeed be shown, associations between genotype and disease outcome or genotype and intermediate phenotype may elucidate the importance of environmentally modifiable factors as causes of disease [112, 113]. These findings are crucial for the understanding of genetic mechanisms associated with diseases and for the development of therapeutic strategies [11, 72].

Causal networks describing the regulatory interactions between different genes are called gene regulatory networks (GRNs) and have been inferred from both observational and interventional gene expression data [77]. Approaches for reverse engineering GRNs from purely observational data (i.e., data collected without any biological or experimental interference on the level of individual genes) need a large sample size and capture only parts of biologically relevant networks [77]. However, it has been shown that it is possible to greatly improve accuracy and performance in network reconstruction by incorporating data from experimental interventions and perturbations (e.g., from gene knockout or knockdown experiments) [91].

It is widely accepted that the most trustworthy method to infer causal relationships from data are experimental studies such as randomized controlled trials. However, the number of variables in biological systems is usually very large, so that it is unfeasible to carry out randomized experiments to discover all possible causal relationships. Nonetheless, throughout this chapter, we show that it is possible to discover the structure of causal networks and to infer causal effects based on observational data alone if certain assumptions are met. In practical context, some of these assumptions may be very restrictive and may not necessarily hold in biological data. However, causal analysis may still be useful, since the conclusions might be indicative of some causal connections in the data [37].

Causal inference from observational studies is a complicated task that encompasses two major challenges: (1) ensuring accuracy of discovered results and (2) reducing computational complexity.

The absence or inadequacy of randomization, combined with the presence of (measured and unmeasured) confounding factors, often leads to spurious conclusions in observational studies. More reliable causal relations can be extracted from data by relying on two imperative pillars: proper randomization and instrumental variables. In systems genetics, Mendelian randomization plays an important role in causal inference. The random segregation of alleles from parents to offspring during meiosis closely resembles the random allocation of treatments (exposure variables or interventions) in randomized controlled trials. In other words, the genotype can be considered as an effect of a randomized intervention, allowing tests of causal hypotheses. Thus, a robust association from a genetic variant to a phenotype that is allegedly free of confounding factors (e.g., behavioral and environmental factors) can be interpreted as a causal relationship. Moreover, when certain assumptions are met, genetic variants can be used as instrumental variables, allowing causal inferences about the effect of phenotypes on outcomes. Causal inference can be greatly improved by exploiting instruments, since biases and effects of confounding factors are minimized [102]. In Sect. 2, we discuss in detail the Mendelian randomization approach, and particularly the conditions on which genetic variants can be defined as instrumental variables.

The computational complexity challenge of causal structure learning problem arises out of the need to develop efficient algorithms to handle large amounts of data. There is no available algorithm for finding an entire causal model in polynomial time. In other words, the causal structure learning problem is NP-hard. Since the possible number of causal networks is exponential in the number of variables, currently, some heuristics are used to limit the search space. The most commonly used approaches for reducing computational complexity assume certain network structure properties, such as sparsity and acyclicity. In Sect. 6, we review some current approaches being used for causal structure learning.

Structural equation models (SEMs) and probabilistic graphical models (PGMs) are widespread adopted methodologies for representing causal associations among variables. While SEM provides a functional representation of the causal mechanisms by which a variable's value is generated, PGM provides an equivalent but graphical representation of these causal mechanisms by using graph theory and probability theory. Thus, by combining elements from SEMs and PGMs, it is possible to model causal relationships using a mathematically rigorous and intuitive language. In Sect. 3, we present the direct correspondence between both representations.

We will address two classes of models which differ in their ability to accommodate feedback loops: (1) the non-recursive SEMs, which are capable of modeling cycles involving both just two variables (direct feedback loops or reciprocal associations) and three or more variables (indirect feedback loops); and (2) the recursive SEMs, which assume all causal effects as unidirectional, so that no two variables are causes of each other. The SEMs can be graphically represented by directed graphs in such a way that there is a direct equivalence between the two representations. More specifically, non-recursive SEMs can be graphically represented by directed cyclic graphs (DCGs), and recursive SEMs can be graphically represented by directed acyclic graphs (DAGs).

Biological systems have been extensively modeled using DAGs because algorithm development is facilitated by using results that are valid under the assumption that the causal structure is acyclic. However, for modeling cyclic phenomena, which are the most prevalent in biological systems, acyclic structures are very restrictive [130]. DCGs can be used as a more appropriate alternative for modeling data in the steady-state (equilibrium state of a time invariant dynamic system), since feedback loops can capture the redundancy and stability of the underlying system. Genetic regulatory networks have been modeled as DCGs, since they are capable of reproducing the stable cyclic pattern of gene expressions [22, 133].

Algorithms for discovering causal structures are often based on a functional representation (a recursive or non-recursive SEM) or on a graphical representation (a directed acyclic or cyclic graph) of causal processes. SEM-based structure learning approaches use optimization methods for estimating the model parameters and techniques for improving efficiency and interpretability such as sparsity-enforcing regularization. In this chapter, we focus mainly on the fundamental concepts used by causal structure learning algorithms that are based on a graphical model. In Sect. 4, we cover the main definitions and properties connecting graphical structure and probability distributions, including the concept of d-separation which allows to derive the conditional independencies entailed by a causal structure. This theory was developed mainly by Judea Pearl [87], Peter Spirtes, Clark Glymour, Richard Scheines [97, 118], and Thomas Richardson [96].

One of the main issues of the causal structure learning theory is the identifiability problem. There are some models which encode precisely the same set of conditional independence relations. Thus, they are considered statistically equivalent and indistinguishable from observational data. In this case, it is not possible to uniquely identify the true underlying causal model. By including information on genetic variants causally associated with phenotypes [e.g., quantitative trait loci (QTL) or quantitative trait nucleotide (QTN)], new conditional independence relations are created, and statistically equivalent phenotype networks may become identifiable. In Sect. 5, these concepts will be discussed in detail.

There are several algorithms available in the literature that were designed to solve the specific problem of discovering the structure of a genotype–phenotype–outcome network. Among these, in chronological order, are [5, 22, 24, 25, 32, 46, 71, 73, 74, 108, 126]. In Sect. 7, we describe in detail four of the most popular algorithms: QTL-directed dependency graph (QDG) [24], QTL + Phenotype supervised orientation (QPSO) [126], QTL-driven phenotype network (QTLnet) [25], and sparsity-aware maximum likelihood (SML) [22]. These are recent algorithms with source code freely available and easily accessible to the users.

# 2 Mendelian Randomization

## 2.1 Randomized Controlled Trial

A widely accepted approach for finding causal relationships is to perform intervention experiments, also known as randomized controlled trials. Such experiments are critically based on randomization and confounding factors control. Treatments (or interventions) are randomly assigned to the subjects and statistical tests are performed to verify whether differences between treatment and control groups are significant. For instance, to verify whether a new medication is superior in comparison with placebo, randomized controlled trials are usually conducted and the randomization is imperative to verify whether the treatment has a causal effect on the disease.

Experiments with randomization have three important implications, namely elimination of selection bias between groups, assurance of allocation concealment, and justification of randomization-based statistical tests. Under random assignment of treatments, selection biases are removed since confounding factors are more likely to be distributed evenly among groups, and statistical tests can be properly performed.

Note that the determination of a causal effect critically depends on whether all confounding factors are properly randomized. A proper randomization can indeed increase the chances of evenly distribute known and unknown confounding factors among groups. However, considering that there is a non-zero probability that confounding factors are not fully balanced among groups, it is recommended to perform some form of restricted randomization (e.g., randomization within homogeneous blocks with respect to a specific known confounding factor) when it is crucial that biases from a particular confounding factor are avoided [17].

## 2.2 Randomized Allocation of Allelic Variation in Genes

Unfortunately, in observational studies (where the data is collected without any intervention) an association between a phenotype and a disease (or other outcome of interest) may not be causal. The main reasons are [20, 36]:

- **Confounding variables:** suppose e are interested in the association between a phenotype and an outcome. Measured or unmeasured factors that affect both variables may create spurious associations when not considered in the model. They are called confounding variables. For instance, let $X$ be a phenotype and $Y$ be an outcome of interest. Consider a variable $Z$ that directly affects both $X$ and $Y$. The causal relationships among these variables can be represented by the scheme: $X \leftarrow Z \rightarrow Y$. If a pairwise correlation analysis is performed, there may be a significant association between the phenotype and the outcome, even when

there is not a direct influence between them. This spurious association vanishes only if the variable $Z$ is considered in the analysis. In this case, we say that $Z$ is a confounder of the relationship between $X$ and $Y$.
- **Reverse causation:** when two variables are causally related, but in the contrary direction to a common presumption (outcome affecting the phenotype), we say that there is a reverse causation. It is a type of misinterpretation in which the effect is allowed to occur before its cause. For instance, given a strong association between low circulating cholesterol levels and risk of cancer, one could suspect that low cholesterol levels increases the risk of cancer. However, it is possible that the causality goes in the opposite direction, i.e., early stages of cancer may, many years before diagnosis, lead to a lowering in cholesterol levels [62].
- **Various biases:** unobserved or imprecisely measured factors can bias estimates of the association between two variables even if the causal direction is correctly specified. Studies with small sample size are more affected by such biases.

These issues show that causal inference can be hard to achieve, or even an impossible task, if only observational associations are considered [112, 114]. However, it is possible to provide evidence for or against a causal relationship and, usually, to quantify causal effect by making specific assumptions and by using additional information.

The region of the genome affecting variation in a quantitative trait (phenotype) is known as quantitative trait locus (QTL), and QTLs have essentially been detected by using panels of microsatellites, mainly for population-based studies of plants and animals, and for family-based studies of humans. Quantitative trait nucleotides (QTNs) are often identified through genome-wide association studies by using single nucleotide polyomrphism (SNP) markers. Genetic variants (QTLs or QTNs) have been used to distinguish causation from association in biological studies.

Based on the Mendel's second law, alleles are randomly assigned from parents to offspring during gamete formation. This random allocation of alleles provides a design analogous to an intervention experiment. Thus, Mendelian randomization can be interpreted as a natural randomized controlled trial, in which different genotypes, rather than treatments, are randomly allocated to individuals. Considering that the variation in genotypes always precedes the differences in phenotype, this natural randomization allows us to use statistical tests to determine whether there is a causal relation from a genetic variant and a phenotype. Since the influence of genotype on phenotype is, in general, independent of any confounding, reverse causation or other biases, causal interpretation may be appropriate.

In the following, we summarize some concepts that provide a foundation for causal inference based on Mendelian randomization.

- **The law of segregation (Mendel's first law):** states that during the gamete formation the members of the allelic pair of each hereditary factor (some gene or genetic locus) segregate from each other independently so that each gamete carries only one allele for each factor and offspring acquire one allele randomly chosen from each parent. Since genetic variants segregation occurs randomly and independently of environmental factors, causal studies are less susceptible to confounding;

- **The law of independent assortment (Mendel's second law):** states that genetic variants segregate independently of other traits. In other words, alleles related to different traits are transmitted independently of one another from parents to offspring. Note that the independent assortment law is violated when two loci are linkage, i.e., when they are on the same chromosome and their genetic distance is small. In this case, the recombination fraction is less than 50 % in a single generation, that is, allele combinations in different loci are not inherited independently of each other [68].
- **Unambiguous causal direction:** since the randomization of marker alleles during meiosis precedes their effect on phenotypes, reverse causality is not an issue. In other words, the direction of the causal effect is always from the genotype to the phenotype.
- **Life-long effects:** genetic variants have life-long effects on exposures as opposed to interventions in randomized controlled trials which only occur over short periods of time.

The same advantages of a randomized controlled trial may be achieved from natural experimental setting provided by Mendelian randomization when there is no interaction with uncontrolled external confounders, such as maternal genotype and environmental perturbation. In this regard, the canalization or developmental compensation phenomenon need to be emphasized. When a genetic or environmental factor is expressed during fetal development or post-natal growth, the expression of other genetic variants may be influenced leading to changes that may alter development in such a way that the effect of the factor is damped (or buffered). This resistance of phenotypes to environmental or genetic perturbation can bias causal inferences and makes it difficult to relate randomized controlled trials and Mendelian randomization studies. In randomized controlled trials, the randomization of the intervention to subjects often occurs during their middle-age. On the other hand, in Mendelian randomization approaches, the randomization occurs before birth. Thus, we must be aware that some findings of studies using Mendelian randomization approach may be unrepresentative of clinical interventions on the exposure performed in a mature population.

Mendelian randomization has been particularly useful for investigating causal effects of an exposure of interest (phenotype) on an outcome (e.g., disease or other clinical outcome), when a genetic variant robustly associated with the exposure is not associated with any confounding factor and it is not associated with the outcome through any other path than through the exposure of interest [112]. In a statistical point of view, causal inference is improved whenever a genetic variant meets the requirements to be used as an instrumental variable [31, 68]. In this case, it is possible to estimate the long-term causal effects of exposures on outcomes. Genetic variants robustly associated with phenotypes have been used in several studies as instrumental variables, improving causal inference in a non-experimental setting [19, 20, 101, 114].

Instrumental variable analysis within the Mendelian randomization context is particularly powerful for experimental crosses (e.g., F2, backcrosses, intercrosses, etc.), which are conducted in controlled conditions and closely mimic randomized experiments. For studies in humans and other natural populations, a more careful analysis is needed, since the population structure and cryptic relatedness might still act as confounders [4]. When a genetic variant is in linkage disequilibrium with another genetic variant (i.e., alleles at the two loci are non-randomly associated), both affecting the outcome or the same metabolic pathway, the instrumental variable assumption that the genetic variant is associated with the outcome only through the exposure of interest may be violated. In the human genome, linkage disequilibrium can occur even between completely unlinked loci (e.g., alleles on separate chromosomes) due to population structure, natural selection, genetic drift, and mutation [111]. Thus, it can be quite a challenge to identify violations of the instrumental variable assumptions.

The instrumental variable approach and its assumptions are described in detail in the next section.

## 2.3 Genetic Variants as Instrumental Variables

Suppose a study for investigating a causal relationship between an exposure (e.g., a phenotype) $X$ and an outcome (e.g., a clinical trait or disease) $Y$, when it is known a genetic variant $M$ which is associated with $X$ as illustrated in Fig. 1.

Considering linear relationships among the variables, the true causal equation for the outcome $Y$ is $Y = \alpha + \beta_2 X + \beta_3 Z + \varepsilon$, where $\alpha$ is the regression intercept, $\beta_2$ and $\beta_3$ are direct causal effects from, respectively, $X$ and $Z$, and $\varepsilon$ is the error term.

Suppose that the simple regression model $Y = \alpha + \beta_2 X + e$, where $e = \beta_3 Z + \varepsilon$, is used in order to estimate the causal effect of $X$ on $Y$, possibly because $Z$ is an unobserved confounding factor. In this case, the ordinary least squares (OLS) estimator

$$\hat{\beta}_2 = \frac{Cov(Y,X)}{Var(X)} = \frac{Cov(\beta_2 X + \beta_3 Z + \varepsilon, X)}{Var(X)},$$

with expectation

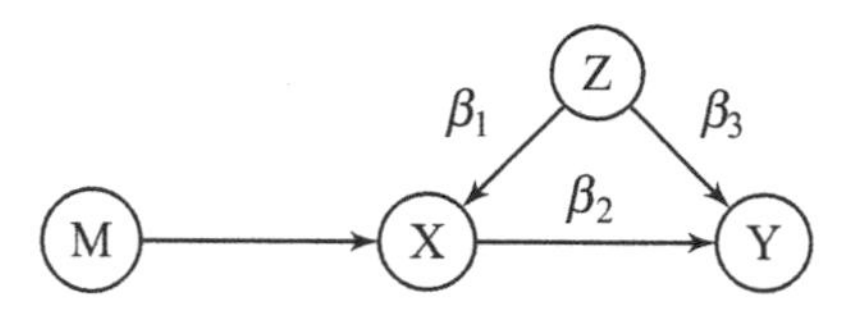


**Fig. 1** Usual scenario investigated in Mendelian randomization. The causal network illustrates the assumed relationships among genetic variant $M$, exposure $X$, outcome $Y$ and confounders $Z$

$$\mathbb{E}(\hat{\beta}_2) = \beta_2 + \beta_3 \frac{Cov(Z,X)}{Var(X)},$$

will have a bias of $\beta_3 \frac{Cov(Z,X)}{Var(X)}$, if $\beta_3 \neq 0$ and $Cov(Z,X) \neq 0$.

In other words, when $X$ and $Z$ are correlated, there is a violation of the assumption that the error $e$ is uncorrelated with the covariate $X$, i.e., $cov(X, \varepsilon) = 0$, and the OLS estimator will be not asymptotically unbiased and consistent.

However, according to the instrumental variable technique introduced by the geneticist Sewal Wright [129], it is possible to determine the causal effect of $X$ on $Y$, $\beta_2$, if there is a variable $M$ (called instrumental variable) which is correlated with $X$, but is uncorrelated with $Z$.

Under the instrumental variable assumptions, i.e., $Cov(X,M) \neq 0$ and $Cov(Z,M) = 0$, the covariance of $Y$ and $Z$ is

$$Cov(Y,M) = Cov(\beta_2 X + \beta_3 Z + \varepsilon, M) = \beta_2 Cov(X,M).$$

Thus, we can obtain the instrumental variable (IV) estimator of $\beta_2$:

$$\hat{\beta}_2^{IV} = \frac{Cov(Y,M)}{Cov(X,M)}.$$

The instrumental variable estimator is also known as two-stage least squares (2SLS) estimator, since it can be obtained in two stages. In the first stage, $X$ is regressed on $M$, and, since $M$ is alleged to be uncorrelated with $Z$, the OLS estimator of the slope coefficient will be consistent and unbiased. In the second stage, $Y$ is regressed on $\hat{X}$, which is obtained in the first stage and represents the predict value of $X$ explained by $M$, but not by $Z$. Since the covariate $\hat{X}$ is uncorrelated with the error term, the OLS estimator is used for concluding the estimation of $\beta_2$ [12, 16].

In the Mendelian randomization scenario illustrated in Fig. 1, the genetic variant $M$, which is affecting the outcome $Y$ only through its effect on the exposure $X$ and it is not associated with confounding factors $Z$, is acting as an instrumental variable, allowing inferences on the causal relation of $X$ to $Y$. When an intervention is made at the instrumental variable $M$ and a significant change is detected in the outcome $Y$, since $M$ is not directly associated with the outcome, the only way of explaining the indirect effect of $M$ on $Y$ is by a causal effect of the exposure $X$ on the outcome $Y$. By using a genetic variant as an instrumental variable, it is not possible to do an intervention experiment, however, the randomization of genotypes to individuals ensures the causal inference similarly.

In order to use a genetic variant as instrumental variable, a number of assumptions must be met:

1. The genetic variant $M$ must be associated with the exposure $X$;
2. The genetic variant $M$ must be independent of measured and unmeasured factors (represented by $Z$) that confound the relationship between the exposure $X$ and the outcome $Y$.

3. Exclusion restriction: the genetic variant $M$ cannot affect the outcome $Y$ by no other way than by the exposure $X$.

We will provide details on each of these assumptions in the following sections.

### 2.3.1 Statistical Association with the Exposure

The first assumption states that the genetic variant $M$ and the exposure $X$ must be statistically associated.

It is important that the association between $M$ and $X$ is strong, otherwise $M$ is considered a weak instrumental variable and can bias estimates of the causal effects even if all other core instrumental variable assumptions are satisfied [119]. The implicit idea is that the genetic variant must be the strongest factor which genetically divides the sample into subgroups according to the exposure of interest, similar to a randomized controlled trial. It is expected that the genetic variant effect is not inhibited by the effect of any confounder.

The genetic variant is not required to be the true functional variant that produces a subsequent effect on the exposure. However, it is necessary that the chosen instrument is in linkage disequilibrium with the functional variant, i.e., they must be statistically associated.

In order to verify the association magnitude between two variables, the Pearson product moment correlation coefficient can be used if the relationship is linear. If the data present non-linear or more complicated relationships, other measures, such as the Spearman's rank correlation coefficient and the mutual information, can be better suited.

### 2.3.2 Independence with Exposure–Outcome Confounders

The assumption that the genetic variant $M$ is independent of any confounder $Z$ for the exposure–outcome relationship is specially hard to verify considering that it must hold even for unmeasured confounders.

However, it is highly recommended to examine whether any statistical association between the genetic variant and the observed confounders of the exposure exists. The absence of such statistical associations does not guarantee that the assumption is fulfilled but, at least, increases the chances of it being true.

### 2.3.3 Exclusion Restriction

The third and the most troubling assumption for using genetic variants as instrumental variables is sometimes referred as exclusion restriction. It states that the genetic variant only affects the outcome through the exposure. More precisely, the genetic variant must be independent of the outcome given the exposure and all confounders (measured and unmeasured) of the exposure–outcome association.

The scheme of Fig. 1 would be an exclusion restriction violation if the genetic variant $M$ is not independent of the outcome $Y$ conditional on exposure $X$ and confounders $Z$. For example, it would be a violation if $M$ is affecting $Y$ through a direct edge and through the exposure $X$. In this case, estimates for the association between $X$ and $Y$ would be biased.

Since it is always possible that the genetic variant affects the outcome via a biological pathway other than the exposure of interest, it may be very difficult to guarantee that this assumption holds. In addition, pleiotropy (i.e., phenomenon of a genetic variant influencing multiple phenotypes) and linkage disequilibrium can be violations of the exclusion restriction if such associations imply the existence of another pathway by which the genetic variant is associated with the outcome [20].

A recommendation is to use only strong instruments, i.e., genetic variants whose functionality and relationship to the exposure are well biologically understood [36]. For instance, if the exposure is a protein, then the best strategy is generally to use a marker in the gene which is responsible for encoding the protein itself.

## 3 Causal Model

There are many definitions of causality in the philosophical and statistical literature [44, 50, 51, 55, 70, 87, 98, 104, 105]. Throughout this chapter we adopt the Pearl's definition of causality. It is a notion of causality which relies mainly on conditional probability and interventions. In order to mathematically represent interventions and distinguish them from observations, Pearl introduced the do-operator. Denoting by $do(X = x)$ the hypothetical intervention in which the variable $X$ is manipulated to be set to the value $x$, and denoting by $P(y|do(x))$ the probability of the response event $Y = y$ under the hypothetical intervention $X = x$, we say that $X$ is a cause of $Y$ if [38, 87]:

$$P(Y = y|do(X = x)) \neq P(Y = y|do(X = x')),$$

when all background variables remain constant. Thus, Pearl's representation of causality has close resemblance to a randomized controlled experiment, in which any change in the outcome variable must be due to the intervention, if all factors influencing their association are either constant, or vary at random.

The Pearl's comprehensive theory of causation resulted from the unification of several approaches to causation, such as the graphical, potential outcome, and SEMs.

As proved by Pearl [87], SEMs provide a language for causality which is mathematically equivalent to the potential-outcome framework, developed by Jerzy Neyman [84] and Donald B. Rubin [105]. Significant contributions to the generalization of the potential-outcome framework as a general mathematical language for causal inference were also given by James Robins [98, 99].

The potential-outcome model is based on randomized experiments and counterfactual variables. By conducting randomized controlled experiments, in general, only one potential outcome is observed for each individual, which is the one corresponding to the exposure value that actually occurred for the individual. An outcome that would have occurred if, contrary to the fact, the exposure had been assigned another value is considered a counterfactual quantity according to Rubin's notation. Within the potential-outcome framework, causal inferences are made by deriving probabilistic properties of these counterfactual quantities as in a missing data problem. The equivalence between SEMs and potential-outcome models could be demonstrated by Pearl by treating counterfactual quantities as random variables. He showed that the consistency rule [98]—which states that an individual's potential outcome under a hypothetical intervention that happened to materialize is precisely the outcome experienced by that individual—is automatically satisfied in the structural model. Thus, expressions involving probabilities of counterfactuals can be converted to expressions involving conditional probabilities of measured variables [88].

The connection between SEMs and graphical models will be described in detail in the following sections. While SEMs provide a functional representation of the causal processes relating the variables, graphical models provide a visual and, thus, more intuitive representation of these relationships.

## 3.1 Functional Causal Representation

In a general context, a causal model is an SEM representing the causal relationships between random variables [87].

Consider random variables $V_i,\ i = 1, \ldots, n$. Let $\mathrm{pa}(V_i)$ denote the set of the parents of $V_i$, that is, the set of random variables that directly determine the value of $V_i$. Also, let $\varepsilon_i,\ i = 1, \ldots, n$ be random variables representing errors due to unknown causes.

The mechanism by which the value of each variable $V_i$ is selected can be defined according to some function $f_i$ (usually, but not necessarily, linear) of the parent variables and of the error variable:

$$V_i = f_i(\mathrm{pa}(V_i), \varepsilon_i), \quad i = 1, \ldots, n. \tag{1}$$

In the context of genotype–phenotype causal networks and Mendelian randomization, the random variables are quantitative phenotypes and variant genetics associated with these phenotypes (QTLs or QTNs). In addition, it is commonly assumed that the functional relationships of the SEM shown in Eq. (1) are linear, and that the effects of genetic variants are fixed.

Before introducing the model, let us first discuss the involved notation and objects. Let:

- $\mathbf{Y}$ be a $p \times n$ matrix where each element $y_{ij}$ represents the observed value of the $i$th quantitative phenotype for the $j$th individual;
- $\mathbf{M} = \mathbf{1}' \otimes \boldsymbol{\mu}$ be a $p \times n$ matrix resulting from the Kronecker product between the transposed n-dimensional unity vector, and $\boldsymbol{\mu}$ be the vector with the expected values of each quantitative phenotype;
- $\mathbf{Q}$ be a $p \times q$ matrix of effects of genetic variants on the phenotypes. The element $q_{ij}$ represents the effect of the $j$th genetic variant on the $i$th phenotype;
- $\mathbf{X}$ be a $q \times n$ matrix where each element $x_{ij}$ represents the predicted genotype of the $i$th genetic variant in the $j$th individual. For SNPs the observed genotype state is used instead of the predicted values;
- $\mathbf{P}$ be a $p \times p$ matrix containing the direct causal effects of the phenotypes on each other. The element $p_{ij}$ corresponds to the effect of the $j$th phenotype on the $i$th phenotype;
- $\mathbf{E}$ be a $p \times n$ matrix where each $e_{ij}$ represents the measurement error of the $i$th phenotype for the $j$th individual.

The causal linear SEM representing the expected pattern of associations among $p$ observed phenotypes and $q$ genetic variants for $n$ individuals is

$$\mathbf{Y} = \mathbf{M} + \mathbf{PY} + \mathbf{QX} + \mathbf{E}. \tag{2}$$

Note that the assumption that variations in QTL or QTN genotypes precede variation in the phenotypes is expressed by the fact that the $\mathbf{Q}$ matrices is always in the right side in Eq. (2), along with the parent phenotype effects, represented in $\mathbf{P}$ matrix.

In the case of biallelic genetic variants, the total genetic effect (represented in $Q$ and $X$ matrices) can be partitioned in order to explicit particular effects, such as additive and dominance effects of each genetic locus and the possible interaction effects among them (epistatic effects). Typically biallelic genetic variants have three genotype states, denoted usually as AA (homozygous dominant), Aa (heterozygous), and aa (homozygous recessive). These genotypes must be encoded according to the type of effect by using two degrees of freedom.

We will illustrate this extension in order to take into account additive and dominance effects. For simplicity, we will not consider interactions between genetic variants (epistasis).

Let us precisely define the components of dominance and additive effects using biallelic genetic variants. Let:

- $\mathbf{Q^a}$ be a $p \times q$ matrix of the additive effects. The element $q^a_{ij}$ represents the additive effect of the $j$th genetic variant on the $i$th phenotype;
- $\mathbf{X^a}$ be a $q \times n$ matrix where each element $x^a_{ij}$ represents the predicted genotype of the $i$th genetic variant for the $j$th individual, properly encoded to represent

additive effects. For instance, we can encode the genotypes aa, Aa, and AA as $-1, 0$, and 1, respectively.

- $\mathbf{Q^d}$ be a $p \times q$ matrix of the dominance effects. The element $q^d_{ij}$ represents the dominance effect of $j$th genetic variant on the $i$th phenotype;
- $\mathbf{X^d}$ be a $q \times n$ matrix where each element $x^a_{ij}$ represents the predicted genotype of the $i$th genetic variant for the $j$th individual, properly encoded to represent dominance effects. Since the dominance effects are due the interaction between the alleles, a possible encoding for the genotypes is 1 for a heterozygous individual, and 0 otherwise.

In this specific case, the causal model for the genotype–phenotype network is

$$\mathbf{Y} = \mathbf{M} + \mathbf{PY} + \mathbf{Q^aX^a} + \mathbf{Q^dX^d} + \mathbf{E}. \tag{3}$$

The interaction effects between genetic variants (describing epistasis) were not illustrated in Model 3. However, we could easily incorporate in the model matrices $\mathbf{Q^{aa}}$, $\mathbf{Q^{dd}}$, and $\mathbf{Q^{ad}}$ representing epistatic interaction effects [27].

Since dominance effects (interaction between alleles) and epistasis (interaction between loci) are higher order effects, it is possible that they have little impact on the inferences about the response variable. Supporting this idea, Burgess et al. (2011) [21] suggest to include only the most important instrumental variables (genetic variants), based on biological knowledge, for a parsimonious modeling of the genetic association (i.e., per allele additive genetic model, rather than using the total degrees of freedom in terms of effects).

However, the literature has shown that many genetic variants are not precisely identified because of the simplicity of the adopted models [75, 120, 132]. So, while lower order effects (additive effects) may be sufficient for genetic mapping, interaction effects may be decisive to analyze complex diseases (as opposed to the Mendelian diseases).

In order to draw causal inferences from observational studies, it has been suggested to select the most relevant instrumental variables (genetic determinants of the exposure) attempting to be as parsimonious as possible. However, when only a small proportion of the variability in the exposure is explained by the genetic variant, it is possible to improve the precision of estimates by using multiple genetic variants [85, 90].

### *3.2 Graphical Causal Representation*

The causal model in Eq. (1) can be graphically represented by a directed graph.

**Definition 1.** Let $\mathbf{V} = \{V_1, \ldots, V_n\}$ be a finite set and $\mathbf{E} \subseteq \{(V_i, V_j) : V_i, V_j \in \mathbf{V}\}$ a set of ordered pairs of vertices. Each element of the set $\mathbf{V}$ is called **vertex** and each element of the set $\mathbf{E}$ is called **directed edge**. The edge $(V_i, V_j)$ represents a direct connection from $V_i$ to $V_j$. The ordered pair $G = (\mathbf{V}, \mathbf{E})$ is called **directed graph** or **digraph**.

We can always build a graphical representation of an SEM by using a directed graph.

In this representation, each vertex of a directed graph corresponds to a distinct random variable. In addition, each edge $(V_i, V_j)$, if it exists, represents a direct functional relationship from the variable $V_i$ to the variable $V_j$. In this case, $V_i$ is called parent of $V_j$ and $V_j$ is a child of the vertex $V_i$. The absence of an edge indicates both variables are not directly associated. Thus, if we draw, for each variable $V_i$, an edge pointing to it from each of its parents, we can build the directed graph which represents their causal mechanisms.

The error terms are not represented in the graph. However, when error terms are correlated, the corresponding pairs of variables must be connected by a bidirected (double-headed) edge.

When the relationships imply causality, the graphical representation of a causal model is referred as *causal graph* or *causal diagram* of the system. The goal of the causal structure learning methods is to discover the causal graph of a system often from observational data.

In the graphical representation of the functional model shown in Eq. (2), the vertices can represent phenotypes or genetic variants and the edges represent the causal relationships among them.

The next definitions introduce concepts that distinguish two classes of graphs according to whether or not the graph structure has cyclic patterns. This distinction is important because many results and procedures for inferring causal relationships using observational data are dependent on the type of graph structure.

**Definition 2.** A **directed path** between two vertices is a sequence of directed edges that begins at one vertex and ends at another vertex, with the restriction that all the edges are oriented in the same direction. Whenever there is a path that begins and ends at the same vertex we have a **cycle**. Cycles of length one are called **self-loops** and cycles of length two corresponds to a **bidirectional influence** or **reciprocal association**.

**Definition 3.** An SEM with uncorrelated error terms and which does not contain cyclic relationships is called **recursive SEM** and its graphical representation is called **directed acyclic graph (DAG)**.

Real biological systems such as GRNs often have natural cyclic behavior [28]. Thus, DAGs can be very restrictive to model such biological data. Directed graphs that can accommodate cycles and reciprocal associations have been used to model feedback processes that have reached equilibrium. For instance, equilibrium expression patterns can be modeled in reverse engineering GRNs from multiple gene expression measurements [26].

An alternative interpretation for cycles is that each feedback relation represents an infinite sequence of variables indexed by time. Thus, a cyclic graph can be viewed as a compact representation of an infinite acyclic graph [96, 117].

**Definition 4.** An SEM which contains at least one cycle is called **non-recursive SEM** and its graphical representation is called **directed cyclic graph (DCG)**. Note that systems with correlated error are also non-recursive, since the corresponding pair of variables are connected by a bidirected edge.

DCGs have been used to represent GRNs. In this representation, vertices are gene expression levels of genes and directed edges indicate regulation processes. The expression of a gene can be controlled by the presence of proteins called activators and repressors (or inhibitors). Thus, the gene in the tail of the edge produces a protein that regulates the gene in the head of the edge. In this case, the genome itself consists in a complex network [26, 83].

From an algebraic point of view, the edges connecting phenotypes correspond to the non-zero elements in $\mathbf{P}$ and the edges pointing from a QTL or QTN to a phenotype exist if the corresponding entries in $\mathbf{Q^a}$ or in $\mathbf{Q^d}$ are non-zero. If the $\mathbf{P}$ matrix can be rearranged as a lower triangular matrix, then we have a recursive model and, consequently, it can be represented as a DAG. Otherwise, the system contains cycles and a non-recursive SEM and a DCG are used to represent it.

Any SEM can be represented by directed graphs, even when the system involves cycles, self-loops, dependent errors, and nonlinearities. In biological context, causal models often represent linear relationships among phenotypes and genetic variants. It is commonly assumed that the system does not contain self-loops and the error terms are uncorrelated.

# 4 Properties Relating Functional and Graphical Models

To provide a statistical connection between the functional and the graphical representation of a causal model, some concepts are fundamental, namely conditional independence, d-separability, directed Markov property, and causal faithfulness.

The graphical model (a directed acyclic or cyclic graph) that precisely encodes the conditional independence relations among the variables of the system is called PGM. When that precise connection between graphical and functional representations can be established, some theoretical results can be used for causal inference and network structure learning. These concepts are presented in the following sections.

## *4.1 d-Separability*

The concept called d-separation is a fundamental criterion used in network structure discovery algorithms. In fact, it can determine whether or not a directed edge exists between two variables. Under d-separation criterion, it is even possible to determine the direction of some edges.

Before giving a precise definition of d-separation, it will be introduced some related concepts: conditional and unconditional independence and undirected path. These concepts can be defined on random variables or on sets of random variables as follows:

**Definition 5.** Let $\mathbf{V} = \{V_1, V_2, \ldots V_n\}$ be a set of random variables. Consider $\mathbf{X}$, $\mathbf{Y}$, and $\mathbf{Z}$ three subsets of $\mathbf{V}$ and $P$ the joint probability distribution function over the variables in $\mathbf{V}$. We say that the sets $\mathbf{X}$ and $\mathbf{Y}$ are **conditionally independent** given $\mathbf{Z}$ if, for any configuration $x$ of the variables in the set $\mathbf{X}$ and for any configurations $y$ and $z$ of the variables in the sets $\mathbf{Y}$ and $\mathbf{Z}$ satisfying $P(\mathbf{Y} = y, \mathbf{Z} = z) > 0$, we have

$$P(\mathbf{X} = x|\mathbf{Y} = y, \mathbf{Z} = z) = P(\mathbf{X} = x|\mathbf{Z} = z).$$

This relationship is denoted by $(\mathbf{X} \perp\!\!\!\perp \mathbf{Y}|\mathbf{Z})_P$ or simply $\mathbf{X} \perp\!\!\!\perp \mathbf{Y}|\mathbf{Z}$.

**Definition 6.** Using the same notations of the Definition 5, $\mathbf{X}$ and $\mathbf{Y}$ are **unconditionally independent** or **marginally independent** if

$$P(\mathbf{X} = x|\mathbf{Y} = y) = P(\mathbf{X} = x)$$

for any configurations $x$ and $y$ of the variables in the sets $\mathbf{X}$ and $\mathbf{Y}$ satisfying $P(\mathbf{Y} = y) > 0$.

This relationship is denoted by $\mathbf{X} \perp\!\!\!\perp \mathbf{Y}|\emptyset$ or simply $\mathbf{X} \perp\!\!\!\perp \mathbf{Y}$.

The conditional and unconditional independencies encoded by a given directed cyclic or acyclic graph can be determined by a graphical criterion based on the definitions of undirected path and collider:

**Definition 7.** Let $G = (\mathbf{V}, \mathbf{E})$ be a directed graph. A sequence of distinct edges $\{E_1, \ldots, E_k\}$ in $G$ is an **undirected path** if there exists a sequence of vertices $\{V_i, \ldots, V_{k+1}\}$ such that for $1 \leq i \leq k$ either $(V_i, V_{i+1}) = E_i$ or $(V_{i+1}, V_i) = E_i$, and $E_i \neq E_{i+1}$. An **acyclic undirected path** is an undirected path in which every vertex in the path occurs no more than once.

In words, an undirected path is a sequence of connected edges ignoring their directions. It is also common to define undirected path as an ordered sequence of vertices that must be transversed, ignoring the direction of the edges. However, this definition is only valid for acyclic graphs, since a pair of vertices can uniquely identify an edge. A proper definition of undirected path which is valid for structures with reciprocal associations uses a sequence of edges rather than a sequence of vertices [117].

**Definition 8.** Let $X$, $Y$, and $Z$ be vertices of a graph and $U$ be an undirected path containing $X$, $Y$, and $Z$ in this order.

$Y$ is a **collider** in $U$ if there are edges pointing into it from both $X$ and $Z$ (i.e., $Y$ is common effect of $X$ and $Z$), preventing transmission of causal effects along such a path.

When $Y$ is a collider and, additionally, $X$ and $Z$ are not connected by an edge, $Y$ is called **unshielded collider** [118]. In addition, the formation $X \rightarrow Y \leftarrow Z$ is called **v-structure** (using Pearl's notation [87]) or **immorality** (using Koller and Friedman's notation [64]).

Having introduced such fundamental concepts, d-separation can be defined as follows [87]:

**Definition 9.** Let $G = (\mathbf{V}, \mathbf{E})$ be a causal graph and $\mathbf{X}$, $\mathbf{Y}$, and $\mathbf{Z}$ be disjoint sets of vertices of $\mathbf{V}$. $\mathbf{X}$ and $\mathbf{Y}$ are **d-separated** given $\mathbf{Z}$ in $G$, if for any undirected acyclic path $U$ between a vertex in $\mathbf{X}$ and a vertex in $\mathbf{Y}$:

- $U$ contains an unshielded collider such that neither the middle vertex (the collider) nor any of its descendants is in $\mathbf{Z}$; or
- $U$ contains a vertex which is not a collider and it is in $\mathbf{Z}$.

When $\mathbf{X}$ and $\mathbf{Y}$ are d-separated by $\mathbf{Z}$ in the graph $G$ we write $(\mathbf{X} \perp\!\!\!\perp \mathbf{Y}|\mathbf{Z})_G$.

Using d-separation criterion, the premise of conditional independence can be observed under two assumptions: global directed Markov and causal faithfulness. We will precisely define them in the following sections, but they assure us that two vertices are conditionally independent given a set of variables $\mathbf{Z}$ if and only if $\mathbf{Z}$ d-separates both variables. Thus, the graphical property of d-separation enables us to determine what conditional independence relations are entailed by a given graphical causal model. For each pair of variables, we can test whether they are independent given all sorts of conditioning variables sets.

Sometimes it is possible to prune away edges that represent spurious associations or even to orient edges using observational data alone. Whenever both variables become independent by conditioning on other variables, we can rule out the edge between them. In addition, unshielded collider formations can be identified testing if two variables become dependent by conditioning on the collider. As the edges going into colliders are oriented, orientations of other edges can be induced.

For instance, consider the following d-separation statements present in the graph of Fig. 2:

$$B \perp\!\!\!\perp C \mid A; \quad A \perp\!\!\!\perp D \mid \{B, C\}; \quad B \perp\!\!\!\perp E \mid D; \quad C \perp\!\!\!\perp E \mid D; \quad A \perp\!\!\!\perp E \mid D.$$

These pairs of vertices are d-separated because every path between them is blocked, that is: (1) in every path containing a collider, the collider and its descendants are not in the conditioning set; and (2) in every path containing only non-colliders and non-descendant of colliders, at least one vertex of the path is in the conditioning set. Therefore, we can conclude that there is not an edge connecting them.

We can orient some of the edges if more tests are made. For instance, we can identify the unshielded collider $C \rightarrow D \leftarrow B$, knowing that $B \not\!\perp\!\!\!\perp C \mid \{A, D\}$. After orienting these edges, we can also identify the true direction between $D$ and $E$. By conditioning to $D$, neither the path between $B$ and $E$ ($B \perp\!\!\!\perp E \mid D$) nor the path

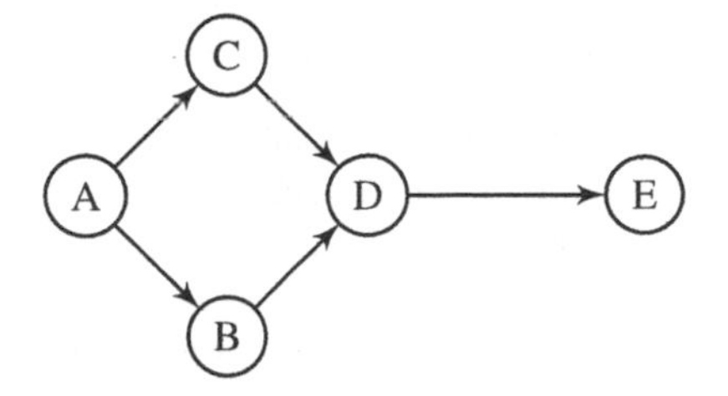


**Fig. 2** A DAG representing causal relationships among five variables. By using d-separation criterion, only the direction of the edges $A \rightarrow C$ and $A \rightarrow B$ cannot be recovered

between $C$ and $E$ ($C \perp\!\!\!\perp E \mid D$) can be blocked, implying that $D$ is not a collider. Thus, the edge $D \rightarrow E$ can be recovered.

Since conditional independence is a symmetric relationship, we cannot orient the two remaining edges. Even knowing that $A$, $B$, and $C$ are not colliders, we cannot discard any of the three following orientations because for all of them we have the d-separation statement $B \perp\!\!\!\perp C \mid A$:

- A chain: $B \rightarrow A \rightarrow C$;
- Another chain: $B \leftarrow A \leftarrow C$;
- A fork: $B \leftarrow A \rightarrow C$.

Graphs with the same set of d-separation statements usually correspond to observationally equivalent models and their structures cannot be fully recovered using observational data alone. In Sect. 5, we will discuss the equivalence problem in more detail.

## *4.2 Global Directed Markov Property*

In order to provide a probabilistic interpretation of the graphs, it is necessary to introduce a property to ensures that the graph with a set of vertices $\mathbf{V}$ can also represent a set of probability distributions over $V$.

If the graph accurately describes the structure entailed by a causal model, then the separation properties of the graph can be associated with conditional independencies and causality relations among variables. In other words, we can use d-separation criterion as a graphical tool to recover the underlying causal mechanisms relating the variables.

**Definition 10.** Let $G$ be a directed acyclic or cyclic graph with a probability distribution $P$. We say that $P$ satisfies the **global directed Markov property** for $G$ if for all disjoints sets of variables $\mathbf{X}$, $\mathbf{Y}$, and $\mathbf{Z}$ the following statement is true: if $\mathbf{X}$ is d-separated from $\mathbf{Y}$ given $\mathbf{Z}$ in $G$, then $\mathbf{X}$ is conditionally independent from $\mathbf{Y}$ given $\mathbf{Z}$ in $P$.

In other words, we say that $P$ satisfies the global directed Markov property for $G$ when:

$$(\mathbf{X} \perp\!\!\!\perp \mathbf{Y}|\mathbf{Z})_G \Rightarrow (\mathbf{X} \perp\!\!\!\perp \mathbf{Y}|\mathbf{Z})_P, \quad \text{foralldisjointsets}\mathbf{X}, \mathbf{Y}\text{and}\mathbf{Z}.$$

Assuming the global directed Markov property holds, the conditional independencies found by d-separation in a given graph $G$ hold for every causal model that can be represented graphically by $G$.

A wide range of statistical models, including recursive linear SEMs with independent errors, regression models, and factor analytic models, satisfy the global directed Markov condition for its associated DAG.

Considering recursive and non-recursive SEMs with independent errors, if the relationships are linear, then the following result guarantees that the probability distribution $P$ of the SEM satisfies the global directed Markov property [65, 117]:

**Theorem 1.** *Let $L$ be a* linear *recursive (or non-recursive SEM) with jointly independent error variables and $G$ be the directed acyclic (or cyclic) graph naturally associated with $L$. Consider the probability distribution $P$ over the variables of $L$.*

*Under these conditions, $P$ satisfies the global directed Markov property for $G$.*

Thus, the Theorem 1 allows us to use d-separation criterion in a graph to read off the conditional independence relations entailed by the associated linear SEM. In addition, all conditional independence relations which hold in a linear SEM are precisely encoded by its natural graphical representation.

The natural graphical representation of a non-recursive SEM is a DCG. It has been shown that there is no DAG that is capable of encoding the conditional independence relations entailed by a non-recursive SEM [94].

### 4.2.1 Local Directed Markov Property in DAGs

By construction, in DAGs, every variable is d-separated from its non-descendants given its parents. Thus, for DAGs, there is a local property equivalent to the global directed Markov property [67]:

**Definition 11.** Consider a directed **acyclic** graph $G$ with a probability distribution $P$, both defined over a set of random variables $\mathbf{V} = \{V_1, \ldots, V_n\}$.

We say that $P$ satisfies the **local directed Markov property** with respect to $G$ if for every variable $V_i \in \mathbf{V}$, in the probability distribution $P$, $V_i$ is independent of all other non-descendants variable (all other vertices except its parents and descendants), given its parents in $G$.

In Pearl's terminology [86] we say that $G$ is an **independency map** (or **I-map**) of $P$ when all the Markov assumptions implied by $G$ are satisfied by $P$.

Thus, the local directed Markov property is sufficient to relate an acyclic graphical representation $G$ to a probability distribution $P$. The equivalence of the global and local directed Markov properties in DAGs holds even when the probability distributions represented by the graph have no density function [67].

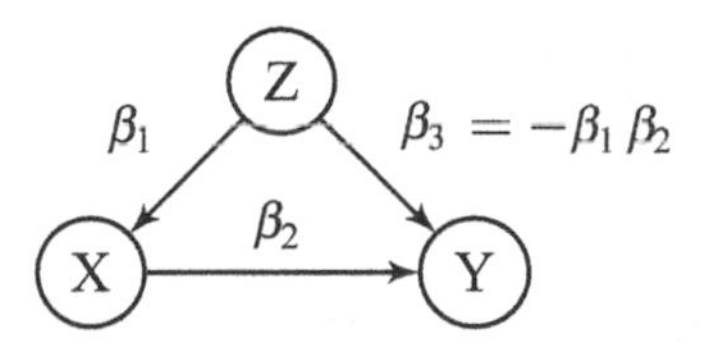


**Fig. 3** Example of an unfaithful distribution to the graph. The direct effect of $Z$ on $Y$ is exactly the additive inverse of the indirect effect through $X$, leaving no total effect

According to Theorem 1, the probability distribution $P$ of a linear recursive SEM $L$ with jointly independent error variables satisfies the global Markov property for the DAG $G$ that provides the natural graphical representation of $L$. Moreover, given the equivalence of the global and local directed Markov properties, $P$ also satisfies the local directed Markov property for $G$.

We point out, however, that the local and global Markov properties are not equivalent for directed *cyclic* graphs (DCGs) [117].

## *4.3 Causal Faithfulness*

Assuming that a probability distribution $P$ satisfies the global directed Markov property for a graph, all d-separation statements obtained graphically hold as conditional independence statements in $P$. However, this assumption does not ensure that all conditional independence statements implied by $P$ are represented in the graph.

Consider, for instance, the system in Fig. 3 represented by an DAG with three vertices, $X$, $Y$, and $Z$, such that $Z$ directly affects $Y$, but it also affects $Y$ indirectly, mediated by $X$.

Assuming that the relations are linear, the total effect of one variable on another is the sum of its direct effect and indirect effects [15]. Moreover, the indirect effect can be calculated by using Sewall Wright's multiplication rule [34, 128], i.e., by multiplying the structural coefficients on the corresponding path.

In Fig. 3, the direct effect is given by $\beta_3$, and the indirect effect is given by the product of the coefficients on path through $X$, that is, $\beta_1\beta_2$. By summing the direct and indirect effects, we have the total effect of $Z$ on $Y$ is equal to $\beta_3 + \beta_1\beta_2$. Since in this specific case $\beta_3$ is defined as $-\beta_1\beta_2$, the total effect is equal to zero.

In order to clarify a bit more the calculation of the total effect under linearity condition, consider the corresponding recursive linear SEM with independent errors that can be derived by describing each variable as a linear function of its parents and of an error variable:

$$
\begin{aligned}
Y &= \beta_3 Z + \beta_2 X + \varepsilon_Y \\
X &= \beta_1 Z + \varepsilon_X.
\end{aligned}
\tag{4}
$$

By substituting the expression for $X$ into the expression for $Y$, we can express $Y$ as a function of $Z$ and error variables:

$$\begin{aligned} Y &= \beta_3 Z + \beta_2(\beta_1 Z + \varepsilon_X) + \varepsilon_Y \\ &= (\beta_3 + \beta_2\beta_1)Z + \beta_2\varepsilon_X + \varepsilon_Y. \end{aligned} \quad (5)$$

Thus, the total effect of $Z$ on $Y$ is given by $\beta_3 + \beta_1\beta_2$ and it vanishes when the parameter $\beta_3$ is set to exactly $-\beta_1\beta_2$.

Although there are no conditional independencies entailed for all values of free parameters, with that specific choice of the $\beta_3$ parameter, the direct effect is cancelled out by the indirect effect and $Z$ and $Y$ will be apparently not associated. In such a case, we say that the population is unfaithful to the graph of the causal structure that generated it.

Under the assumption that a probability distribution $P$ is faithful to a graph, we have the guarantee that the conditional independencies entailed by $P$ can be read off from the graph by applying d-separation criterion.

**Definition 12.** Let $G$ be a directed acyclic or cyclic graph $G$ with a probability distribution $P$. We say that $P$ satisfies the **causal faithfulness condition** for $G$ if for all disjoints sets of variables $\mathbf{X}$, $\mathbf{Y}$, and $\mathbf{Z}$, the following statement is true: if $\mathbf{X}$ is conditionally independent from $\mathbf{Y}$ given $\mathbf{Z}$ in $P$, then $\mathbf{X}$ is d-separated from $\mathbf{Y}$ given $\mathbf{Z}$ in $G$.

In other words, we say that $P$ satisfies the causal faithfulness condition for $G$ when:

$$(\mathbf{X} \perp\!\!\!\perp \mathbf{Y}|\mathbf{Z})_P \Rightarrow (\mathbf{X} \perp\!\!\!\perp \mathbf{Y}|\mathbf{Z})_G, \quad \text{for all disjoint sets } \mathbf{X}, \mathbf{Y} \text{ and } \mathbf{Z}.$$

The following theorem allow us to test if a probability distribution is faithful to a directed *acyclic* graph [118]:

**Theorem 2.** *Let G be a DAG with a probability distribution P. If P is faithful to some DAG, then P is faithful to G if and only if*

1. *for any vertices X and Y of G, X, and Y are adjacent if and only if X and Y are dependent conditional on every set of vertices of G that does not include X or Y; and*
2. *for any vertices X, Y, and Z, such that X is adjacent to Y, Y is adjacent to Z, and X is not adjacent to Z, $X \rightarrow Y \leftarrow Z$ is a subgraph of G if and only if X and Z are dependent conditional on every set containing Y but not X or Z.*

## 4.4 Factorization of Joint Probability Distribution Functions

Consider an arbitrary probability distribution function $P$, defined on $n$ random variables $V_1, \dots V_n$. By successive application of the chain rule of probability, $P$ can be factorized as a product of $n$ conditional probability distribution functions:

$$P(V_1, V_2, \dots, V_n) = \prod_j P(V_j | V_1, \dots, V_{j-1}).$$

The probability distribution function $P$ can be either specified by a probability mass function (for qualitative random variables) or by a probability density function (for quantitative random variables).

We will present a definition for the factorization of a joint probability *density* function according to a directed graph which may be either acyclic or cyclic. It was first proposed by Lauritzen et al. [67] for DAGs and its generalization was due to Thomas Richardson and Peter Spirtes [97, 117].

To precisely define that factorization property, let us first introduce the concept of ancestor of a vertex:

**Definition 13.** Let $G = (\mathbf{V}, \mathbf{E})$ be a directed graph. A vertex $V_i \in \mathbf{V}$ is an **ancestor** of a vertex $V_j \in \mathbf{V}$, if there is an acyclic directed path from $V_i$ to $V_j$ or $V_i = V_j$.

With the concept of ancestor, we can now define the factorization property of joint densities according to directed cyclic or acyclic graph.

**Definition 14.** Let $G = (\mathbf{V}, \mathbf{E})$ be a directed graph and $\mathbf{X}$ be a subset of $\mathbf{V}$. Consider the probability density function $f(\mathbf{V})$ for a probability distribution $P$ with respect to a product measure $\mu$ over $\mathbf{V}$ (i.e., $P = f \cdot \mu$). Denote by $f(\mathbf{Y})$ the marginal of $f(\mathbf{V})$ for a subset $\mathbf{Y}$ of $\mathbf{V}$. Also, denote the set of ancestors of members of $\mathbf{X}$ by $An(\mathbf{X}, G)$ and the set of parents of a vertex $X_i$ in $G$ by $\text{pa}(X_i, G)$.

For a non-negative function $g$, we say that $P$ **factors according to the directed graph** $G$ if for every subset $\mathbf{X}$ of $\mathbf{V}$,

$$f(An(\mathbf{X}, G)) = \prod_{X_i \in An(\mathbf{X}, G)} g(X_i, \text{pa}(X_i, G)).$$

Directed *acyclic* graphs are built in such a way that each variable is d-separated from its non-descendants given its parents. For this reason, it is possible to use a simpler form of Definition 14. Moreover, the factorization property according to a DAG is defined for any probability distribution function (defined either over quantitative or qualitative random variables):

**Definition 15.** Let $G$ be a directed **acyclic** graph and $P$ be a probability distribution function, both defined over a set of random variables $\mathbf{V} = \{V_1, \dots, V_n\}$. Denote by $\text{pa}(V_j, G)$ the set of parents of a vertex $V_j$ in $G$.

We say that $P$ **factors according to the directed acyclic graph** $G$ if $P$ can be written as the product of the individual distribution functions, conditional on their parent variables:

$$P(V_1, \dots, V_n) = \prod_j P(V_j | \operatorname{pa}(V_j, G)).$$

### 4.4.1 Factorization and Global Markov Property

The following results relate factorization of probability density functions and global directed Markov property. Their proofs are due to Thomas Richardson and Peter Spirtes [97] and are based on the proofs by Lauritzen et al. [67] for DAGs.

**Theorem 3.** *Let $P$ be a probability distribution that is absolutely continuous with respect to a product measure $\mu$ (and, thus, it has a non-negative probability density function).*

*If $P$ factors according to a directed graph $G$, then $P$ satisfies the global directed Markov property for $G$.*

The Theorem 3 states that if a probability distribution complies with the Definition 14, then it also satisfies the global directed Markov property for the corresponding directed (acyclic or cyclic) graph.

The reverse direction holds for acyclic graphs under the same hypotheses. However, to extend this result for cyclic graphs, a further constraint on the probability distribution $P$ is necessary: it must has a *strictly positive* probability density function $f$ [97, 117].

**Theorem 4.** *Let $P$ be a probability distribution, defined over a set of random variables $\mathbf{V} = \{V_1, \dots, V_n\}$, that is absolutely continuous with respect to a product measure $\mu$, and has a positive probability density function $f(\mathbf{V})$.*

*If $P$ satisfies the global directed Markov property for a directed (cyclic or acyclic) graph $G$, then $f(\mathbf{V})$ factors according to $G$.*

The proof of Theorem 4 by Thomas Richardson and Peter Spirtes uses some ideas of Lauritzen et al. [67], which are mainly based on the moralized version of a graph (i.e., the undirected version of the graph obtained after connecting or marrying the parents of each immorality). The positivity assumption is needed to use the Hammersley–Clifford theorem. It gives necessary and sufficient conditions under which a *positive* probability distribution factorizes according to an undirected graph. A discussion on the problems involved is given by Terry Speed [115].

To conclude this section, we want to emphasize three statements that are equivalent for a probability distribution function $P$ associated with the directed *acyclic* graph $G$ [67]:

- $P$ satisfies the global directed Markov property with respect to the DAG $G$;
- $P$ satisfies the local directed Markov property with respect to the DAG $G$;

- $P$ factors according to the DAG $G$, i.e., the joint probability distribution can be expressed by a product of conditional distributions for each variable given its parents. For instance, considering the system shown in Fig. 2,

$$P(A,B,C,D,E) = P(E|D)\,P(D|C,B)\,P(C|A)\,P(B|A)\,P(A).$$

## 4.5 Linear Entailment and Partial Correlations

In practice, algorithms for causal structure learning assume a d-separation oracle which precisely tell us whether two variables are d-separated in a directed graph given a set of other variables. Thus, we can ask the oracle whether two variables are d-separated given every possible conditioning set. If it is possible to find a conditioning set that makes two variables d-separated, then we can conclude that does not exist an edge connecting these two variables. That is a fundamental idea for reconstructing association networks (an undirected graph representing only direct associations among variables). For instance, this idea is used in the first step of the classical PC-algorithm, called PC-skeleton algorithm [118]. The direction of each edge connecting two variables that could not be d-separated by any conditioning set is determined in subsequent steps of structure learning algorithms.

Under the assumption that all involved variables have a joint multivariate normal distribution, a zero partial correlation ties with conditional independence. In addition, as shown in Definition 12, the faithfulness assumption ensures that conditional independence implies d-separability. Thus, under faithfulness and normality assumptions, it is possible to apply d-separation criterion by using a statistical test for zero partial correlations. The level of statistical significance of the test is decisive for the determination of direct associations between variables. The choice of the significance level depends on the maximum acceptable probability of making a type I error, but the most commonly used significance levels are 1 % and 5 %. In the PC-skeleton algorithm of the R package pcalg [60] (which is used in the first step of the QDG [24] and QPSO [126] genotype–phenotype discovery algorithms), the default significance level for individual partial correlation tests is 1 %. However, in the R package QTLnet [82], which implements the QDG algorithms, the suggested significance level is very small (equals to 0.05 %), probably to compensate for multiple comparisons.

The partial correlation measures the strength of the linear association between two continuous variables when the effect of a set of other random variables is controlled. We define the partial correlation coefficient in the following [59]:

**Definition 16.** Let $X$ and $Y$ be two random variables and $\mathbf{Z} = (Z_1, \ldots, Z_p)$ a set of $p$ other random variables. Let $\mu_X$ and $\mu_Y$ be the means of $X$ and $Y$, respectively, and $\boldsymbol{\mu}_{\mathbf{Z}}$ the mean vector of $\mathbf{Z}$. Also, let $\Sigma$ be the covariance matrix of $(X, Y, Z_1, \ldots, Z_p)$ with the following partition notations:

$$\boldsymbol{\Sigma} = \begin{pmatrix} \sigma_{XX} & \sigma_{XY} & \sigma_{XZ_1} & \cdots & \sigma_{XZ_p} \\ \sigma_{YX} & \sigma_{YY} & \sigma_{YZ_1} & \cdots & \sigma_{YZ_p} \\ \sigma_{XZ_1} & \sigma_{YZ_1} & \sigma_{Z_1Z_1} & \cdots & \sigma_{Z_pZ_1} \\ \vdots & \vdots & \vdots & \ddots & \vdots \\ \sigma_{XZ_p} & \sigma_{YZ_p} & \sigma_{Z_1Z_p} & \cdots & \sigma_{Z_pZ_p} \end{pmatrix} = \begin{pmatrix} \sigma_{XX} & \sigma_{XY} & \boldsymbol{\Sigma}_{X\mathbf{Z}} \\ \sigma_{YX} & \sigma_{YY} & \boldsymbol{\Sigma}_{Y\mathbf{Z}} \\ \boldsymbol{\Sigma}'_{X\mathbf{Z}} & \boldsymbol{\Sigma}'_{Y\mathbf{Z}} & \boldsymbol{\Sigma}_{\mathbf{ZZ}} \end{pmatrix}.$$

The prediction errors of $X$ and $Y$ given $\mathbf{Z}$ when using the best linear predictors (which minimize the mean square error) are, respectively,

$$X - \mu_X - \boldsymbol{\Sigma}_{X\mathbf{Z}}\boldsymbol{\Sigma}_{\mathbf{ZZ}}^{-1}(\mathbf{Z} - \boldsymbol{\mu}_{\mathbf{Z}})$$
$$Y - \mu_Y - \boldsymbol{\Sigma}_{Y\mathbf{Z}}\boldsymbol{\Sigma}_{\mathbf{ZZ}}^{-1}(\mathbf{Z} - \boldsymbol{\mu}_{\mathbf{Z}}),$$

with error covariance matrix that can be calculated as

$$\Sigma_{XY\cdot\mathbf{Z}} \doteq \begin{pmatrix} \sigma_{XX\cdot\mathbf{Z}} & \sigma_{XY\cdot\mathbf{Z}} \\ \sigma_{YX\cdot\mathbf{Z}} & \sigma_{YY\cdot\mathbf{Z}} \end{pmatrix} = \begin{pmatrix} \sigma_{XX} & \sigma_{XY} \\ \sigma_{YX} & \sigma_{YY} \end{pmatrix} - \begin{pmatrix} \boldsymbol{\Sigma}_{X\mathbf{Z}} \\ \boldsymbol{\Sigma}_{Y\mathbf{Z}} \end{pmatrix} \boldsymbol{\Sigma}_{\mathbf{ZZ}}^{-1} \left(\boldsymbol{\Sigma}'_{X\mathbf{Z}} \; \boldsymbol{\Sigma}'_{Y\mathbf{Z}}\right).$$

The partial correlation coefficient between $X$ and $Y$ eliminating the effect of $\mathbf{Z}$ is defined by the correlation between the prediction errors of $X$ and $Y$ given $\mathbf{Z}$, determined from $\Sigma_{XY\cdot\mathbf{Z}}$:

$$\rho_{XY\cdot\mathbf{Z}} = \frac{\sigma_{XY\cdot\mathbf{Z}}}{\sqrt{\sigma_{XX\cdot\mathbf{Z}}}\sqrt{\sigma_{YY\cdot\mathbf{Z}}}}.$$

The partial correlation coefficient shown in Definition 16 can be estimated using the sample covariance matrices. In the case of the variables having a joint multivariate normal distributed, that sample partial correlation coefficient is the maximum-likelihood estimator.

It has also been shown that zero partial correlation and conditional independence are equivalent only in Gaussian distribution [6]. However, the d-separation oracle does not necessarily need to be a statistical test for conditional independence. It can be any statistical constraint that provides the d-separability relations in a graph.

In the following, we will show some conditions linking zero partial correlation with d-separation, without any normality assumption. The main assumption is linearity of the relations among the variables.

It is noteworthy that randomization can provide the basis for making inferences without assuming a particular distribution [81]. Thus, randomization-based hypothesis tests can be used within Mendelian randomization approach when the assumption of normality is not met.

To state precisely the conditions that must be satisfied, let us first introduce some notation. Consider $L$ a *linear* SEM with jointly independent error terms and $G$ the directed graph corresponding to $L$. Note that $L$ can be a recursive or non-recursive SEM. Thus, it can be associated with a directed acyclic or cyclic graph $G$. Consider also the notation $(X \perp\!\!\!\perp Y|Z)_L$ to say that $X$ is independent of $Y$ given $Z$ in $L$, the notation $(X \perp\!\!\!\perp Y|Z)_G$ to say that $X$ is d-separated of $Y$ in $G$, and the notation $\rho_{XY \cdot Z}$ for the partial correlation of $X$ and $Y$ given $Z$.

**Definition 17.** Let $\mathbf{X}$, $\mathbf{Y}$, and $\mathbf{Z}$ be disjoint sets of random variables.

We say that $L$ **linearly entails** the conditional independence relation between $\mathbf{X}$ and $\mathbf{Y}$ given $\mathbf{Z}$ (i.e., $(\mathbf{X} \perp\!\!\!\perp \mathbf{Y}|\mathbf{Z})_L$) when $(\mathbf{X} \perp\!\!\!\perp \mathbf{Y}|\mathbf{Z})_L$ for all values of the non-zero linear coefficients and all distributions of the exogenous variables in which they are jointly independent and have positive variances.

**Theorem 5.** *Using the same notation of the definition 17,* $(\mathbf{X} \perp\!\!\!\perp \mathbf{Y}|\mathbf{Z})_G$ *if and only if* $L$ *linearly entails that* $(\mathbf{X} \perp\!\!\!\perp \mathbf{Y}|\mathbf{Z})_L$.

Even if the relations between the variables are *non-linear*, d-separation is a necessary condition for a conditional independence claim to be entailed by a recursive or non-recursive SEM. Thus, whenever two variables are conditionally independent according to an SEM, they can be d-separated in the associated directed graph. However, d-separation is a sufficient condition for conditional independence only in recursive SEMs and linear non-recursive SEMs. There are non-linear non-recursive SEMs in which a d-separation relation exists in the naturally associated graph, but their conditional independence is not entailed by the model. Peter Spirtes showed that in a modified graphical representation of the SEM, called collapsed graph, d-separation statements imply conditional independence relations [117].

In the following, we show some concepts linking d-separation and partial correlation in a linear SEM.

**Definition 18.** Let $\mathbf{Z}$ be a set of random variables and $X$ and $Y$ be random variables such that $X \neq Y$ and $X$ and $Y$ are not in $\mathbf{Z}$.

We say that $L$ **linearly entails** the zero partial correlation between $X$ and $Y$ given $\mathbf{Z}$ (i.e., $\rho_{XY \cdot \mathbf{Z}} = 0$) when $\rho_{XY \cdot \mathbf{Z}} = 0$ for all values of the non-zero linear coefficients and all distribution of the exogenous variables in which each pair of exogenous variables has zero correlation, each exogenous variable has positive variance, and in which $\rho_{XY \cdot \mathbf{Z}}$ is defined.

**Theorem 6.** *Using the same notation of the Definition 18,* $(X \perp\!\!\!\perp Y|\mathbf{Z})_G$ *if and only if* $L$ *linearly entails that* $\rho_{XY.\mathbf{Z}} = 0$.

By the Theorem 6, when the model is linear with jointly independent errors, partial correlation marks d-separability. In other words, the partial correlations identified by applying d-separation criterion in acyclic or cyclic graphs are guaranteed to vanish [116–118]. Thus, in this case, we can use a statistical test for zero partial correlation as d-separation oracle. Actually, under these assumptions, tests for any statistic that vanishes when partial correlations vanish would suffice [118].

In DAGs, it is possible to test a small number of partial correlations that constitute a basis for the entire set. A possible basis is the one which reflects the local directed Markov property of DAGs, i.e., the set of zero partial correlations between each variable and its predecessors (non-parental non-descendant variables) given its parents [87]. The cardinality of the basis is equal to the number of missing edges in the graph. Thus, the sparser the graph, more tests are required to reconstruct the structure. The PC-algorithm [118] is a fundamental algorithm to recover DAG structures. It runs in the worst case in exponential time with respect to the number of vertices, but if the true underlying DAG is sparse this reduces to a polynomial runtime.

In linear cyclic models, it is also possible to discover features of the graph performing statistical tests of zero partial correlation in a subset of the set of all d-separation relations. The CCD algorithm [97] is a discovery algorithm for linear cyclic models that contain no latent variables. It can infer features of sparse directed graphs from a probability distribution in polynomial time.

# 5 Equivalent Models

It may happen that two or more causal models or structures share the same conditional independence relations. For instance, a chain ($A \rightarrow B \rightarrow C$), a reverse chain ($A \leftarrow B \leftarrow C$), and a fork ($A \leftarrow B \rightarrow C$) share the following set of conditional and unconditional independence relations:

$$I = \{A \perp\!\!\!\perp C \mid B;\ A \not\perp\!\!\!\perp B;\ B \not\perp\!\!\!\perp C;\ A \not\perp\!\!\!\perp C\}.$$

In this case, the same probability distribution satisfies the global directed Markov and the causal faithfulness conditions for all these graphs. We say that these three models are members of the same equivalence class. In this case, it is not possible to distinguish one from another without any other information or assumption. In other words, they are indistinguishable by observational data alone.

These concepts are formalized in Definitions 19, 20, and 21:

**Definition 19.** Let $S_1$ and $S_2$ be two different SEMs.

We say that $S_1$ and $S_2$ are **observationally equivalent** if every probability distribution that is generated by one of the models can also be generated by the other.

**Definition 20.** Let $G_1$ and $G_2$ be two directed cyclic or acyclic graphs.

We say that $G_1$ and $G_2$ are **Markov equivalent** or **faithful indistinguishable** if any probability distribution $P$ which satisfies the global directed Markov and faithful conditions with respect to $G_1$ also satisfies these conditions with respect to $G_2$, and vice-versa.

Since the global directed Markov and causal faithfulness conditions only places conditional independence constraints on distributions, the following equivalent definition can be established:

**Definition 21.** Let $G_1$ and $G_2$ be two directed cyclic or acyclic graphs.

We say that $G_1$ and $G_2$ are **Markov equivalent** or **faithful indistinguishable** if the same d-separation relations hold in both graphs, or, equivalently, if they both linearly entail the same set of conditional independencies.

The Theorem 7 is an important result that holds only for DAGs:

**Theorem 7.** *Let $G_1$ and $G_2$ be two DAGs.*
*The two DAGs are Markov equivalent if and only if they have the same skeleton (the undirected version of the graph) and the same unshielded colliders.*

We can verify that the two first graphs in Fig. 4 are equivalent using the Theorem 7. Both graphs have acyclic structures, the same skeleton and the same set of unshielded colliders (empty set in this case). Note that $A$ and $B$ are connected in both graphs. Thus, neither the collider $C$ of the first graph nor the collider $B$ of the second graph is an unshielded collider. The third graph in Fig. 4 shows that the acyclicity hypothesis is important for the Theorem 7. Even though the third graph does not have the same skeleton and the same set of unshielded colliders than the first two graphs ($B$ is an unshielded collider in the third graph), these three graphs are Markov equivalent because they hold no d-separation relations [95].

A more complex theory has been developed by Thomas Richardson and Peter Spirtes [95–97, 117] to completely characterize cyclic Markov equivalence classes. We will not discuss these results in this chapter. The more interested reader may refer to [96, 97].

Logsdon et al. [74] demonstrate some results that characterize the set of perturbations (e.g., driving QTLs or QTNs) that minimizes the equivalence classes. Moreover, the authors demonstrate an important theorem, namely "Recovery Theorem," describing how the set of equivalent DCGs can be recovered from the corresponding moralized graph. As mentioned by Logsdon et al. [74], their results can also be proven by using Thomas Richardson's work [94].

As a result of the Recovery Theorem, it is possible to guarantee identifiability of both cyclic and acyclic models when each vertex (e.g., phenotype) has at least one unique perturbation associated (e.g., a QTL or QTN) and the genetic architecture is known. That result generalizes the assumption made by Chaibub Neto et al. in the QDG algorithm [24]. By providing a genetic mapping where every phenotype is associated with an unique genetic variant, a directed acyclic or cyclic network can be uniquely recovered [22, 74].

In studies of genetic associations, Mendelian randomization can reduce the size of equivalence classes of phenotypes by using driving genetic variants. This is

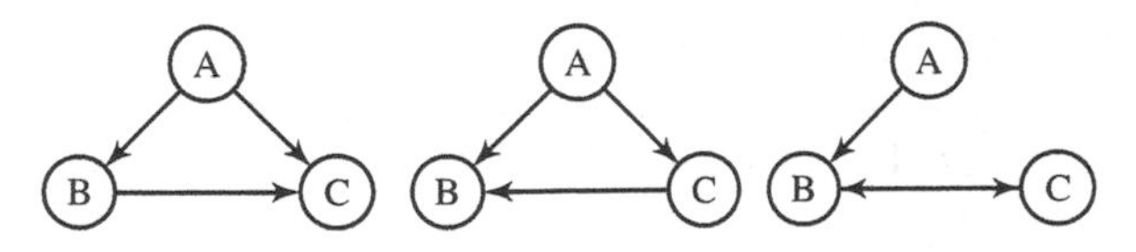


**Fig. 4** Two DAGs and one DCG which are Markov equivalent. Every pair of vertices in these graphs cannot be d-separated

because the additional information of the genetic variants causing phenotypes can create new conditional independence relationships among the vertices, getting rid of some Markov-equivalence. Background knowledge can also be used to rule out some equivalent graphs. However, these approaches may not always narrow the set of possibilities to a single graph.

# 6 Causal Structure Learning

Inferring the causal effects may be a problem of estimation of a SEM if the causal structure is known. For a genotype–phenotype network, the structure is specified indicating the non-zero entries of the matrices $\mathbf{P}$ and $\mathbf{Q}$ of the model shown in Eq. (2). The values of these entries can be estimated using an appropriate method of SEM estimation.

When the causal structure is not known a priori, we can use an algorithm to recover or discover it. This problem is known as *causal structure learning* or *causal structure discovery*.

In Sect. 5, we showed that causal structures can be statistically indistinguishable from each other. This means we cannot distinguish equivalent models based on observational data. The use of QTL or QTN genotype information can break statistical equivalence. However, it is still possible that the true causal structure cannot be uniquely recovered. When the exact identifiability is not possible, the algorithms often report suitable summary statistics of all graphs within the equivalence class.

There are two main approaches to infer the network structure that is more compatible with the joint distribution of the data: (1) an SEM-based approach, in which the structure is determined by fitting an SEM with regularization and variable selection techniques for identifying relevant associations; and (2) an approach based on the graphical representation of the causal processes, in which d-separation criterion and greedy strategies are often used to determine an adequate structure network.

## 6.1 *Learning Structural Equation Models*

In SEM-based approaches, the network structure learning problem is reduced to the estimation of the model parameters, often using a regularizer that controls the model complexity. The regularizers have a key role in the learning of the network structure because of the variable selection effect. Additional constraints, such as sparsity and smoothness, are incorporated in the likelihood function of an SEM model. Structure learning algorithms based on SEM infer both the causal structure and the parameters of the model.

Most methods for learning genotype–phenotype networks adopt the model shown in Eq. (2) (or a slightly modified version) in which genetic variants are incorporated. The network structure is specified by the non-zero entries of the matrices **P** (causal relations among phenotypes) and **Q** (causal relations from genotypes to phenotypes). Thus, the causal structure learning problem is equivalent to estimate which entries of these matrices are non-zero. Some methodologies require more assumptions than others for estimating these parameters. The most common assumptions are listed in the following:

- The response variables (phenotypes) are assumed to follow a multivariate normal distribution;
- There are no self-loops, so that all diagonal entries of **P** are zero; Self-loops are often used in modeling of time series, implying that a variable depends on its own path. In this case, the model is static, and the state of a phenotype at any instant of time $t$ is assumed to be not dependent on its state at the past time step $t-1$;
- The error term corresponding to the $j$th individual (column $j$ of the matrix **E**) is modeled as a zero-mean Gaussian vector with covariance $\sigma^2\mathbf{I}$, where **I** denotes the $p \times p$ identity matrix. Thus, the error terms are assumed to be uncorrelated;
- No dominance or epistatic effects are considered in the model. It is assumed that all effects of genetic variants are due to the additive effects of alleles;
- The $q$ QTLs have been predetermined by an existing method, but the magnitude of their effects is unknown.

The most popular regularization methods are those that lead to sparse variable selection. A main reason is that the reduction in the effective number of parameters to be estimated (in sparse models, many parameters are expected to be zero) reduces computational cost, and contributes to the selection of relevant features. Sparsity is often achieved by imposing the $L1$-norm (i.e., the sum of absolute values) on the parameters as a regularization. Examples of sparse regularizers include least absolute shrinkage and selection operator (LASSO) [122] and its extensions, such as adaptive LASSO [136] and Dantzig selector [23]. A limitation of the LASSO is that it tends to select only one variable from a group of highly correlated variables. To overcome this limitation, some regularizers, such as elastic net [137] and its adaptive version [138], combine $L1$-norm with the $L2$-norm (i.e., the sum of squared absolute values) regularization.

In the field of genotype–phenotype network learning, the use of the sparsity constraint is particularly attractive for inferring the structure of GRN. Several studies indicate that biological gene networks, including protein–protein interaction, metabolic, signalling, and transcription-regulatory networks, contain few highly connected vertices (also know as hubs) [1, 10, 58, 69, 93, 124]. Thus, by exploiting the sparsity prior information, it may be possible to both improve computational efficiency and achieve a biologically realistic representation. The Adaptive LASSO (AL)-based algorithm [74], for instance, was designed for determining regulatory relationships underlying observed gene expressions by using the adaptive LASSO procedure for feature selection. Another algorithm intended to infer the structure of GRNs is the SML algorithm, proposed by Cai et al. [22]. The SML algorithm infers sparse SEMs in which an $L1$-norm penalty is incorporated on the entries of the matrix **P** of Eq. (2), inducing a sparsity constraint. Later, Anhui Huang

[54] extended the SML algorithm by incorporating the adaptive elastic net penalty into the SEM likelihood. Other SEM-based structure learning algorithms deserving attention are [32, 71, 73].

## 6.2 Learning Causal Graphical Models

Causal structure learning algorithms which are based on the graphical representation of the causal processes often use the d-separability concept. Thus, they often assume the global Markov and causal faithfulness conditions. These algorithms search for the structure more compatible with the joint distribution of the data. One of the approaches used to reduce complexity of the algorithm is to constrain the search space by imposing specific properties to the structure of the graphs, such as acyclicity and sparsity.

The most common assumptions often made by structure learning algorithms based on graphical representation are

- **Causal sufficiency**: there are no hidden confounders, that is, all common causes of the underlying causal system have been observed and the error variables are jointly independent. That is a problematic assumption, since it is difficult to be confirmed and, in general, depends on factual knowledge. Within Mendelian randomization framework, the improvement of causal inference by using genetic variants as instrumental variables may compensate for biases introduced by small departures from causal sufficiency. Thus, identification of the true network structure may still be achieved even when causal sufficiency condition is not perfectly fulfilled [40].
- **Causal Markov condition**: the distribution generated by a causal structure (represented by a directed graph) satisfies the global directed Markov condition. It permits inference from probabilistic dependence to causal connection. Note that, for linear SEMs, this assumption holds if the error terms are independent.
- **Causal faithfulness**: all conditional independence relations present in a directed graph $G$ are consequences of the global directed Markov condition applied to the true causal structure $G$. This is an assumption that any conditional independence relation holding in $G$ is due to the causal structure rather than a particular parameterization of the model. Thus, it permits inference from probabilistic independence to causal separation.

In order to learn causal graphical models, three approaches are often used: constraint-based approaches, score-based approaches, and hybrid approaches, where techniques from constraint-based and score-based approaches are combined. In the following, we will discuss the main ideas used in constraint-based and score-based approaches.

### 6.2.1 Constraint-Based Approaches

The algorithms in this category are based on significance tests for the null hypothesis that a certain conditional independence statement is true. These individual constraints are used both to decide if a given pair of variables is adjacent or not as well as to orient some edges. In order to read off the implied conditional independencies, d-separation criterion is often used. Thus, the causal sufficiency, causal Markov, and causal faithfulness assumptions must be made in order to safely apply these conditional independence tests.

The most basic causal discovery method is the SGS (Spirtes-Glymour-Scheines) algorithm. The correctness of the SGS algorithm follows from the Theorem 2, which is stated only for DAGs. Thus, acyclicity is assumed.

The SGS algorithm works similarly to the exercise shown in Sect. 4.1, in which the graph of Fig. 2 is reconstructed by using d-separation criterion. The algorithm starts with a complete undirected graph. Then, the skeleton of the graph is inferred: for each pair of variables, it tests whether they are conditionally independent on any set of variables. If so, then the edge connecting the pair can be removed. The reason is that, if the dependence between two vertices can be explained away, then there cannot be a direct causal connection between them. In the next step, the algorithm finds and then orients the edges of the unshielded colliders. The unshielded collider is the only configuration for three vertices and one missing edge that can be uniquely oriented. The orientation of other edges are determined by consistency. That is recursively made until no more edges can be oriented. In this last step, the algorithm checks if any loop is created.

The SGS algorithm is statistically consistent, but it is computationally inefficient. In the edge-removal step, each pair of variables should be conditioned on all possible subsets of the remaining variables. Thus, the number of tests it does grows exponentially in the number of variables.

The PC (Peter and Clark) algorithm is very similar to the SGS algorithm, but it is more efficient, specially for sparse graphs. In the edge-removal step, it tries to condition on as few variables as possible. It only conditions on adjacent variables and the sets are sorted in order of increasing size. The PC algorithm has the same assumptions as the SGS algorithm, and the same consistency properties.

The first step of the PC-algorithm, where an association (undirected) graph is inferred, is called PC-skeleton.

The PC-skeleton is used in the first step of two popular genotype–phenotype structure learning algorithms: the QDG, proposed by Chaibub Neto et al. [24], and the QPSO algorithms, by Huange Wang and Fred van Eeuwijk [126]. Technically, these two algorithms are not constraint-based approaches because they use a score-based approach to orient the edges. Thus, they are considered *hybrid approaches*. Score-based approaches are described in the following section.

### 6.2.2 Score-Based Approaches

Given a score that indicates how well the network fits the data, score-based algorithms search the space of all possible structures for the network with the highest score.

The search for the global optimal network is an NP-hard problem. Thus, the time required to solve the problem increases very quickly as the number of vertices grows. By the solution of the enumeration problem of labeled directed graphs (i.e., graphs in which each vertex has been assigned a different label, so that all vertices are considered distinct), it is known that there are $2^{n(n-1)}$ different causal structures (directed graphs) with $n$ vertices [103]. In asymptotic notation, there are $2^{O(n^2)}$ structures. Just to give an idea of how this number increases, the number of directed graphs with $n$ labeled vertices, for $n$ varying from 1 to 6, is $1, 4, 64, 4096, 1\,048\,576$ and $1\,073\,741\,824$.

Thus, heuristics such as greedy search are used to find a sub-optimal structure. However, local optimal solutions can be far away from the global optimal solutions. This becomes even more critical when the number of sampled configurations is small compared to the number of vertices.

The simplest search algorithm over the structure space is the greedy hill-climbing search. A series of modifications of the local structures are made by adding, removing, or reversing an edge, and the score of the new structure is computed after each modification. The search ends when there are no more modifications that increase the score.

The scores offer model selection criteria for the network structure. There is no consensus on what is the best criterion, since that depends on the objective that one wants to achieve [56]. One of the most used measure is the Bayesian information criterion (BIC) [109], which penalizes complex models and the penalty increases with the sample size. It is an approximation for the posterior predictive distribution with respect to the model parameters. The posterior probability of a structural feature (e.g., the presence of an edge) is the total posterior probability of all models that contain it. By estimating the posterior probability of a feature, we are estimating the strength with which the data indicates the presence of it.

Score-based approaches often assume acyclicity, because every DAG has a topological ordering, that is, an ordering of the vertices as $V_1, \ldots, V_n$ so that for every edge $(V_i, V_j)$ we have $i < j$. In this case, each vertex $V_i$ can have parents only from the set $\{V_1, \ldots, V_{i-1}\}$. That significantly reduces the search space and has implications that can reduce the computational cost of the whole process [121].

Robert W. Robinson derived a recurrence relation to count how many labeled DAGs have $n$ vertices [100]. By applying the recurrence relation for $n$ varying from 1 to 6, we notice that corresponding number of labeled DAGs is $1, 3, 25, 543, 29\,281$, and $3\,781\,503$. Using asymptotic notation there are $2^{O(n \log n)}$ orderings, as opposed to the $2^{O(n^2)}$ structures. Thus, the space of orders is smaller and more regular than the space of structures.

In Bayesian approaches, it is also assumed acyclicity for estimating the probability of a structural feature over the set of all orderings. That is often performed by using a Markov chain Monte Carlo (MCMC) algorithm. It was noted by empirical studies [110] that different runs of MCMC over the structure space typically lead to very different estimates in the posterior probabilities. That poor convergence to the stationary distribution has not been found running MCMC over ordering space.

Two recent methodologies using MCMC to jointly infer the causal structure of a genotype–phenotype network are the QTLnet algorithm, proposed by Chaibub Neto et al. [25], and the Bayesian framework for inference of the genotype–phenotype map for segregating populations, proposed by Hageman et al. [46].

## 7 Algorithms for Causal Discovery in Genetic Systems

In this section, we will describe some of the most popular algorithms to infer the structure of a genotype–phenotype network, namely QDG [24], QPSO [126], QTLnet [25], and SML [22].

We have tested all proposed features by these algorithms in a simulation study. Some simulated networks are shown in Fig. 5. The letters $A$, $B$, $C$, and $D$ were used to identify phenotypes. In all other networks, there is a distinct genetic variant associated with each phenotype. These genetic variants are identified by the letter $M$ followed by the associated phenotype letter in subscript. We simulated genotype and phenotype data for 500 individuals, independently. QTL genotypes were generated from an F2 intercross using the R/QTL package [18], so that QTLs of each simulated network are unlinked and in linkage equilibrium. The phenotypes were generated according to Eq. (2), with error terms following a normal distribution with zero mean and variance 0.01.

It is worth mentioning that the source codes of the four algorithms are freely available. Details about how to obtain the source code of each algorithms are provided in the following sections.

We want to emphasize that the purpose of our simulation studies is only to check the capabilities of the algorithms. Although it would be very interesting to do a comparison study of the algorithms, it is out of the scope of this chapter. By running simulations under different configurations of phenotype networks, we investigated the advantages and limitations of the algorithms, and we noted that they differ mainly in their ability to discover networks with the following properties: genetic variants with pleiotropic effects, phenotypes associated with multiple genetic variants, acyclic structure, feedback loops, and reciprocal associations. In Sects. 7.1–7.4, we show how these issues are addressed. We conclude this section providing a summary of the main features implemented by each of the algorithms in Table 1.

The QDG and QPSO algorithms are closely related. Both are designed to orient edges into a phenotype association network. In other words, the QDG and QPSO algorithms focus on discovering the causal direction among variables which are known to be statistically associated. However, they can only achieve this goal if genetic variants robustly associated with the phenotypes are previously selected.

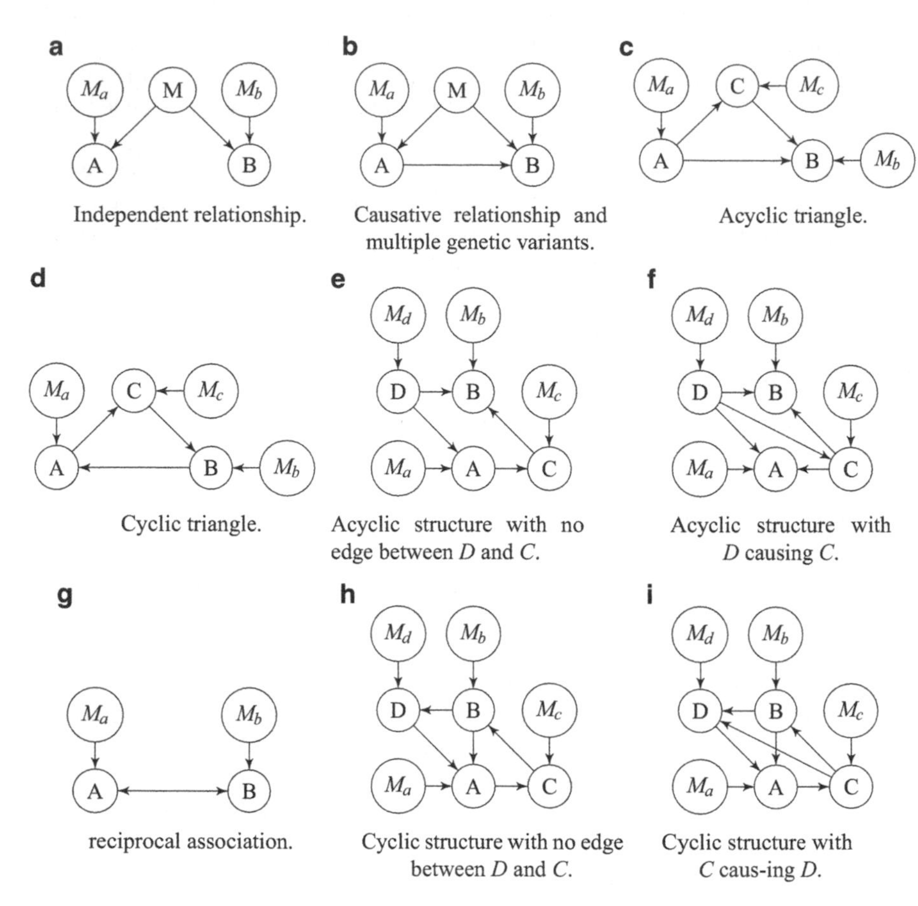


**Fig. 5** Causal networks, in which $A$, $B$, $C$, and $D$ are phenotypic vertices. Each is associated with a distinct genetic variant, $M_a$, $M_b$, $M_c$, and $M_d$, respectively. (**a**) Independent relationship. (**b**) Causative relationship and multiple genetic variants. (**c**) Acyclic triangle. (**d**) Cyclic triangle. (**e**) Acyclic structure with no edge between $D$ and $C$. (**f**) Acyclic structure with $D$ causing $C$. (**g**) Reciprocal association. (**h**) Cyclic structure with no edge between $D$ and $C$. (**i**) Cyclic structure with $C$ causing $D$

Thus, besides the phenotype association network, it is also necessary to provide a genetic map as input to these algorithms.

The association network required by both QDG and QPSO algorithms is an undirected graph among phenotypes constructed in such a way that an edge does not exist between two phenotypes if they are d-separated. The main method to infer an association network is the PC-skeleton algorithm [118], discussed in Sect. 6.2.1. Another recommended method to infer association networks is the undirected dependency graph (UDG) [24, 29].

There are several genetic mapping approaches in the literature, most of them using microsatellite or SNP markers.

Linkage mapping is the conventional genetic mapping technique, mostly used for QTLs mapping using microsatellite markers in inbred populations of plants and animals (e.g., F2, backcross, and recombinant inbred lines), and for family based QTL mapping in humans and natural populations [2]. In the simple interval mapping approach [66], the QTL position is assumed to be in a region flanked by two linked observed markers. The Haley–Knott [47] approach gives a remarkably good approximation of the interval mapping, and it is computationally more efficient. These methods model a single QTL at a time and assume that the phenotypes follow a normal distribution or a mixture of normal distributions [135]. When it is necessary to model multiple QTLs at once, composite interval mapping (CIM) [57, 131] and multiple interval mapping (MIM) [61] can be used. In order to estimate the specific position of a QTN (sometimes within a QTL), fine mapping techniques are performed. Linkage disequilibrium and/or SNP information, as well as imputation in sparse marker panels, are currently used for improving fine mapping [43, 52, 125].

The identification of QTNs that are robustly associated with a particular phenotype is mainly done by genome-wide association (GWA) in the case of complex traits and diseases [3] and increasingly by next-generation sequencing in the case of Mendelian traits [30]. In particular, exome sequencing has been a powerful approach for identifying rare variants that underlie Mendelian disorders in circumstances in which conventional approaches have failed [7]. These studies provide much higher resolution than linkage mapping and often involve studying human population.

The QDG algorithm can discover acyclic and cyclic structures, while the QPSO algorithm can only discover acyclic structures. Reciprocal associations cannot be detected by any of these two algorithms.

Unlike the QDG and QPSO, the QTLnet algorithm [25] jointly infers the phenotype network and the associated genetic architecture. Thus, it is not necessary to provide as input to this algorithm a phenotype association network and a genetic map. It assumes that the structure is acyclic and that the phenotypes are normally distributed.

These three aforementioned methods are intended to infer network structures based on concepts of graphical models. The size of the causal effects is not estimated. If it is also necessary to estimate the magnitude of the causal effects, once the model structure is inferred, it may be done by estimating the parameters of the corresponding SEM.

Some SEM-based structure learning algorithms can infer both the phenotype network structure and the magnitude of the causal effects. An example is the SML [22] algorithm which will also be discussed in this section.

As the QDG and QPSO algorithms, the SML algorithm assumes that the genetic variants (e.g., QTLs or QTNs) directly associated with the phenotypes (and preferably not violating the instrumental variable assumptions), must be known a priori and provided as input to the algorithm. It can discover acyclic and cyclic structures, even when reciprocal associations are present.

The output of these algorithms is the most likely structure network according to the data. Except for the SML method, these network reconstruction algorithms also report a goodness-of-fit statistic indicating how well the model fits the data. In particular, this statistic used is the BIC score. The causal model with the lowest BIC score is determined as the best solution.

The QDG and QTLnet algorithms report the BIC score of each fitted model and the score comparing the possible directions of each edge. Thus, the output contains information not only on the best solution, but also on all fitted models. By comparing information on different solutions, one can decide whether there are Markov equivalent networks (possibly models with similar BIC scores) and spurious edges.

In the next sections, each of these four algorithms will be described in more detail.

## *7.1 QTL-Directed Dependency Graph Algorithm*

The QDG algorithm was proposed by Chaibub Neto et al. [24]. The goal of this algorithm is to infer causal directions into a phenotype association network. That is achieved by using information of genetic variants (e.g., QTLs) associated with each phenotype. The algorithm is designed to discover causal structures that can contain cycles but not reciprocal associations.

Both a phenotype association network and a genetic map must be provided as input to the algorithm.

The phenotype association network can be an UDG [29] or the graph skeleton built in the first step of the PC (Peter-Clark) [118] algorithm. Both options are implemented in the QDG algorithm.

In the genetic mapping (e.g., QTL mapping), each phenotype must be associated with at least one unique genetic variant. However, the algorithm can deal with both multiple and pleiotropic genetic variants.

Since the genetic mapping and the determination of the phenotype association network are independently performed, spurious edges and indirectly associated genetic variants can be removed when both are put together. In a pre-processing step of the QDG algorithm, tests are conducted to remove associations identified due to only a pleiotropic effect. In this case, the algorithm verifies whether an association between two phenotypes vanishes by conditioning to the common genetic variant. If so, the edge connecting them is removed.

The orientation edge step is a score-based procedure, similar to those described in Sect. 6.2.2. We will define the logarithm of odds (LOD) score that is used to compare the two possible orientations of each edge.

Consider $A$ and $B$ a pair of connected phenotypes, each with a non-empty set of associated genetic variants, $M_a$ and $M_b$, respectively. Let $pa(A)$ and $pa(B)$ represent the set of phenotypes that are parents of $A$ and $B$, respectively. Also, let $f$ be the predictive density, that is, the sampling model with parameters replaced by the corresponding maximum-likelihood estimates. We use a subscript $i$ to represent the phenotype or genotype values of each individual.

The LOD score corresponding to the direction $A \to B$ is given by:

$$LOD_{A\to B} = LOD_A + LOD_{B|A} \tag{6}$$
$$= \log_{10}\left\{\prod_{i=1}^{n}\frac{f(A_i|M_{a_i}, pa_i(A))}{f(A_i)}\right\} + \log_{10}\left\{\prod_{i=1}^{n}\frac{f(B_i|A_i, M_{b_i}, pa_i(B))}{f(B_i)}\right\}.$$

The LOD score corresponding to the direction $B \to A$ is given by the formula (6) changing the roles of $A$ and $B$.

The LOD score is defined as $LOD = LOD_{A\to B} - LOD_{B\to A}$. If it is positive, then the edge is oriented in favor of the direction $A \to B$. Otherwise, it is oriented in the opposite direction. The genetic variants associated with the phenotypes play an important role in this step. As discussed in Sect. 5, they can break likelihood equivalences by creating new conditional independence relationships.

The edges are oriented in a greedy strategy. For each edge, it is computed the LOD score, and the chosen direction is the one with the higher likelihood. This process follows a randomly chosen ordering of the edges. Whenever the orientation of an edge changes, the graph is updated before moving to the next edge.

Different solutions can be obtained by running the algorithm from a different edge ordering. Thus, it is recommended to rerun the algorithm using different edge orderings to get all possible solutions. If more than one solution is obtained in this process, the graph with the lowest BIC score is selected as the best solution.

The output of the algorithm is a detailed report of all possible solutions. It contains the BIC score of each model and the LOD score of each edge. That information can be useful when deciding between equivalent models and whether a certain edge is spurious. For instance, an LOD score very close to zero means that there is not a strong evidence for any direction. In other words, that edge may be spurious.

In Fig. 5a, phenotypes $A$ and $B$ are correlated only due to pleiotropic effect of the common causal genetic variant $M$. The QDG algorithm could not remove the edge connecting $A$ and $B$ in our simulations for this network, but its LOD score was very close to zero (−0.04).

In the simulation studies reported by the Chaibub Neto et al. [24], the QDG algorithm could infer correctly feedback loops, but not reciprocal associations. When there is a reciprocal association, Chaibub Neto et al. noted that the direction detected corresponds to the one with highest regression coefficient. These observations are consistent with the results obtained in our simulations, since the QDG algorithm failed in recovering the network shown in Fig. 5g.

In our simulations, the QDG algorithm was able to precisely infer simple acyclic and cyclic networks, including those shown in Fig. 5b, c, d, e, h.

However, the QDG algorithm only performs well if the association network is inferred correctly. For instance, the PC-skeleton algorithm was successful in inferring the skeleton of the graphs in Fig. 5e, h, but it failed in recovering a bit more complex version of these networks, such as those shown in Fig. 5f, i. Consequently, the QDG also failed to recover these structures.

The QDG algorithm is implemented in the R/QTLnet package [82], available at https://cran.r-project.org/web/packages/qtlnet/. For reference, the QTLnet package version we used was 1.3.6, published in March, 2014. We run the tests using R environment, version 3.2.3 [92].

## 7.2 QTL+Phenotype Supervised Orientation Algorithm

The QPSO algorithm was proposed by Wang and van Eeuwijk [126]. It was designed to infer causal connections between pairs of phenotypes from an association network.

Likewise the QDG algorithm described in Sect. 7.1, a phenotype association network and a genetic map must be determined a priori. The genetic map can contain both multiple and pleiotropic genetic variants causally associated with phenotypes. However, the QDG and the QPSO algorithms differ in the sense that the QPSO algorithm does not assume that every phenotype has at least a unique causal genetic variant.

In order to orient the edges between pairs of phenotypes, two important assumptions are made: phenotypes follow a Gaussian distribution and the structure is locally acyclic. Because of the acyclicity assumption for local networks, the algorithm may not perform well in detecting cyclic structures. Moreover, it cannot detect reciprocal associations.

The edge orientation step is a score-based procedure, similar to those shown in Sect. 6.2.2. In each step of a heuristic search, the algorithm chooses a pair of connected phenotypes and extracts its local network. This local network is denoted by local generalized phenotype network (LGPN) and consists of the two connected phenotypes, their parents (both genetic variants and phenotypes) and other phenotypes connected by undirected edges. It is assumed that the LGPN is a conditional linear Gaussian model, in which discrete variables (QTLs or QTNs) are not allowed to have continuous parents (phenotypes).

This local network is thoroughly investigated by using the log-likelihood score which will be defined in the following.

Let $A$ and $B$ be two connected phenotypes. Denote by $pa(A)$ and $pa(B)$ the set of parent vertices of $A$ and $B$, respectively, including genetic variants and other phenotypes. Let $f$ be the probability density function with parameters replaced by the corresponding maximum-likelihood estimates. A subscript $i$ is used to indicate values for the $i$th individual.

The log-likelihood score of the local structure is given by:

$$\sum_{i=1}^{n} \log_{10}(f(A_i|pa_i(A))\, f(B_i|pa_i(B))).$$

Using this score, both genetic variants and phenotypes identified as parent vertices can break Markov equivalence among phenotype networks.

Under the assumption that the local structure is acyclic, Wang and van Eeuwijk [126] showed a result that allows all undirected edges of the LGPN be oriented simultaneously.

According to Theorem 7, if two DAGs have different sets of unshielded colliders, then they are not Markov equivalent. Wang and van Eeuwijk [126] showed that all candidate DAGs derived from an LGPN have a distinct set of unshielded colliders if two conditions are satisfied: (1) the pair of connected phenotypes must have at least one parent vertex, and (2) each phenotype connected to the pair by an undirected edge must be nonadjacent to at least one of the parents of the pair's phenotype to which it is connected. Thus, the problem is identifiable under acyclicity assumption and the two aforementioned conditions.

The phenotype network is inferred in a greedy strategy. For each pair of phenotypes satisfying the two identifiability conditions, it is obtained the log-likelihood score of each possible configuration of the respective local network. If there are $k$ undirected edges in the LGPN, then there are $2^k$ directed graphs to be tested. The configuration with the highest log-likelihood score is considered as the locally optimal directed graph (LODG). This process can be computationally very expensive, since the number of candidate directed graphs increases exponentially.

To prevent the algorithm from converging to a network that is a locally optimal solution, it is recommended that the edge orientation procedure is repeated several times from different starting points. The BIC score is used as a global evaluation metric to determine the most likely solution among those obtained in multiple runs.

Despite the acyclicity assumption in determining the LODG, it is possible that the algorithm builds a cyclic structure when combining the LODGs. However, since the algorithm was not designed with the purpose of discovering cyclic networks, the correct structure can be recovered only by chance.

The accuracy of the QPSO algorithm depends on the association network provided as input to it in a similar way to the QDG. If the association network is not the true skeleton of the network, then the algorithm will fail in recovering the true causal structure.

In our simulation studies, the QPSO algorithm could not recover the true structure of most networks we simulated. Out of the networks shown in Fig. 5, it correctly recover only the networks shown in Fig. 5b, c.

In addition, since the QPSO algorithm only provides information on the network that is solution of the problem, it is difficult to do further analysis in order to identify equivalent networks and spurious edges.

The QPSO algorithm is implemented in Matlab, and it is available upon request to the authors. All simulations were performed using Matlab 8.1.0.604 (R2013a).

## 7.3 QTL-Driven Phenotype Network Algorithm

The QTLnet algorithm was proposed by [25] to jointly infer causal relationships between genotypes and phenotypes. Thus, unlike the QDG and QPSO algorithms, it is not necessary to provide a phenotype association network and a genetic map as input to the QTLnet algorithm.

The algorithm infers a genotype–phenotype network by a Bayesian procedure under the assumptions that the phenotype network structure is acyclic and the phenotypes follow a normal distribution.

The phenotypes are modeled by a set of structural equations similar to the model shown in Eq. (2).

Let $\mathbf{Y}_i = (Y_{ti})_{t=1}^T$, for each $i = 1, \ldots, n$, be a vector with the measurements of $T$ phenotypes for the $i$th individual, and $\varepsilon_{ti}$ represent the corresponding independent normal error terms. Denote by $\mu_t$ the overall mean for the phenotype $t$. Consider a row vector $\mathbf{X}_{ti}$ with the QTL genotypes and observed values of other covariates associated with the phenotype $t$ for the individual $i$, and a column vector $\boldsymbol{\theta}_t$ with their linear (additive) effects. Thus, the genetic architecture is defined by the elements of $\boldsymbol{\theta}_t$. The effect of the phenotype $k$ on the phenotype $t$ is represented by the coefficient $\beta_{tk}$. The notation $pa(Y_t)$ represents the set of phenotypic parent vertices of $Y_t$.

Each phenotype $t$ of the $i$th individual is modeled by the following SEM denoted as homogeneous conditional Gaussian regression (HCGR) model:

$$Y_{ti} = \mu_t + \mathbf{X}_{ti}\boldsymbol{\theta}_t + \sum_{Y_k \in pa(Y_t)} \beta_{tk} Y_{ki} + \varepsilon_{ti}, \quad \varepsilon_{ti} \sim \mathcal{N}(0, \sigma_t^2). \tag{7}$$

Though the parametric family HCGR can accommodate cyclic and acyclic networks, only DAGs can be recovered by using the QTLnet algorithm.

Let $\mathcal{M}$ represent a specified network structure and $\Gamma$ represent all parameters of the model. Also, let $\mathbf{q}_i = \{\mathbf{q}_{1i}, \ldots, \mathbf{q}_{Ti}\}$ be the QTL map of the $i$th individual, in which $\mathbf{q}_{ti}$, for $t = 1, \ldots, T$, represents the set of QTLs associated with the phenotype $t$. The likelihood of the HCGR model is equal to the probability of the observed phenotypes conditional to the QTL genotypes, with respect to the parameters $\mathcal{M}$ and $\Gamma$. Under the acyclicity assumption, the factorization property shown in Definition 15 holds. Thus, considering the individual likelihood functions

$$p(Y_{ti}|\mathbf{q}_{ti}, pa(Y_t)) \sim \mathcal{N}(\mu_t + \mathbf{X}_{ti}\boldsymbol{\theta}_t + \sum_{Y_k \in pa(Y_t)} \beta_{tk} Y_{ki}, \sigma_t^2),$$

we can write the likelihood of the HCGR model as a product of the all individual likelihood functions:

$$p(\mathbf{Y}_i|\mathbf{q}_i; \Gamma, \mathcal{M}) = \prod_t p(Y_{ti}|\mathbf{q}_{ti}, pa(Y_t)). \tag{8}$$

Thus, by using the DAG factorization property, the likelihood function can be written as a product of normal distributions, one for each value of the data. It means we can use the maximum-likelihood estimation (MLE) technique to estimate the parameters of the HCGR model, in the same way as the classical linear regression model.

By taking the product of the likelihood function and a prior density, up to a normalizing constant, we have the posterior probability distribution. The QTLnet algorithm estimates this posterior probability by using an MCMC algorithm. Thus, the QTLnet algorithm is similar to the Bayesian score-based approaches presented in Sect. 6.2.2.

Specifically, a modified Metropolis–Hastings algorithm was proposed to integrate the sampling of network structures and QTL mapping. It searches across the model space sampling from the derived posterior distribution. In each step of the search, it is proposed a single modification, such as an edge deletion, addition, or reversion, so that the resulting network does not contain cycles. In addition to this simple approach (described in the paper and initially implemented in the software), the R/QTLnet package now supports a more effective M-H sampler [45], which improves a lot the mixing of the Markov chain.

The algorithm does not consider the network with the highest posterior probability as solution of the problem. Instead, the solution is an average network constructed by putting together all causal relationships such that the posterior probability is maximum or above a predetermined threshold.

An advantage of the Bayesian approach is its ability to incorporate prior information in the analysis. In an extension of the QTLnet algorithm [78], it is possible to specify a prior density using, for instance, biological knowledge or sparsity to produce a more predictive network.

Since genotype and phenotype information are jointly analyzed by the QTLnet algorithm, common genetic variants are no longer hidden confounders, reducing the possibility of inferring networks with spurious edges. For instance, the QTLnet could test for the independence between the vertices $A$ and $B$, conditioned to the common QTL $M$, correctly inferring the structure of the network shown in Fig. 5a. Additionally, it precisely inferred all acyclic networks we simulated, including those shown in Fig. 5b, c, e, f. On the other hand, since QTLnet assumes acyclicity, it could not discover any cyclic structure.

The output of the algorithm contains the set of the posterior probabilities for each possible network structure and the averaged probabilities for each edge direction. Thoroughly analyzing this information, it is possible to identify equivalent networks. In addition, averaged probabilities close to 0.5 indicate suspicious directions and should be further investigated.

The QTLnet algorithm uses the R/QTL package [18] for performing a single-QTL genome scan, by using methods such as the Haley–Knott regression. This process may involve prediction or imputation of genotypes, requiring information on the experimental cross design. For this reason, the current implementation of the QTLnet is intended for studies in segregating populations. Thus, the source code must be adapted for studies in natural populations.

As the QDG algorithm, it is possible to use the QTLnet algorithm from the R/QTLnet package [82]. We run our simulations using the last version available of QTLnet package (version 1.3.6, published in March, 2014), and the R environment, version 3.2.3 [92].

## 7.4 Sparsity-Aware Maximum Likelihood Algorithm

The SML algorithm was proposed by Cai et al. [22]. It is an SEM-based approach to infer sparse SEMs integrating genotypic and phenotypic information.

The network is postulated to obey the SEM shown in Eq. (2). The only assumption placed on the phenotype structure network is that there is no self-loops. Thus, the algorithm can infer both acyclic and cyclic networks.

It is assumed that every phenotype is associated with at least one genetic variant. According to the Recovery Theorem [74] discussed in Sect. 5, that restriction guarantees that the network structure can be uniquely identified for both cyclic and acyclic models.

In the paper describing the SML algorithm [22], Cai et al. comment that genetic maps with both multiple and pleiotropic genetic variants affecting phenotypes are supported. However, we could not verify that feature by our simulation studies. That is not yet implemented in the current version of the algorithm (obtained in January 2016 from the supporting information in the online version of the article—doi: 10.1371/journal.pcbi.1003068.s008). In this version, it is only possible to provide genetic variants affecting one phenotype and every phenotype must be associated with only one distinct genetic variant. Considering the Eq. (2), the one-to-one correspondence is forced by placing constraints on the elements of the matrix $\mathbf{Q}$. It is required that all elements on the main diagonal are non-zero and all other elements are equal to zero.

Because of that limitation, the algorithm failed to discover the true structure of the network shown in Fig. 5a. It inferred a reciprocal association between the phenotypes $A$ and $B$, with a causal effect, in both direction, of 0.168. For the graph in Fig. 5b, the algorithm estimated a causal effect of 0.23 from $B$ to $A$ and a causal effect of 1.12 from $A$ to $B$. The effects of these spurious connections are relatively low. However, we cannot decide whether they are significantly non-zero because we were not provided a significance test for these coefficients.

All the common assumptions listed in Sect. 6.1 are required to carry out the SML algorithm. That is, the model assumes that the phenotypic variance is due to the additive effects, but not due to dominance or epistatic effects of genetic variants. In addition, it is assumed that phenotypes are normally distributed with independent and normally distributed error terms.

The network structure inference is achieved by estimating all the off-diagonal entries of the $\mathbf{P}$ matrix (the diagonal has only null entries, since it is assumed that no self-loop are present). The matrix $\mathbf{P}$ may or may not be a triangular matrix, implying that the phenotype network structure is acyclic or cyclic, respectively.

The genetic architecture is specified in the matrix **Q**, but the additive effects need to be estimated. Since the **Q** matrix has only one genetic variant associated with each phenotype, only the diagonal entries of the **Q** matrix need to be estimated by the SML algorithm.

In order to efficiently estimate the parameters, it is assumed that the **P** matrix is sparse, that is, most of its entries are equal to zero. The $l_1$-norm is used to control the number of zeros in the network structure matrix **P**.

Under the normality assumption, the $l_1$-regularized log-likelihood maximization problem is solved by using a non-convex optimization algorithm called *block-coordinate ascent* [13].

Although the Recovery Theorem guarantees the identifiability of the network, the algorithm can converge to a local maximum. Thus, it is recommended to run the algorithm several times from different initial values.

In our simulation studies, it was necessary to use an empirically determined threshold of 0.15 in an attempt to eliminate weak causal effects representing spurious associations. Apart from that, the SML algorithm performed very well under the assumption that there is one, and only one, genetic variant associated with each phenotype. The algorithm precisely recovered all acyclic and cyclic causal structures (with both reciprocal associations and feedback loops), including those illustrated in Fig. 5c–i. We noted, however, that the more reciprocal association, the greater the error of the estimated causal effects.

The output of the algorithm contains the estimated **P** and **Q** matrices and the algorithm does not provide a goodness-of-fit measure.

The Matlab package implementing the SML algorithm is available as supporting information in the online version of the article [22]. We run the simulations using Matlab 8.1.0.604 (R2013a).

## 7.5 Summary

In Sects. 7.1–7.4, we described the algorithms QDG [24], QPSO [126], QTLnet [25], and SML [22].

These algorithms were designed under different assumptions. Thus, one can be more suitable for particular tasks than others. By running simulations, we investigated whether the algorithms are capable to discover the structure of a genotype–phenotype network which was generated according to the expected assumptions. The characterization of the algorithms was based on their ability to discover networks in the following situations: genetic variants with pleiotropic effects, phenotypes associated with multiple genetic variants, acyclic structure, feedback loops, and reciprocal associations. In addition, we investigated their behaviors taking into account the input parameters and the output information.

In Table 1, we summarize the input, output, assumptions, and features of these four algorithms.

**Table 1** Input, output, assumptions, and features of the algorithms QDG, QPSO, QTLnet, and SML

| | QDG | QPSO | QTLnet | SML |
|---|---|---|---|---|
| Input | | | | |
| A phenotype association network | ✓ | ✓ | | |
| A genetic map in which each phenotype | | | | |
| must be associated with a distinct genetic variant | ✓ | | | ✓ |
| can be associated with multiple genetic variants | ✓ | ✓ | | ✓* |
| can share a genetic variant with other phenotypes | ✓ | ✓ | | ✓* |
| Assumptions | | | | |
| Phenotypes are normally distributed | | ✓ | ✓ | ✓ |
| The phenotype network is acyclic | | ✓ | ✓ | |
| Features | | | | |
| Discovers acyclic structures | ✓ | ✓ | ✓ | ✓ |
| Discovers cyclic structures | ✓ | | | ✓ |
| Discovers reciprocal associations | | | | ✓ |
| Performs multiple genetic (QTL) mapping | | | ✓ | |
| Estimates causal effects | | | | ✓ |
| Output | | | | |
| The most likely structure network | ✓ | ✓ | ✓ | ✓ |
| A list of the most likely solutions | ✓ | | ✓ | |
| A goodness-of-fit measure of the solution | ✓ | ✓ | ✓ | |

The check mark indicates that the corresponding algorithm has the property. (*) indicates that the feature could not be verified by our simulation studies

# 8 Conclusions

The best approach for inferring causality is to conduct randomized controlled trials. However, very often not all interventions can be tested because of the large number of variables and resource constraints.

Throughout this chapter, we showed some concepts and algorithms that allow us to discover causal associations among variables based on observational data. However, we need to be aware that many assumptions are made in this process. It is important to check the validity of these assumptions since they may be unrealistic from a biological point of view.

In this chapter, we explained in detail the reasons for those assumptions. The causal sufficiency is probably the most difficult assumption to retain, since it is difficult to design an experiment in which all causes involved are properly measured. However, in systems genetics, causal inference is aided by Mendelian randomization. The random assignment of genotypes to individuals (from parents to offspring during meiosis) mimics a randomized controlled trial on genetic level, assuring the causal association from genotypes to phenotypes. Moreover, it is possible to improve inferences on the causal effects by using genetic variants as instrumental variables when they are not associated with confounders of the exposure–outcome association of interest or with the outcome through a path other than through the exposure. Considering this scenario, it is even possible to draw appropriate causal inferences on the network structure when the causal sufficiency assumption is being slightly violated. What we mean is that the causal assumption is deterministic, but the causal modeling contains error terms and involves estimation of the residuals of the model as well as robust goodness-of-fit statistics. Thus, causal associations may still be identified if departures from causal sufficiency do not overly inflate the estimate of the error term variance.

We also discussed that Mendelian randomization is a powerful tool for causal structure learning from genomic data. By adding information on causal relationships from genotypes to phenotypes, some conditional independence relations among the variables are created. Thereby, QTL or QTN information has been used for reducing Markov equivalence classes.

Some concepts and definitions in Pearl's causality theory [87] were first developed for acyclic graphical models. Therefore, there are many results in causal inference under the acyclicity hypothesis. Throughout this chapter, we prioritized the exposition of the generalized theory which is applicable to both cyclic and acyclic cases and has been developed mainly by Thomas Richardson and Peter Spirtes [95–97, 117]. However, some results that facilitate the development of algorithms, such as the factorization property of the joint probability distribution as shown in Definition 15 and the Markov equivalence theorem for DAGs as shown in Theorem 7, are only stated for the acyclic case. Thus, more efforts are still needed to make the generalized theory more accessible for practical applications.

Algorithms for discovering causal phenotype networks are of great interest in genomic studies. The output of these algorithms is a directed graph in which the direction of the edges indicates the flow of information in the causal processes. The investigation of the inferred network structure allows a better understanding of the mechanisms of the underlying biological system (e.g., a gene regulation network) and identification of causes of interest (e.g., genetic determinants and risk factors in diseases).

We described in detail four algorithms for genotype–phenotype network learning, namely (1) QDG, (2) QPSO, (3) QTLnet, and (4) SML. These algorithms are similar in the sense that they leverage genetic variant information to help in determining causal directions among phenotypes. However, they were designed under different assumptions, and therefore some may be more suitable than others for a particular biological application.

The most common assumptions include acyclicity of the network structure and normality of the phenotype distribution. The acyclic structure assumption may be quite restrictive for some applications. In GRNs, for instance, modeling cyclic phenomena is particularly important. Therefore, algorithms that are capable to recover networks containing feedback loops and reciprocal associations are possibly more attractive for modeling biological networks.

Some algorithms require that every phenotype is associated with a distinct genetic variant. Under these conditions, the Recovery Theorem assures that the network structure is uniquely identifiable. One must be particularly careful when selecting genetic variants associated with phenotypes of interest. Pleiotropy and linkage disequilibrium may violate some instrumental variables assumptions within Mendelian randomization framework, and thus misleading conclusions may be more likely to be drawn.

In order to achieve efficiency in high-dimensional data analysis, sparsity has been exploited in some causal structure discovery algorithms. As shown in Sect. 4.5, the PC-algorithm, for instance, may run in polynomial runtime if the true network is acyclic and sparse. There are results in the literature supporting the claim that biological networks, such as transcriptional regulatory networks, are sparse and with few highly connected vertices [48]. However, that is not always the case and the sparsity assumption must be verified in the research field of interest.

We investigated some properties of the QDG, QPSO, QTLnet, and SML algorithms by running some simulations. This study had no intentions of comparing the algorithms, but a comparative study would be very valuable for the field. Both the accuracy of the inferred causal structure and the computational performance of the algorithms (in terms of computational time and memory requirements) should be evaluated. The accuracy and complexity of the algorithms are mainly dependent on the number of data samples used, on the number of vertices (dimension of the data), on the number of edges (degree of sparsity of the network), and also on the degree distribution of the vertices (that is, number of edges per vertex). In regard to the latter, it would be interesting a comparison between Erdös–Renyi [35] and Barabási–Albert [9] random graphs, since it must be harder to orient edges incident on (rather than emanating from) highly connected vertices. Another issues that should be taken into account in a comparative study of algorithms are programming language and high performance computing techniques to save computational time, such as parallel and distributed processing.

In this chapter, we discussed linear SEMs representing genotype–phenotype networks in which the genetic effects are considered fixed effects. Gianola et al. [42] presented an alternative modeling of the causal network among phenotypes. The authors proposed a linear recursive SEM in which the additive genetic effects are random effects following a multivariate normal distribution. Valente et al. [123] proposed an algorithm based on d-separation tests for recovering a DAG (or a class of observationally equivalent acyclic causal structures) which represents a recursive SEM in the context of mixed models applied to quantitative genetics.

We also want to emphasize that the methodologies presented in this chapter are intended to model time invariant systems. However, a better understanding of dynamic biological processes, such as signaling, metabolic, and regulatory activities, can be provided by using a model that takes into account the temporal patterns in the data. Using dynamic networks it is possible to infer causal networks exploiting the temporal aspect of time series data. In this case, cyclic associations are inferred using time delay information [41, 80]. Dynamic Bayesian networks (DBNs) have been widely used for inferring GRNs from time series gene expression data [8, 63, 79, 89].

We conclude with a recommendation for future research in genotype–phenotype network structure inference. Most of the reconstruction network structure algorithms discussed in this chapter assume that phenotypes are normally distributed. There are special correlations, such as tetrachoric, polychoric, biserial, and polyserial correlations, which measure the strength of association between continuous and categorical variables and are particularly robust to deviations from symmetry and kurtosis [33, 127]. Additionally, it was observed that SEMs with ordinal categorical indicators are best estimated using special correlation matrix instead of using Pearson's correlation matrix [14, 127]. Thus, an interesting direction for future work is to investigate SEM-based approaches for reconstructing phenotypes networks using these more robust association measures.

**Acknowledgements** This work was supported by FAPESP (2013/01715-3 and 2014/09576-5, 2015/01587-0), CNPq (306319/2010-1 and 473063/2013-1), CAPES, and NAP eScience–PRP–USP. We are greatly thankful to Guilherme J. M. Rosa for valuable comments.

## References

1. ALBERT, R. Scale-free networks in cell biology. *Journal of cell science 118*, 21 (2005), 4947–4957.
2. ALMASY, L., AND BLANGERO, J. Human QTL linkage mapping. *Genetica 136*, 2 (2009), 333–340.
3. ALTSHULER, D., DALY, M. J., AND LANDER, E. S. Genetic mapping in human disease. *Science 322*, 5903 (2008), 881–888.
4. ASTLE, W., AND BALDING, D. J. Population structure and cryptic relatedness in genetic association studies. *Statistical Science* (2009), 451–471.
5. ATEN, J. E., FULLER, T. F., LUSIS, A. J., AND HORVATH, S. Using genetic markers to orient the edges in quantitative trait networks: the neo software. *BMC Systems Biology 2*, 1 (2008), 34.
6. BABA, K., SHIBATA, R., AND SIBUYA, M. Partial correlation and conditional correlation as measures of conditional independence. *Australian & New Zealand Journal of Statistics 46*, 4 (2004), 657–664.
7. BAMSHAD, M. J., NG, S. B., BIGHAM, A. W., TABOR, H. K., EMOND, M. J., NICKERSON, D. A., AND SHENDURE, J. Exome sequencing as a tool for Mendelian disease gene discovery. *Nature Reviews Genetics 12*, 11 (2011), 745–755.
8. BAR-JOSEPH, Z., GITTER, A., AND SIMON, I. Studying and modelling dynamic biological processes using time-series gene expression data. *Nature Reviews Genetics 13*, 8 (2012), 552–564.
9. BARABÁSI, A.-L., AND ALBERT, R. Emergence of scaling in random networks. *Science 286*, 5439 (1999), 509–512.
10. BARABÁSI, A.-L., AND OLTVAI, Z. N. Network biology: understanding the cell's functional organization. *Nature reviews genetics 5*, 2 (2004), 101–113.
11. BARYSHNIKOVA, A., COSTANZO, M., MYERS, C. L., ANDREWS, B., AND BOONE, C. Genetic interaction networks: toward an understanding of heritability. *Annual review of genomics and human genetics 14* (2013), 111–133.
12. BAUM, C. F., SCHAFFER, M. E., STILLMAN, S., ET AL. Instrumental variables and GMM: Estimation and testing. *Stata journal 3*, 1 (2003), 1–31.

13. Bazerque, J. A. *Leveraging sparsity for genetic and wireless cognitive networks*. PhD thesis, University of Minnesota, 2013.
14. Bistaffa, B. C. *Incorporação de indicadores categóricos ordinais em modelos de equações estruturais*. PhD thesis, Universidade de São Paulo, 2010.
15. Bollen, K. A. *Structural equations with latent variables*. John Wiley & Sons, 1989.
16. Bowden, R. J., and Turkington, D. A. *Instrumental variables*, vol. 8. Cambridge University Press, 1990.
17. Box, G. E., Hunter, W. G., Hunter, J. S., et al. *Statistics for experimenters*. John Wiley & Sons, New York, 1978.
18. Broman, K. W., Wu, H., Sen, Ś., and Churchill, G. A. R/qtl: Qtl mapping in experimental crosses. *Bioinformatics 19*, 7 (2003), 889–890.
19. Burgess, S., Daniel, R. M., Butterworth, A. S., Thompson, S. G., Consortium, E.-I., et al. Network Mendelian randomization: using genetic variants as instrumental variables to investigate mediation in causal pathways. *International journal of epidemiology* (2014), dyu176.
20. Burgess, S., and Thompson, S. G. *Mendelian Randomization: Methods for using Genetic Variants in Causal Estimation*. CRC Press, 2015.
21. Burgess, S., Thompson, S. G., et al. Avoiding bias from weak instruments in Mendelian randomization studies. *International journal of epidemiology 40*, 3 (2011), 755–764.
22. Cai, X., Bazerque, J. A., and Giannakis, G. B. Inference of gene regulatory networks with sparse structural equation models exploiting genetic perturbations. *PLoS Comput Biol 9*, 5 (2013), e1003068.
23. Candes, E., and Tao, T. The Dantzig selector: statistical estimation when p is much larger than n. *The Annals of Statistics* (2007), 2313–2351.
24. Chaibub Neto, E., Ferrara, C. T., Attie, A. D., and Yandell, B. S. Inferring causal phenotype networks from segregating populations. *Genetics 179*, 2 (2008), 1089–1100.
25. Chaibub Neto, E., Keller, M. P., Attie, A. D., and Yandell, B. S. Causal graphical models in systems genetics: a unified framework for joint inference of causal network and genetic architecture for correlated phenotypes. *The annals of applied statistics 4*, 1 (2010), 320.
26. Chu, T., Glymour, C., Scheines, R., and Spirtes, P. A statistical problem for inference to regulatory structure from associations of gene expression measurements with microarrays. *Bioinformatics 19*, 9 (2003), 1147–1152.
27. Cordell, H. J. Epistasis: what it means, what it doesn't mean, and statistical methods to detect it in humans. *Human molecular genetics 11*, 20 (2002), 2463–2468.
28. Danks, D., Glymour, C., and Spirtes, P. The computational and experimental complexity of gene perturbations for regulatory network search. In *Proceedings of IJCAI-2003 Workshop on Learning Graphical Models for Computational Genomic* (2003), pp. 22–31.
29. De La Fuente, A., Bing, N., Hoeschele, I., and Mendes, P. Discovery of meaningful associations in genomic data using partial correlation coefficients. *Bioinformatics 20*, 18 (2004), 3565–3574.
30. DePristo, M. A., Banks, E., Poplin, R., Garimella, K. V., Maguire, J. R., Hartl, C., Philippakis, A. A., Del Angel, G., Rivas, M. A., Hanna, M., et al. A framework for variation discovery and genotyping using next-generation DNA sequencing data. *Nature genetics 43*, 5 (2011), 491–498.
31. Didelez, V., and Sheehan, N. Mendelian randomization as an instrumental variable approach to causal inference. *Statistical methods in medical research 16*, 4 (2007), 309–330.
32. Dong, Z., Song, T., and Yuan, C. Inference of gene regulatory networks from genetic perturbations with linear regression model. *PloS one 8*, 12 (2013), e83263.
33. Drasgow, F. Polychoric and polyserial correlations. *Encyclopedia of Statistical Sciences* (1988).
34. Duncan, O. D. *Introduction to structural equation models*. Elsevier, 2014.
35. Erdős, P., and Rényi, A. On random graphs. *Publicationes Mathematicae Debrecen 6* (1959), 290–297.

36. EVANS, D. M., AND DAVEY SMITH, G. Mendelian randomization: New applications in the coming age of hypothesis-free causality. *Annual review of genomics and human genetics*, (2015).
37. FRIEDMAN, N., LINIAL, M., NACHMAN, I., AND PE'ER, D. Using Bayesian networks to analyze expression data. *Journal of computational biology 7*, 3–4 (2000), 601–620.
38. GALLES, D., AND PEARL, J. Axioms of causal relevance. *Artificial Intelligence 97*, 1 (1997), 9–43.
39. GARDNER, T. S., DI BERNARDO, D., LORENZ, D., AND COLLINS, J. J. Inferring genetic networks and identifying compound mode of action via expression profiling. *Science 301*, 5629 (2003), 102–105.
40. GERIS, L., AND GOMEZ-CABRERO, D. *Uncertainty in biology: a computational modeling approach*, vol. 17. Springer, 2015.
41. GHAHRAMANI, Z. Learning dynamic Bayesian networks. In *Adaptive processing of sequences and data structures*. Springer, 1998, pp. 168–197.
42. GIANOLA, D., AND SORENSEN, D. Quantitative genetic models for describing simultaneous and recursive relationships between phenotypes. *Genetics 167*, 3 (2004), 1407–1424.
43. GODDARD, M., ET AL. Fine mapping of quantitative trait loci using linkage disequilibria with closely linked marker loci. *Genetics 155*, 1 (2000), 421–430.
44. GRANGER, C. W. Investigating causal relations by econometric models and cross-spectral methods. *Econometrica: Journal of the Econometric Society* (1969), 424–438.
45. GRZEGORCZYK, M., AND HUSMEIER, D. Improving the structure MCMC sampler for bayesian networks by introducing a new edge reversal move. *Machine Learning 71*, 2–3 (2008), 265–305.
46. HAGEMAN, R. S., LEDUC, M. S., KORSTANJE, R., PAIGEN, B., AND CHURCHILL, G. A. A bayesian framework for inference of the genotype–phenotype map for segregating populations. *Genetics 187*, 4 (2011), 1163–1170.
47. HALEY, C. S., AND KNOTT, S. A. A simple regression method for mapping quantitative trait loci in line crosses using flanking markers. *Heredity 69*, 4 (1992), 315–324.
48. HAO, D., REN, C., AND LI, C. Revisiting the variation of clustering coefficient of biological networks suggests new modular structure. *BMC systems biology 6*, 1 (2012), 34.
49. HERNÁN, M. A., HERNÁNDEZ-DÍAZ, S., WERLER, M. M., AND MITCHELL, A. A. Causal knowledge as a prerequisite for confounding evaluation: an application to birth defects epidemiology. *American journal of epidemiology 155*, 2 (2002), 176–184.
50. HOCUTT, M. Aristotle's four becauses. *Philosophy 49*, 190 (1974), 385–399.
51. HOLLAND, P. W. Causal inference, path analysis and recursive structural equations models. *ETS Research Report Series 1988*, 1 (1988), i–50.
52. HORIKOSHI, M., PASQUALI, L., WILTSHIRE, S., HUYGHE, J. R., MAHAJAN, A., ASIMIT, J. L., FERREIRA, T., LOCKE, A. E., ROBERTSON, N. R., WANG, X., ET AL. Transancestral fine-mapping of four type 2 diabetes susceptibility loci highlights potential causal regulatory mechanisms. *Human molecular genetics* (2016), ddw048.
53. HOULE, D., GOVINDARAJU, D. R., AND OMHOLT, S. Phenomics: the next challenge. *Nature Reviews Genetics 11*, 12 (2010), 855–866.
54. HUANG, A. *Sparse model learning for inferring genotype and phenotype associations*. PhD thesis, University of Miami, 2014.
55. HUME, D., AND BEAUCHAMP, T. L. *An enquiry concerning human understanding: A critical edition*, vol. 3. Oxford University Press, 2000.
56. IACOBUCCI, D. Structural equations modeling: Fit indices, sample size, and advanced topics. *Journal of Consumer Psychology 20*, 1 (2010), 90–98.
57. JANSEN, R. C. Interval mapping of multiple quantitative trait loci. *Genetics 135*, 1 (1993), 205–211.
58. JEONG, H., TOMBOR, B., ALBERT, R., OLTVAI, Z. N., AND BARABÁSI, A.-L. The large-scale organization of metabolic networks. *Nature 407*, 6804 (2000), 651–654.
59. JOHNSON, R. A., WICHERN, D. W., ET AL. *Applied multivariate statistical analysis*, vol. 4. Prentice hall Englewood Cliffs, NJ, 1992.

60. KALISCH, M., MÄCHLER, M., COLOMBO, D., MAATHUIS, M. H., AND BÜHLMANN, P. Causal inference using graphical models with the r package pcalg. *Journal of Statistical Software 47*, 11 (2012), 1–26.
61. KAO, C.-H., ZENG, Z.-B., AND TEASDALE, R. D. Multiple interval mapping for quantitative trait loci. *Genetics 152*, 3 (1999), 1203–1216.
62. KATAN, M. B. Apolipoprotein e isoforms, serum cholesterol, and cancer. *International journal of epidemiology 33*, 1 (2004), 9–9.
63. KIM, S. Y., IMOTO, S., MIYANO, S., ET AL. Inferring gene networks from time series microarray data using dynamic bayesian networks. *Briefings in bioinformatics 4*, 3 (2003), 228.
64. KOLLER, D., AND FRIEDMAN, N. *Probabilistic graphical models: principles and techniques.* MIT press, 2009.
65. KOSTER, J. T., ET AL. Markov properties of nonrecursive causal models. *The Annals of Statistics 24*, 5 (1996), 2148–2177.
66. LANDER, E. S., AND BOTSTEIN, D. Mapping Mendelian factors underlying quantitative traits using RFLP linkage maps. *Genetics 121*, 1 (1989), 185–199.
67. LAURITZEN, S. L., DAWID, A. P., LARSEN, B. N., AND LEIMER, H.-G. Independence properties of directed Markov fields. *Networks 20*, 5 (1990), 491–505.
68. LAWLOR, D. A., HARBORD, R. M., STERNE, J. A., TIMPSON, N., AND DAVEY SMITH, G. Mendelian randomization: using genes as instruments for making causal inferences in epidemiology. *Statistics in medicine 27*, 8 (2008), 1133–1163.
69. LECLERC, R. D. Survival of the sparsest: robust gene networks are parsimonious. *Molecular systems biology 4*, 1 (2008).
70. LEWIS, D. *Counterfactuals.* Harvard University Press, 1973.
71. LI, R., TSAIH, S.-W., SHOCKLEY, K., STYLIANOU, I. M., WERGEDAL, J., PAIGEN, B., AND CHURCHILL, G. A. Structural model analysis of multiple quantitative traits. *PLoS Genet 2*, 7 (2006), e114.
72. LIANG, Y., AND MIKLER, A. R. Big data problems on discovering and analyzing causal relationships in epidemiological data. In *Big Data (Big Data), 2014 IEEE International Conference on* (2014), Washington, DC, pp. 11–18.
73. LIU, B., DE LA FUENTE, A., AND HOESCHELE, I. Gene network inference via structural equation modeling in genetical genomics experiments. *Genetics 178*, 3 (2008), 1763–1776.
74. LOGSDON, B. A., AND MEZEY, J. Gene expression network reconstruction by convex feature selection when incorporating genetic perturbations. *PLoS Comput Biol 6*, 12 (2010), 429–435.
75. LOPES, M. S., BASTIAANSEN, J. W., HARLIZIUS, B., KNOL, E. F., AND BOVENHUIS, H. A genome-wide association study reveals dominance effects on number of teats in pigs. *PloS one 9*, 8 (2014), e105867.
76. LUSIS, A. J., ATTIE, A. D., AND REUE, K. Metabolic syndrome: from epidemiology to systems biology. *Nature Reviews Genetics 9*, 11 (2008), 819–830.
77. MARKOWETZ, F., AND SPANG, R. Inferring cellular networks–a review. *BMC bioinformatics 8*, Suppl 6 (2007), S5.
78. MOON, J. Y., CHAIBUB NETO, E., YANDELL, B., AND DENG, X. Bayesian causal phenotype network incorporating genetic variation and biological knowledge. *Probabilistic Graphical Models in Genetics, Genomics, and Postgenomics* (2014), 165–195.
79. MURPHY, K., MIAN, S., ET AL. Modelling gene expression data using dynamic bayesian networks. Tech. rep., Computer Science Division, University of California, Berkeley, CA, 1999.
80. MURPHY, K. P. *Dynamic bayesian networks: representation, inference and learning*. PhD thesis, University of California, Berkeley, 2002.
81. NETER, J., KUTNER, M. H., NACHTSHEIM, C. J., AND WASSERMAN, W. *Applied linear statistical models*, vol. 4. Irwin Chicago, 1996.
82. NETO, E. C., AND YANDELL, M. B. S. Package 'qtlnet'. *Genetics 179* (2014), 1089–1100.
83. NEWMAN, M. E. The structure and function of complex networks. *SIAM review 45*, 2 (2003), 167–256.

84. NEYMAN, J. Sur les applications de la thar des probabilities aux experiences agaricales: Essay des principle. excerpts reprinted (1990) in English. *Statistical Science 5* (1923), 463–472.
85. PALMER, T. M., LAWLOR, D. A., HARBORD, R. M., SHEEHAN, N. A., TOBIAS, J. H., TIMPSON, N. J., SMITH, G. D., AND STERNE, J. A. Using multiple genetic variants as instrumental variables for modifiable risk factors. *Statistical methods in medical research 21*, 3 (2012), 223–242.
86. PEARL, J. *Probabilistic reasoning in intelligent systems: networks of plausible inference*. Morgan Kaufmann Publishers Inc, San Francisco, CA, 1988.
87. PEARL, J. *Causality: models, reasoning, and inference*. Cambridge University Press, Cambridge, UK, 2000.
88. PEARL, J. On the consistency rule in causal inference: axiom, definition, assumption, or theorem? *Epidemiology 21*, 6 (2010), 872–875.
89. PERRIN, B.-E., RALAIVOLA, L., MAZURIE, A., BOTTANI, S., MALLET, J., AND D'ALCHE BUC, F. Gene networks inference using dynamic bayesian networks. *Bioinformatics 19*, suppl 2 (2003), ii138–ii148.
90. PIERCE, B. L., AHSAN, H., AND VANDERWEELE, T. J. Power and instrument strength requirements for mendelian randomization studies using multiple genetic variants. *International journal of epidemiology* (2010), dyq151.
91. POURNARA, I., AND WERNISCH, L. Reconstruction of gene networks using bayesian learning and manipulation experiments. *Bioinformatics 20*, 17 (2004), 2934–2942.
92. R CORE TEAM. *R: A Language and Environment for Statistical Computing*. R Foundation for Statistical Computing, Vienna, Austria, 2014.
93. RAVASZ, E., SOMERA, A. L., MONGRU, D. A., OLTVAI, Z. N., AND BARABÁSI, A.-L. Hierarchical organization of modularity in metabolic networks. *Science 297*, 5586 (2002), 1551–1555.
94. RICHARDSON, T. A discovery algorithm for directed cyclic graphs. In *Proceedings of the Twelfth international conference on Uncertainty in artificial intelligence* (1996), Morgan Kaufmann Publishers Inc., pp. 454–461.
95. RICHARDSON, T. A polynomial-time algorithm for deciding Markov equivalence of directed cyclic graphical models. In *Proceedings of the Twelfth international conference on Uncertainty in artificial intelligence* (1996), Morgan Kaufmann Publishers Inc., pp. 462–469.
96. RICHARDSON, T. A characterization of markov equivalence for directed cyclic graphs. *International Journal of Approximate Reasoning 17*, 2 (1997), 107–162.
97. RICHARDSON, T., AND SPIRTES, P. Automated discovery of linear feedback models. Tech. rep., CMU-PHIL-75, Dept. of Philosophy, Carnegie Mellon University, 1996.
98. ROBINS, J. A new approach to causal inference in mortality studies with a sustained exposure period—application to control of the healthy worker survivor effect. *Mathematical Modelling 7*, 9 (1986), 1393–1512.
99. ROBINS, J. M. Addendum to "a new approach to causal inference in mortality studies with a sustained exposure period—application to control of the healthy worker survivor effect". *Computers & Mathematics with Applications 14*, 9 (1987), 923–945.
100. ROBINSON, R. W. Counting labeled acyclic digraphs. In *New Directions in the Theory of Graphs*. Academic Press New York, 1973.
101. ROCKMAN, M. V. Reverse engineering the genotype–phenotype map with natural genetic variation. *Nature 456*, 7223 (2008), 738–744.
102. ROSA, G. J., VALENTE, B. D., DE LOS CAMPOS, G., WU, X.-L., GIANOLA, D., AND SILVA, M. A. Inferring causal phenotype networks using structural equation models. *Genet Sel Evol 43*, 6 (2011).
103. ROSEN, K. H. *Handbook of discrete and combinatorial mathematics*. CRC press, 1999.
104. ROSENBAUM, P. R., AND RUBIN, D. B. The central role of the propensity score in observational studies for causal effects. *Biometrika 70*, 1 (1983), 41–55.
105. RUBIN, D. B. Estimating causal effects of treatments in randomized and nonrandomized studies. *Journal of educational Psychology 66*, 5 (1974), 688.

106. Sachs, K., Perez, O., Pe'er, D., Lauffenburger, D. A., and Nolan, G. P. Causal protein-signaling networks derived from multiparameter single-cell data. *Science 308*, 5721 (2005), 523–529.
107. Schadt, E. E. Molecular networks as sensors and drivers of common human diseases. *Nature 461*, 7261 (2009), 218–223.
108. Schadt, E. E., Lamb, J., Yang, X., Zhu, J., Edwards, S., GuhaThakurta, D., Sieberts, S. K., Monks, S., Reitman, M., Zhang, C., et al. An integrative genomics approach to infer causal associations between gene expression and disease. *Nature genetics 37*, 7 (2005), 710–717.
109. Schwarz, G., et al. Estimating the dimension of a model. *The annals of statistics 6*, 2 (1978), 461–464.
110. Segal, E., Shapira, M., Regev, A., Pe'er, D., Botstein, D., Koller, D., and Friedman, N. Module networks: identifying regulatory modules and their condition-specific regulators from gene expression data. *Nature genetics 34*, 2 (2003), 166–176.
111. Slatkin, M. Linkage disequilibrium—understanding the evolutionary past and mapping the medical future. *Nature Reviews Genetics 9*, 6 (2008), 477–485.
112. Smith, G. D., and Ebrahim, S. 'mendelian randomization': can genetic epidemiology contribute to understanding environmental determinants of disease? *International journal of epidemiology 32*, 1 (2003), 1–22.
113. Smith, G. D., and Ebrahim, S. Mendelian randomization: prospects, potentials, and limitations. *International journal of epidemiology 33*, 1 (2004), 30–42.
114. Smith, G. D., and Hemani, G. Mendelian randomization: genetic anchors for causal inference in epidemiological studies. *Human molecular genetics 23*, R1 (2014), R89–R98.
115. Speed, T. P. A note on nearest-neighbour gibbs and markov probabilities. *Sankhyā: The Indian Journal of Statistics, Series A* (1979), 184–197.
116. Spirtes, P. Directed cyclic graphs, conditional independence, and non-recursive linear structural equation models. Tech. rep., CMU-PHIL-35, Dept. of Philosophy, Carnegie Mellon University, 1993.
117. Spirtes, P. Directed cyclic graphical representations of feedback models. In *Proceedings of the Eleventh conference on Uncertainty in artificial intelligence* (1995), Morgan Kaufmann Publishers Inc., pp. 491–498.
118. Spirtes, P., Glymour, C. N., and Scheines, R. *Causation, prediction, and search*, vol. 81. MIT press, 2000.
119. Staiger, D. O., and Stock, J. H. Instrumental variables regression with weak instruments. *Econometrica: Journal of the Econometric Society 65*, 3 (1997), 557–586.
120. Sun, C., VanRaden, P. M., Cole, J. B., and O'Connell, J. R. Improvement of prediction ability for genomic selection of dairy cattle by including dominance effects. *PloS one 9*, 8 (2014), e103934.
121. Teyssier, M., and Koller, D. Ordering-based search: A simple and effective algorithm for learning bayesian networks. *arXiv preprint arXiv:1207.1429* (2012).
122. Tibshirani, R. Regression shrinkage and selection via the lasso. *Journal of the Royal Statistical Society. Series B (Methodological)* (1996), 267–288.
123. Valente, B. D., Rosa, G. J., de los Campos, G., Gianola, D., and Silva, M. A. Searching for recursive causal structures in multivariate quantitative genetics mixed models. *Genetics 185*, 2 (2010), 633–644.
124. Wagner, A., and Fell, D. A. The small world inside large metabolic networks. *Proceedings of the Royal Society of London B: Biological Sciences 268*, 1478 (2001), 1803–1810.
125. Wang, H., Haiman, C. A., Burnett, T., Fortini, B. K., Kolonel, L. N., Henderson, B. E., Signorello, L. B., Blot, W. J., Keku, T. O., Berndt, S. I., et al. Fine-mapping of genome-wide association study-identified risk loci for colorectal cancer in African Americans. *Human molecular genetics 22*, 24 (2013), 5048–5055.
126. Wang, H., and van Eeuwijk, F. A. A new method to infer causal phenotype networks using qtl and phenotypic information. *PloS one 9*, 8 (2014), e103997.

127. WEST, S. G., FINCH, J. F., AND CURRAN, P. J. Structural equation models with nonnormal variables. *Structural equation modeling: Concepts, issues, and applications* (1995), 56–75.
128. WRIGHT, S. Correlation and causation. *Journal of agricultural research 20*, 7 (1921), 557–585.
129. WRIGHT, S. *Appendix to The Tariff on Animal and Vegetable Oils, by P. G. Wright.* New York: MacMillan, 1928.
130. YU, J., SMITH, V. A., WANG, P. P., HARTEMINK, A. J., AND JARVIS, E. D. Advances to bayesian network inference for generating causal networks from observational biological data. *Bioinformatics 20*, 18 (2004), 3594–3603.
131. ZENG, Z.-B. Theoretical basis for separation of multiple linked gene effects in mapping quantitative trait loci. *Proceedings of the National Academy of Sciences 90*, 23 (1993), 10972–10976.
132. ZHANG, L., LI, H., LI, Z., AND WANG, J. Interactions between markers can be caused by the dominance effect of quantitative trait loci. *Genetics 180*, 2 (2008), 1177–1190.
133. ZHAO, W., SERPEDIN, E., AND DOUGHERTY, E. R. Recovering genetic regulatory networks from chromatin immunoprecipitation and steady-state microarray data. *EURASIP Journal on Bioinformatics and Systems Biology 2008*, 1 (2008), 248747.
134. ZHU, J., ZHANG, B., SMITH, E. N., DREES, B., BREM, R. B., KRUGLYAK, L., BUMGARNER, R. E., AND SCHADT, E. E. Integrating large-scale functional genomic data to dissect the complexity of yeast regulatory networks. *Nature genetics 40*, 7 (2008), 854–861.
135. ZHU, X., ZHANG, S., ZHAO, H., AND COOPER, R. S. Association mapping, using a mixture model for complex traits. *Genetic epidemiology 23*, 2 (2002), 181–196.
136. ZOU, H. The adaptive lasso and its oracle properties. *Journal of the American statistical association 101*, 476 (2006), 1418–1429.
137. ZOU, H., AND HASTIE, T. Regularization and variable selection via the elastic net. *Journal of the Royal Statistical Society: Series B (Statistical Methodology) 67*, 2 (2005), 301–320.
138. ZOU, H., AND ZHANG, H. H. On the adaptive elastic-net with a diverging number of parameters. *Annals of statistics 37*, 4 (2009), 1733.

# Genomic Applications of the Neyman–Pearson Classification Paradigm

**Jingyi Jessica Li and Xin Tong**

**Abstract** The Neyman–Pearson (NP) classification paradigm addresses an important binary classification problem where users want to minimize type II error while controlling type I error under some specified level $\alpha$, usually a small number. This problem is often faced in many genomic applications involving binary classification tasks. The terminology Neyman–Pearson classification paradigm arises from its connection to the Neyman–Pearson paradigm in hypothesis testing. The NP paradigm is applicable when one type of error (e.g., type I error) is far more important than the other type (e.g., type II error), and users have a specific target bound for the former. In this chapter, we review the NP classification literature, with a focus on the genomic applications as well as our contribution to the NP classification theory and algorithms. We also provide simulation examples and a genomic case study to demonstrate how to use the NP classification algorithm in practice.

**Keywords** Classification • Genomic applications • Neyman–Pearson • Statistical learning • Methodology

## 1 Introduction

As an important statistical and machine learning method, classification has been widely used in genomic studies. Binary classification is the basis of all types of classification problems, and there exist many approaches to ensemble binary classifiers to solve multi-class classification problems or to reduce multi-class to binary classification. Important genomic applications of binary classification include labeling microarray data as tumor or non-tumor samples [18, 53], dividing

J.J. Li (✉)
Department of Statistics, University of California, Los Angeles, Los Angeles, CA, USA
e-mail: jli@stat.ucla.edu

X. Tong
Department of Data Sciences and Operations, University of Southern California, Los Angeles, CA, USA
e-mail: xint@marshall.usc.edu



K.-C. Wong (ed.), *Big Data Analytics in Genomics*,
DOI 10.1007/978-3-319-41279-5_4

genes into housekeeping and single-tissue specific groups [54], classifying genomic hairpin structures into precursor microRNAs and pseudo hairpins (i.e., genomic inverted repeats that are not precursor microRNAs) [32], and predicting transcription factor binding sites and other DNA regulatory elements based on genomic features [7, 20].

The aim of binary classification is to accurately predict binary (i.e., 0 or 1) labels for new observations on the basis of labeled training data. There are two types of errors: *type I error* (the conditional probability that the predicted label for a new observation is 1 given that the observation has a true label 0) and *type II error* (the conditional probability that the predicted label for a new observation is 0 given that the observation has a true label 1). For more than half a century, significant advances have been made in the development of binary classification theory and methods to construct good classifiers with various desirable properties [24]. Most existing binary classification methods aim to optimize the *risk*, which is the expected classification error (the probability that the predicted label is different from the true label) and can be expressed as a weighted sum of the type I and II errors, where the two weights are the marginal probabilities of the true label being 0 and 1, respectively. In real-world applications, however, users' priorities for type I and type II errors may differ from these weights, and then minimizing the risk may lead to unsatisfactory classifiers. For example, in tumor diagnosis, suppose that we label a tumor sample as 0 and a normal sample as 1, the risk minimization approach fails if it leads to a classifier with type I error (i.e., the conditional probability that a tumor sample is misclassified as a normal sample) equal to 0.3 but doctors prefer to constrain the type I error under 0.05.

There are many scenarios where users need asymmetric error control, and they often occur when the two types of classification errors lead to vastly different consequences. Again in the example of tumor diagnosis, mispredicting a normal sample as a tumor sample may increase a patient's anxiety and impose additional medical costs, but misclassifying a tumor sample as a normal sample may delay a patient's treatment and even cause a life loss. Hence, the latter type of error—type I error—is more severe and should be controlled at a low level. In another example of classifying genes into housekeeping ones (say class 0) and cell-specific ones (say class 1), if the research aim is to identify novel cell-specific genes for a cell type (e.g., human embryonic stem cells) and the identified genes will be validated by experiments, researchers would generally prefer to control the type I error (the conditional probability of misclassifying a housekeeping gene as a cell-specific gene) at a low level to reduce experimental costs.

One common approach to addressing asymmetric classification errors is cost-sensitive learning, which allows users to assign two different costs as weights for type I and type II errors [14, 56]. Although this approach has many merits, its effectiveness is largely limited when there lacks a consensus way to assign costs. Cost-sensitive learning is also unable to serve the purpose when users desire a specific high probabilistic bound $\alpha$ on the type I or II error [e.g., $\mathbb{P}(\text{type I error} \leq \alpha) > 1 - \delta$, the probability that a chosen classifier has type I error not exceeding $\alpha$ is greater than $1 - \delta$ for some small positive $\alpha$ (e.g., 0.05) and $\delta$ (e.g., 0.05)], even though users may vary the two costs to achieve a small type I or type II error. There are several other classification approaches that target on small type I errors. Examples include asymmetric support vector machines (SVM) [52] and $p$-values for classification [13]. But like cost-sensitive learning, these approaches also provide no probabilistic guarantee for the type I error bound and could lead to non-negligible probability of large type I errors. In practice, there has been a long-time intuitive and straightforward common practice, that is to tune the observed type I error (also called empirical type I error) on the training data to the desired type I error bound $\alpha$, for example, by adjusting the costs of errors or by changing the classification threshold. However, this approach cannot control the type I error of the chosen classifier on a new data set to be under $\alpha$ with high probability; in fact, a classifier chosen in this way will have type I error greater than $\alpha$ for approximately half the chance. Figure 1 illustrates this phenomenon with a simple two-class Gaussian example.

Unlike the above approaches, the Neyman–Pearson (NP) classification, which was motivated by the century-long Neyman–Pearson paradigm in hypothesis testing, specifically aims to bound one type of error with high probability and meanwhile minimize the other type error. The main advantage of the NP classification is that it provides high probability guarantee on controlling one type of error under a user desired level.

This chapter is organized as follows. Section 2 provides a review of the Neyman–Pearson classification paradigm, including its theoretical and algorithmic advances. Section 3 presents three simulation examples to demonstrate how to implement the NP classification with popular classification algorithms (logistic regression, SVM, and random forests) that are widely used in genomic applications. Section 4 implements the NP classification on a genomic case study, where the goal is to classify DNA regions containing transcription factor motifs into two classes: transcription factor binding sites and non-binding sites, using two genomic features (absolute DNase-seq tag counts and DNase-seq footprint scores). Section 5 describes future research directions and potential genomic applications of the NP classification.

## 2 Neyman–Pearson Paradigm

A few commonly used notations are set up to facilitate our discussion. Let $(X, Y)$ be random variables where $X \in \mathscr{X} \subset \mathbb{R}^d$ is a vector of features and $Y \in \{0, 1\}$ is a class label. A classifier $h(\cdot)$ is a mapping $h : \mathscr{X} \to \{0, 1\}$ that returns the predicted class given $X$. An error occurs when $h(X) \neq Y$, and the binary loss is $\mathbb{I}(h(X) \neq Y)$, where $\mathbb{I}(\cdot)$ denotes the indicator function. The risk is the expected loss with respect to the joint distribution of $(X, Y)$: $R(h) = \mathrm{E}\left[\mathbb{I}(h(X) \neq Y)\right] = \mathbb{P}\left(h(X) \neq Y\right)$, which can be expressed as a weighted sum of type I and II errors: $R(h) = \mathbb{P}(Y = 0)R_0(h) + \mathbb{P}(Y = 1)R_1(h)$, where $R_0(h) = \mathbb{P}\left(h(X) \neq Y | Y = 0\right)$ denotes the type I error, and $R_1(h) = \mathbb{P}\left(h(X) \neq Y | Y = 1\right)$ denotes the type II error. While the classical binary classification aims to minimize the risk $R(\cdot)$, the NP classification aims to mimic the NP oracle classifier $\phi^*$, which is defined as

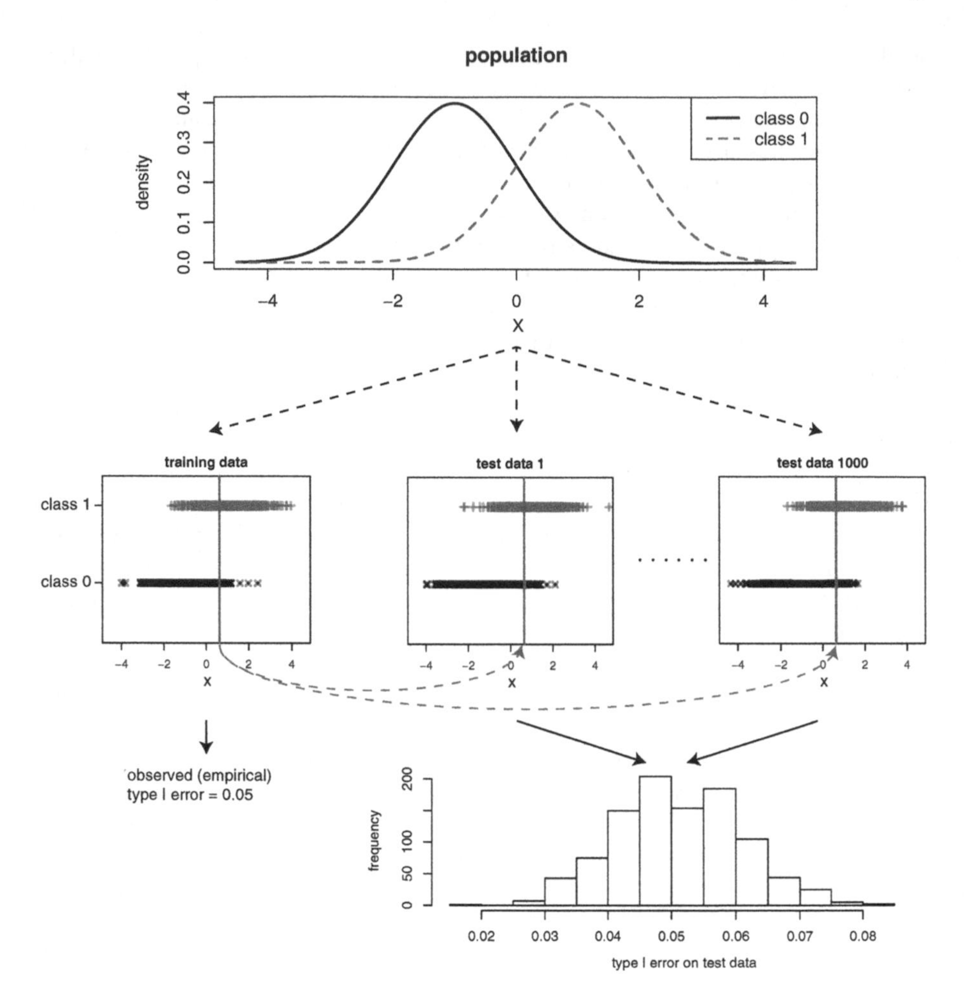


**Fig. 1** An example to illustrate that tuning the empirical type I error on training data to $\alpha$ cannot control the type I error on test data under $\alpha$ with high probability. The population is a two-class Gaussian distribution, where $X$ follows $N(-1, 1)$ under class 0 and $N(1, 1)$ under class 1. The two classes have equal probabilities. A training data set with size $n = 1000$ is generated from this population, and a threshold $t = 0.635$ (the *dark blue vertical line*) is chosen so that the resulting classifier $\mathbb{I}(X \geq t)$ has the observed (empirical) type I error equal to $\alpha = 0.05$ on the training data. This classifier is then applied to $B = 1000$ test data sets from the same population, and the resulting empirical type I errors on each of these test data sets are summarized in the histogram, which shows that approximately 50 % of the type I errors are greater than $\alpha$ and 18.1 % of the errors are even greater than 0.06

$$\phi^* = \underset{\phi:\, R_0(\phi) \leq \alpha}{\arg\min}\; R_1(\phi)\,,$$

where the user-specified level $\alpha$ reflects a conservative attitude (priority) towards the type I error. Figure 2 shows a toy example that demonstrates the difference between a classical classifier that aims to minimize the risk and an NP classifier.

Earlier work on the NP classification came from the engineering community. Earlier theoretical development for the NP classification includes traditional statistical learning results such as probably approximately correct bounds and oracle type inequalities [8, 38, 40]. Then performance measures for the NP classification were proposed [39]. More recently, a general solution to semi-supervised novelty detection via reduction to NP classification was developed [3]. There are also other related work [9, 19]. All these work follow an empirical risk minimization (ERM) approach, and suffer a common limitation: a relaxed empirical type I error constraint is used in the optimization program, and as a result, the type I error can only be shown to satisfy some relaxed upper bound, which is bigger than $\alpha$.

We have worked extensively on NP classification using ERM and plug-in approaches [37, 46, 58]. We initiated a significant departure from the previous NP classification literature in [37] by arguing that a good classifier $\hat{\phi}$ under the NP paradigm should respect the chosen significance level $\alpha$, rather than some relaxation of it. More concretely, the following two properties should both be satisfied with high probability.

(1) The type I error constraint is respected, i.e., $R_0(\hat{\phi}) \leq \alpha$.
(2) The excess type II error $R_1(\hat{\phi}) - R_1(\phi^*)$ diminishes with an explicit rate (w.r.t. sample size).

We say a classifier satisfies the NP oracle inequalities if it has properties (1) and (2) with high probability. The NP oracle inequalities measure the theoretical performance of classifiers under the NP paradigm, as well as define a new NP counterpart of the well-established oracle inequalities under the classical paradigm (see [23] and the references within). In contrast, for a classifier $\hat{h}$, the classical oracle inequality insists that with high probability,

$$\mathrm{the excess risk R(\hat{h}) - R(h^*) diminishes with an explicit rate,}$$

where $h^*(x) = \mathbb{I}(\eta(x) \geq 1/2)$ is the Bayes classifier under the classical paradigm, with $\eta(x) = \mathrm{E}[Y|X = x] = \mathbb{P}(Y = 1|X = x)$ denoting the regression function of $Y$ on $X$.

Using a more stringent empirical type I error constraint (less than $\alpha$), we established NP oracle inequalities for their proposed classifiers under convex loss functions (as opposed to the binary loss) [37]. We also proved a negative result by constructing a counterexample: under the binary loss, ERM approaches cannot guarantee diminishing excess type II error if one insists that type I error of the classifier be bounded from above by $\alpha$ with high probability. This negative result motivated us to develop a plug-in approach to NP classification, described in [46].

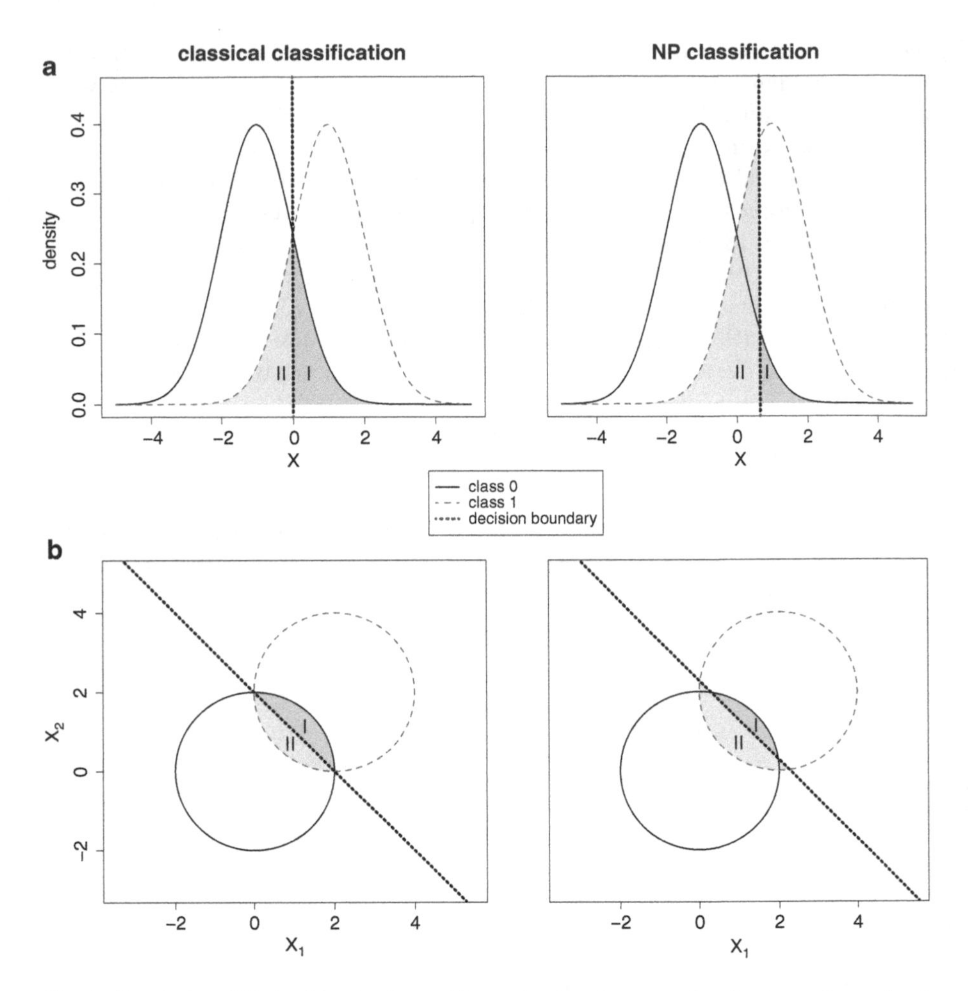


**Fig. 2** Classical vs. NP classification in two binary classification examples. (**a**) A one-dimensional toy example where $X$ has a two-class Gaussian distribution. $X$ follows $N(-1, 1)$ under class 0 and $N(1, 1)$ under class 1. The two balanced classes have equal marginal probabilities. Suppose that a user prefers a type I error $\leq 0.05$. The classical classifier $\mathbb{I}(X \geq 0)$ that minimizes the risk would result in a type I error $= 0.16 > 0.05$. On the other hand, the NP classifier $\mathbb{I}(X \geq 0.65)$ that minimizes the type II error under the type I error constraint ($\leq 0.05$) delivers the desirable type I error. (**b**) A two-dimensional toy example where $(X_1, X_2)$ has a two-class uniform distribution over circles. $(X_1, X_2)$ follows a uniform distribution on $\{X_1^2 + X_2^2 \leq 4\}$ under class 0 and a uniform distribution on $\{(X_1 - 2)^2 + (X_2 - 2)^2 \leq 4\}$ under class 1. The two balanced classes have equal marginal probabilities. Suppose that a user prefers a classifier that is linear in $X_1$ and $X_2$ and has type I error $\leq 0.05$. The classical classifier $\mathbb{I}(X_1 + X_2 \leq 2)$ that minimizes the risk would result in a type I error $= 0.29 > 0.05$. On the other hand, the NP classifier $\mathbb{I}(X_1 + X_2 \leq 2.28)$ that minimizes the type II error under the type I error constraint ($\leq 0.05$) delivers the desirable type I error

In classical binary classification, plug-in methods that target the Bayes classifier $\mathbb{I}(\eta(x) \geq 1/2)$ have been studied. The earliest works cast doubt on the efficacy of the plug-in approach to classification. For example, it was shown that plug-in estimators cannot achieve excess risk with rates faster than $O(1/\sqrt{n})$ under certain assumptions [55], while direct methods can achieve fast rates up to $O(1/n)$ under margin assumption [30, 45, 49, 50]. However, a more recent work combined a smoothness condition on $\eta$ with the margin assumption and showed that plug-in classifiers $\mathbb{I}(\hat{\eta}_n \geq 1/2)$ based on local polynomial estimators can achieve rates faster than $O(1/n)$ [1].

The oracle classifier under the NP paradigm arises from its close connection to the Neyman–Pearson Lemma in statistical hypothesis testing. Hypothesis testing bears a strong resemblance to binary classification if we assume the following model. Let $P_1$ and $P_0$ be two *known* probability distributions on $\mathcal{X} \subset \mathbb{R}^d$. Let $\zeta \in (0, 1)$ and assume that $Y \sim Bernouli(\zeta)$. Assume further that the conditional distribution of $X$ given $Y$ is denoted by $P_Y$. Given such a model, the goal of statistical hypothesis testing is to determine whether $X$ was generated from $P_1$ or from $P_0$. To this end, we construct a randomized test $\phi : \mathcal{X} \to [0, 1]$ and the conclusion of the test based on $\phi$ is that $X$ is generated from $P_1$ with probability $\phi(X)$ and from $P_0$ with probability $1 - \phi(X)$. Two types of errors arise: type I error occurs when $P_0$ is rejected given $X \sim P_0$, and type II error occurs when $P_0$ is not rejected given $X \sim P_1$. The Neyman–Pearson paradigm in hypothesis testing amounts to choosing $\phi$ that

$$\text{maximizes} E[\phi(X)|Y = 1]\,, \text{subjectto} E[\phi(X)|Y = 0] \leq \alpha\,,$$

where $\alpha \in (0, 1)$ is the significance level of the test. A solution to this constrained optimization problem is called a most powerful test of level $\alpha$. The Neyman–Pearson Lemma gives mild sufficient conditions for the existence of such a test.

**Theorem 1 (Neyman–Pearson Lemma).** *Let $P_0$ and $P_1$ be probability distributions possessing densities $q$ and $p$, respectively, with respect to some measure $\mu$. Let $r(x) = p(x)/q(x)$ and $C_\alpha$ be such that $P_0(r(X) > C_\alpha) \leq \alpha$ and $P_0(r(X) \geq C_\alpha) \geq \alpha$. Then for a given level $\alpha$, the most powerful test of level $\alpha$ is defined by*

$$\phi^*(X) = \begin{cases} 1 & \text{if } r(X) > C_\alpha \\ 0 & \text{if } r(X) < C_\alpha \\ \frac{\alpha - P_0(r(X) > C_\alpha)}{P_0(r(X) = C_\alpha)} & \text{if } r(X) = C_\alpha \end{cases}.$$

Therefore, our plug-in target under the NP paradigm is the **oracle classifier**

$$\phi^*(x) = \mathbb{I}(r(x) \geq C_\alpha) = \mathbb{I}(\eta(x) \geq D_\alpha), \text{where} D_\alpha = \frac{\mathbb{P}(Y = 1)C_\alpha}{\mathbb{P}(Y = 1)C_\alpha + \mathbb{P}(Y = 0)}\,.$$

Note that under the classical paradigm, the oracle classifier $\mathbb{I}(\eta(x) \geq 1/2)$ puts a threshold on the regression function $\eta$ at precisely $1/2$, so plug-in methods do not involve estimating the threshold level. In contrast, the NP paradigm poses more challenges because the threshold level $C_\alpha$ or $D_\alpha$ needs to be estimated in addition to $r(x)$ or $\eta(x)$.

Also note that in practice, the threshold on $\hat{\eta}(x)$ is often not set to $1/2$ but is chosen by data-driven approaches such as cross validation and bootstrap. In contrast to the NP classification, these data-driven approaches aim to minimize an estimated classification risk, not the type II error with a type I error constraint.

## 2.1 An Estimate of $C_\alpha$

Pinning down a good estimate of $C_\alpha$ is of central importance for classifiers under the NP paradigm. Contrary to common intuition, naïvely tuning the empirical type I error to $\alpha$ does not deliver a desirable classifier, as we have shown in Fig. 1. To facilitate our discussion, we assume that our sample contains $n$ i.i.d. observations $\mathcal{S}^1 = \{U_1, \cdots, U_n\}$ from class 1 with density $p$, and $m$ i.i.d. observations $\mathcal{S}^0 = \{V_1, \cdots, V_m\}$ from class 0 with density $q$. The sample $\mathcal{S}^0$ is decomposed as follows: $\mathcal{S}^0 = \mathcal{S}_1^0 \cup \mathcal{S}_2^0$, where $|\mathcal{S}_1^0| = m_1$ and $|\mathcal{S}_2^0| = m_2$. Below is a generic procedure introduced in our recent paper [58].

**General Neyman–Pearson Plug-In Procedure**

*Step 1*: Use $\mathcal{S}^1$ and $\mathcal{S}_1^0$ to construct a density ratio estimate $\hat{r}$.
*Step 2*: Given $\hat{r}$, choose a threshold estimate $\widehat{C}_\alpha$ from the set $\hat{r}(\mathcal{S}_2^0) = \{\hat{r}(V_{i+m_1})\}_{i=1}^{m_2}$. Denote by $\hat{r}_{(k)}(\mathcal{S}_2^0)$ the $k$th order statistic of $\hat{r}(\mathcal{S}_2^0)$, $k \in \{1, \cdots, m_2\}$. The corresponding plug-in classifier by setting $\widehat{C}_\alpha = \hat{r}_{(k)}(\mathcal{S}_2^0)$ is

$$\hat{\phi}_k(x) = \mathbb{I}\{\hat{r}(x) \geq \hat{r}_{(k)}(\mathcal{S}_2^0)\}. \tag{1}$$

The general strategy is that for any given estimate $\hat{r}$, we want to find a proper order statistic $\hat{r}_{(k)}(\mathcal{S}_2^0)$ to estimate the threshold $C_\alpha$, so that type I error of the classifier defined in (1) will be controlled from above by $\alpha$ with high probability $1 - \delta$. To achieve this, it is necessary to study the distribution of order statistics, which we find to be beta-distributed. Based on a concentration inequality for beta distributed variables, we have derived the following high probability bound for $R_0(\hat{\phi}_k)$:

**Proposition 1.** *Suppose $\hat{r}$ is such that $F_{0,\hat{r}}(t) = P_0(\hat{r}(X) \le t)$ is continuous almost surely. For any $\delta \in (0, 1)$ and $k \in \{1, \cdots, m_2\}$, it holds that*

$$\mathbb{P}\left(R_0(\hat{\phi}_k) > g(\delta, m_2, k)\right) \le \delta,$$

*where*

$$g(\delta, m_2, k) = \frac{m_2 + 1 - k}{m_2 + 1} + \sqrt{\frac{k(m_2 + 1 - k)}{\delta(m_2 + 2)(m_2 + 1)^2}}.$$

Let $\mathcal{K} = \mathcal{K}(\alpha, \delta, m_2) = \{k \in \{1, \cdots, m_2\} : g(\delta, m_2, k) \le \alpha\}$. Proposition 1 implies that $k \in \mathcal{K}(\alpha, \delta, m_2)$ is a sufficient condition for the classifier $\hat{\phi}_k$ to satisfy the NP oracle inequality (1). The next step is to characterize $\mathcal{K}$. The smallest $k \in \mathcal{K}$ accommodates small excess type II error for $\hat{\phi}_k$; for details, please see [58].

**Proposition 2.** *The minimum $k$ that satisfies $g(\delta, m_2, k) \le \alpha$ is $k_{\min} := \lceil A_{\alpha,\delta}(m_2) \cdot (m_2 + 1) \rceil$, where $\lceil z \rceil$ denotes the smallest integer larger than or equal to $z$ and*

$$A_{\alpha,\delta}(m_2) = \frac{1 + 2\delta(m_2 + 2)(1 - \alpha) + \sqrt{1 + 4\delta(1 - \alpha)\alpha(m_2 + 2)}}{2\left[\delta(m_2 + 2) + 1\right]}.$$

The choice $k_{\min}$ coupled with a good estimate of $r$ or $\eta$ delivers a plug-in NP classifier that satisfies the NP oracle inequalities. We have worked out estimates based on parametric and nonparametric naïve Bayes models [58], but estimates for more complex model assumptions are not yet developed. While these directions are interesting to explore, we would like to note a limitation in the use of theoretical estimates for the threshold $C_\alpha$. That is, the theoretical results require concentration inequalities, which are not specific to certain types of data distributions and sometimes give threshold estimates that are too conservative in practice. Therefore, we have developed an alternative route to implement the NP paradigm. This route makes the NP classification more adaptable to popular classification algorithms and thus more useful in practice.

## 2.2 The NP Umbrella Algorithm

Here we present the alternative route, the NP umbrella algorithm we developed in [47], as pseudocodes in Algorithm 1. The essential idea is to use bootstrap to approximate the distribution of type I errors and determine a threshold such that the corresponding classifier has type I errors bounded by a predefined level with high probability. This algorithm is widely applicable to the scoring type of classification methods, which include a wide range of popular methods, such as logistic regression [11], SVM [10], random forests [6], naïve Bayes [26], and neural networks [43]. Methods of the scoring type output a numeric value, i.e., a classification score, to represent the degree to which a test data point belongs to class 1. The classification scores can be strict probabilities or uncalibrated numeric values, as long as a higher score indicates a higher probability of an observation belonging to class 1. Many other classification methods that only output class labels can be converted to the scoring type via bagging to generate an ensemble of classifiers, each of which predicts a class label for a test data point, and the proportion of predicted labels being 1 serves as a classification score. Since almost all the state-of-the-art classification methods belong to or can be converted to the scoring type, this NP umbrella algorithm is easily adaptable in practice, though its theoretical properties are difficult to establish.

**Algorithm 1:** The NP umbrella classification algorithm

1: **input**:
  training data with two parts: a mixed i.i.d. sample $\mathscr{S}$ and a class 0 sample $\mathscr{S}^0 = \{X_1, \ldots, X_m\}$
  type I error upper bound $\alpha \in [0, 1]$
  small tolerance level $\delta \in (0, 1)$
  number of bootstrap samples $B$
2: **procedure** NPTHRESHOLD($\mathscr{S}, \mathscr{S}^0, \alpha, \delta, B$)
3: $f \leftarrow$ `classificationalgorithm`($\mathscr{S}$)
  ▷ *train a classification scoring function f by inputting $\mathscr{S}$ into the classification algorithm; let f have a larger expected value for class* 1 *data*
4: $T_0 = (T_{0,1}, \ldots, T_{0,m})^T \leftarrow (f(X_1), \ldots, f(X_m))^T$
  ▷ *apply the scoring function f to $\mathscr{S}^0$ to obtain a set of threshold candidates*
5: **for** $b$ *in* $\{1, \ldots, B\}$ **do** ▷ *bootstrap $T_0$ for B times*
6: $T_0^{(b)} = \left(T_{0,1}^{(b)}, \ldots, T_{0,m}^{(b)}\right)^T \leftarrow$ `sample`($T_0$, `size` $= m$, `replace` $=$ `TRUE`) ▷ *sample m points with replacement from $T_0$*
7: **for** $t$ *in* $T_0$ **do** ▷ *for each threshold candidate t*
8: **for** $b$ *in* $\{1, \ldots, B\}$ **do** ▷ *for each bootstrap sample b*
9: $e^{(b)}(t) \leftarrow \frac{1}{m}\sum_{i=1}^{m} \mathbb{I}\left(T_{0,i}^{(b)} > t\right)$ ▷ *calculate the type I error of threshold t in bootstrap sample b*
10: $v(t) \leftarrow \frac{1}{B}\sum_{b=1}^{B} \mathbb{I}\left(e^{(b)}(t) > \alpha\right)$ ▷ *calculate the violation rate of threshold t*
11: $t^* \leftarrow \min\{t : v(t) \leq \delta\}$ ▷ *pick the minimal threshold whose violation rate is under $\delta$*
12: **output**:
  an NP classifier $\phi(X) = \mathbb{I}\left(f(X) > t^*\right)$

# 3 Simulation

In this section, we demonstrate the use of NP classification with three popular classification algorithms: logistic regression [11], SVM [10], and random forests [6]. The three simulation examples, each employing one algorithm, are implemented by calling the R package nproc [15] we developed in recent work [47].

The nproc package can be installed by calling

```
> install.packages("nproc")
```

and loaded into the R environment using the next command. All the following numerical results were generated by the nproc package version 0.1.

```
> library(nproc)
```

We first simulate a training data set from a logistic regression model. The training data have 1000 observations with binary responses and two-dimensional features.

```
> # training data
> set.seed(1)
> x1 <- rnorm(1000)          # feature 1
> x2 <- rnorm(1000)          # feature 2
> x <- cbind(x1, x2)         # matrix of features
> z <- 1 + 2*x1 + 3*x2       # linear combination of the
                               two features
> pr <- 1/(1+exp(-z))        # logisitic function to
                             generate probability
> y <- rbinom(1000,1,pr) # response as Bernoulli
                             variable
> df <- data.frame(x1=x1, x2=x2, y=y)
```

Figure 3 shows a scatterplot of the training data. We also simulate 1000 test data sets with 1000 observations from the same model, to evaluate the performance of classifiers.

```
> # test data
> test_data <- lapply(1:1000, FUN=function(i) {
+   set.seed(i+1)
+   x1 <- rnorm(1000)
+   x2 <- rnorm(1000)
+   x <- cbind(x1, x2)
+   z <- 1 + 2*x1 + 3*x2
+   pr <- 1/(1+exp(-z))
+   y <- rbinom(1000, 1, pr)
+   df <- data.frame(x1=x1, x2=x2, y=y)
+   return(list(x=x, y=y, df=df))
+ })
```

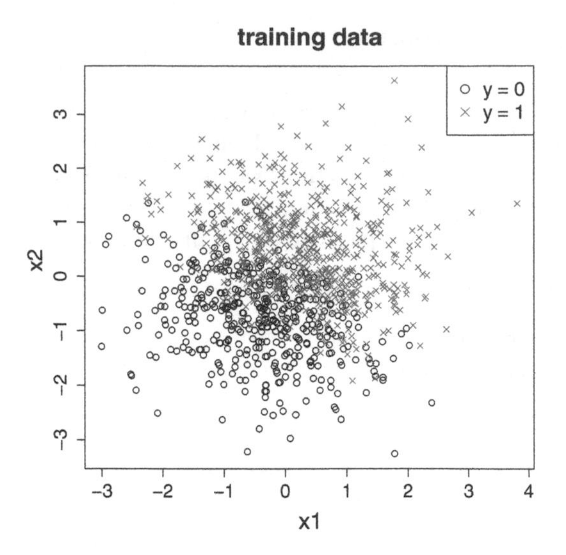


**Fig. 3** Scatterplot of the training data with 1000 observations and two-dimensional features. *Black circles* and *red crosses* represent class 0 and class 1, respectively

## 3.1 Logistic Regression

Logistic regression is a type of generalized linear model. For binary classification purposes, logistic regression can serve as a classification method. One way to interpret logistic regression is that it models

$$p := \mathbb{P}(Y = 1) = \frac{1}{1 + e^{-(\beta_0 + \beta^T X)}}$$

for a binary response $Y$ with $d$ features $X \in \mathbb{R}^d$. Given training data $\{(x_i, y_i)\}_{i=1}^n$, logistic regression estimates $\beta_0$ and $\beta$ as maximum likelihood estimates $\hat{\beta}_0$ and $\hat{\beta}$, and then estimates $p$ as $\hat{p} = \left(1 + e^{-(\hat{\beta}_0 + \hat{\beta}^T X)}\right)^{-1}$. This estimated probability can be interpreted as a classification score, as logistic regression predicts $Y$ by a linear decision rule, i.e., $\hat{Y} = \mathbb{I}(\hat{\beta}^T X \geq c) = \mathbb{I}\left(\hat{p} \geq \left(1 + e^{-\hat{\beta}_0 - c}\right)^{-1}\right)$ for some threshold $c \in \mathbb{R}$. Under the classical classification paradigm, $c = -\hat{\beta}_0$, which corresponds to $\hat{p} \geq 1/2$. Under the NP paradigm, we potentially have different choices for $c$. By regarding $\hat{p}$ as a classification scoring function of $X$, we use the umbrella algorithm to find a threshold on $\hat{p}$ so that the resulting classifier will have type I error below the desired level $\alpha$ with high probability close to $1 - \delta$.

We use the simulated data to demonstrate the use of NP classification with logistic regression and compare it with the classical paradigm. We first train a logistic regression model on the training data under the classical paradigm.

```
> lr_model1 <- glm(y~x1+x2, data=df, family="binomial")
```

Then we apply the trained model `lr_model1` to the 1000 test data sets to evaluate the distribution of its empirical type I errors on test data.

```
> lr_model1_err <- sapply(test_data, FUN=function(tdat)
{ +    pred <- predict(lr_model1, tdat$df,
type="response") > 0.5 +    ind0 <- which(tdat$y==0)
+    typeI <- mean(pred[ind0]!=tdat$y[ind0])
+ })
> summary(lr_model1_err)

   Min. 1st Qu.  Median    Mean 3rd Qu.    Max.
 0.1290  0.1723  0.1846  0.1849  0.1980  0.2368
```

We next train a logistic regression model on the training data under the NP paradigm with type I error bound $\alpha = 0.05$, using the npc function.

```
> set.seed(1001)
# for reproducible purposes, because npc()
  involves bootstrap,
# whose results will be reproducible with a fixed seed
> lr_model2 <- npc(x=x, y=y, method='logistic',
  alpha=0.05)
```

Then we also applied the trained model `lr_model2` to the test data.

```
> lr_model2_err <- sapply(test_data, FUN=function(tdat){
+    pred <- predict(lr_model2, tdat$x)
+    ind0 <- which(tdat$y==0)
+    typeI <- mean(pred$pred.label[ind0]!=tdat$y[ind0])
+ })
> summary(lr_model2_err)

    Min.  1st Qu.   Median     Mean  3rd Qu.     Max.
0.009732 0.027230 0.033530 0.033950 0.040400 0.062330

> sum(lr_model2_err <= 0.05) / 1000

[1] 0.948
```

Comparing the empirical type I errors of the two logistic regression classifiers found under the classical and the NP paradigm, respectively, we can see that the NP classifier gives much smaller type I errors, 94.8 % of which are under $\alpha = 0.05$.

## 3.2 *Support Vector Machines*

Similar to logistic regression, SVM is also a scoring type of classification method, for which approximate posterior probabilities of class labels proposed by Platt [35] can be used as classification scores. We demonstrate the use of NP classification with SVM on the simulated data as follows.

We first train an SVM model on the training data under the NP paradigm.

```
> set.seed(1001)
> svm_model <- npc(x=x, y=y, method='svm', alpha=0.05)
```

Then we apply the trained model `svm_model` to the 1000 test data sets to evaluate the distribution of its empirical type I errors on test data.

```
> svm_model_err <- sapply(test_data, FUN=function(tdat)
{
+    pred <- predict(svm_model, tdat$x)
+    ind0 <- which(tdat$y==0)
+    typeI <- mean(pred$pred.label[ind0]!=tdat$y[ind0])
+ })
> summary(svm_model_err)
    Min.  1st Qu.   Median     Mean  3rd Qu.     Max.
0.007772 0.023530 0.029560 0.029900 0.035350 0.062330
> sum(svm_model_err <= 0.05) / 1000
[1] 0.987
```

We see that the SVM classifier found by the NP algorithm has empirical type I errors under $\alpha = 0.05$ with high probability.

### 3.3 Random Forests

Random forests is another popular and powerful classification method. It is an ensemble method of tree-based classifiers. We can also interpret it as a scoring type of method, if we consider the proportion of output votes for class 1 (i.e., the proportion of trees that predict an observation as class 1) as classification scores. We demonstrate the use of NP classification with random forests on the toy data as follows.

We first train a random forest model on the training data under the NP paradigm.

```
> set.seed(1001)
> rf_model <- npc(x=x, y=y, method='randomforest',
  alpha=0.05)
```

Then we apply the trained model `rf_model` to the 1000 test data sets to evaluate the distribution of its empirical type I errors on test data.

```
> rf_model_err <- sapply(test_data, FUN=function(tdat) {
+   pred <- predict(rf_model, tdat$x)
+   ind0 <- which(tdat$y==0)
+   typeI <- mean(pred$pred.label[ind0]!=tdat$y[ind0])
+ })
> summary(rf_model_err)

    Min.  1st Qu.   Median     Mean  3rd Qu.     Max.
0.002451 0.021000 0.026280 0.026810 0.032260 0.054460

> sum(rf_model_err <= 0.05) / 1000

[1] 0.996
```

We see that the random forest classifier found by the NP algorithm has empirical type I errors under $\alpha = 0.05$ with high probability.

## 4 Case Study

We demonstrate the use of NP classification in a genomic case study on the prediction of transcription factor binding sites. A recent study [20] found that DNase-seq signals can well predict whether genomic regions containing transcription factor sequence motifs are transcription factor binding sites. DNase-seq is a recent high-throughput technology that combines traditional DNaseI footprinting [17] and next-generation DNA sequencing to identify genomic regions where regulatory factors interact with DNA to modify chromatin structure [5, 12, 31, 42]. An important question investigated in this study is which DNase-seq features can well predict binding sites of CTCF, a transcription factor that acts as an insulator to regulate the 3D structure of chromatin [34].

The study [20] formulated this question as a binary classification problem, where the goal is to classify genomic regions that contain CTCF sequence motifs into CTCF binding sites (i.e., class 1) and non-binding sites (i.e., class 0). For this task, two one-dimensional genomic features extracted from DNase-seq data are compared: (1) the number of DNase-seq tags in a 200 base pair window centered in each CTCF motif site and (2) the DNaseI footprint score calculated using the formula $f = -[(n_C + 1)/(n_R + 1) + (n_C + 1)/(n_L + 1)]$, where $n_C$, $n_R$, and $n_L$ represent, respectively, the tag count in the motif region and the flanking regions to the right and left of the motif (the lengths of the flanks are both the same as that of the motif). In the data, there are $n = 216,929$ genomic regions, each with one tag count and one footprint score. Among these regions, 27,220 regions were found as CTCF binding sites (class 0), and the rest 189,709 regions were considered as non-binding sites (class 1). By varying the threshold on each feature, the study showed that the footprint score outperforms the tag count at low FPR and underperforms at higher FPR. In other words, if users desire a small type I error, for example, if they prefer to predict fewer but more confident CTCF binding sites, the footprint score is a better genomic feature; otherwise if users prefer a small type II error, for example, if they prefer to predict more potential CTCF binding sites, the tag count is a better feature.

However, this study only reported the observed (empirical) type I and II errors on one data set, without assessing the randomness of these empirical errors. If a user is interested in knowing which feature is better when the type I error (or FDR) is constrained under $\alpha$ (e.g., 5 %) with high probability, this analysis cannot provide a good answer. Here we address this question using the NP classification algorithm described in Sect. 2.2. We also compare the performance of the NP classification with the common practice, which is to tune the empirical type I error on the training data to $\alpha$. At three different type I error bounds $\alpha = 0.01, 0.05$, and 0.1, we find their corresponding thresholds on the number of DNase-seq tags or the footprint scores via the common practice or the NP approach. Specifically, in the common practice we find the thresholds as the 99th, 95th, and 90th percentiles of the number of DNase-seq tags or the footprint scores of the 189, 709 non-binding sites; in the NP approach we use the NP algorithm with the number of bootstrap runs $B = 1000$ and the violation tolerance level $\delta = 0.05$ to find the thresholds on the number of DNase-seq tags or the footprint scores of the 189, 709 non-binding sites. Since values of each feature serve as natural classification scores in this case, we do not need to train a classification scoring function, and all the class 0 data points can be used to find the NP thresholds. That is, in the algorithm, $\mathscr{S}^0$ contains all the class 0 data points, and the $f$ function in Step 3 is just the identity map.

We evaluate the thresholds and their corresponding classifiers found by the NP approach or the common practice via bootstrap. We generate $B' = 1000$ sets of $n = 216,929$ bootstrap regions with corresponding tag counts and footprint scores from the original data via random sampling with replacement. Then we evaluate the empirical type I and type II errors of each threshold on each bootstrap data set, and summarize the distribution of the empirical type I and type II errors of all bootstrap data sets. The results in Table 1 show that the classifiers found by the NP approach have empirical type I errors bounded by $\alpha$ with high probability, while the classifiers found by the common practice have large portions of empirical type I errors above $\alpha$. Comparing the means of the empirical classification errors, we can see that the classifiers found by the NP approach have slightly smaller mean type I errors and slightly larger mean type II errors, a reasonable result given its high probability bound on the type I errors. The standard deviations are similar for the two approaches. Back to the question about which feature is better when the type I error is under $\alpha$ with high probability, the NP classification results suggest that at $\alpha = 0.01$ and 0.05, the footprint score is a better feature, while the tag count is a better feature at $\alpha = 0.1$.

**Table 1** Comparison of classifiers established by the common practice vs. the NP approach

| | | | Type I errors | | | Type II errors | |
|---|---|---|---|---|---|---|---|
| Features | $\alpha$ | Approaches | Mean[a] | sd[b] | $\% > \alpha$[c] | Mean | sd |
| Tag counts | 0.01 | Common | 0.010 | $2.32e^{-4}$ | 46.5 | 0.976 | $9.34e^{-4}$ |
| | | NP | 0.009 | $2.25e^{-4}$ | 0.00 | 0.979 | $8.84e^{-4}$ |
| | 0.05 | Common | 0.050 | $4.83e^{-4}$ | 48.4 | 0.707 | $2.78e^{-3}$ |
| | | NP | 0.049 | $4.78e^{-4}$ | 1.10 | 0.716 | $2.78e^{-3}$ |
| | 0.10 | Common | 0.100 | $6.89e^{-4}$ | 39.8 | 0.335 | $2.86e^{-3}$ |
| | | NP | 0.099 | $6.83e^{-4}$ | 3.90 | 0.340 | $2.87e^{-3}$ |
| Footprint scores | 0.01 | Common | 0.010 | $2.25e^{-4}$ | 48.6 | 0.766 | $2.59e^{-3}$ |
| | | NP | 0.009 | $2.07e^{-4}$ | 0.00 | 0.775 | $2.59e^{-3}$ |
| | 0.05 | Common | 0.050 | $4.96e^{-4}$ | 33.5 | 0.596 | $2.97e^{-3}$ |
| | | NP | 0.049 | $4.90e^{-4}$ | 1.90 | 0.598 | $2.96e^{-3}$ |
| | 0.10 | Common | 0.100 | $6.58e^{-4}$ | 49.5 | 0.493 | $3.07e^{-3}$ |
| | | NP | 0.099 | $6.54e^{-4}$ | 1.40 | 0.494 | $3.06e^{-3}$ |

[a] Mean of the empirical classification errors over the $B' = 1000$ bootstrap runs
[b] Standard deviation of the empirical classification errors over the $B' = 1000$ bootstrap runs
[c] Percentage of the empirical type I errors that are greater than $\alpha$ in the $B' = 1000$ bootstrap runs

# 5 Future Research and Genomic Applications of the Neyman–Pearson Classification

Neyman–Pearson classification paradigm handles binary class classification problem, but it can be extended to address multi-class problems where errors are asymmetric in nature. Neyman–Pearson classification has wide application potentials in genomics. In Sects. 5.2–5.4 we describe three potential applications as future research directions: sample size determination, automatic disease diagnosis, and disease marker detection. The latter two applications were also discussed in our recent review paper [48].

## 5.1 *Extension to Multi-class*

Originating from binary trade-offs, the NP classification methods can also be applied to multi-class ($Y \in \{1, \cdots, K\}$, $K \geq 3$) problems using the following two strategies:

- **[Strategy 1]** Missing class 1 has more severe consequences than missing other classes. A two-step procedure can be implemented: Apply an NP method to classify a subject into class 1 versus the other classes. Stop if the subject is assigned to class 1. Otherwise, continue and apply a (multi-class) classification algorithm to assign this subject to one of the other classes $\{2, \cdots, K\}$.

- **[Strategy 2]** There is a hierarchical order (class $1 > \cdots >$ class $K$) of class priorities (i.e., severity of missing each class). A possible procedure is to first apply an NP method to classify a subject into class 1 versus other classes $\{2, \cdots, K\}$. Stop if this subject is assigned to class 1. Otherwise, apply again the NP method to classify it into class 2 versus classes $\{3, \cdots, K\}$. Continue along this line until this subject is assigned to a class.

## 5.2 Sample Size Determination

In clinical trials and other experimental designs, sample acquisition can be quite expensive. Therefore, how to determine the minimal sufficient sample size is an important question. Admittedly, different criteria would lead to different sample size selection procedures; so there is no hope of finding a universally golden rule. The Neyman–Pearson paradigm inspires one procedure to choose the sample size based on the theoretical upper bound of the excess type II error in the NP oracle inequalities.

We describe a contrived version of this procedure in the following. Suppose a classifier $\hat{\varphi}$ satisfies the NP oracle inequalities, and the excess type II error of $\hat{\varphi}$ is bounded from above by $f(m, n)$, where $m$ is the sample size of class 0, and $n$ is the sample size of class 1, and $f(m, n) \to 0$ as $m$ and $n$ go to infinity. If the user has a target to control the excess type II error at some $\gamma > 0$. Then $m$ and $n$ can be selected such that $f(m, n) \leq \gamma$. As this procedure is based on concentration inequalities, it is conservative and might overestimate the sample size. Yet this procedure provides a valid upper bound of the needed sample size. Future statistical research is need for more accurate sample size determination.

## 5.3 Automatic Disease Diagnosis

Another application is a long-time challenge in clinical research: automatic disease diagnosis from patient genomic data. This challenge involves a classification problem, where diseases correspond to different classes, and the goal is to predict the diseases that are most likely associated with a patient's genomic sample. Thanks to the development of high-throughput genomic technologies [e.g., microarray and next-generation sequencing (NGS)], a large amount of disease related genomic data can serve as training data in this classification problem. Taking gene expression data as an example, the National Center for Biotechnology Information (NCBI) Gene Expression Omnibus (GEO) contains more than 690,000 human gene expression samples that are related to hundreds of diseases, such as heart diseases, mental illnesses, infectious diseases, and various cancers.

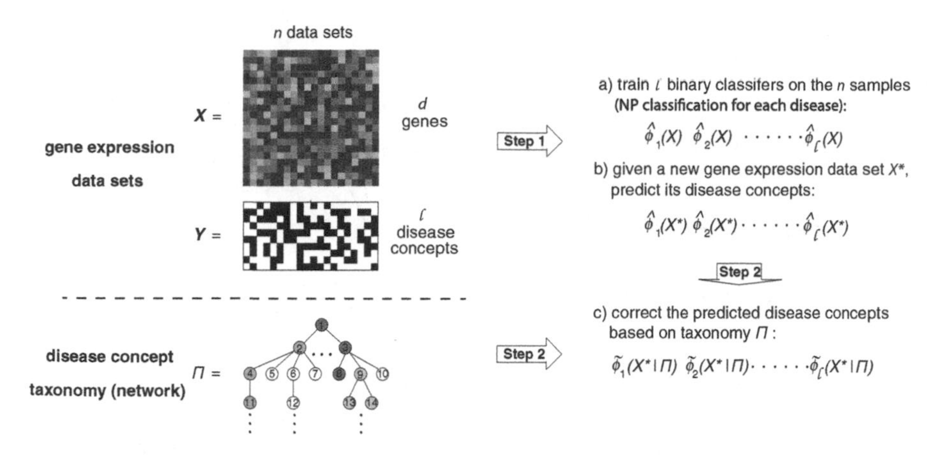


**Fig. 4** Automatic disease diagnosis via NP classification and network-assisted correction

We can study automatic disease diagnosis by NP classification and network-assisted correction using a two-step approach (Fig. 4). Step 1: using public microarray and NGS gene expression data sets with disease labels (e.g., $>$ 100 Unified Medical Language System standardized disease concepts), we can (a) use NP classification to build a binary classifier for each disease class, and (b) classify a patient's microarray gene expression sample into these disease classes. Step 2: (c) correct the predicted diseases based on the disease taxonomy (network). In Step 1, since the disease classes are non-exclusive (one data set may have multiple disease class labels), this multi-label classification problem is inherently composed of multiple binary classification problems, where every disease class needs a binary decision. In previous works [21, 27], binary classifiers such as SVM and naïve Bayes classifiers were used, and all disease classes were treated as interchangeable. This raises an important issue, though: some diseases are more life-threatening than others, e.g., lung cancer vs. arthritis. Therefore, it is important to allow doctors to have different levels of conservativeness, i.e., different thresholds $\alpha$ on the type I error (the chance of missing a disease when a patient in fact has it), for different diseases. Although previous researchers have attempted to address this trade-off between false positives and false negatives in disease diagnosis [16], they failed to control false negative rates under a desired threshold with high probability. Given the pressing need for precise disease diagnosis, the developed NP classification algorithms are in high demand to address this issue.

## 5.4 Disease Marker Detection

The multi-class extension of NP classification has application potentials in detecting and screening for key markers (i.e., genes and genomic features) to aid disease diagnosis as well as to understand molecular mechanisms of diseases. In early cancer diagnosis studies that aimed to determine which genes should be included

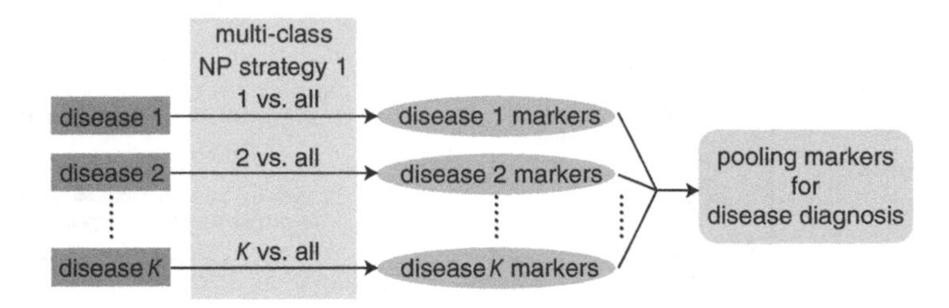


**Fig. 5** Marker detection via multi-class NP strategy 1

as features (markers) [16, 36, 41], classification error of each disease class versus others was used as a criterion. In other words, "the smallest set" of genes that results in low classification error for a disease class was retained as markers for that disease. However, this criterion lacks consideration of asymmetric classification errors, and as a result, the selected markers for a disease could lead to high false negative rates in the diagnosis—a dangerous situation for severe diseases such as cancers. Therefore, in the diagnosis of severe diseases, a more reasonable criterion would be to minimize the FPR given a pre-specified false negative rate control. The multi-class NP classification (Strategy 1) serves the purpose: key markers are selected so that low FPR are attained while the false negative rates are constrained below a threshold (see Fig. 5). Markers selected by this new detection strategy can be pooled to make disease prediction, in the hope of increasing the sensitivity of disease diagnosis. To implement and evaluate this strategy, we need to compare it with the more recent state-of-the-art disease prediction methods, which are for example based on multi-task learning [2, 28, 57, 59], group lasso [29], multicategory support vector machines [25], partial least squares regression [4, 33], neural networks [22, 51], and others [44].

**Acknowledgements** Dr. Jingyi Jessica Li's work was supported by the start-up fund of the UCLA Department of Statistics and the Hellman Fellowship. Dr. Xin Tong's work was supported by Zumberge Individual Award from University of Southern California and summer research support from Marshall School of Business. We thank Dr. Yang Feng in Department of Statistics at Columbia University and Ms. Anqi Zhao in Department of Statistics at Harvard University for their help in developing the Neyman–Pearson classification algorithms. We also thank Dr. Wei Li and Mr. Sheng'en Shawn Hu in Dr. X. Shirley Liu's group in Department of Biostatistics and Computational Biology at Dana-Farber Cancer Institute and Harvard School of Public Health for kindly sharing the data for our genomic case study in Sect. 4.

## References

1. Audibert, J., Tsybakov, A.: Fast learning rates for plug-in classifiers under the margin condition. Annals of Statistics **35**, 608–633 (2007)
2. Bi, J., Xiong, T., Yu, S., Dundar, M., Rao, R.B.: An improved multi-task learning approach with applications in medical diagnosis. In: Machine Learning and Knowledge Discovery in Databases, pp. 117–132. Springer (2008)

3. Blanchard, G., Lee, G., Scott, C.: Semi-supervised novelty detection. Journal of Machine Learning Research **11**, 2973–3009 (2010)
4. Booij, B.B., Lindahl, T., Wetterberg, P., Skaane, N.V., Sæbø, S., Feten, G., Rye, P.D., Kristiansen, L.I., Hagen, N., Jensen, M., et al.: A gene expression pattern in blood for the early detection of Alzheimer's disease. Journal of Alzheimer's Disease **23**(1), 109–119 (2011)
5. Boyle, A.P., Song, L., Lee, B.K., London, D., Keefe, D., Birney, E., Iyer, V.R., Crawford, G.E., Furey, T.S.: High-resolution genome-wide in vivo footprinting of diverse transcription factors in human cells. Genome research **21**(3), 456–464 (2011)
6. Breiman, L.: Random forests. Machine learning **45**(1), 5–32 (2001)
7. Bulyk, M.L., et al.: Computational prediction of transcription-factor binding site locations. Genome biology **5**(1), 201–201 (2004)
8. Cannon, A., Howse, J., Hush, D., Scovel, C.: Learning with the Neyman-Pearson and min-max criteria. Technical Report LA-UR-02-2951 (2002)
9. Casasent, D., Chen, X.: Radial basis function neural networks for nonlinear fisher discrimination and Neyman-Pearson classification. Neural Networks **16**(5–6), 529–535 (2003)
10. Cortes, C., Vapnik, V.: Support-vector networks. Machine learning **20**(3), 273–297 (1995)
11. Cox, D.R.: The regression analysis of binary sequences. Journal of the Royal Statistical Society. Series B (Methodological) pp. 215–242 (1958)
12. Degner, J.F., Pai, A.A., Pique-Regi, R., Veyrieras, J.B., Gaffney, D.J., Pickrell, J.K., De Leon, S., Michelini, K., Lewellen, N., Crawford, G.E., et al.: DNase I sensitivity QTLs are a major determinant of human expression variation. Nature **482**(7385), 390–394 (2012)
13. Dümbgen, L., Igl, B., Munk, A.: P-values for classification. Electronic Journal of Statistics **2**, 468–493 (2008)
14. Elkan, C.: The foundations of cost-sensitive learning. In Proceedings of the Seventeenth International Joint Conference on Artificial Intelligence pp. 973–978 (2001)
15. Feng, Y., Li, J., Tong, X.: nproc: Neyman-Pearson Receiver Operator Curve (2016). URL http://CRAN.R-project.org/package=nproc. R package version 0.1
16. Furey, T.S., Cristianini, N., Duffy, N., Bednarski, D.W., Schummer, M., Haussler, D.: Support vector machine classification and validation of cancer tissue samples using microarray expression data. Bioinformatics **16**(10), 906–914 (2000)
17. Galas, D.J., Schmitz, A.: DNase footprinting a simple method for the detection of protein-DNA binding specificity. Nucleic acids research **5**(9), 3157–3170 (1978)
18. Golub, T.R., Slonim, D.K., Tamayo, P., Huard, C., Gaasenbeek, M., Mesirov, J.P., Coller, H., Loh, M.L., Downing, J.R., Caligiuri, M.A., et al.: Molecular classification of cancer: class discovery and class prediction by gene expression monitoring. science **286**(5439), 531–537 (1999)
19. Han, M., Chen, D., Sun, Z.: Analysis to Neyman-Pearson classification with convex loss function. Anal. Theory Appl. **24**(1), 18–28 (2008). DOI 10.1007/s10496-008-0018-3
20. He, H.H., Meyer, C.A., Chen, M.W., Zang, C., Liu, Y., Rao, P.K., Fei, T., Xu, H., Long, H., Liu, X.S., et al.: Refined DNase-seq protocol and data analysis reveals intrinsic bias in transcription factor footprint identification. Nature methods **11**(1), 73–78 (2014)
21. Huang, H., Liu, C.C., Zhou, X.J.: Bayesian approach to transforming public gene expression repositories into disease diagnosis databases. Proceedings of the National Academy of Sciences **107**(15), 6823–6828 (2010)
22. Khan, J., Wei, J.S., Ringner, M., Saal, L.H., Ladanyi, M., Westermann, F., Berthold, F., Schwab, M., Antonescu, C.R., Peterson, C., et al.: Classification and diagnostic prediction of cancers using gene expression profiling and artificial neural networks. Nature medicine **7**(6), 673–679 (2001)
23. Koltchinskii, V.: Oracle Inequalities in Empirical Risk Minimization and Sparse Recovery Problems (2008)
24. Kotsiantis, S.B., Zaharakis, I., Pintelas, P.: Supervised machine learning: A review of classification techniques. Informatica **31**, 249–268 (2007)
25. Lee, Y., Lee, C.K.: Classification of multiple cancer types by multicategory support vector machines using gene expression data. Bioinformatics **19**(9), 1132–1139 (2003)

26. Lewis, D.D.: Naive (Bayes) at forty: The independence assumption in information retrieval. In: Machine learning: ECML-98, pp. 4–15. Springer (1998)
27. Liu, C.C., Hu, J., Kalakrishnan, M., Huang, H., Zhou, X.J.: Integrative disease classification based on cross-platform microarray data. BMC Bioinformatics **10**(Suppl 1), S25 (2009)
28. Liu, F., Wee, C.Y., Chen, H., Shen, D.: Inter-modality relationship constrained multi-modality multi-task feature selection for Alzheimer's disease and mild cognitive impairment identification. NeuroImage **84**, 466–475 (2014)
29. Ma, S., Song, X., Huang, J.: Supervised group lasso with applications to microarray data analysis. BMC bioinformatics **8**(1), 1 (2007)
30. Mammen, E., Tsybakov, A.: Smooth discrimination analysis. Annals of Statistics **27**, 1808–1829 (1999)
31. Neph, S., Vierstra, J., Stergachis, A.B., Reynolds, A.P., Haugen, E., Vernot, B., Thurman, R.E., John, S., Sandstrom, R., Johnson, A.K., et al.: An expansive human regulatory lexicon encoded in transcription factor footprints. Nature **489**(7414), 83–90 (2012)
32. Ng, K.L.S., Mishra, S.K.: De novo svm classification of precursor microRNAs from genomic pseudo hairpins using global and intrinsic folding measures. Bioinformatics **23**(11), 1321–1330 (2007)
33. Park, P.J., Tian, L., Kohane, I.S.: Linking gene expression data with patient survival times using partial least squares. Bioinformatics **18**(suppl 1), S120–S127 (2002)
34. Phillips, J.E., Corces, V.G.: Ctcf: master weaver of the genome. Cell **137**(7), 1194–1211 (2009)
35. Platt, J., et al.: Probabilistic outputs for support vector machines and comparisons to regularized likelihood methods. Advances in large margin classifiers **10**(3), 61–74 (1999)
36. Ramaswamy, S., Tamayo, P., Rifkin, R., Mukherjee, S., Yeang, C.H., Angelo, M., Ladd, C., Reich, M., Latulippe, E., Mesirov, J.P., Poggio, T., Gerald, W., Loda, M., Lander, E.S., Golub, T.R.: Multiclass cancer diagnosis using tumor gene expression signatures. Proceedings of the National Academy of Sciences **98**(26), 15,149–15,154 (2001)
37. Rigollet, P., Tong, X.: Neyman-Pearson classification, convexity and stochastic constraints. Journal of Machine Learning Research **12**, 2831–2855 (2011)
38. Scott, C.: Comparison and design of Neyman-Pearson classifiers. Unpublished (2005)
39. Scott, C.: Performance measures for Neyman-Pearson classification. IEEE Transactions on Information Theory **53**(8), 2852–2863 (2007)
40. Scott, C., Nowak, R.: A Neyman-Pearson approach to statistical learning. IEEE Transactions on Information Theory **51**(11), 3806–3819 (2005)
41. Segal, N.H., Pavlidis, P., Antonescu, C.R., Maki, R.G., Noble, W.S., DeSantis, D., Woodruff, J.M., Lewis, J.J., Brennan, M.F., Houghton, A.N., Cordon-Cardo, C.: Classification and subtype prediction of adult soft tissue sarcoma by functional genomics. The American Journal of Pathology **163**(2), 691–700 (2003)
42. Song, L., Zhang, Z., Grasfeder, L.L., Boyle, A.P., Giresi, P.G., Lee, B.K., Sheffield, N.C., Gräf, S., Huss, M., Keefe, D., et al.: Open chromatin defined by DNaseI and faire identifies regulatory elements that shape cell-type identity. Genome research **21**(10), 1757–1767 (2011)
43. Specht, D.F.: Probabilistic neural networks. Neural networks **3**(1), 109–118 (1990)
44. Statnikov, A., Aliferis, C.F., Tsamardinos, I., Hardin, D., Levy, S.: A comprehensive evaluation of multicategory classification methods for microarray gene expression cancer diagnosis. Bioinformatics **21**(5), 631–643 (2005)
45. Tarigan, B., van de Geer, S.: Classifiers of support vector machine type with l1 complexity regularization. Bernoulli **12**, 1045–1076 (2006)
46. Tong, X.: A plug-in approach to Neyman-Pearson classification. Journal of Machine Learning Research **14**, 3011–3040 (2013)
47. Tong, X., Feng, Y., Li, J.J.: Neyman-pearson (np) classification algorithms and np receiver operating characteristic (np-roc) curves Manuscript
48. Tong, X., Feng, Y., Zhao, A.: A survey on Neyman-Pearson classification and suggestions for future research. Wiley Interdisciplinary Reviews: Computational Statistics **8**, 64–81 (2016)
49. Tsybakov, A.: Optimal aggregation of classifiers in statistical learning. Annals of Statistics **32**, 135–166 (2004)

50. Tsybakov, A., van de Geer, S.: Square root penalty: Adaptation to the margin in classification and in edge estimation. Annals of Statistics **33**, 1203–1224 (2005)
51. Wei, J.S., Greer, B.T., Westermann, F., Steinberg, S.M., Son, C.G., Chen, Q.R., Whiteford, C.C., Bilke, S., Krasnoselsky, A.L., Cenacchi, N., et al.: Prediction of clinical outcome using gene expression profiling and artificial neural networks for patients with neuroblastoma. Cancer research **64**(19), 6883–6891 (2004)
52. Wu, S., Lin, K., Chen, C., M., C.: Asymmetric support vector machines: low false-positive learning under the user tolerance (2008)
53. Xing, E.P., Jordan, M.I., Karp, R.M., et al.: Feature selection for high-dimensional genomic microarray data. In: ICML, vol. 1, pp. 601–608. Citeseer (2001)
54. Yanai, I., Benjamin, H., Shmoish, M., Chalifa-Caspi, V., Shklar, M., Ophir, R., Bar-Even, A., Horn-Saban, S., Safran, M., Domany, E., et al.: Genome-wide midrange transcription profiles reveal expression level relationships in human tissue specification. Bioinformatics **21**(5), 650–659 (2005)
55. Yang, Y.: Minimax nonparametric classification-part i: rates of convergence. IEEE Transaction Information Theory **45**, 2271–2284 (1999)
56. Zadrozny, B., Langford, J., Abe, N.: Cost-sensitive learning by cost-proportionate example weighting. IEEE International Conference on Data Mining p. 435 (2003)
57. Zhang, D., Shen, D., Initiative, A.D.N., et al.: Multi-modal multi-task learning for joint prediction of multiple regression and classification variables in Alzheimer's disease. NeuroImage **59**(2), 895–907 (2012)
58. Zhao, A., Feng, Y., Wang, L., Tong, X.: Neyman-Pearson classification under high dimensional settings (2015). URL http://arxiv.org/abs/1508.03106
59. Zhou, J., Yuan, L., Liu, J., Ye, J.: A multi-task learning formulation for predicting disease progression. In: Proceedings of the 17th ACM SIGKDD international conference on Knowledge discovery and data mining, pp. 814–822. ACM (2011)

# Part II
# Computational Analytics

# Improving Re-annotation of Annotated Eukaryotic Genomes

**Shishir K. Gupta*, Elena Bencurova*, Mugdha Srivastava*, Pirasteh Pahlavan, Johannes Balkenhol, and Thomas Dandekar**

**Abstract** In the age of post-genomics, the task of improving existing annotation is one of the major challenge. The sequenced transcriptome allows to revisit the annotated sequenced genome of the corresponding organism and improve the existing gene models. In addition, misleading annotations propagate in multiple databases by comparative approaches of annotation, automatic annotation, and lack of curating power in the face of large data volume. In this pursuit, re-annotated improved gene models can prevent misleading structural and functional annotation of genes and proteins. In this chapter, we will highlight annotation and re-annotation procedures and will explain how annotations can be improved using computational methods. Our integrative workflow can be used to re-annotate genomes of any sequenced eukaryotic organism. We describe the annotation of splice sites, open reading frames, encoded proteins and peptides, hints for functional annotation including phylogenetic and domain analysis as well as critical evaluation of data transfer procedures, and the genome annotation process.

---

*Authors contributed equally with all other contributors.

S.K. Gupta
Department of Bioinformatics, Biocenter, University of Würzburg, Am Hubland, 97074 Wuerzburg, Germany

Department of Microbiology, Biocenter, University of Würzburg, Am Hubland, 97074 Wuerzburg, Germany
e-mail: shishir.gupta@uni-wuerzburg.de

E. Bencurova
Department of Bioinformatics, Biocenter, University of Würzburg, Am Hubland, 97074 Wuerzburg, Germany

University of Veterinary Medicine and Pharmacy, Komenskeho 73, Kosice, 04181 Slovakia
e-mail: elena.bencurova@uni-wuerzburg.de

M. Srivastava • J. Balkenhol
Department of Bioinformatics, Biocenter, University of Würzburg, Am Hubland, 97074 Wuerzburg, Germany
e-mail: mugdha.srivastava@uni-wuerzburg.de; johannes.balkenhol@uni-wuerzburg.de



K.-C. Wong (ed.), *Big Data Analytics in Genomics*,
DOI 10.1007/978-3-319-41279-5_5

## 1 Background

In 1975, the new era of the biological research started with the development of DNA sequencing. Only 15 years later the discovery of DNA double helix, Sanger sequencing method, had predestinated further direction of the nature sciences. Since the first sequenced genome—phage ΦX174 [1], till date, more than 4000 genomes have been sequenced (1081 eukaryotic genomes, KEGG Genome database, December 2015) including the genomes of several bacteria, followed by yeasts, archaeans, virus, plants, animals and, of course, the human genome with its over 3 billion bases [2]. The development of new techniques, improvement of the methods, and, at least, the hunger for new knowledge significantly changed our view on the structure and function of genes. As more and more genomes are sequenced, there is an increasing need for the fast, precise, and accurate sequencing and data analysis. The simple Sanger sequencing is currently not fast enough and the novel and enhanced sequencing platforms becoming the daily routine in the research—the development of pyrosequencing in the 1990s, followed by 454 pyrosequencing, semiconductor and epigenetic sequencing, the RNA and transcriptome sequencing, or new generation sequencing (NGS) entering the modern phase of the biological research. The new technologies allow to analyze not only the isolated in vitro cultures, but also the ecological and environmental samples.

One of the most popular sequencing platform, NGS, generates thousands and thousands of sequencing reads in parallel. Presently, there are several NGS platforms families; e.g., 454, Illumina, SOLiD, PacBio, Ion Torrent, and many more, which differ in the type of clonal amplification, used chemistry, and read length. For example, Ion Torrent produces output with up to 50 million reads per run with the reads length of approximately 200 bases, while SOLiD is able to generate approximately three billion reads per run with 75-bases long reads. The longest length of read, up to 1000 bases with one million reads per run, is able to obtain by 454 NGS sequencing platform that is based on pyrosequencing method [3]. Typically, the reads can be acquired from a pool of PCR-amplicons, cDNA libraries, or fragmented libraries. However, there are several crucial factors influencing the accuracy of the sequencing results: the read length and precision

P. Pahlavan
Department of Bioinformatics, Biocenter, University of Würzburg, Am Hubland, 97074 Wuerzburg, Germany

Leibniz-Institute, German Collection of Microorganisms and Cell Cultures Gmbh, Braunschweig, Germany
e-mail: pirasteh.pahlavan@uni-wuerzburg.de

T. Dandekar (✉)
Department of Bioinformatics, Biocenter, University of Würzburg, Am Hubland, 97074 Wuerzburg, Germany

EMBL Heidelberg, BioComputing Unit, Meyerhofstraße 1, 69117 Heidelberg, Germany
e-mail: dandekar@biozentrum.uni-wuerzburg.de

of the sequencing, cost of the experiments and the formation of the PCR bias, chimeric sequences, and secondary structure-related matters [4]. Nevertheless NGS is a powerful and valuable tool with significant benefaction for the medical, pharmacological, environmental and ecological, and forensics studies.

The obtaining of the DNA sequences of the genome is just one side of the coin. Reads represent the huge dataset containing the fragments of genes, which must be annotated in order to obtain the overall view of the genomes. Gene annotation, in a broad sense, is the description and localization of the genes in the genome by computational approach. This comprehends structural and functional characterization of protein-coding and non-coding genes, as well as other genomic features. However, it seems as an unpretentious topic, the genome contains a various repetitive sequences—transposable elements, simple sequence repeats, as well as single nucleotide polymorphisms, that can vary in the gene expression. Those factors, together with the infrequent phenotype of individuals, presence of non-coding regions and short reads of highly duplicated sequences, make the genome annotation and analysis one of the most challenging tasks in biology. As the raw data contain several thousands of the sequences, the choice of the annotation process is crucial. The manual annotation, based on information originated from sequence homology searches, is enormously time-consuming and labor-intense. Even the annotation by experts, albeit presumably most accurate, is not always affordable, and highly automated computational methods are called upon to fill the gap [5, 6]. The second option, automated annotation, provides the practical and faster way to obtain precise information; however, the part of the genome annotation is still based on the researcher. Despite very perspective results from manual or automated approach, it is very hard to annotate all the genes. Currently, well-studied genomes contain hundreds of the genes coding "hypothetical" proteins, without predicted function or structure. Those parts of the genomes remain still unannotated or the level of the annotation is very low, even compared to evolutionary close species.

As the amount of the obtained data is increasing every day, the need for the useful databases and bioinformatics tools is enormous. The first program for the DNA sequence analysis was written in COBOL by McCallum and Smith [7], and rapidly new programs to interpret and annotate sequence data became more and more available.

A short time after, development of searching program BLAST in 1990 brings the new view to identification genes and comparison to other sequences [8]. The first swallow for the storage of the obtained data was the creation of the nucleotide and protein sequence database, which was established in 1981 [9]. Only 5 years after, the National Institute of Health created the most used and most influential database GenBank, which is collection of all available nucleotide sequences and its protein translations [10].

The aim of this chapter is to describe major modern tools for genome annotations, discusses some of the basic errors, and how to avoid them. Also, we would like to offer the reader deep understanding to the processes, which precluding the whole process of genome annotation.

## 2 Genome Annotation

The tremendous amount of data generated by advances in NGS projects has motivated more reliable annotation of the genome. In general, the process of interpreting raw sequence data into biological information is the process of annotation [11]. We earlier stressed the need for continuously improving genome annotation, for instance, by reanalyzing the annotation of a genome again after 5 years. The growth of databases, improvement of software, and meticulous re-annotation then led in the cases examined to improved annotation for the major part of the genome annotations reanalyzed, including both RNA and protein genes [12, 13].

The primary step of genome annotation is to find the sequence that codes the gene. Recent sophisticated software is using the various algorithms (e.g., sensor algorithm, WeederH algorithm) to identify the key structural features, e.g., transcriptional start sites, and the sequence of the potential gene is then discovered by scanning the reads in all existing open reading frames (ORFs). On sequence-labeling effort in genome annotation, discriminative strategies have been shown to outperform generative hidden Markov models (HMMs) [14, 15]. Once the genome sequence is completed, gene is identified according to sequence similarities and ab initio methods, and then further annotated functionally by searching the protein families and motifs, their corresponding binding sites on the DNA, or by application of the phylogenetic footprinting. Moreover, two levels of computational genome annotation can be applied: "single species: many genes" or "single gene: many species" aspects, which is utilized mainly for the identification of the conserved motifs [16, 17].

The annotation is a complex process. It includes various methods to gain the accurate localization and function of genes: from the gapped BLAST (structural approach of annotation) through motif analysis to the detection of homologs (functional approach of annotation). The structural annotation relies mostly on the identification of ORFs and gene candidates in DNA sequence using a computational algorithm, and usually is less precise. The functional annotation, which uses sequence similarities searches against the genes of know functions (templates), is more precise; however, it is not suitable for the genomes, such as phylogenetically distant genome. The typical genome annotation has been made with functional annotation, usually from 54 to 79 % of protein-coding genes.

Briefly, the annotation denotes and demarcates the genomic elements in the genome and subsequently link these genomic elements to biological function. Therefore, the annotation process can be referred to as the combination of two different process, i.e., gene structure prediction and functional annotation. The main difference between structural and functional annotation will be described later in this chapter.

## 2.1 Genome Re-annotation

The annotation projects of large eukaryotic genomes often provide a broad but shallow view of structure and function of genes as their central perspective is to offer overview of entire genome rather than defining individual genes [11]. The re-annotation can be defined as the improved structural and functional annotation of previously annotated genome by integrating the advanced computational analysis, auxiliary biological data, and biological expertise to improve the existing annotation. The key issue of genome annotation is its accuracy. Generally, a first genome annotation still contains large numbers of errors, which may cause wrong conclusions during further analysis. Critical is detection of sequence conservation, as otherwise functional sites are overlooked, as well as sufficient repetitions in the raw data. An erroneous annotation can arise from numerous sources; for instance, errors in the gene model caused by incorrect predictions of splicing junctions, erroneous, and inconsistent gene naming owing to transferred annotation from the gene based on sequence similarity where the original gene name is itself incorrect. The flawed functional annotation is inevitably propagated by BLAST based annotations if not followed by the extensive and careful manual curations. In this scenario, the re-annotation procedure improves the improper original annotations; for instance, the recent update of *Apis mellifera* genome resulted in annotation of ~5000 more protein-coding genes that were not previously reported [18].

Extending own previous earlier efforts in prokaryotes [12, 13], we will focus on how the re-annotation process in eukaryotes can improve the original annotation and some tools which can be used for this process. Several re-annotation procedures have been conducted for eukaryotic organisms in last 6 years (Table 1), which in general, are quite less in compare to re-annotation efforts for prokaryotes. These studies have shown increase in the number of protein-coding genes, missing gene identification, increase in the overall total coding length, more number of coding regions, increase in untranslated region (UTR) coverage, splice site corrections in the gene models and detection of alternative splicing events that further allows more accurate gene models, and predicted protein sequence and annotations. The reader is referred to the literature pointed in Table 1 for the details of improvements as per individual re-annotation efforts. For a gene, good re-annotation procedure can improve both the gene structure and its functional annotation together. For instance, in a recent study, Gupta and coworkers found the presence of pattern recognition receptor PGRP-LC in ant *Camponotus floridanus* genome [19], which was wrongly annotated as PGRP-LE in the official version of the ant genome cflo3.3 [20]. It was found that PGRP-LE consists of a long N-terminal domain and a transmembrane domain. Before the re-annotation it was named as PGRP-LE. The gene structure of PGRP-LE was improved during re-annotation and corrected by addition of the missing N-terminal (Fig. 1) further, the correct name of the protein was adapted from high similarity with *A. mellifera* PGRP-LE during function annotation. This wrong annotation occurred probably because of the high similarity between PGRP-LC and PGRP-LE. In fly these two proteins even have

**Table 1** List of re-annotated eukaryotes

| Organism | Common name | References |
|---|---|---|
| *Anolis carolinensis* | Lizard (green anole) | [36] |
| *Apis mellifera* | Honey bee | [18] |
| *Arabidopsis lyrata* | (Plant) | [80] |
| *Camponotus floridanus* | Carpenter ant | [19] |
| *Cryptosporidium parvum, C. hominis* | (Protozoa) | [81] |
| *Drosophila melanogaster* | Fruit fly | [82] |
| *Entamoeba histolytica* | Intestinal parasitic protozoan | [37] |
| *Fragaria vesca* | Wild strawberry | [39] |
| *Gallus gallus domesticus* | Chicken | [83] |
| *Rattus norvegicus* | Rat | [84] |
| *Trichoderma reesei* | (Fungus) | [85] |

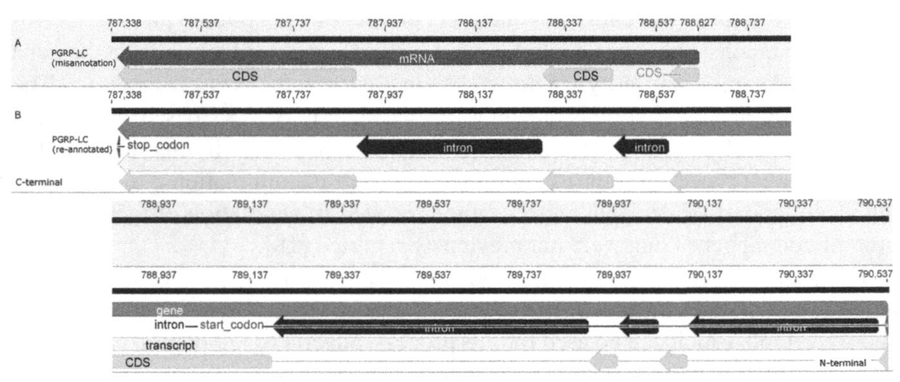


**Fig. 1** Re-annotation of carpenter ant PGRP-LC gene. (**a**) Original annotation PGRP-LC (wrongly annotated as PGRP-LE; accession number EFN63542.1/Cflo_03358) in cflo_OGSv3.3 and (**b**) Re-annotated version of carpenter ant PGRP-LC (accession number Cflo_N_g10272t1)

high similarity at the structural level with root mean square deviation of ~0.75 A. In such cases, conclusions based on similarity and whether the annotation is correct or incorrect are cumbersome, therefore computationally the strategies like orthologous clustering, domain analysis, and gene bases phylogeny could be implemented for improvement in function annotation. Expertise in biology also serves with a great potential for re-annotation such as here the absence of extracellularly located pattern recognition receptor encoded by *C. floridanus* was quite surprising which directed the re-annotator to verify the annotation.

One functional annotation error often occurs in the annotation of nitric oxide synthase (NOS) enzymes due to their significant similarity with NADPH-cytochrome P450 reductase. Both may have the similar flavodoxin/NOS and NADPH-cytochrome p450 reductase and FAD-binding alpha-helical domains, hence, the BLAST based annotation may carry artifact. This could be corrected by observing the presence of N-terminal NOS domain in the sequences as cytochrome p450 lacks this domain.

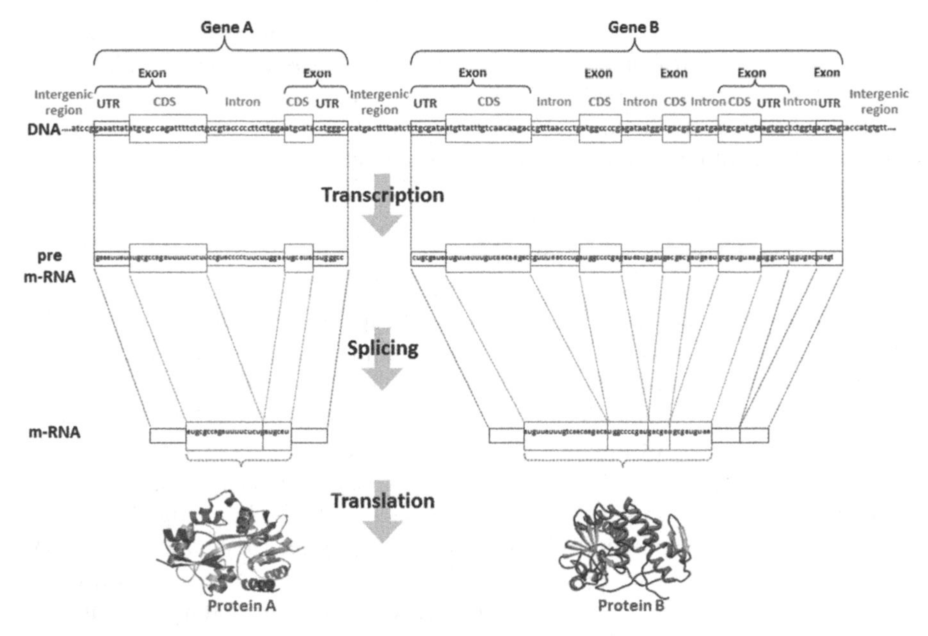


**Fig. 2** Graphical depiction of the structure of eukaryotic gene and splicing process

## 2.2 *Structural Annotation*

Compared to prokaryotes the structure of a eukaryotic gene is more complex as the exons are separated by the scattered introns which are removed during splicing and multiple exons can join together in different arrangements to code the single or alternative form of proteins. Figure 2 graphically depicts the typical structure of a eukaryotic gene and its splicing process.

The correct determination of splicing sites is critical for improved gene prediction in eukaryotic genomes. The accuracy of gene prediction can be well defined in terms of sensitivity and specificity.

For a feature (coding base, exon, transcript, and gene) the sensitivity is defined as the number of correctly predicted features divided by the number of annotated features. Specificity is the number of correctly predicted features divided by the number of predicted features [21].

$$Accuracy = \frac{Sensitivity + Specificity}{2} \tag{1}$$

Genome annotation ASsessment Project (GASP) provides the overview of the accuracy of the gene predictions tools. The two independent gene prediction tools assessments occurred in the year 2006 and 2008 which indicated Augustus, Fgenesh++, and mGene as robust gene predictors with comparatively high accuracy than alternative tools in the assessment. EGASP on human ENCODE regions proposed Augustus as best ab initio predictor with overall prediction accuracy ~51 % [22]. The nGASP on *Caenorhabditis elegans* proposed mGene and Augustus as best ab initio predictors both with overall prediction accuracy ~65 % and in the category of hybrid predictors proposed Fgenesh++ and Augustus as best hybrid predictors both with overall prediction accuracy ~76 % [23].

### 2.2.1 Tools for Structural Annotation

Multiple methods can be used for gene prediction such as ab initio methods, evidence based methods (e.g., EST alignment and protein alignment), comparative genomics methods, and the hybrid approaches which combine ab initio methods and evidence based methods. The hybrid methods predict and perform better than individual predictors. All the individual gene prediction methods have their own advantages and limitations [24]. In the probabilistic approaches, the prediction accuracy is limited by the quality of training sets while the homology based and comparative methods are constrained by the similarity and annotation of closely related species. Notably, mapping of transcriptome data (such as RNA-seq reads or ESTs) over genome promises major advances for gene finding. The hybrid method takes advantages of different methods to allow a more accurate and gene prediction over complete genome, however, hybrid approaches are also limited, for instance, by incomplete evidence or insufficient training data but overall their prediction performance in terms of accuracy is much better than individual predictors [25, 26]. The predictors are trained to recognize the pattern of splice sites which helps the predictor of ab initio gene finding. Furthermore, the ORFs are identified by the presence of start and stop codons. Table 2 lists the tools and software available for gene prediction and the methods associated with them. Below, we introduce some of the good software according to GASP that can be used for better structural annotations.

**Table 2** Tools for gene prediction with associated methods

| Software | Method |
|---|---|
| GRAIL, GeneID, GeneParser, Fgeneh, GeneFinder, GENSCAN, HMMGene, GeneMark.hmm, Gene-Zilla, GlimmerHMM, Augustus, SNAP, Conrad, Contrast | ab initio |
| PROCRUSTES, GeneWise, Ensembl | Evidence based |
| Genie, Fgenes++, Augustus+, mGene, BRAKER1, CodingQuarry | Hybrid |
| GenomeScan, TWINSCAN, GAZE, N-SCAN, Plaza, GSA-MPSA | Comparative genomics |
| TWINSCANEST, N-SCANEST | Comparative genomics + evidence based |

## Augustus

Augustus is a generalized hidden Markov model (GHMM) based predictor which is flexible for incorporating extrinsic information such as EST alignment, protein alignment, and RNA-seq data to define the accurate gene models [27]. Like most of the machine learning method based ab initio gene predictor, in the first step the algorithm has to be trained with accurate and refined set of genes consisting of curated annotations. The trained algorithm can be exploited to predict the gene structure over complete genome. Augustus package comes with optimization script that is used to retrain the algorithm with tenfold cross validation for improvement in the prediction parameters. If the user decides not to do training it is possible to use the prediction parameters of phylogenetically closer organisms if available with Augustus package. Currently, the program package comes with gene prediction parameters of >50 organisms. However, we do not recommend to use the closer organism parameters if the user objective is to perform re-annotation as the sequence features such as codon bias and splicing signals vary from organism to organism and even the nearest phylogenetic neighbor does not necessarily possess compatible parameters. The flexibility of incorporating different data with Augustus provides the many opportunities to improve the annotations. These data can be used to generate evidence of hints of introns, exons, and exon parts. The evidences are further used during the run of Augustus over repeat-masked genome with species-specific parameters for final structural re-annotation. Nowadays, the bloom in NGS generates transcriptome data which is also highly valuable for re-annotation. The mapping of transcriptome sequence reads on corresponding genome provides the clue of exact exon–intron boundaries which is attributed for improving gene models in re-annotation procedure. For purely biological researchers Augustus also provides a web version for training and gene predictions. The different combination of datasets can be used for training, for example, genome and cDNA/ESTs, genome and RNA-seq transcripts, genome and protein file, genome and gene structure file, genome, cDNA file, protein file or genome, cDNA file, and gene structure file. Augustus can perform the prediction of alternative transcripts, 5′UTR and 3′UTR including introns. A recent study has shown the successful re-annotation of ant *Camponotus floridanus* genome with Augustus [19].

## Fgenesh++

Fgenesh++ [28] is the automatic eukaryotic genome annotation pipeline that includes several programs come with the Fgenesh++ suite; for instance, it uses the HMM based gene prediction program Fgenesh++ that uses the information of the gene structure of existing homologous proteins for more accurate gene assembly from predicted exons. Currently, the web version of Fgenesh++ consists of genome-specific parameters for gene predictions for more than 250 different organisms Fgenesh++ yielded the most accurate gene predictor for predicting maize genes among the five gene predictors [29]. Similar to Augustus, Fgenesh++ can incorporate the supporting data such as mRNA or homologous protein sequences for improved gene predictions. In an evaluation of three gene prediction pipelines for ability to reproduce the results of gene predictions for 44 selected ENCODE sequences, Fgenesh++ pipeline performs best and could identify 91 % of coding nucleotides with a specificity of 90 % [30]. Briefly, the pipeline initiates with the mapping of known mRNAs/cDNAs to genome then the genes are predicted based on homology to known proteins and finally ab initio predictions are made on the genomic regions having neither mapped mRNAs nor genes predicted based on protein homology.

### mGene

mGene is a computational tool for the genome-wide prediction of protein-coding genes. For the recognition of genes training techniques are used that use eukaryotic sequences as a testing set. A combination of HMM and support vector machines (SVM) is utilized to differ coding regions from non-coding regions [5, 6]. Thus, the machine learning algorithm of mGene is called hidden semi-Markov support vector machine (HSMSVM). During the training process the state-of-the-art kernel machines refine their parameters for the recognition of signals from, e.g., donor and acceptor splice sites, transcription start site. These iterative processes improve the determination of distinct DNA motifs, such as exons and introns The signals of the input DNA sequences are analyzed by a string kernel or weighted degree kernel in order to establish a weighted degree subgraph. In the second step the different signal outputs are assembled to detect whole gene structures.

The weighted output of the HMSVM is contributing to a global score. While testing the algorithm the difference between the score and the true segmentation and wrong segmentation is maximized. In the order to compare the performance of mGene with other gene detection tools, specificity and sensitivity of gene detection are specified (Table 3; [5, 6]). The program has shown excellent performance for predicting genes of the nematode model system *C. elegans* [23]. The Galaxy based webserver for gene prediction with mGene is also available.

**Table 3** Performance of different gene finder software regarding mGene [5, 6]

| | Nucleotide | | | Exon | | | Transcript | | |
|---|---|---|---|---|---|---|---|---|---|
| Method | Sn | Sp | $\frac{Sn+Sp}{2}$ | Sn | Sp | $\frac{Sn+Sp}{2}$ | Sn | Sp | $\frac{Sn+Sp}{2}$ |
| mGene.init | 96.8 | 90.9 | 93.8 | 85.1 | 80.2 | 82.6 | 49.6 | 42.3 | 45.9 |
| mGene.init (dev) | 96.9 | 91.6 | 94.2 | 84.2 | 78.6 | 81.4 | 44.3 | 38.7 | 41.5 |
| Craig | 95.5 | 90.9 | 93.2 | 80.3 | 78.2 | 79.2 | 35.7 | 35.4 | 35.6 |
| Fgenesh | 98.2 | 87.1 | 92.7 | 86.4 | 73.6 | 80.0 | 47.1 | 34.1 | 40.6 |
| Augustus | 97.0 | 89.0 | 93.0 | 86.1 | 72.6 | 79.3 | 52.9 | 28.6 | 40.8 |

*Sn* sensitivity, *Sp* specificity, $\frac{Sn+Sp}{2}$—accuracy

**Table 4** Results of different gene finders applied to EuGene and FGenesH

| Gene finder | Sne | Spe | Sng | Spg |
|---|---|---|---|---|
| GenScan (*A. thaliana*) | 69.6 | 78 | 25.8 | 29 |
| GeneMark.HMM (*A. thaliana*) | 73.1 | 76.6 | 32.4 | 31.6 |
| FGenesH (*A. thaliana*) | 85.3 | 81.4 | 47 | 46.5 |
| FGenesH (*M. truncatula*) | 85.1 | 80.7 | 52.8 | 47.8 |
| EuGène (ab initio, *M truncatula*) | 84.7 | 85.4 | 55.5 | 50.5 |
| + FGenesH | 90 | 86.9 | 63.2 | 56.4 |
| + Protein similarity | 92.4 | 88 | 69.2 | 61.8 |
| + Transcript similarity | 94.4 | 94.6 | 80.2 | 79.4 |

*Sne* sensitivity for exone, *Spe* specificity for exon, *Sng* gene level sensitivity, *Spg* gene level specificity [31]

## EuGene

EuGene is an integrative gene finder for eukaryotic and prokaryotic genomes. The underlying algorithm is a HMM that is adjusting state transition probabilities to a sample set of sequences. In addition EuGene utilizes information of homology prediction, RNA-seq analysis, structural proteins similarities, and other statistical source. The results of existing gene finders tool contribute to the global score of predicted genes. Additionally, EuGene offers the possibility to use plugins, such as software for detecting frameshift, EST similarity, translation start, and anti-sense genes. The plugins are either ready to use software, or can be coded in C++, for more experienced users. EuGene is a command line tool. The results of the integrative analysis intend to obtain a maximum score with maximal consistent information provided (Table 4). For comparability with other gene finders software results of a test set of eukaryotic genes are stated with sensitivity and specificity values [31].

## GeneScan

GenScan determines coding gene regions by the application of the mathematical model of GHMM. The software establishes a probabilistic model of the gene structure of human genomic sequences. It incorporates description of the basic transcriptional and splicing signals, as well as length distributions and compositional features of exons, introns, and intergenic regions of the human. Hence, a probabilistic model of the gene structure of human genomic sequences is provided. In contrast to mGene the tool identifies distinct pattern in C + G content of DNA region and also focuses on strand information of genes. The output of the software identifies complete exon/intron structures of genes in genomic DNA. Around 75–80 % of exons are identified with GenScan [32]. Sample output of GenScan tool is illustrated in Fig. 3.

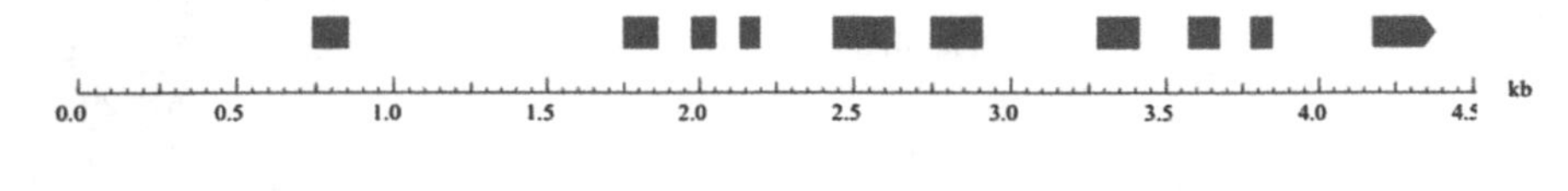


## *Legend:*

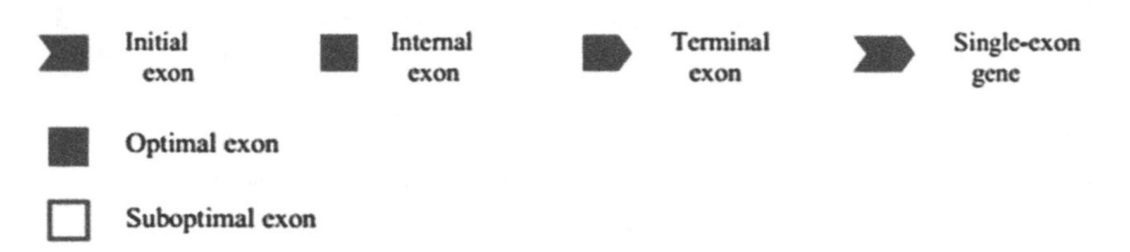


**Fig. 3** The graphical output of GeneScan shows an example of a predicted gene HS307871

**Table 5** Performance on an *A. thaliana* cDNAs (*a.t.*) and *Aspergillus fumigatus* CDSs (*a.f.*) by GeneZilla [34]

| | % Nucl. accuracy | | % Exon sensitivity | | % Exon specificity | | % Exact genes | |
|---|---|---|---|---|---|---|---|---|
| | *a.t.* | *a.f.* | *a.t.* | *a.f* | *a.t.* | *a.f.* | *a.t.* | *a.f.* |
| TigrScan | 96 | 90 | 77 | 37 | 81 | 47 | 43 | 19 |
| GlimmerHMM | 96 | 91 | 71 | 36 | 79 | 49 | 33 | 21 |
| GenScan+ | 95 | 87 | 75 | 23 | 82 | 4 | 35 | 11 |

### GeneZilla

GeneZilla, former known as TigrScan, is a eukaryotic gene finder that is utilizing a GHMM. The software can be trained by the user with an own set of annotated genes. GeneZilla is highly reconfigurable and is programmed in C++ and is able to optimize the parameters by retraining. Owing to specific decoding algorithm the run time and memory requirements are linear in the sequence length, and are in general better than those of competing gene predictors [33]. Skilled user can adopt the open source code for their own purposes. Furthermore, the algorithm is retrained by the end user. The runtime of GeneZilla is linear in length. The graph-theoretic representation of the high scoring ORF is provided. In this perspective, the information of states for signal peptides, branch point, TATA boxes, CAP sites, and CpG islands is integrated. Likewise, GeneZilla is capable of recognizing various types of exons (e.g., initial/internal/single) by the hand of distinctly adapted sensors. Interpolated Markov models and maximal dependence decomposition are two of several underlying submodels for calculating the output. The specificity and sensitivity are calculated for *Aspergillus fumigatus*, *A. thaliana* (Table 5), and *Toxoplasma gondii* genomes [34].

EVidenceModeler

EVidenceModeler (EVM) is a tool for the automated eukaryotic gene annotation, including the ab initio gene prediction, prediction of spliced protein, and gene transcription alignments using different weights. The EVM algorithm considers initial exon, internal exon, terminal exon, and single exon corresponding to intronless gene to assess the gene structure. The coordinates, position and length of each exon are stored and further validated. Introns with the minimal length of 20 bp are also stored as discrete features and together with exons are contributed to assessment of the protein and transcript spliced alignment data. All candidates are evaluated by unique scoring mechanism and predictions with low suppers for each locus are filter out. To set the accuracy and sensitivity of the prediction, several evaluation tools to estimate optimal evidence weights are applied. Except the automated processes, expected evidence error rates can also be set manually by user, or can be trained from a training set separately [35]. EVM belongs to poplar software for the genome re-annotation and was used, e.g., in *Anolis carolinensis* [36] and *Entamoeba histolytica* [37] re-annotation projects.

MAKER

MAKER [38] is a popular configurable genome annotation pipeline written in the Perl+. It operates from the preassembled RNA-seq data or Cufflinks outputs. Despite originally was this tool developed for de novo annotation of emerging model organism, MARKER was used in more than 200 eukaryotic and prokaryotic genome studies and re-annotation projects, such as woodland strawberry [39] and *Pogonomyrmex barbatus* [40]. This tool provides several features, such as identification of the repeats, alignments of ESTs and protein to a genome, or ab initio prediction of the genes, as well as re-evaluation of legacy annotation sets. The quality control is assessed by annotation edit distance [41] and the outputs are automatically synthetized into gene annotations. Nevertheless the complete annotation pipeline can take several weeks, the simply usage and reproducible data make MAKER one of the most using annotation tool and several extensions are already available, for example, MAKER-P—for the plant genome projects and MAKER2, toolkit with extended features.

## 2.3 *Free Cloud Infrastructures*

In the last decade, the popularity of the cloud infrastructure (CI) grows rapidly. As processing of the genome data requires expensive computer equipment, CI provides all necessary assumes for each register users. The CI platform utilizes the clouds to handle with the data storage, analysis, and access to the databases. It provides

**Table 6** Selected iPlant application

| Application | Description |
|---|---|
| Atmosphere module | Virtual hosting of web applications, sites, and databases |
| Discovery environment | Computing, storage, and analysis application for clustering and network analysis, mapping of QTL and GWAS, sequence alignments, phylogenetic analysis, and many more applications |
| MyPlant tool | Allows social networking and group formation |
| DNA subway | Tool for prediction and annotation of genes, analysis of RNA-seq data and DNA barcoding, phylogenetic analysis |
| PhytoBisque | Analysis of plant-related images |
| Taxonomic name resolution service | Computer-assisted standardization of plant scientific names |
| TreeViewer | Phylogenetic analysis |

a complex suite of high-performance and high-throughput computing, access to the computational data management, networking tools, as well as overall cloud computing approach.

CI is becoming increasingly popular among the users and various platforms are used as a daily routine in many laboratories. One of the most used CI for the analysis of the plant and animal genome is iPlant Collaborative (http://www.iplantcollaborative.org/; [42]). The Data Store is placed on the network of iRODS servers (University of Arizona, USA) and replicated at the TACC using the iPlant fAPI job services. The platform allows the work with several applications, which are highlighted in Table 6. iPlant provides huge number of analytic tools, such as BEDTools, SAMTools or EMBOSS for bioinformatic analysis, QTL Cartographer and TASSEL for genetyping and Trinity, and Newblemet or SOAPdenovo for the de novo assembly. Despite the name of this CI, it was also successfully used in the animal science [43]. Data storage can be accessed with several web interfaces and desktop applications, e.g., iPlantDE, FUSE interface, Davis web application or iDROP, and each user with get universal unique identifier for each file. Initially, the user gets 100 GB space in the iPlant Data Store, but the data can be increased upon a request. Furthermore, iPlant also provides work with web service platform BioExtract, Gates Integrated Breeding Platform, and CiPRES computation tool, as well as many other underlying infrastructure.

Apart from iPlant, many other CI platforms provide high-performance computing, data analysis, virtual organization, learning, and workforce. The Mercury platform represents an automated approach for the all-step analysis of NGS data using Amazon Web services on the DNAnexus platform [44]. The Cloud BioLinux serves as a platform for variety of bioinformatics analysis, for instance, the processing of NGS data (provided by toolsets Fastx utilities SAM, BAM), databases tools for similarity searches (HMMER, BLAST) as well as genome assembly by SSAKE, NEWBLER, or the Velvet application [45]. Among the popular CI platforms, GALAXY portal (http://galaxyproject.org/) and Google cloud platform (https://cloud.google.com/genomics/) are also a good choice for the re-annotation projects.

## 2.4 Functional Annotation

Most of the gene prediction suites provide their own identifier for the predicted genes or proteins. In such a case, it is not possible to recognize the actual gene name or the gene function. Here the functional annotation comes in act and provide the name and function of all the annotated proteins. Generally, the name for a particular protein is derived based on the similarity with its well-annotated homolog sequence. The similarity search for annotation transfer is performed by using powerful search algorithms such as BLAST. Nevertheless, the fuzzy criteria of BLAST thresholds for functional annotation may cause the misannotations which further may propagate the annotation errors by homology based annotation transfers. Indeed, 10–30 % of all the functional annotations may be wrong [46, 47]. Generally, above levels of 50 % pairwise sequence identity enzyme functions are strongly conserved and less than 30 % of the pair fragments above 50 % sequence identity have entirely identical Enzyme Commission (EC) numbers [48]. However, such criteria are not always true, for example, on considering the full length sequence the human and *C. elegans* p53 transcription factor does not show significant similarity for annotation transfer, but due to the presence of similar domains both are annotated as p53. In many cases, annotation can be erroneously transferred in proteins families that belong to same superfamily [48].

Such misannotation occurs because BLAST is not very discriminative with fuzzy thresholds. Misclassifications are also originated from similarities in relatively short regions and/or from transferring annotations for different domains in multidomain proteins. To avoid the errors the extensive manual curations are required although using robust algorithm. Moreover, the selecting sensitive parameters for functional annotation can be applied to reduce the annotation errors.

The encoded short peptides by gene prediction efforts can also be annotated. However, for short peptide the HMM based or comparative analysis based annotations are more suitable. In case of missing proper annotation for proteins by homology based annotation transfer, annotation of functional domains can be performed with Pfam [49], Smart [50], and InterPro ([51, 52]) like tools. Occasionally phylogenic analysis can be used to dissociate the members of gene family that belongs to same superfamily into appropriate groups, for instance, the classification of Glutathione *S*-transferases (GSTs) into multiple groups after the re-annotation [19].

### 2.4.1 Tools for Functional Annotation

Most of the functional annotation pipelines are based on homology based annotation transfer were user can tune the parameters to minimize the annotation error. Here we have summarized some good and state-of-the-art programs for functional annotation.

### AutoFact

AutoFact [53] is one of the good classical tools for functional annotation. AutoFact is written in Perl and is a BLAST based annotation tool which considers a nice hierarchical system for annotation. Users can give their preference number to the databases for the annotation. Initially, the hits for annotation transfer are verified in first rank database such as Uniprot90 or Uniprot100 and so on. If the hits are not satisfactory with the sequence databases, then the hits against Pfam database are analyzed and in case of significant similarity protein is annotated as "x-domain containing protein" where x is the name of the functional domain in target sequence homolog in Pfam. This process is quite useful to minimize the number of proteins annotated as hypothetical proteins.

### BLAST2GO

BLAST2GO (B2G) represents one of the most frequently used tools for functional annotation, in either basic or pro frameworks, available free or commercial, respectively. Besides the user-friendly interface for gene ontology (GO) annotations, B2G interface also provides the facility to KEGG metabolic pathway enrichment, prediction of InterPro motifs, and enzyme code (EC), which allow the users to get to graphical function of data mining on annotation procedure, without the need for computational knowledge. The user can set up their own BLAST database downloaded from NCBI portal for the faster calculations. B2G transfers the annotations based on the similarity with top hits. The analysis process starts with BLAST (either BLASTN, BLASTP, BLASTX, or TBLASTX) through either public database (such as NCBI nr and EST using QBLAST) or a local database (such as single species DB and GO annotated sequence set). Retrieving significant hits is performed based on provided *E*-values and hit number thresholds, as well as additional filters such as minimal alignment length (hsp-length). The resulted hit gene identifiers (gi) and gene accessions are used in order to get GO annotation and evidence code, which is evaluated based on annotation rule (AR) including highest hit similarity and parent node annotation. The statistical assessment of functionality of GO terms is performed by comparing with a reference group by Gossip and calculating false discovery rate (with Fisher Exact test) and ranking by one-test *p*-values or corrected ones. These procedures will be visualized step-by-step through graphs and charts with informative statistics data, besides of GoPAG possibility, which highlights the most relevant biological meaning of a set of sequence [54, 55]. With all these good features B2GO ensures the improving possibilities in functional annotation.

### Plaza

Plaza is an online platform for plant comparative genomics, it includes functional as well as structural annotations of available plant genomes, from a wide range of species of green plant lineage. It provides various interactive tools for extracting biological information about gene function, as well as gene and genome evolution. Protein-coding genes are assigned to homologous gene families based on sequence similarity and their paralogs and orthologs are shown in phylogenetic trees. A web interface toolkit allows access to various tools. These include local synteny plots, which describe positional orthologs, Skyline plots, which calculate gene based collinearity between all chromosomes belong to different species, and WGDotplot, which facilitates the study of large-scale duplicates within a genome. Currently, Plaza hosts different organisms from eudicots, monocots, lycopods, moss, and algae. Plaza is a useful application for browsing gene families, functional gene clustering, studying multispecies collinearity, as well as for the prediction and evaluation of complex gene orthology [56, 57].

### Trapid

Trapid is a web based tool, providing transcriptome analysis derived from de novo assembly of RNA-seq or EST sequencing. The process starts with a sequence similarity search using PAPSearc [58] with a user-selectable $E$-value cutoff, followed by detection of ORFs in order to identify coding sequences and gene family assignment. This is achieved by building a graph with genes for nodes and BLAST bitscores for edges, performing transcript quality control and annotating homology based functional. The analysis process is followed by detailed sequence analysis of the transcriptome. This includes multiple sequence alignment using MUSCLE, detection of homologs from related species using JalView [59] and construction of a phylogenetic tree using FastTree2 [60] in order to define relationships between genes (orthologs and paralogs) and the allelic transcript variant from Plaza V2.5 [57] or OrthoMCL-DB V5 [61]. Although TRAPID utilizes ~175 reference proteomes, it does not process data from large-scale metagenomics studies. Output overview of the rapid analysis will be shown in a Toolbox, including general statistical information on input quality, meta annotation, similarity search annotation, gene family, and GO functional annotation [62].

### BLASTKOALA and GhostKOALA

These KOALA (KEGG Orthology and Links Annotation) tools are automatic annotation servers that perform BLASTP and GhostX searches against nonredundant KEGG DGENES or KEGG MGENES, respectively. This is achieved by computing CD-Hit clusters, with 50 % cutoff identity of species/genus/family group. The K number assignment to the query sequence is made by KOALA algorithm, through weighted sum of SW (Smith-Waterman) score calculation. BLASTKOALA performs best to annotate fully sequenced genomes, whereas GhostKOALA is more suitable for large amount of metagenome sequences annotation. Implementation of KEGG Map-per tool will be done in order to reconstructing of KEGG pathway maps, BRITE functional hierarchical, KEGG modules, as well as KEGG taxonomy or NCBI taxonomy. The general overview of annotation data will be shown in the result page as pie charts [63].

MapMan

MapMan is a software tool to compare biological responses of different gene categories. It is applicable for a very wide range of crops and wild species, which is useful for the identification of potential orthologs between different plants. This is achieved by using reciprocal BLAST and a range of *E*-value cutoffs. MapMan structure is based on a set of modules. The first module is Scavenger, which generates the functional categories of genes, enzymes, proteins, and metabolites. The resulting ontologies (mapping files) will be imported into the ImageAnnotator module to visualize the biological pathways or processes, by using a Wilcoxon test. In case of multiple experiments, the mapping files will be visualized by PageMan through compressing of all the pathways. This tool facilitates the process of investigating analogous responses in any plant species with existing gene models or Unigenes [64]. This procedure is also available via an automated sequence annotation pipeline (Mercator), a web application of MapMan, in order to make functional annotation based on BLAST and protein domain search for NGS sequences [65].

## 3 Steps for Re-annotation

### *3.1 Genome Assembly*

Genome assembly process stitches together a genome sequence from the short sequenced pieces of DNA. Although the advancement in NGS technology, generating the large, continuous regions of DNA sequences by high quality assembly of reads is still challenging mainly because of the presence of repetitive elements. Repeats not only cause gaps in assembly procedure but can also erroneously collapsed on top of one another and subsequently can cause complex, misassembled rearrangements [66, 67]. The misassembled genome further attributes to wrong interpretation of absence or presence of genes. Korf et al. showed that an assembly of the chicken genome lacks 36 genes that are conserved across plants, yeast, and other organisms. However, these genes were only missed from the assembly rather than the organism as the focused re-analysis of the raw data showed the presence of most of these genes in sequences that had not been included in the assembly [68].

It is possible that number of genes in the different assembly may not be similar for the same organism. Therefore, the features missing from the assembly should not be assumed missed from the organism without the follow-up studies and validations. Although, re-assembly and re-annotation both are independent procedures, but for the sake of completeness we have briefly summarize the genome assembly in reference to missing genes.

## 3.2 *Repeat-Masking*

The first requirement for re-annotation is genome itself. The annotation programs are generally run on the repeat-masked version of the genome. Therefore, first the genome should be repeat-masked. The program RepeatMasker [69] and Repeat-Modeler [70] can be used for the repeat identification. RepeatMasker uses the library containing those repetitive elements for masking the homologous regions and RepeatModeler performs the ab initio modeling of the genome. Finally, the combined library should be used for the repeat-masking. When preparing repeat-masked sequences, we recommend not masking low complexity regions and simple repetitive motifs, as they may also be the parts of coding sequences.

## 3.3 *Training-Set Creation*

The training set for gene predictors should be made with caution as the quality of the training set greatly affects the final gene predictions. Tools such as Pasa [71], BLAT [72], and Scipio [73] can be used for the creation of good quality training sets structure using a combination of GeneWise [74], HMMER [75], and GeneID [76] program. Pasa performs the alignment of assembled transcripts on the test genome using GMAP (Wu et al. 2005) aligner for modeling the gene structure. Scipio can be used to derive gene models from the well-annotated proteins of test genome or close species by aligning them to the test genome using spliced alignment tool Blat, followed by determination of the correct exon–intron junctions by hit refinements and filtering with Scipio. In case of the availability of multiple well-annotated proteins, we recommend the use of protein belongs to different GO categories for creating training set with Scipio. The consensus or non-overlapping set of gene models for training procedure can be obtained from combination of Cegma [77], Pasa, and Scipio based training sets.

## 3.4 Training and Optimization

The gene prediction parameter for the test species should be optimized to get the best sensitivity and specificity. Here, we recommend the cross-fold validation during the training to achieve the best optimized parameters. It should be noted that the prediction parameters can significantly influence the number of exons and/or genes predicted by the gene finder. These parameters often include the probabilities of transition between the given states of the GHMM.

## 3.5 Integration of Extrinsic Data

After the training-set creation step, the next step is to collect the additional supporting data, for example, the EST/cDNA, RNA-seq data to improve the ab initio gene predictions. The intron, exon, or exon part hints for evidence supported gene structure could be derived from EST, cDNA, and raw reads as well as assembled transcripts. Different gene predictors provide their own tutorials for integrating the RNA-seq data. Generally, the raw or assembled reads are mapped to the reference genome using aligners such as Tophat2/Bowtie2 [78, 79] to generate the evidences of splice sites, introns, and exons. For example, in Augustus evidences of introns can be generated by a two-step process, first the raw reads are mapped over repeat-masked genome of the annotating species using Tophat2/Bowtie2 to create the preliminary intron hints then Augustus performs the gene predictions with the created preliminary intron hints to generate intermediate gene models and exon–exon junction database. Finally, the reads are again mapped now on exon–exon junction database using Bowtie2 to generate the final intron evidences. Moreover, to generate the evidences of exons, part of exon and introns, the assembled reads or EST data can also be used which can be mapped over the repeat-masked or unmasked annotating genome using the splice aligner such as BLAT [72]. The generated evidences serve as hints for hybrid predictors and it could be accessed that how much percentage of the genes are supported by transcriptome data. More coverage of genome by the extrinsic data increases the reliability of the annotation.

## 3.6 Structural and Functional Annotation

Finally, to predict the gene structures and estimate the alternative transcripts, gene predictions with extrinsic events should be performed on the repeat-masked test genome using the tuned optimized parameters of the test species. Our recommendation is to use the robust tool Augustus [27] as the gene prediction platform and later using AutoFact [53] for annotating the function of the re-annotated genes.

## 4 Conclusion

If our chapter at least alerted you to the problem of constantly improving annotation of eukaryotic genomes, we are already rewarded. Clearly, in our Big Data era, a good eukaryotic genome annotation is one of the first victims of data growth as data grow so rapidly. Transferring annotation mistakes from one genome to the new one often happen and manual annotation requires anyway a lot of biological insights. We also focused in this chapter on protein gene annotation, the most important and direct task. Bear in mind that the area of annotating repetitive DNA and intergenic regions or annotation of encoded RNA would be full grown chapters of their own. However, our second intent was to give you automatic tools in hand which certainly cannot take all the burden from you of carefully reanalyze what you achieved regarding annotation, but they should give you the momentum and confidence to really do it with your favorite genome.

## References


1. Sanger, F., Coulson, A. R., Friedmann, T., Air, G. M., Barrell, B. G., Brown, N. L., Fiddes, J.C., Hutchison, C.A., Slocombe, P.M and Smith, M. (1978). The nucleotide sequence of bacteriophage φX174. *Journal of molecular biology* (2),225–246.
2. Venter, J. C., Adams, M. D., Myers, E. W., Li, P. W., Mural, R. J., Sutton, G. G., Smith, H. O., Yandell, M., Evans, C.A., Holt, R.A. and Gocayne, J. D. (2001). The sequence of the human genome. *Science, 291*(5507), 1304–1351.
3. Hodkinson, B. P., and Grice, E. A. (2015). Next-generation sequencing: a review of technologies and tools for wound microbiome research. *Advances in wound care, 4*(1), 50–58.
4. Shokralla, S., Spall, J. L., Gibson, J. F., and Hajibabaei, M. (2012). Next generation sequencing technologies for environmental DNA research. *Molecular ecology, 21*(8), 1794–1805.
5. Schweikert, G., Behr, J., Zien, A., Zeller, G., Ong, C. S., Sonnenburg, S., & Rätsch, G. (2009). mGene. web: a web service for accurate computational gene finding. *Nucleic acids research, 37*(suppl 2), W312–W316.
6. Schweikert, G., Zien, A., Zeller, G., Behr, J., Dieterich, C., Ong, C. S., Philips, P., De Bona, F., Hartmann, L., Bohlen, A. and Krüger, N. (2009). mGene: accurate SVM-based gene finding with an application to nematode genomes. *Genome research.*
7. McCallum, D., and Smith, M. (1977). Computer processing of DNA sequence data. *Journal of Molecular. Biology, 116*, 29–30
8. Altschul, S. F., Gish, W., Miller, W., Myers, E. W., and Lipman, D. J. (1990). Basic local alignment search tool. *Journal of molecular biology, 215*(3), 403–410.
9. Dayhoff, M. O., Schwartz, R. M., Chen, H. R., Barker, W. C., Hunt, L. T., and Orcutt, B. C. (1981). Nucleic acid sequence database. *DNA, 1*(1), 51–58.
10. Bilofsky, H. S., Burks, C., Fickett, J. W., Goad, W. B., Lewitter, F. I., Rindone, W. P., Swindell, C.D and Tung, C. S. (1986). The GenBank genetic RNA sequence databank. *Nucleic acids research, 14*(1), 1–4.
11. Lewis, S., Ashburner, M., and Reese, M. G. (2000). Annotating eukaryote genomes. *Current opinion in structural biology, 10*(3), 349–354.
12. Dandekar T, Huynen M, Regula JT, Ueberle B, Zimmermann CU, Andrade MA, Doerks T, Sánchez-Pulido L, Snel B, Suyama M, Yuan YP, Herrmann R, Bork P. (2000) Re-annotating the *Mycoplasma pneumoniae* genome sequence: adding value, function and reading frames. *Nucleic Acids Res. 28*(17), 3278–88.

13. Gaudermann P, Vogl I, Zientz E, Silva FJ, Moya A, Gross R, Dandekar T. (2006) Analysis of and function predictions for previously conserved hypothetical or putative proteins in *Blochmannia floridanus*. *BMC Microbiology*. 6, 1.
14. DeCaprio, D., Vinson, J. P., Pearson, M. D., Montgomery, P., Doherty, M., and Galagan, J. E. (2007). Conrad: gene prediction using conditional random fields. *Genome research*, *17*(9), 1389–1398.
15. Tsochantaridis, I., Hofmann, T., Joachims, T., and Altun, Y. (2004, July). Support vector machine learning for interdependent and structured output spaces. In *Proceedings of the twenty-first international conference on Machine learning* (p. 104). ACM.
16. Bulyk, M. L. (2004). Computational prediction of transcription-factor binding site locations. *Genome biology*, *5*(1), 201–201.
17. Pavesi, G., Mauri, G., and Pesole, G. (2004). *In silico* representation and discovery of transcription factor binding sites. *Briefings in Bioinformatics*, *5*(3), 217–236.
18. Elsik, C. G., Worley, K. C., Bennett, A. K., Beye, M., Camara, F., Childers, C. P., de Graaf, D.C., Debyser, G., Deng, J., Devreese, B. and Elhaik, E. (2014). Finding the missing honey bee genes: lessons learned from a genome upgrade. *BMC genomics*, *15*(1), 86.
19. Gupta, S. K., Kupper, M., Ratzka, C., Feldhaar, H., Vilcinskas, A., Gross, R., Dandekar, T. and Förster, F. (2015). Scrutinizing the immune defence inventory of *Camponotus floridanus* applying total transcriptome sequencing. *BMC genomics*, *16*(1), 540.
20. Bonasio, R., Zhang, G., Ye, C., Mutti, N. S., Fang, X., Qin, N., Donahue, G., Yang, P., Li, Q., Li, C. and Zhang, P. (2010). Genomic comparison of the ants *Camponotus floridanus* and *Harpegnathos saltator*. *Science*, *329*(5995), 1068–1071.
21. Sokolova, M., Japkowicz, N., & Szpakowicz, S. (2006). Beyond accuracy, F-score and ROC: a family of discriminant measures for performance evaluation. In AI 2006: *Advances in Artificial Intelligence* (pp. 1015–1021). Springer Berlin Heidelberg.
22. Guigó Serra, R., Flicek, P., Abril Ferrando, J. F., Reymond, A., Lagarde, J., Denoeud, F., Antonarakis, S., Ashburner, M., Bajic, V.B., Birney, E. and Castelo, R. (2006). EGASP: the human ENCODE Genome Annotation Assessment Project. *Guigó R, Reese MG, editors. EGASP'05: ENCODE genome annotation assessment Project. Genome biology. 2006; 7 Suppl 1.*
23. Coghlan, A., Fiedler, T. J., McKay, S. J., Flicek, P., Harris, T. W., Blasiar, D., and Stein, L. D. (2008). nGASP–the nematode genome annotation assessment project. *BMC bioinformatics*, *9*(1).
24. Mathé, C., Sagot, M. F., Schiex, T., and Rouzé, P. (2002). Current methods of gene prediction, their strengths and weaknesses. *Nucleic acids research*, 30(19), 4103–4117.
25. Keller, O., Kollmar, M., Stanke, M., and Waack, S. (2011). A novel hybrid gene prediction method employing protein multiple sequence alignments. *Bioinformatics*, 27(6), 757–763.
26. Zickmann, F., and Renard, B. Y. (2015). IPred-integrating ab initio and evidence based gene predictions to improve prediction accuracy. *BMC genomics*, 16(1).
27. Stanke, M., Schöffmann, O., Morgenstern, B., and Waack, S. (2006). Gene prediction in eukaryotes with a generalized hidden Markov model that uses hints from external sources. *BMC bioinformatics*, *7*(1), 62.
28. Solovyev V.V. (2007) Statistical approaches in Eukaryotic gene prediction. In Handbook of Statistical genetics (eds. Balding D., Cannings C., Bishop M.), *Wiley-Interscience*; 3d edition, 1616 p
29. Yao, H., Guo, L., Fu, Y., Borsuk, L. A., Wen, T. J., Skibbe, D. S., Cui, X., Scheffler, B.E., Cao, J., Emrich, S.J. and Ashlock, D. A. (2005). Evaluation of five ab initio gene prediction programs for the discovery of maize genes. *Plant molecular biology*, *57*(3), 445–460.
30. Solovyev, V., Kosarev, P., Seledsov, I., & Vorobyev, D. (2006). Automatic annotation of eukaryotic genes, pseudogenes and promoters. *Genome Biology,* 7(Suppl 1), S10.
31. Foissac, S., Gouzy, J., Rombauts, S., Mathé, C., Amselem, J., Sterck, L., de Peer, Y.V., Rouzé, P.& Schiex, T. (2008). Genome annotation in plants and fungi: EuGene as a model platform. *Current Bioinformatics*, *3*(2), 87–97.

32. Burge, C., & Karlin, S. (1997). Prediction of complete gene structures in human genomic DNA. *Journal of molecular biology*, *268*(1), 78–94.
33. Allen, J. E., Majoros, W. H., Pertea, M., and Salzberg, S. L. (2006). JIGSAW, GeneZilla, and GlimmerHMM: puzzling out the features of human genes in the ENCODE regions. *Genome Biology*, *7*(Suppl 1), 1–13.
34. Majoros, W. H., Pertea, M., & Salzberg, S. L. (2004). TigrScan and GlimmerHMM: two open source *ab initio* eukaryotic gene-finders. *Bioinformatics*, *20*(16), 2878–2879.
35. Haas, B. J., Salzberg, S. L., Zhu, W., Pertea, M., Allen, J. E., Orvis, J., & Wortman, J. R. (2008). Automated eukaryotic gene structure annotation using EVidenceModeler and the Program to Assemble Spliced Alignments. *Genome biology*, 9(1), R7.
36. Eckalbar, W. L., Hutchins, E. D., Markov, G. J., Allen, A. N., Corneveaux, J. J., Lindblad-Toh, K., Di Palma, F., Alföldi, J., Huentelman, M.J. and Kusumi, K. (2013). Genome reannotation of the lizard *Anolis carolinensis* based on 14 adult and embryonic deep transcriptomes. *BMC genomics*, *14*(1), 49.
37. Lorenzi, H. A., Puiu, D., Miller, J. R., Brinkac, L. M., Amedeo, P., Hall, N., and Caler, E. V. (2010). New assembly, reannotation and analysis of the *Entamoeba histolytica* genome reveal new genomic features and protein content information. *PLoS Neglected Tropical Diseases*, *4*(6), e716.
38. Cantarel, B. L., Korf, I., Robb, S. M., Parra, G., Ross, E., Moore, B., Holt, C., Alvarado, A.S & Yandell, M. (2008). MAKER: an easy-to-use annotation pipeline designed for emerging model organism genomes. *Genome research*, 18(1), 188–196.
39. Darwish, O., Shahan, R., Liu, Z., Slovin, J. P., and Alkharouf, N. W. (2015). Re-annotation of the woodland strawberry *(Fragaria vesca)* genome. *BMC genomics*, *16*(1), 29.
40. Smith, C. R., Smith, C. D., Robertson, H. M., Helmkampf, M., Zimin, A., Yandell, M., Holt, C., Hu, H., Abouheif, E., Benton, R. & Cash, E. (2011). Draft genome of the red harvester ant *Pogonomyrmex barbatus*. *PNAS*, 108(14), 5667–5672.
41. Eilbeck, K., Moore, B., Holt, C., & Yandell, M. (2009). Quantitative measures for the management and comparison of annotated genomes. *BMC bioinformatics*, 10(1), 1.
42. Merchant, N., Lyons, E., Goff, S., Vaughn, M., Ware, D., Micklos, D., & Antin, P. (2016). The iPlant Collaborative: Cyberinfrastructure for Enabling Data to Discovery for the Life Sciences. *PLoS Biol,* 14(1), e1002342.
43. Soderlund, C. A., Nelson, W. M., & Goff, S. A. (2014). Allele Workbench: transcriptome pipeline and interactive graphics for allele-specific expression. *PloS one*, 9(12), e115740.
44. Reid, J. G., Carroll, A., Veeraraghavan, N., Dahdouli, M., Sundquist, A., English, A., & Yu, F. (2014). Launching genomics into the cloud: deployment of Mercury, a next generation sequence analysis pipeline. *BMC bioinformatics*, 15(1), 1.
45. Krampis, K., Booth, T., Chapman, B., Tiwari, B., Bicak, M., Field, D., & Nelson, K. E. (2012). Cloud BioLinux: pre-configured and on-demand bioinformatics computing for the genomics community. *BMC bioinformatics,* 13(1), 1.
46. Brenner, S. E. (1999). Errors in genome annotation. *Trends in Genetics*, *15*(4), 132–133.
47. Devos, D., and Valencia, A. (2001). Intrinsic errors in genome annotation. *TRENDS in Genetics*, *17*(8), 429–431.
48. Rost, B. (2002). Enzyme function less conserved than anticipated. *Journal of molecular biology*, *318*(2), 595–608.
49. Finn, R. D., Coggill, P., Eberhardt, R. Y., Eddy, S. R., Mistry, J., Mitchell, A. L., Potter, S.C., Punta, M., Qureshi, M., Sangrador-Vegas, A. and Salazar, G. A. (2015). The Pfam protein families database: towards a more sustainable future. *Nucleic acids research*, gkv1344.
50. Letunic, I., Doerks, T., and Bork, P. (2015). SMART: recent updates, new developments and status in 2015. *Nucleic acids research*, *43*(D1), D257–D260.
51. Jones, P., Binns, D., Chang, H. Y., Fraser, M., Li, W., McAnulla, C., McWilliam, H., Maslen, J., Mitchell, A., Nuka, G. and Pesseat, S. (2014). InterProScan 5: genome-scale protein function classification. *Bioinformatics*, *30*(9), 1236–1240.

52. Mitchell, A., Chang, H. Y., Daugherty, L., Fraser, M., Hunter, S., Lopez, R., McAnulla, C., McMenamin, C., Nuka, G., Pesseat, S and Sangrador-Vegas, A. (2014). The InterPro protein families database: the classification resource after 15 years. *Nucleic acids research*, gku1243.
53. Koski, L. B., Gray, M. W., Lang, B. F., & Burger, G. (2005). AutoFACT: an automatic functional annotation and classification tool. *BMC bioinformatics*, *6*(1), 151.
54. Conesa, A., and Götz, S. (2008). BLAST2GO: A comprehensive suite for functional analysis in plant genomics. *International journal of plant genomics*, *2008*.
55. Conesa, A., Götz, S., García-Gómez, J. M., Terol, J., Talón, M., and Robles, M. (2005). BLAST2GO: a universal tool for annotation, visualization and analysis in functional genomics research. *Bioinformatics*, *21*(18), 3674–3676.
56. Proost, S., Van Bel, M., Sterck, L., Billiau, K., Van Parys, T., Van de Peer, Y., and Vandepoele, K. (2009). PLAZA: a comparative genomics resource to study gene and genome evolution in plants. *The Plant Cell*, *21*(12), 3718–3731.
57. Van Bel, M., Proost, S., Wischnitzki, E., Movahedi, S., Scheerlinck, C., Van de Peer, Y., and Vandepoele, K. (2011). Dissecting plant genomes with the PLAZA comparative genomics platform. *Plant physiology*, pp 111.
58. Zhao, Q. Y., Wang, Y., Kong, Y. M., Luo, D., Li, X., and Hao, P. (2011). Optimizing de novo transcriptome assembly from short-read RNA-seq data: a comparative study. *BMC bioinformatics*, *12*(Suppl 14), S2.
59. Waterhouse, A. M., Procter, J. B., Martin, D. M., Clamp, M., and Barton, G. J. (2009). Jalview Version 2—a multiple sequence alignment editor and analysis workbench. *Bioinformatics*, *25*(9), 1189–1191.
60. Price, M. N., Dehal, P. S., and Arkin, A. P. (2010). FastTree 2–approximately maximum-likelihood trees for large alignments. *PloS one*, *5*(3), e9490.
61. Chen, F., Mackey, A. J., Stoeckert, C. J., and Roos, D. S. (2006). OrthoMCL-DB: querying a comprehensive multi-species collection of ortholog groups. *Nucleic acids research*, *34*(suppl 1), D363–D368.
62. Van Bel, M., Proost, S., Van Neste, C., Deforce, D., Van de Peer, Y., and Vandepoele, K. (2013). TRAPID: an efficient online tool for the functional and comparative analysis of de novo RNA-seq transcriptomes. *Genome Biology*, *14*(12), R134.
63. Kanehisa, M., Sato, Y., and Morishima, K. (2015). BLASTKOALA and GhostKOALA: KEGG Tools for Functional Characterization of Genome and Metagenome Sequences. *Journal of Molecular Biology*.
64. Usadel, B., Poree, F., Nagel, A., Lohse, M., Czedikeysenberg, A and Stitt, M. (2009). A guide to using MapMan to visualize and compare Omics data in plants: a case study in the crop species, Maize. *Plant, cell* and *environment*, *32*(9), 1211–1229.
65. Lohse, M., Nagel, A., Herter, T., May, P., Schroda, M., Zrenner, R., Tohge, T., Fernie, A.R., Stitt, M. and Usadel, B. (2014). Mercator: a fast and simple web server for genome scale functional annotation of plant sequence data. *Plant, cell* and *environment*, *37*(5), 1250–1258.
66. Phillippy, A. M., Schatz, M. C., and Pop, M. (2008). Genome assembly forensics: finding the elusive mis-assembly. *Genome Biology*, 9(3), R55.
67. Pop, M., and Salzberg, and S. L. (2008). Bioinformatics challenges of new sequencing technology. *Trends in Genetics*, 24(3), 142–149.
68. Parra, G., Bradnam, K., Ning, Z., Keane, T., and Korf, I. (2009). Assessing the gene space in draft genomes. *Nucleic acids research*, 37(1), 289–297.
69. Smit, A. F., Hubley, R., & Green, P. (1996). RepeatMasker. *Published on the web at* http://www.repeatmasker.org.
70. Smit, A. F. A., & Hubley, R. (2010). RepeatModeler Open-1.0. *Repeat Masker Website*.
71. Haas, B. J., Delcher, A. L., Mount, S. M., Wortman, J. R., Smith Jr, R. K., Hannick, L. I., Maiti, R., Ronning, C.M., Rusch, D.B., Town, C.D & Salzberg, S. L. (2003). Improving the Arabidopsis genome annotation using maximal transcript alignment assemblies. *Nucleic acids research*, *31*(19), 5654–5666.
72. Kent, W. J. (2002). BLAT—the BLAST-like alignment tool. *Genome research*, 12(4), 656–664.

73. Keller, O., Odronitz, F., Stanke, M., Kollmar, M., & Waack, S. (2008). Scipio: using protein sequences to determine the precise exon/intron structures of genes and their orthologs in closely related species. *BMC bioinformatics*, *9*(1), 278.
74. Birney, E., Clamp, M., and Durbin, R. (2004). GeneWise and Genomewise. *Genome research*, *14*(5), 988–995.
75. Finn, R. D., Clements, J., and Eddy, S. R. (2011). HMMER web server: interactive sequence similarity searching. *Nucleic acids research*, gkr367.
76. Parra, G., Bradnam, K., & Korf, I. (2007). CEGMA: a pipeline to accurately annotate core genes in eukaryotic genomes. *Bioinformatics*, *23*(9), 1061–1067.
77. Parra, G., Blanco, E., and Guigó, R. (2000). GeneID in drosophila. *Genome research*, *10*(4), 511–515.
78. Kim, D., Pertea, G., Trapnell, C., Pimentel, H., Kelley, R., and Salzberg, S. L. (2013). TopHat2: accurate alignment of transcriptomes in the presence of insertions, deletions and gene fusions. *Genome Biology*, 14(4), R36.
79. Langmead, B., & Salzberg, S. L. (2012). Fast gapped-read alignment with Bowtie 2. *Nature methods*, 9(4), 357–359.
80. Rawat, V., Abdelsamad, A., Pietzenuk, B., Seymour, D. K., Koenig, D., Weigel, D., Pecinka, A. and Schneeberger, K. (2015). Improving the Annotation of *Arabidopsis lyrata* Using RNA-seq Data. *PloS one*, *10*(9), e0137391.
81. Isaza, J. P., Galván, A. L., Polanco, V., Huang, B., Matveyev, A. V., Serrano, M. G., Manque, P., Buck, G.A. and Alzate, J. F. (2015). Revisiting the reference genomes of human pathogenic *Cryptosporidium* species: reannotation of *C. parvum* Iowa and a new *C. hominis* reference. *Scientific reports*, *5*.
82. Misra, S., Crosby, M. A., Mungall, C. J., Matthews, B. B., Campbell, K. S., Hradecky, P., Huang, Y., Kaminker, J.S., Millburn, G.H., Prochnik, S.E. and Smith, C. D. (2002). Annotation of the *Drosophila melanogaster* euchromatic genome: a systematic review. *Genome biology*, *3*(12), research0083
83. van den Berg, B. H., McCarthy, F. M., Lamont, S. J., and Burgess, S. C. (2010). Re-annotation is an essential step in systems biology modeling of functional genomics data. *PLoS One*, *5*(5), e10642.
84. Li, L., Chen, E., Yang, C., Zhu, J., Jayaraman, P., De Pons, J., Kaczorowski, C.C., Jacob, H.J., Greene, A.S., Hodges, M.R. & Cowley, A. W. (2015). Improved rat genome gene prediction by integration of ESTs with RNA-Seq information. *Bioinformatics*, *31*(1), 25–32.
85. Häkkinen, M., Arvas, M., Oja, M., Aro, N., Penttilä, M., Saloheimo, M., and Pakula, T. M. (2012). Re-annotation of the CAZy genes of *Trichoderma reesei* and transcription in the presence of lignocellulosic substrates. *Microbial Cell Factories*, *11*(1), 134.

# State-of-the-Art in Smith–Waterman Protein Database Search on HPC Platforms

**Enzo Rucci, Carlos García, Guillermo Botella, Armando De Giusti, Marcelo Naiouf, and Manuel Prieto-Matías**

**Abstract** Searching biological sequence database is a common and repeated task in bioinformatics and molecular biology. The Smith–Waterman algorithm is the most accurate method for this kind of search. Unfortunately, this algorithm is computationally demanding and the situation gets worse due to the exponential growth of biological data in the last years. For that reason, the scientific community has made great efforts to accelerate Smith–Waterman biological database searches in a wide variety of hardware platforms. We give a survey of the state-of-the-art in Smith–Waterman protein database search, focusing on four hardware architectures: central processing units, graphics processing units, field programmable gate arrays and Xeon Phi coprocessors. After briefly describing each hardware platform, we analyse temporal evolution, contributions, limitations and experimental work and the results of each implementation. Additionally, as energy efficiency is becoming more important every day, we also survey performance/power consumption works. Finally, we give our view on the future of Smith–Waterman protein searches considering next generations of hardware architectures and its upcoming technologies.

**Keywords** Bioinformatics • Computational acceleration • Database search • Smith–Waterman algorithm • Protein sequence

---

E. Rucci (✉) • A. De Giusti
Instituto de Investigación en Informática LIDI (III-LIDI), CONICET, Universidad Nacional de La Plata, Buenos Aires, Argentina
e-mail: erucci@lidi.info.unlp.edu.ar; degiusti@lidi.info.unlp.edu.ar

M. Naiouf
Instituto de Investigación en Informática LIDI (III-LIDI), CONICET, Facultad de Informática, Universidad Nacional de La Plata, Buenos Aires, Argentina
e-mail: mnaiouf@lidi.info.unlp.edu.ar

C. García • G. Botella • M. Prieto-Matías
Dpto. Arquitectura de Computadores y Automática Universidad Complutense de Madrid, Madrid, Spain
e-mail: garsanca@ucm.es; gbotella@fdi.ucm.es; mpmatias@ucm.es



K.-C. Wong (ed.), *Big Data Analytics in Genomics*,
DOI 10.1007/978-3-319-41279-5_6

# 1 Introduction

Searching biological sequence database is a common and repeated task in bioinformatics and molecular biology. In a typical search operation, biological sequences with unknown functionalities (usually referred as query sequences) are aligned to a database of known sequences to find similarities. The alignment process computes a score that represents the degree of similarity between each pair of query and database sequences. Sequence alignment methods are classified as either global or local. Global alignments try to maximise the number of matches between the two sequences along their entire lengths and are useful when the sequences are similar. On the other hand, local alignments try to maximise the number of matches between small portions of the two sequences. This kind of alignment exposes much better similarity between unrelated sequences and, at the same time, leads to more biologically relevant results [18]. The Smith–Waterman (SW) algorithm is the most accurate method for local sequence alignment. This algorithm is based on dynamic programming approach and its high sensitivity comes from exploring all the possible alignments between two sequences. Unfortunately, this method is computationally demanding and the situation gets worse due to the exponential growth of biological data in the last years. One frequently used approach to speed up this time demanding operation is to introduce heuristics in order to reduce the search space. Heuristics usually produce considerably good results. However, they are deficient in searching the best match subsequences and, in consequence, are not guaranteed to discover the optimal alignment. For that reason, the scientific community has made great efforts to accelerate SW protein database searches through high-performance computing (HPC) in a wide variety of hardware platforms. This chapter gives a survey of the state-of-the-art in SW protein database search, focusing on four hardware architectures: central processing units (CPU), graphics processing units (GPU), field programmable gate arrays (FPGA) and Xeon Phi coprocessors. After briefly describing each hardware platform, we analyse temporal evolution, contributions, limitations and experimental work and the results of each implementation. Additionally, as energy efficiency is becoming more important every day, we also survey performance/power consumption works. Finally, we give our view on the future of SW protein searches considering next generations of hardware architectures and its upcoming technologies.

The rest of the chapter is organised as follows: Sect. 2 briefly describes the considered hardware platforms. Section 3 introduces the basic concepts of the SW algorithm. Section 4 reviews hardware acceleration of SW protein database search. Section 5 overviews performance-power consumption evaluations on SW context. Section 6 gives our view on the future of SW protein searches considering next generations of hardware architectures. Finally, Sect. 7 presents the conclusions of this chapter.

## 2 Hardware Platforms

Scientific community has made great efforts to accelerate Smith–Waterman biological database searches in a wide variety of hardware architectures. Next there is a brief description for each hardware platform considered in this work.

### *2.1 Central Processing Units*

Traditional chip design was guided by increasing transistor count and clock speed, which enabled designers to implement many advanced techniques that permitted to increment instruction level parallelism (ILP) and, in consequence, led to improved application performance. However, at the beginning of this century, this design process got stuck due to two reasons:

- extracting more ILP from programmes became a hard task and
- increasing clock frequency reached unsustainable power consumption and heat generation levels.

Multi-core processors arose as a solution to this problem. Hardware vendors decided to integrate two or more computational cores within the same chip. Even though these cores are simpler and slower, when combined, they permit enhancing the global performance of the processor while making an efficient use of energy [28]. Its introduction also affected application programmers because explicit parallelism should be exploited to take advantage of multi-core hardware; in particular, both data and task parallelism. The first multi-core CPUs were simply two processors on the same die but later generations incorporated more cores, additional cache levels and better interconnection networks, among other features.

Currently, the main CPU vendor is Intel followed by AMD. In 2015, Intel presented the Skylake micro-architecture introducing the first processors of this family and more models were announced to the next 2 years. In particular, the high-performance Xeon line will incorporate several improvements, such as support for more sockets, channels of DDR4 memory and PCIe slots. Additionally, these processors will include AVX-512 vectorial instruction set[1] (a 512-bit extension of the current Advanced Vector Extensions with 256-bit width) and will give support to integrated FPGAs [36]. According to Intel, the next two micro-architectures will be available in 2016 (Kaby Lake) and 2017 (Cannonlake). Kaby Lake will be an upgraded version of Skylake while Cannonlake will shrink the fabrication technology to 10 nm [14].

AMD introduced three different micro-architectural families in 2011. The Fusion family corresponds to the Accelerated Processing Units that integrate CPUs and GPUs on the same chip. On the other hand, the Bobcat family was designed for low-power and low-cost devices while the Bulldozer family is oriented to desktop

[1] AVX-512 Extensions: https://software.intel.com/en-us/blogs/additional-avx-512-instructions.

computers and servers. Finally, the next AMD micro-architecture is named Zen and will be available by the end of 2016. Zen's main purpose consists in improving performance per core more than increasing number of cores or hardware threads. In particular, some preliminary reports state up to 40 % more instructions per clock cycle [30]. Unlike previous micro-architectural families, Zen will adopt simultaneous multi-threading capabilities and will be developed using 14 nm fabrication technology.

## *2.2 GPU*

GPUs were originally developed for computer games and its designs were orientated for that purpose. The first non-graphic applications were programmed adapting primitives from graphic languages like OpenGL or DirectX. In the last decade, GPU architectures were modified and several programming libraries were introduced that permitted avoiding graphic primitives. These changes increased GPU usage in significant manner. Nowadays, they have consolidated as general purpose accelerators in HPC community due to the increasing compute power and energy efficiency.

Currently, most popular programming languages for GPUs are CUDA [33], OpenCL [45] and, in lesser extent, OpenACC [35]. While these languages reduce programming cost compared to initial graphic languages, they still represent a hard task because they significantly differ from traditional CPU's programming model. Therefore, programmers must learn specific GPU knowledge to achieve high-performance applications. For example, common optimisation techniques comprise increasing hardware occupancy, exploiting memory hierarchy, organising memory accesses, avoiding divergent branches, among others.

The two main GPU vendors are NVIDIA and AMD. In 2011, AMD introduced Graphics Core Next (GCN) architecture, which is the basis for its individual and integrated GPUs. GCN was designed to achieve high performance not only in graphic applications but also in general purpose tasks [42]. One of the main AMD innovations in the last years is high-bandwidth memory (HBM) technology, a new type of *3D memory* that can be used in CPUs and GPUs. This kind of memory has several advantages: significant space savings and increased bandwidth and energetic efficiency [3]. HBM has been incorporated in AMD GPUs codenamed Fiji from GCN architecture, introduced in 2015, and more AMD cards with this technology will be available in 2016.

Current NVIDIA GPUs are based on Maxwell architecture, which was presented in 2014. Maxwell family redesigned Streaming Multiprocessor architecture, the heart of each NVIDIA GPU, and also the memory hierarchy [12]. These changes allowed Maxwell to improve performance and power efficiency in relation to its predecessor Kepler. NVIDIA has announced its next architecture codenamed Pascal for 2016, which will include HBM adoption and 16 nm manufacturing process. According to NVIDIA, Pascal will improve performance, performance per watt, memory capacity and bandwidth of Maxwell [31].

## 2.3 FPGA

FPGAs are reconfigurable integrated circuits comprising programmable interconnections that join programmable logic blocks, embedded memory blocks and digital signal processor blocks. Communication to the outside is performed through I/O blocks, which are arranged in a ring form around the circumference of these devices. As opposed to CPUs and GPUs, FPGA resources may be configured and linked together to create custom instruction pipelines through which data is processed. Also, they work at lower clock frequencies and have lower peak performances. However, since FPGAs can configure its hardware for each specific application, they usually reach better performance efficiencies. Additionally, they are normally more efficient from energetic point of view as there is no silicon waste [41, 48].

Since its development, FPGAs have significantly evolved continuously incrementing its available resources and incorporating features like standards for interconnection networks and high-speed I/O. At the beginning, FPGAs were used for digital signal processing. However, in the last few years, there are two clear trends to enlarge FPGA usage in other application domains. The first comprises the increasing integration of FPGAs with CPUs due to accelerators consolidation in HPC community as a way of improving performance while keeping power efficiency. In particular, the two main FPGA makers Xilinx and Altera have established different agreements with important CPU vendors to develop hybrid CPU-FPGA architectures. IBM has announced a strategic partnership with Xilinx to enable higher performance and energy-efficient applications through FPGA-enabled workload acceleration on IBM POWER-based systems [15]. On its behalf, Altera has been recently acquired by Intel and they plan to combine Altera's FPGA products with Intel Xeon processors as highly customised, integrated products [16]. The second trend consists in reducing FPGA programming cost. Generally, digital design verification and creation have involved the use of hardware description languages (HDLs), like Verilog and VHDL. However, HDLs are tedious, error prone and affected by an extra abstraction layer as they contain the additional concept of time. Currently, both Altera and Xilinx are working on high-level tools that seek to reduce the programming cost of these devices; in particular, through OpenCL standard [2, 52].

## 2.4 Xeon Phi

The Xeon Phi is a recent many-core coprocessor developed by Intel for HPC applications. In its current generation, the Xeon Phi features up to 61 x86 pentium cores with extended vector processing units (512-bit) named Knight Corner (KNC) and simultaneous multi-threading capabilities (four hardware threads per core). Each core integrates an L1 cache and has an associated fully coherent L2 cache.

Additionally, a high-speed ring interconnection allows data transfer among all the L2 caches and the memory subsystem.

The Xeon Phi offers two execution modes: *offload* and *native*. In the offload mode, the Xeon Phi acts as a coprocessor. It takes on computationally demanding parts of programmes delegated by the CPU. In the native mode, the Xeon Phi runs as a completely standalone computing system. In this mode, applications can use solely the resources of the coprocessor.

From a programming point of view, one of the main advantages of this platform is the support of existing parallel programming models traditionally used on HPC systems such as the OpenMP or MPI paradigms, which simplifies code development and improves portability over other alternatives based on accelerator-specific programming languages such as CUDA or OpenCL.

With regard to the future of Intel many-core coprocessors, Intel has announced the next generation, called Knights Landing, which is planned to run HPC systems in 2016. Among the main differences posted, the chip will be built with 14 nm technology and be able to operate as a standalone CPU rather than as a coprocessor. It will also incorporate Intel Silvermont processors with AVX-512 vector capabilities, unifying in this way vector extensions with general purpose Intel Xeon Skylake processors. Lastly, main memory will have a stacked organisation, similar to HBM proposal [36].

## 3 Smith–Waterman Algorithm

In 1970, Saul Needleman and Christian Wunsch introduced an algorithm to compute optimal global alignment between two biological sequences, known as the Needleman–Wunsch algorithm [32]. Later, in 1981, Temple Smith and Michael Waterman proposed a variant of the Needleman–Wunsch algorithm to find the optimal local alignment of two sequences [43]. The SW method has been used as the basis for many subsequent algorithms and is often employed as a benchmark when comparing different alignment techniques [13]. Its strength comes from the guarantee of discovering optimal alignment because it explores all possible alignments between the pair of sequences.

The SW algorithm computes the optimal local alignment between two sequences following a dynamic programming approach and can be divided into two stages: (1) similarity matrix (also called alignment matrix) filling, to obtain optimal alignment score; and (2) traceback, to obtain optimal alignment.

1. *Similarity matrix filling:* given two sequences $q = q_1q_2q_3 \dots q_m$ and $d = d_1d_2d_3 \dots d_n$, SW fills a matrix $H$ which keeps track of the degree of similarity between them. The recurrence relations for the SW algorithm with the modifications of Gotoh [11] for handling multiple sized gap penalties are shown below:

$$H_{i,j} = max\Big\{0,\ \ H_{i-1,j-1} + SM(q_i, d_j),\ \ E_{i,j},\ \ F_{i,j}\Big\} \tag{1}$$

$$E_{i,j} = max\left\{H_{i,j-1} - G_{oe},\ E_{i,j-1} - G_e\right\} \tag{2}$$

$$F_{i,j} = max\left\{H_{i-1,j} - G_{oe},\ F_{i-1,j} - G_e\right\} \tag{3}$$

The residues of sequence $q$, usually called *query sequence*, label the rows. In similar way, the residues of sequence $d$, usually called *database sequence*, label the columns. $H_{i,j}$ represents the score for aligning the prefixes of $q$ and $d$ ending at position $i$ and $j$, respectively. $E_{i,j}$ and $F_{i,j}$ are the scores ending with a gap involving the first $i$ residues of $q$ and the first $j$ residues of $d$, respectively. *SM* is the *substitution matrix* which defines the substitution scores for all residue pairs. Generally *SM* rewards with a positive value when $q_i$ and $d_j$ are identical or relatives, and punishes with a negative value otherwise. Common substitution matrices for protein alignment are BLOSUM or PAM families. $G_{oe}$ is the sum of gap open and gap extension penalties while $G_e$ is the gap extension penalty. The recurrences should be calculated with $1 \le i \le m$ and $1 \le j \le n$, after initialising $H$, $E$ and $F$ with 0 when $i = 0$ or $j = 0$. The maximal alignment score in the matrix $H$ is the optimal local alignment score $S$.

2. *Traceback:* Based on the position in matrix $H$ where the value $S$ was found, a traceback procedure is performed to obtain the pair of segments with maximum similarity, until a position whose value is zero is reached (this being the starting alignment point of the segments). These two segments represent the best local alignment.

The SW algorithm has quadratic time complexity. To compute optimal alignments, this method has quadratic spatial complexity. However, computing optimal alignment scores do not require storing full similarity matrix and can be computed in linear space complexity.

Figure 1 shows the calculation of four cells ($H_{1,1}$, $H_{1,2}$, $H_{2,1}$ and $H_{2,2}$) in the similarity matrix corresponding to the SW alignment between protein sequences CAWHEAET ($q$) and CITAGWHEE ($d$). BLOSUM62 was selected as the scoring matrix, and gap insertion and extension penalties were set to 6 and 2, respectively. After initialising $H$, $E$ and $F$ with zero when $i = 0$ or $j = 0$, the other cells in the similarity matrix are computed according to Eq. (1). For example, the cell $H_{1,1}$ in Fig. 1 is 9 because that is the maximum of 0, 9 (the upper-left neighbour plus the similarity score from BLOSUM62 substitution matrix, $0 + 9 = 9$), $-2$ (the alignment score ending with a gap in the query sequence, $max\{0-8, 0-2\} = -2$) and $-2$ (the alignment score ending with a gap in the database sequence, $max\{0-8, 0-2\} = -2$).

It is important to note that no cell value can be less than zero. Additionally, there is a strict order of computation in matrix $H$ due to the data dependences inherent to this problem. Any cell in matrix $H$ has a dependence on three cells: the one to the left, the one above and the one from the upper-left diagonal. This dependence is illustrated in Fig. 1 through arrows.

Once all values in matrix $H$ were computed, a traceback procedure is performed to obtain the best local alignment. Figure 2 illustrates the complete similarity matrix corresponding to the SW alignment shown in Fig. 1. The traceback starts in $S$

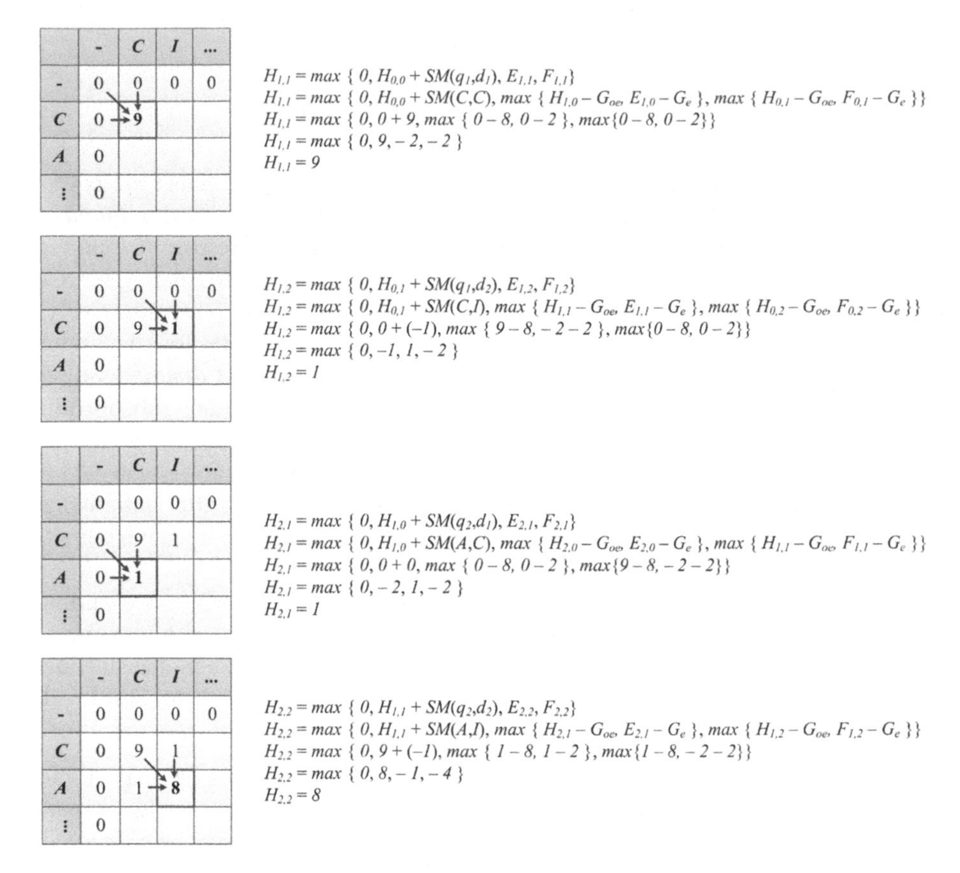


**Fig. 1** Calculation of four cells in the similarity matrix corresponding to the SW alignment between protein sequences CAWHEAET and CITAGWHEE

position and then works backwards (upwards and to the left). Moving upwards inserts a gap in the database sequence while moving to the left inserts a gap in the query sequence. The traceback procedure terminates when a position whose value is zero is reached (this being the starting alignment point of the segments). These two segments represent the best local alignment. In this example, the optimal alignment score is 20 while the best local alignment is shown under the similarity matrix.

## 4 Acceleration of SW Protein Database Search

As explained in Sect. 1, dynamic programming algorithms can be too computationally expensive to be used in biological database searches. In fact, due to the exponential growth in biological data, heuristic approaches are not enough in some occasions [18]. This is where parallelism exploitation becomes fundamental to accelerate this kind of searches. This section focuses on the state-of-the-art of

**Fig. 2** Illustration of Smith–Waterman alignment between protein sequences CAWHEAET and CITAGWHEE

| | - | C | I | T | A | G | W | H | E | E |
|---|---|---|---|---|---|---|---|---|---|---|
| - | 0 | 0 | 0 | 0 | 0 | 0 | 0 | 0 | 0 | 0 |
| C | 0 | 9 | 1 | 0 | 0 | 0 | 0 | 0 | 0 | 0 |
| A | 0 | 1 | 8 | 1 | **4** | 0 | 0 | 0 | 0 | 0 |
| W | 0 | 0 | 0 | 6 | 0 | **15** | 7 | 5 | 3 | 1 |
| H | 0 | 0 | 0 | 0 | 4 | 7 | **13** | ←**15** | 7 | 5 |
| E | 0 | 0 | 0 | 0 | 0 | 5 | 5 | 13 | **20** | 12 |
| A | 0 | 0 | 0 | 0 | 4 | 3 | 5 | 5 | 12 | 19 |
| E | 0 | 0 | 0 | 0 | 0 | 1 | 1 | 5 | 10 | 17 |
| T | 0 | 0 | 0 | 5 | 0 | 0 | 0 | 1 | 8 | 9 |

*Best local alignment* A W H – E
A G W H E

SW algorithm acceleration. First of all, we study the data dependences of SW algorithm and the possible ways of parallelise it. Next, we describe the available implementations for four different hardware platforms: CPU, GPU, FPGA and Xeon Phi.

## *4.1 Data Dependences and Parallelism*

The most computational expensive part of SW algorithm is the similarity matrix filling. Even though the data dependences described in Sect. 3 restrict the ways in that the similarity matrix can be computed, the SW computation has some inherent parallelism that can be exploited to reduce its computational cost. In general, accelerated implementations adopt one of the following two approaches:

- In the intra-task parallelism approach, the parallelism within a single pair of sequences is exploited. The implementations following this approach compute several anti-diagonal cells in parallel, since these computations are independent among them. It is also possible to compute several cells in a row or a column at the same time; however, a subsequent adjustment mechanism is required to maintain algorithm coherency due to data dependence ignorance.
- Inter-task parallelism is based on performing several pairwise alignments concurrently. Its backbone is based on null data dependence between alignments, which turns the problem into an embarrassingly parallel one.

Both approaches have been extensively explored by scientific community in a wide variety of hardware platforms. In the following subsection, we describe the works based on CPU, GPU, FPGA and Xeon Phi.

## 4.2 Available Implementations

*Cell updates per second* (CUPS) is a commonly used performance measure in the Smith–Waterman context, because it allows removal of the dependence on the query sequences and the databases utilised for the different tests as well as the hardware device. A CUPS represents the time for a complete computation of one cell in matrix $H$, including all memory operations and the corresponding computation of the values in the $E$ and $F$ arrays. Given a query sequence $Q$ and a database $D$, the GCUPS (billion cell updates per second) value is calculated by:

$$\frac{|Q| \times |D|}{t \times 10^9} \tag{4}$$

where $|Q|$ is the total number of residues in the query sequence, $|D|$ is the total number of residues in the database and $t$ is the runtime in seconds [23].

Next, we describe the most notable implementations according to the hardware architecture employed.

### 4.2.1 CPU Implementations

The first efforts to accelerate SW algorithm date back to the 1990s. Alpern et al. [1] proposed several techniques including a parallel implementation that used microparallelism by dividing the 64-bit Z-buffer registers of the Intel Paragon i860 processor into four parts. This approach allowed comparing the query sequence and four database sequences at the same time. As a result, they reached a 5× speed-up compared to a conventional implementation.

The intra-register parallelism of the previous approach would be simpler and easier with the future introduction of small vectorial capabilities from processor vendors. With the rise of multimedia applications, general purpose processors incorporated SIMD technology, like the MMX, SSE, AVX or AltiVec extensions [37]. In general, four different approaches can be identified to vectorisation in similarity matrix computations. Figure 3 illustrates these approaches.

Most efforts focused on intra-task parallelism. In 1997, Wozniak [51] presented a parallel implementation for a Sun Ultra Sparc processor that exploited the SIMD video instructions to compute several anti-diagonal cells in parallel, since they have no dependences among them. Unfortunately, getting substitution scores for anti-diagonal cells resulted complex and hard to resolve efficiently, because each amino acid pair requires different indexation in substitution matrix. Even so, Wozniak achieved a 2× acceleration over the fastest serial implementation of the time.

Three years later, in 2000, Rognes and Seeberg [38] introduced a SIMD version using the MMX/SSE extensions, becoming the first to take advantage of these instruction sets. They found that vectorising along the query sequence was faster than vectorising along the anti-diagonals, in spite of having to make more calculations due to ignoring some of the data dependences mentioned in Sect. 4.1.

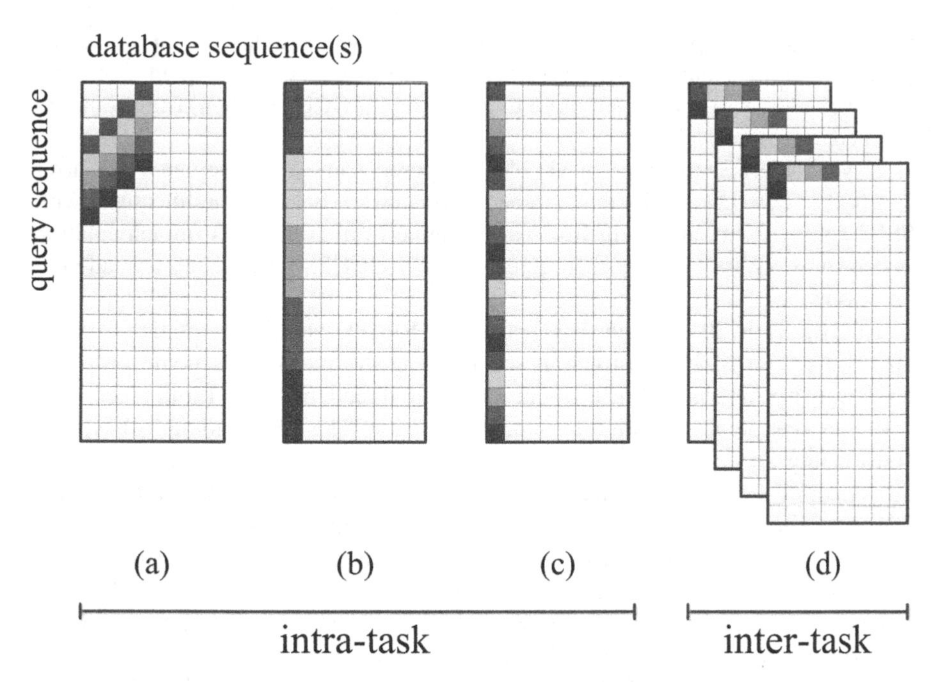


**Fig. 3** Approaches to vectorisation in similarity matrix computations (adapted from [37]): (**a**) Vectorisation along the anti-diagonals, proposed by Wozniak [51]; (**b**) Vectorisation along the query sequence, proposed by Rognes and Seeberg [38]; (**c**) Vectorisation along the query sequence (striped approach), proposed by Farrar [9]; (**d**) Vectorisation along multiple database sequences, proposed by Alpern et al. [1]

Another difference with previous solutions lies on Rognes and Seeberg used 8-bit integer data. This fact allowed them to compute a higher number of alignments in parallel at the cost of reducing score representation range to 0–255. This scheme results beneficial because overflow only occurs when sequences are long and/or very similar between them. When overflow occurs, a 16-bit integer version of the algorithm is employed to guarantee correct results. In addition, they introduced the Query Profile (QP) technique, which consists in building an auxiliary two-dimensional array of size $|q| \times |\sum|$, where $q$ is the query sequence and $\sum$ is the alphabet. Each row of this array contains the scores of the corresponding query residue against each possible residue in the alphabet. QP technique improves data locality by replacing a random access to the substitution matrix with a linear access to the QP matrix in the algorithm innermost loop. The SIMD instructions usage together with QP technique allowed Rognes and Seeberg to reach a 6× speed-up over an optimised serial implementation.

Later, in 2007, Farrar [9, 10] used SSE2 extensions to develop an improved version of the Rognes and Seeberg implementation. Just as the previous approach, Farrar vectorised along the query sequence. However, the alignment matrix computations were re-organised to do them in a striped manner. This access pattern minimises data dependences impact and reduces misaligned vector accesses. Additionally, Farrar proposed the *lazy F* technique, which helps to minimise the number of conditional jumps inside the innermost loop. As a consequence of these improvements, Farrar reported more than 11 and 20 GCUPS when using four and eight cores, respectively.

A year later, Szalkowski et al. [44] proposed some improvements to Farrar approach. They presented a multi-threaded version for both x86 architectures with SSE2 compatibility and Cell/Broadband Engine from IBM. This implementation is known as SWPS3 and reached a peak of 15.7 GCUPS when using a quad-core processor.

Afterwards, in 2011, Rognes [37] presented SWIPE, an implementation for Intel processors with SSSE3 instruction set adopting the inter-task scheme proposed previously by Alpern et al. [1]. Rognes also introduced the Score Profile technique (SP) to accelerate the substitution scores extraction. The SP technique is based on constructing an auxiliary $n \times l \times |\sum|$ two-dimensional score array, where $n$ is the length of the database sequence, $l$ is the number of vector lanes and $\sum$ is the alphabet. Since each row of the auxiliary matrix forms an $l$-lane score vector, its values can be loaded at the same time through a single load vectorial instruction. The main SP disadvantage is that score array must be constructed for each database sequence. Using two six-core processors, SWIPE achieved a peak of 106.2 GCUPS, being up to six times faster than SWPS3 and Farrar approach.

In 2013, Zhao et al. [54] presented SSW, a library developed in C/C + + to facilitate SW integration to other genomic applications. SSW adopts Farrar approach to compute optimal alignment as well as optimal score. As SSW is also available as an autonomous tool, the authors measured its performance while searching Swiss-Prot database; SSW reported up to 2.53 GCUPS using an AMD x86 64 2.0Ghz processor.

Two years later, in 2015, Rucci et al. [39] introduced SWIMM, an implementation for Intel heterogeneous systems combining Xeon and Xeon Phi processors. SWIMM adopts the inter-task scheme and offers three execution mode: (1) Xeon, (2) Xeon Phi and (3) concurrent Xeon and Xeon Phi. Unlike previous implementations, SWIMM is able to take advantage of AVX2 as well as SSE extensions. The AVX2 exploitation allows SWIMM to compute up to 32 alignments in parallel in place of 16 (SSE case). In their work, the authors showed that SWIMM can be comparable with SWIPE when using SSE instruction set. However, when taking advantage of AVX2 extensions, SWIMM demonstrated to be superior to SWIPE. In the Xeon mode with AVX2 exploitation, SWIMM reached up to 360 GCUPS using two Intel Xeon E5-2695 v3 2.3 GHz processors when searching Environmental NR database.

Finally, in 2016, Daily [7] presented Parasail, a C-based library containing implementations of different pairwise sequence alignment algorithms. The intent of this library is to be integrated into other software packages, not necessarily to replace already highly performing database search tools. In that sense, Parasail implements most known algorithms for vectorised SW sequence alignment that follows intra-task parallelism scheme (including Wozniak [51], Rognes and Seeberg [38] and Farrar [9, 10] approaches described above). Additionally, the Parasail library imple-

**Table 1** Performance summary of CPU implementations

| Year | Implementation | Hardware used | No. of threads | GCUPS |
|---|---|---|---|---|
| 1997 | Wozniak [51] | Sun Ultra Sparc Enterprise 6000 167 MHz | 12 | 0.2 |
| 2000 | Rognes and Seeberg [38] | Intel Pentium III 500 MHz | 1 | 0.15 |
| 2007 | Farrar [9] | Intel Xeon Core 2 Duo 2.0 GHz | 1 | 2.9 |
| 2008 | SWPS3, Szalkowski et al. [44] | Intel Core 2 Quad Q6600 2.4 GHz | 4 | 15.07 |
| 2011 | SWIPE, Rognes [37] | 2×Intel Xeon X5650 2.67 GHz | 24 | 106.2 |
| 2013 | SSW, Zhao et al. [54] | AMD x86 64 2.0Ghz | 1 | 2.53 |
| 2015 | SWIMM, Rucci et al. [39] | 2×Intel Xeon E5-2695 v3 2.3 GHz | 28<br>56 | 309.3<br>360 |
| 2016 | Parasail, Jeff Daily [7] | 2×Intel Xeon E5-2670 2.3 GHz | 24 | 291.5 |

ments each of these methods for different instructions sets: SSE2, SSE4.1, AVX2 and KNC. In his work, the author showed that Parasail's AVX2 implementation is able to outperform SWIPE for query sequences longer than approximately 500 amino acids. Parasail reported up to 291.5 GCUPS using two Intel Xeon E5-2670 2.3 GHz processors when searching UniProt Knowledgebase database.

Table 1 summarises the performance of CPU implementations.

### 4.2.2 GPU Implementations

The first GPU implementations date back to 2006 and they were proposed by Weiguo Liu et al. [21] and Yang Liu et al. [22]. Both proposals are very similar: they are based on the OpenGL library, compute alignment matrices by anti-diagonals and store sequences as well as auxiliary buffers in texture memory. Weiguo Liu et al. implementation only processes protein sequences of length shorter than 4096 amino acids due to limitations imposed by the texture memory of that time, and for that reason, the experimental work was carried out with a reduced version of Swiss-Prot database (99.8 % of original sequences). This implementation reached 0.65 GCUPS using a NVIDIA GeForce 6800 GT, which represented a speed-up of 16× over a serial CPU implementation. On its behalf, the Yang Liu et al. implementation does not impose restrictions on the sequence length and it offers two execution modes: (1) optimal alignment and (2) optimal alignment score. Using a NVIDIA GeForce 7800 GTX, this implementation achieved 0.18 and 0.24 GCUPS in (1) and (2) execution modes, respectively.

In 2008, Manavski and Valle [27] presented the first CUDA implementation for SW protein database search, which was named SW-CUDA. As opposed to previous proposals, SW-CUDA adopts inter-task parallelism scheme because each CUDA thread computes a complete alignment between the query sequence and a particular database sequence. Another distinctive characteristic of SW-CUDA is the QP technique usage to obtain scores from substitution matrix. SW-CUDA reported 1.85 GCUPS using a NVIDIA GeForce 8800 GTX card when searching Swiss-Prot database. Additionally, it showed good scalability with the amount of GPUs owing to a dynamic workload balance technique that considers the compute power of each particular device.

In 2009, Yongchao Liu et al. [23] introduced CUDASW++, an implementation for CUDA-enabled GPUs that combines both intra-task and inter-task parallelism approaches. The inter-task scheme usually reports better performance than intra-task scheme; however, it requires more memory resources. Therefore, CUDASW++ sets a configurable length threshold to compute the alignments. Database sequences of length less than or equal to the threshold are computed according to the inter-task scheme. The alignments of database sequences of length greater than the threshold are carried out following the intra-task scheme. CUDASW++ also carefully arranges memory accesses to get coalesced access. Besides, it exploits memory hierarchy by storing query sequence and substitution matrix in constant and shared memories, respectively. This set of optimisations allowed CUDASW++ to reach 9.63 and 16.09 when searching Swiss-Prot database using a single-GPU NVIDIA GeForce GTX 280 and a dual-GPU NVIDIA GeForce GTX 295, respectively.

A year later, the same authors of CUDASW++ presented an improved version of this tool, known as CUDASW++ 2.0 [25]. This version offers two execution modes: the first one adopts the original mode but includes QP technique and replaces scalar data with packed data; the second one follows the Farrar approach [9] through SIMD instruction virtualisation on graphic cards. Searching Swiss-Prot database, CUDASW++ 2.0 reported similar behaviour between both execution modes, reaching 16.9 and 29.6 GCUPS on a single-GPU NVIDIA GeForce GTX 280 and on a dual-GPU NVIDIA GeForce GTX 295, respectively.

GASW is another tool for CUDA-compatible GPUs [18]. This software was introduced in 2010 and among its optimisation we can mention elimination of memory bottlenecks and the conversion of the database to a format convenient for GPU usage. GASW achieved up to 21.36 GCUPS on a NVIDIA GeForce GTX 275 when searching Swiss-Prot database.

In 2011, Zou et al. [55] presented a CUDA-based implementation that combines different optimisations: global memory accesses in coalescent manner, hierarchy memory exploitation and loop unrolling. This implementation reached 28.35 GCUPS on a NVIDIA GeForce GTX 470 when searching a synthetic database of 107520 sequences (each sequence contains 1024 random residues).

There are few known implementations based on OpenCL for GPUs. Khalafallah et al. [19] follow inter-task approach and reuse several optimisations from previous implementations like global memory accesses arrangement to obtain coalescent access and texture memory usage to store QP matrix. This implementation achieved 12.29 and 65.99 GCUPS when searching a reduced version of Swiss-Prot database on a NVIDIA GeForce 9800 GT and an ATI HD 5850, respectively. On its behalf, Borovska and Lazarova proposal [6] also adopts inter-task scheme, although it does not provide enough implementation details. With Swiss-Prot database, this software reported 1.6 and 7.8 GCUPS on a NVIDIA Quadro FX3600M and on a NVIDIA GeForce GTX 295 (dual-GPU), respectively.

Finally, in 2013, Yongchao Liu et al. [26] presented a third version of CUDASW++, which was named CUDASW++ 3.0. This implementation targets NVIDIA GPUs based on the Kepler architecture and combines concurrent CPU and GPU computations adopting the inter-task approach on both cases. The workload is distributed dynamically using an heuristic based on hardware characteristics and constants derived from empirical evaluations. For the GPU computation, CUDASW++ 3.0 employs CUDA PTX video instructions. For the CPU computation, this algorithm uses SSE extensions; specially the host code is based on SWIPE code. Alignments are computed using 8-bit integers on both CPU and GPU devices. Once all alignments are processed, CPU detects overflow cases and recomputes them using 16-bit integer. It is important to mention that, because GPU adopts inter-task scheme, it only processes those database sequences of length less than, or equal to, the threshold. Longer sequences are obligatory computed in CPU. With Swiss-Prot as benchmark database, CUDASW++ 3.0 reached 119 GCUPS using a personal computer based on a quad-core Intel i7 2700k 3.5Ghz processor and a NVIDIA GeForce GTX 680. Because NVIDIA decided to cut down the capability of the PTX video instructions in Maxwell architecture, CUDASW++ 3.0 could not run at full speed on these GPUs. Beyond the previous limitation, CUDASW++ 3.0 is considered the fastest SW implementation for CUDA-compatible systems as of today.

Table 2 summarises the performance of GPU implementations.

### 4.2.3 FPGA Implementations

Acceleration of sequence alignment using FPGAs is a widely studied topic in HPC community. However, most of these implementations focus on DNA alignment because it is simpler than protein alignment from an algorithmic perspective (DNA alignment has a reduced alphabet and adopts a simpler scoring scheme).

Beyond sequence type, FPGA implementations are usually based on creating basic building blocks that can compute a matrix cell in a clock cycle. Next, multiple instances of these blocks are combined at the same time to create systolic arrays capable of processing large amounts of data in parallel. A systolic array is an arrangement of processing units in array form, where data flows synchronously among the units, usually in a specific direction. This kind of array works like vectorial units in modern CPUs (e.g. SSE units) but, instead of having a fixed length, systolic arrays can configure its length [47].

Unfortunately, analytic comparison among these implementations is quite difficult due to different causes [8, 13]:

- There is a wide variety of FPGAs and each of them implements its circuitry in a different way, which complicates direct comparison.

**Table 2** Performance summary of GPU implementations

| Year | Implementation | Programming model | Hardware used | Database | GCUPS |
|---|---|---|---|---|---|
| 2006 | Weiguo Liu et al. [21] | OpenGL | NVIDIA GeForce 6800 GT | Reduced Swiss-Prot (99.8 %) | 0.65 |
| 2006 | Yang Liu et al. [22] | OpenGL | NVIDIA GeForce 7800 GTX | 983 sequences (462862 amino acids) | 0.24 |
| 2008 | SW-CUDA, Manavski and Valle [27] | CUDA | NVIDIA GeForce 8800 GTX | Swiss-Prot | 1.85 |
| | | | 2×NVIDIA GeForce 8800 GTX | | 3.61 |
| 2009 | CUDASW++, Yongchao Liu et al. [23] | CUDA | NVIDIA GeForce GTX 280 | Swiss-Prot | 9.63 |
| | | | NVIDIA GeForce GTX 295 (dual-GPU) | | 16.09 |
| 2010 | CUDASW++ 2.0, Yongchao Liu et al. [25] | CUDA | NVIDIA GeForce GTX 280 | Swiss-Prot | 16.9 |
| | | | NVIDIA GeForce GTX 295 (dual-GPU) | | 29.6 |
| 2010 | GASW, Kentie [18] | CUDA | NVIDIA GeForce GTX 275 | Swiss-Prot | 21.36 |
| 2010 | Khalafallah et al. [19] | OpenCL | NVIDIA GeForce 9800 GT | Reduced Swiss-Prot (45 %) | 12.29 |
| | | | ATI HD 5850 | | 65.99 |

| Year | Implementation | Programming model | Hardware used | Database | GCUPS |
|---|---|---|---|---|---|
| 2011 | Zou et al. [55] | CUDA | NVIDIA GeForce GTX 280 | 107520 sequences of 1024 amino acids each | 13.71 |
| | | | NVIDIA GeForce GTX 470 | | 28.35 |
| 2011 | Borovska and Lazarova [6] | OpenCL | NVIDIA Quadro FX3600M | Swiss-Prot | 1.6 |
| | | | NVIDIA GeForce GTX 295 (dual-GPU) | | 7.8 |
| 2013 | CUDASW++ 3.0, Yongchao Liu et al. [26] | Pthreads + CUDA | Intel i7 2700k 3.5Ghz + NVIDIA GeForce GTX 680 | Swiss-Prot | 119 |
| | | | Intel i7 2700k 3.5Ghz + NVIDIA GeForce GTX 690 (dual-GPU) | | 185.6 |

- There are very few fully functional tools. The implementations that report the best performance only contemplate synthetic tests with the aim to show the potential of this kind of accelerator although its usage in real world is very limited. Some implementations effectively employ real data but also present some limitations: query sequence is embedded in the design, sequences have a fixed or limited length, search parameters (gap penalisations or substitution matrix) are fixed or they require hardware reconfiguration to change them, among others.
- Lack of documentation on implementation details. For example, the implementation performance depends strongly on the data width used. Normally, performance improves as data width reduces, although a very small size could be insufficient to compute all alignments. Researchers tend to omit this kind of detail.

Among the fully functional implementations, we can found Isa et al. [17], Benkrid et al. [4] and Rucci et al. [40]. In 2011, Isa et al. [17] proposed a linear systolic array implementation for a Xilinx Virtex-5 XC5VLX110 FPGA although programming language was not specified. This implementation consists of a pipeline of basic processing elements, each holding one query residue whereas the subject sequence is shifted systolically through it. The alignments are computed using 11-bit integers. When searching Swiss-Prot database, this implementation reached up to 28 GCUPS.

A year later, Benkrid et al. [4] introduced an implementation similar to Isa et al. proposal [17]. This implementation is also based on a linear systolic array. However, alignments are computed using 16-bit integers and the corresponding FPGA design was captured in a C-based high-level hardware language, called Handel-C [29]. Using a Xilinx Virtex-4 LX160-11 FPGA, the authors reported up to 19.4 GCUPS when searching Swiss-Prot database.

In 2016, Rucci et al. [40] presented OSWALD, a tool developed with Altera OpenCL SDK for Altera FPGA-based systems. Unlike the rest of the implementations, this tool does not follow a linear systolic array fashion, on the contrary, it adopts the inter-task scheme. OSWALD computes alignments in FPGA using 8-bit integers and the host recomputes overflowed alignments using wider integer data. Also, it is able to combine concurrent CPU computations through multi-threading and SIMD exploitation. On a heterogeneous platform based on two Xeon E5-2670 and a single Altera Stratix V GSD5 FPGA, OSWALD reached up to 58.4 GCUPS on FPGA mode and 178.9 GCUPS on hybrid mode (host+FPGA), while searching Environmental NR database.

Table 3 summarises the performance of known FPGA implementations.

#### 4.2.4 Xeon Phi Implementations

Xeon Phi coprocessors can also be employed to accelerate SW alignments. In 2014, Liu and Schmidt [24] introduced SWAPHI, a tool for similarity searches based on OpenMP. This implementation adopts the offload model and is capable

**Table 3** Performance summary of FPGA implementations

| Year | Implementation | Programming language | Hardware used | GCUPS | Comments |
|---|---|---|---|---|---|
| 2004 | Dydel and Bala [8] | VHDL | Xilinx Virtex II XC2VP70-5 | 11.18 | Synthetic tests. Fixed sequence lengths |
| 2005 | Oliver et al. [34] | Verilog | Xilinx Virtex II XC2V6000 | 10.6 | Query sequence of limited length. Require hardware reconfiguration to change search parameters |
| 2007 | Zhang et al. [53] | Not specified | Altera Stratix II EP25180 | 25.6 | Require hardware reconfiguration to change search parameters |
| 2007 | Van Court and Herbordt [46] | VHDL | Xilinx Virtex II Pro XC2VP70-5 | 5.41 | Sequences of limited length. Require hardware reconfiguration to change search parameters |
| 2009 | Benkrid et al. [5] | Handel-C | Xilinx Virtex II XC2V6000-4 | 7.66* | Require hardware reconfiguration to change search parameters. *Theoretical peak performance |

(continued)

**Table 3** (continued)

| Year | Implementation | Programming language | Hardware used | GCUPS | Comments |
|---|---|---|---|---|---|
| 2011 | Isa et al. [17] | Not specified | Xilinx Virtex-5 XC5VLX110 | 28 | |
| 2011 | Zou et al. [55] | Not specified | Xilinx Virtex-5 XC5VLX330 | 47 | Synthetic tests |
| 2012 | Benkrid et al. [4] | Handel-C | Xilinx Virtex-4 LX160-11 | 19.4 | |
| 2016 | OSWALD, Rucci et al. [40] | OpenMP + OpenCL | 2×Intel Xeon E5-2670 2.60Ghz + Altera Stratix V GSD5 | 58.4 | FPGA mode |
| | | | | 178.9 | Hybrid mode |

of taking advantage of several coprocessors at the same time. In this work, the authors explored the benefits of intra-task and inter-task approaches as well as QP and SP techniques. In particular, SWAPHI is able to compute 16 cells in parallel due to KNC instruction set. Using a Xeon Phi 5110P, SWAPHI achieved up to 45.6 and 58.8 when searching TrEMBL database with intra-task and inter-task schemes, respectively. When using four coprocessors, performance increased to 164.9 (intra-task) and 228.4 GCUPS (inter-task) for the same database searches.

XSW is another similarity search tool based on Xeon Phi coprocessors [49]. Like SWAPHI, XSW employs inter-task scheme, SP technique and the KNC instruction set. Unlike SWAPHI, XSW works in native mode. XSW reported up to 70 GCUPS when searching Environmental NR database using a Xeon Phi 7110P.

An extended version of XSW, known as XSW 2.0, was developed subsequently by the same authors [50]. This implementation follows the offload model and combines concurrent CPU computations through multi-threading and SSE extensions. Using a Intel Xeon E5-2620 processor and a Xeon Phi 7110P coprocessor, XSW 2.0 reached up to 100 GCUPS when searching Environmental NR database.

In 2015, Rucci et al. [39] presented the previously described SWIMM tool. In their work, the authors state that, despite having more cores and wider vectorial processing units, the poor performance (in terms of GCUPS) of the Xeon Phi compared to Xeon is due mainly to the absence of low-range vector capabilities on the Xeon Phi. Beyond that, SWIMM showed to be comparable with SWAPHI in the Xeon Phi mode and significantly superior to XSW 2.0 in the hybrid mode. When searching Environmental NR database, SWIMM reported 160 GCUPS using two Intel Xeon E5-2670 2.60Ghz processors and an Intel Xeon Phi 3120P.

Finally, XSW 2.0 was replaced by another tool named LSBDS [20]. The main difference between XSW 2.0 and LSBDS is that the latter adopted a multi-pass method to compute the alignment matrices that solved the performance drop problem for long query sequences of the former. LSBDS reported up to 220 GCUPS when searching a merged database (Environmental NR + TrEMBL) using two Intel Xeon E5-2620 v2 2.0Ghz processors and two Intel Xeon Phi 7110P.

Table 4 summarises the performance of Xeon Phi implementations.

## 5 Performance/Power Consumption Evaluations

Energy efficiency is becoming more important every day in the HPC community. There is a wide availability of works exploring performance and power consumption of different hardware devices. However, only three do it in the SW context. All of these works have similar coarse-grain results but differ in the improvement coefficients due to different methodological aspects.

In 2011, Zhou et al. [55] evaluated performance and power consumption of several implementations for CPU, GPU and FPGA. Considering energy efficiency as GCUPS/Watt, they found that FPGA outperforms CPU and GPU by factors of

**Table 4** Performance summary of Xeon Phi implementations

| Year | Implementation | Programming model | Hardware used | Database | GCUPS |
|---|---|---|---|---|---|
| 2014 | SWAPHI, Liu and Schmidt [24] | OpenMP (offload) | Intel Xeon Phi 5110P<br>4× Intel Xeon Phi 5110P | TrEMBL | 58.8<br>228.4 |
| 2014 | XSW, Wang et al. [49] | OpenMP + Pthreads (native) | Intel Xeon Phi 7110P | Environmental NR | 70 |
| 2014 | XSW 2.0, Wang et al. [50] | OpenMP + Pthreads (offload) | Intel Xeon E5-2620 + Intel Xeon Phi 7110P | Environmental NR | 100 |
| 2015 | SWIMM, Rucci et al. [39] | OpenMP (offload) | 2×Intel Xeon E5-2670 2.60Ghz + Intel Xeon Phi 3120P | Environmental NR | 160 |
| 2015 | LSBDS, Lan et al. [20] | OpenMP + Pthreads (offload) | 2×Intel Xeon E5-2620 v2 2.0Ghz + 2×Intel Xeon Phi 7110P | TrEMBL + Environmental NR | 220 |

50× and 26×, respectively. These significant differences can be explained according to three reasons. In the first place, host consumption was not considered in GPU and FPGA implementations to the detriment of CPU implementations. In the second place, CPU and GPU implementations are sub-optimal. These implementations reported less GCUPS than others previously presented in literature. In the third place, they employed synthetic data benefiting FPGA implementations, besides being non-representative of real world biological searches.

Benkrid et al. [4] is another performance/power consumption evaluation in the SW context. In this work, the authors presented some implementations for different hardware devices: CPU, GPU and FPGA. They found that FPGA is the most energy-efficient platform being up to 500× and 23× better than CPU and GPU, respectively. These impressive results are explained due to several reasons. Firstly, host consumption was not considered in FPGA and GPU cases as in the previous work. Secondly, the authors state that they chose specific hardware platforms based on the same fabrication technology (90 nm) to enable a fair comparison between devices. This fact clearly benefits FPGA and GPU implementations since CPUs are more advanced in this aspect. Lastly, they employed sub-optimal implementations for all devices. As the work also evaluates programming cost of each implementation, they chose not to use the results of the fastest implementations reported in the literature, but instead to perform their own experiments using solutions developed by a set of Ph.D. students with relatively equal experience on each platform.

Rucci et al. [40] is probably the most realistic SW performance/power consumption evaluation, since the authors considered host consumption in accelerator versions and employed powerful hardware platforms, the best implementations available in literature so far for each device and real biological data. Taking CPU-based systems as baseline, these authors compared performance and energy efficiency of hybrid systems using all its available computational resources (host and accelerators). They found that CPUs offer a good balance between performance and power consumption, especially those with AVX2 instruction set. Xeon Phi-based systems are not a good choice for this problem from energy efficiency perspective principally due to the absence of low-range vector capabilities on this coprocessor. The performance gain is smaller than the increase in power consumption, which translates into less GCUPS/Watt. Both CPU-FPGA and CPU-GPU systems are able to improve energy efficiency, being the first step forward to the second. GPU accelerated systems offer higher performance rates but at the expense of higher power consumption rates too. CPU-FPGA systems offer less GCUPS than GPU-based platforms. However, because its power consumptions are lower, the energy efficiency rates are higher. In particular, GPU incorporation performed up to 1.6× and 1.22× in performance and energy efficiency points of view, respectively. In the FPGA case, its addition produced improvements of up to 1.4× in performance and 1.28× in energy efficiency.

## 6 Future View

Biological data continue increasing its size and accelerating SW database searches still remains as a challenging task. Fortunately, several new hardware and software technologies will be upcoming in the near future that will help to mitigate this issue:

- Multi-core processors offer a good balance between performance and energy efficiency for SW database searches, in particular those with AVX2 instruction set. Future processors of this kind will have more cores, extended vector units and more complex and larger memory hierarchies. As explained in Sect. 4.2.1, CPU-based implementations can be benefited directly from these features.
- GPUs have demonstrated to be powerful platforms to accelerate SW algorithm. Next generation GPUs will have more computational power and better memory performance. These characteristics can lead to faster implementations.
- FPGAs have also proved to be a good option for accelerating SW protein searches, specially from energy efficiency perspective. In this way, new hybrid CPU-FPGA architectures appear as a promising opportunity.
- Current generation of Xeon Phi is not a good alternative for accelerating SW protein searches due mainly to the absence of low-range vector capabilities on this coprocessor. Fortunately, next generation of Xeon Phi (Knights Landing) will solve this issue incorporating the AVX-512 instruction set. Therefore, better Xeon Phi performances are expected considering the number of cores and the vector capabilities.

Some of these technologies will deliver improvements in a transparent manner to programmers. For example, all hardware vendors plan to reduce the manufacturing process technology of their devices, which can turn into faster communications and more available computational resources. Or also stacked memory adoption that will increase bandwidth and energetic efficiency. However, other technologies will require programmer's intervention to take advantage of them, like new hybrid CPU-FPGA architectures or also the AVX-512 instruction set that will be available in Xeon Skylake processors and next generation Xeon Phi coprocessors. Therefore, programming efforts will be necessary to develop new computational tools capable of taking advantage of these upcoming technologies.

## 7 Conclusions

In this chapter we gave a survey of the state-of-the-art in SW database protein search, focusing on four widespread hardware architectures: CPU, GPU, FPGA and Xeon Phi. First, we presented a brief description of each platform. Next, we explained the SW algorithm followed by the study of its data dependences and the possible parallelism schemes. We reviewed the existing implementations for the hardware platforms under study including temporal evolution, contributions, limitations and experimental work and results of each of them. Additionally, as energy efficiency is becoming more important every day, we also surveyed performance/power consumption works in SW context. Finally, we gave our view on the future of SW protein searches considering next generations of hardware architectures and its upcoming technologies.

Biological data continue increasing its size and, as a consequence, increasing SW search time. Upcoming technologies will help to mitigate this issue but will also present new challenges to programmers in order to take advantage of them. We expect that this chapter can serve as a good starting point to future acceleration of SW protein database searches.

**Acknowledgements** Enzo Rucci holds a PhD CONICET Fellowship from the Argentinian Government, and this work has been partially supported by Spanish research project TIN 2012-32180.

## References

1. Alpern B, Carter L and Gatlin KS (1995) Microparallelism and High-performance Protein Matching. SC95, doi:10.1109/SUPERC.1995.242795
2. Altera Corporation (2016) Altera SDK for OpenCL. Available at https://www.altera.com/products/design-software/embedded-software-developers/opencl/overview.html Cited 08 Jan 2016
3. AMD (2016) High-Bandwidth Memory. Available at http://www.amd.com/en-us/innovations/software-technologies/hbm Cited 08 Jan 2016
4. Benkrid K, Akoglu A, Ling C, Song Y, Liu Y and Tian X (2012) High performance biological pairwise sequence alignment: FPGA versus GPU versus cell BE versus GPP. Int. J. Reconfig. Comput., doi:10.1155/2012/752910
5. Benkrid K, Ying L and Benkrid A (2009) A Highly Parameterized and Efficient FPGA-Based Skeleton for Pairwise Biological Sequence Alignment. IEEE Transactions on Very Large Scale Integration (VLSI) Systems, doi:10.1109/TVLSI.2008.2005314
6. Borovska P and Lazarova M (2011) Parallel models for sequence alignment on CPU and GPU. CompSysTech 2011, doi:10.1145/2023607.2023644
7. Daily J (2016) Parasail: SIMD C library for global, semi-global, and local pairwise sequence alignments. BMC Bioinformatics, doi: 10.1186/s12859-016-0930-z
8. Dydel S and Bala P (2004) Large Scale Protein Sequence Alignment Using FPGA Reprogrammable Logic Devices. LNCS, doi:10.1007/978-3-540-30117-2_ 5
9. Farrar M (2007) Striped Smith–Waterman speeds database searches six time over other SIMD implementations. Bioinformatics, doi:10.1093/bioinformatics/btl582
10. Farrar M (2008) Optimizing Smith–Waterman for the Cell Broad-band Engine. Available at http://farrar.michael.googlepages.com/SW-CellBE.pdf Cited 21 Mar 2009
11. Gotoh O (1982) An improved algorithm for matching biological sequences. J. Mol. Biol., doi:10.1016/0022-2836(82)90398-9
12. Harris M (2014) Maxwell: The Most Advanced CUDA GPU Ever Made. Available at http://devblogs.nvidia.com/parallelforall/maxwell-most-advanced-cuda-gpu-ever-made/ Cited 08 Jan 2016
13. Hasan L and Al-Ars Z (2011) An Overview of Hardware-Based Acceleration of Biological Sequence Alignment. In: Lopes H (ed) Computational Biology and Applied Bioinformatics. InTech

14. Howse B and Smith R (2015) Tick Tock On The Rocks: Intel Delays 10nm, Adds 3rd Gen 14nm Core Product Kaby Lake. Available at http://www.anandtech.com/show/9447/intel-10nm-and-kaby-lake Cited 08 Dec 2015
15. IBM (2015) IBM and Xilinx Announce Strategic Collaboration to Accelerate Data Center Applications. Available at https://www-03.ibm.com/press/us/en/pressrelease/48074.wss Cited 18 Jan 2016
16. Intel (2016) Intel Acquisition of Altera. Available at intelacquiresaltera.transactionannouncement.com Cited 18 Jan 2016
17. Isa MN, Benkrid K, Clayton T, Ling C and Erdogan AT (2011) An FPGA-based parameterised and scalable optimal solutions for pairwise biological sequence analysis. AHS 2011, doi:10.1109/AHS.2011.5963957
18. Kentie M (2010) Biological Sequence Alignment Using Graphics Processing Units. MSc Thesis, TUDelft
19. Khalafallah A, Elbabb HF, Mahmoud O and Elshamy A (2010) Optimizing Smith–Waterman algorithm on Graphics Processing Unit. ICCTD 2010, doi:10.1109/ICCTD.2010.5645976
20. Lan h, Liu W, Schmidt B, and Wang B (2015) Accelerating Large-Scale Biological Database Search on Xeon Phi-based Neo-Heterogeneous Architectures. BIBM 2015, doi:10.1109/BIBM.2015.7359735
21. Liu W, Schmidt B, Voss G, Schroder A and Muller-Wittig W (2006) Bio-sequence database scanning on a GPU. IPDPS 2006, doi:IPDPS.2006.1639531
22. Liu Y, Huang W, Johnson J and Vaidya S (2006) GPU Accelerated Smith–Waterman. LNCS, doi:10.1007/11758549_29
23. Liu Y, Maskell DL and Schmidt B (2009) CUDASW++: optimizing Smith–Waterman sequence database searches for CUDA-enabled graphics processing units. BMC Research Notes, doi:10.1186/1756-0500-2-73
24. Liu Y and Schmidt B (2014) SWAPHI: Smith–Waterman Protein Database Search on Xeon Phi Coprocessors. ASAP 2014, doi:10.1109/ASAP.2014.6868657
25. Liu Y, Schmidt B and Maskell DL (2010) CUDASW++ 2.0: enhanced Smith–Waterman protein database search on CUDA-enabled GPUs based on SIMT and virtualized SIMD abstractions. BMC Research Notes, doi:10.1186/1756-0500-3-93
26. Liu Y, Wirawan A and Schmidt B (2013) CUDASW++ 3.0: accelerating Smith–Waterman protein database search by coupling CPU and GPU SIMD instructions. BMC Bioinformatics, doi:10.1186/1471-2105-14-117
27. Manavski S and Valle G (2008) CUDA compatible GPU cards as efficient hardware accelerators for Smith–Waterman sequence alignment. BMC Bioinformatics, doi:10.1186/1471-2105-9-S2-S10
28. McCool MD (2008) Scalable Programming Models for Massively Multicore Processors. Proceedings of the IEEE, doi: 10.1109/JPROC.2008.917731
29. Mentor Graphics (2015) Handel-C System Methodology. Available at https://www.mentor.com/products/fpga/handel-c/ Cited 08 Jan 2016
30. Moammer K (2015) AMD Zen CPU Microarchitecture Details Leaked In Patch - Doubles Down On IPC And Floating Point Throughput. Available at http://wccftech.com/amd-zen-cpu-core-microarchitecture-detailed/2/ Cited 16 Oct 2015
31. Moammer K (2015) Nvidia : Pascal Is 10X Maxwell, Launching in 2016 - Features 16nm, 3D Memory, NV-Link and Mixed Precision. Available at http://wccftech.com/nvidia-pascal-gpu-gtc-2015/ Cited 16 Jan 2016
32. Needleman SB and Wunsch CD (1970) A general method applicable to the search for similarities in the amino acid sequence of two proteins, J. Mol. Biol., doi:10.1016/0022-2836(70)90057-4
33. NVIDIA Corporation (2016) CUDA. Available at http://www.nvidia.com/object/cuda_home_new.html Cited 08 Jan 2016
34. Oliver TF, Schmidt B and Maskell DL (2005) Reconfigurable architectures for bio-sequence database scanning on FPGAs. IEEE Transactions on Circuits and Systems, doi:10.1109/TCSII.2005.853340

35. OpenACC Organization (2016) OpenACC. Available at http://www.openacc.org/ Cited 08 Jan 2016
36. Pirzada U (2015) Intel's Skylake Purley Family of Microprocessors Will Boast upto 28 Cores and 56 Threads - Next Generation Xeon Platform Landing in 2016. Available at http://wccftech.com/intel-skylake-purley-platform-upto-28-cores-56-threads/ Cited 08 Dec 2015
37. Rognes T (2011) Faster Smith–Waterman database searches with inter-sequence SIMD parallelization. BMC Bioinformatics, doi:10.1186/1471-2105-12-221
38. Rognes T and Seeberg E (2000) Six-fold speed-up of Smith–Waterman sequence database searches using parallel processing on common microprocessors. Bioinformatics, doi:10.1093/bioinformatics/16.8.699
39. Rucci E, García C, Botella, G, De Giusti A, Naiouf M and Prieto-Matías M (2015) An energy-aware performance analysis of SWIMM: Smith–Waterman implementation on Intel's Multicore and Manycore architectures. CPE, doi: 10.1002/cpe.3598
40. Rucci E, García C, Botella, G, De Giusti A, Naiouf M and Prieto-Matías M (2016) OSWALD: OpenCL Smith-Waterman on Altera's FPGA for large protein databases. IJHPCA, doi: 10.1177/1094342016654215
41. Seetle S (2013) High-performance Dynamic Programming on FPGAs with OpenCL. Available at http://ieee-hpec.org/2013/index_htm_files/29-High-performance-Settle-2876089.pdf Cited 08 Jan 2016
42. Smith R (2011) AMD's Graphics Core Next Preview: AMD's New GPU, Architected For Compute. Available at http://www.anandtech.com/show/4455/amds-graphics-core-next-preview-amd-architects-for-compute Cited 08 Jan 2016
43. Smith TF and Waterman MS (1981) Identification of common molecular subsequences. J. Mol. Biol., doi:10.1016/0022-2836(81)90087-5
44. Szalkowski A, Ledergerber C, Krahenbuhl P and Dessimoz C (2008) SWPS3 - fast multi-threaded vectorized Smith–Waterman for IBM Cell/B.E. and x86/SSE2. BMC Research Notes, doi:10.1186/1756-0500-1-107
45. The Khronos Group (2016) OpenCL: The open standard for parallel programming of heterogeneous systems. Available at https://www.khronos.org/opencl/ Cited 08 Jan 2016
46. Van Court T and Herbordt MC (2004) Families of FPGA-based algorithms for approximate string matching. ASAP 2004, doi:10.1109/ASAP.2004.1342484
47. Vermij E (2011) Genetic sequence alignment on a supercomputing platform. MSc Thesis, TUDelft
48. Vestias M and Neto H (2014) Trends of CPU, GPU and FPGA for high-performance computing. FPL 2014, doi:10.1109/FPL.2014.6927483
49. Wang L, Chan Y, Duan X, Lan H, Meng X and Liu W (2014) XSW: Accelerating Biological Database Search on Xeon Phi. IPDPS 2014, doi:10.1109/IPDPSW.2014.108
50. Wang L, Chan Y, Duan X, Lan H, Meng X and Liu W (2014) XSW 2.0: A fast Smith–Waterman Algorithm Implementation on Intel Xeon Phi Coprocessors. Available at http://sdu-hpcl.github.io/XSW/ Cited 16 Nov 2015
51. Wozniak A (1997) Using video-oriented instructions to speed up sequence comparison. CABIOS 13-2:145–150
52. Xilinx Inc. (2016) SDAccel Development Environment. Available at http://www.xilinx.com/products/design-tools/software-zone/sdaccel.html Cited 08 Jan 2016
53. Zhang P, Tan G and Gao GR (2007) Implementation of the Smith-Waterman Algorithm on a Reconfigurable Supercomputing Platform. HPRCTA 2007, doi:10.1145/1328554.1328565
54. Zhao M, Lee W, Garrison E and Marth G (2013)SSW Library: An SIMD Smith-Waterman C/C++ Library for Use in Genomic Applications. PLoS One, doi:10.1371/journal.pone.0082138
55. Zou D, Dou Y and Xia F (2011) Optimization schemes and performance evaluation of Smith-Waterman algorithm on CPU, GPU and FPGA. CPE, doi: 10.1002/cpe.1913

# A Survey of Computational Methods for Protein Function Prediction

**Amarda Shehu, Daniel Barbará, and Kevin Molloy**

**Abstract** Rapid advances in high-throughout genome sequencing technologies have resulted in millions of protein-encoding gene sequences with no functional characterization. Automated protein function annotation or prediction is a prime problem for computational methods to tackle in the post-genomic era of big molecular data. While recent community-driven experiments demonstrate that the accuracy of function prediction methods has significantly improved, challenges remain. The latter are related to the different sources of data exploited to predict function, as well as different choices in representing and integrating heterogeneous data. Current methods predict function from a protein's sequence, often in the context of evolutionary relationships, from a protein's three-dimensional structure or specific patterns in the structure, from neighbors in a protein–protein interaction network, from microarray data, or a combination of these different types of data. Here we review these methods and the state of protein function prediction, emphasizing recent algorithmic developments, remaining challenges, and prospects for future research.

**Keywords** Computational biology • Protein function prediction • Algorithms • Machine learning • Homology

A. Shehu (✉)
Department of Computer Science, Department of Bioengineering, George Mason University, Fairfax, VA 22030, USA
e-mail: amarda@gmu.edu

D. Barbará
Department of Computer Science, George Mason University, Fairfax, VA 22030, USA
e-mail: dbarbara@gmu.edu

K. Molloy
LAAS-CNRS, 7, avenue du Colonel Roche, 31077 Toulouse, France
e-mail: kmolloy@laas.fr



K.-C. Wong (ed.), *Big Data Analytics in Genomics*,
DOI 10.1007/978-3-319-41279-5_7

## 1 Introduction

Molecular biology now finds itself in the era of big data. The focus of the field on high-throughout, automated wet-laboratory protocols has resulted in a vast amount of gene sequence, expression, interactions, and protein structure data [212]. In particular, due to the increasingly fast pace with which whole genomes can be sequenced, we are now faced with millions of protein products for which no functional information is readily available [39, 198]. The December 2015 release of the Universal Protein (UniProt) database [68] contains a little over 55.2 million sequences, less than 1 % of which have reliable and detailed annotations.

The gap between unannotated and annotated gene/protein sequences has exceeded two orders of magnitude. Fundamental information is currently missing for 40 % of the protein sequences deposited in the National Center for Biotechnology Information (NCBI) database; around 32 % of the protein sequences in the comprehensive UniProtKB database are currently labeled "unknown." The missing information includes coarse-grained, low-resolution information such as where protein products are expressed, meta-resolution information, such as what chemical pathways proteins participate in the living cell, and high-resolution information, such as what molecular partners a protein recognizes and binds to directly in the cell.

Getting at what proteins do in the living cell is central to our efforts to understand biology, as proteins are ubiquitous macromolecules [4] involved in virtually every cellular process, from cell growth, maintenance, proliferation, to apoptosis [5]. Understanding what a protein does in the cell is also central to our ability to understand and treat disease [351]. Moreover, computer-aided drug design (CADD) often begins with identifying a protein target whose activity in the diseased cell needs to be altered or regulated via binding compounds to cure or treat disease [322].

Given the exponential increase in the number of protein sequences with no functional characterization and the central role of proteins to human biology and health, predicting where a protein acts in the cell and exactly what it does is a central question to address in molecular biology. Originally, this question was only investigated in the wet laboratory and on a small set of target genes or proteins. Wet-laboratory approaches that elucidate the role of a protein in the cell include gene knockout, targeted mutations, inhibition of gene expressions, mass spectrometry, and RNAi [5].

Gene knockout, targeted mutations, and inhibition of gene expression methods demand considerable effort and time and can only handle one protein product or gene at a time [306]; in other words, these are low-throughput methods. Higher-throughput wet-laboratory annotation initiatives, such as the European Functional Analysis Network [264] have also proven unable to keep up with the pace of whole genome sequencing. In particular, wet-laboratory experiments that use mass spectrometry or RNAi are found to yield biased and less specific information about protein function than the low-throughput methods [309].

Human experts known as biocurators also often peruse published wet-laboratory studies to provide functional information on a protein [38, 308, 309]. For instance, the popular UniProtKB database, which is the central hub for the collection of functional information on protein sequences, consists of two sections: SWISS-PROT, which contains manual and reviewed annotations (less than 1 % of protein sequences in UniProt are annotated in SWISS-PROT), and TrEMBL, which contains computational annotations yet to be validated by experts [34]. About 30–40 % of computational and manual annotations contain errors [308]. The rapid rise of the CRISPR/Cas technology [67, 140], which can edit specific genes and do so rather efficiently [303], proves promising to observe the phenotypic effect of deliberately introduced gene variants [134, 210, 286], but applications for large-scale function prediction are under-pursued at the moment.

In light of the increasing gap between the amount of protein sequence data and the amount of functionally annotated proteins, computational approaches seem poised to tackle protein function prediction and narrow this gap. Before venturing into a comprehensive description of such methods, which is the subject of this review, it is important to formulate exactly what one means by protein function. One can find different definitions in literature, because the function of a protein can be described at different degrees of detail. Information on the cellular localization of a protein can provide important clues towards the processes in which a protein is involved but is not sufficient in itself. Describing function from a physiological aspect entails knowing the biological processes in which a protein participates. From a phenotypical aspect, one is more concerned about the disease or disorders induced by a misbehaving protein. To capture all these different aspects and aid computational approaches, various classification schemes have been proposed. By now, the most broadly accepted and utilized scheme is the Gene Ontology (GO) scheme originally proposed in [12]. GO is a hierarchical description of protein function that describes three different aspects, each one increasing the level of detail.

- **Cellular component** describes the component or anatomical structure in a cell where a gene product operates. Examples include the rough endoplasmic reticulumn, nucleus, ribosome, proteasome, and more.
- **Biological process** captures the physiological description of protein function and allows specifying the processes in which a gene product participates in the cell. A process is defined as a series of events or molecular functions, and examples include membrane fusion, cellular component organization, macromolecular complex assembly, and various distribution processes.
- **Molecular function** is different from the biological processes in which a gene product is involved and instead captures at a finer level of resolution what a protein does in the cell, such as transporting molecules around, binding to molecules, holding molecular systems together, changing systems from one state to another. Examples include ligand binding, catalysis, conformational switching, and more.

Computational methods for protein function prediction are diverse, particularly when one considers methods that limit themselves to prediction of specific aspects of protein function. The focus of this review is on methods that aim to provide GO annotations of protein products. These methods can be organized in distinct categories based on the type of data they employ to predict the GO annotation of an uncharacterized protein.

The first category of computational methods is comprised of methods that make predictions based on observations that sequence similarity is a good indicator of functional similarity. Such methods were among the first to be employed for automatic annotations, and they are often the first tools employed in this regard. They are summarized in Sect. 2. Recent advancements in sequence-based methods concern expanding their applicability beyond proteins of very high sequence similarity, known as close homologs, to remote homologs. These methods consider additional information such as genomic context and evolutionary relationships and are described in Sects. 3 and 4.

Another category is comprised of methods that use more than sequence information and employ information on the three-dimensional, biologically active structure of a protein. This information is often difficult to obtain in the wet laboratory and not available on many proteins. However, advances in structure resolving techniques, both in the wet and dry laboratories, are allowing the application of such methods for protein function prediction. Structure-based methods are mainly distinguished by the representations they choose of protein structure and the amount of protein structure they exploit. These methods are described in Sect. 5.

Yet another category is comprised of methods that employ information on known interactions of a protein product as encapsulated in protein–protein interaction networks. This category is rich in machine learning methods and is the subject of Sect. 6. Methods that exploit gene expression data are summarized in Sect. 7.

Currently, the best-performing methods are those that are enriched with additional information on sequence, structure, and gene expression data, resulting in a category of methods known as hybrid methods, which we describe in Sect. 8. Another category of methods, described in Sect. 9, exclusively mine biomedical literature to annotate query proteins.

Methods for automated function prediction are now evaluated and tested in community-driven experiments and global initiatives, such as the Enzyme Function Initiative (EFI), the COMputational BRidges to EXperiments (COMBREX) initiative, and the Critical Assessment of Function Annotation (CAFA) community-driven experiment. In particular, CAFA is becoming the main venue to objectively compare function prediction methods to one another and highlight the state of the art in automated function prediction [287]. Evaluation is done in two rounds. In CAFA1, several thousands of unannotated query sequences are provided to participants. About 48,298 targets from 18 species were provided in 2014. Participants submit predicted GO terms, and predictions are then evaluated according to community-agreed metrics, such as the top-20, threshold measure, and the maximum F1 score over all recall-precision pairs obtained with the threshold measure, also known as the Fmax score [124] and others.

In the second round of CAFA, CAFA2, twice as many queries are released. For instance, 100,816 target protein sequences from 27 species were provided to participants in 2014. Evaluation of predicted GO terms from different labs is then held as a special interest group meeting at the Intelligent Systems in Molecular Biology (ISMB) conference. Annual reviews are released tracking progress, challenges, and the state of the art in automated function prediction. We summarize the latest review of function prediction methods in the context of their performance in CAFA to conclude our survey of these methods. In particular, the survey concludes with a critical summary of the state of protein function prediction, remaining challenges, and prospects for future research.

## 2 Sequence-Based Methods for Function Prediction

Sequence-based methods transfer onto an uncharacterized target protein sequence the functional annotation of a characterized protein sequence with high sequence similarity to the target. Some of the earliest efforts in bioinformatics focused on understanding the relationship between sequence, structure, and function similarity. This was made possible by the advent of standardized sequence formats and sequence comparison tools, such as FASTA [278], and fast sequence alignment and comparison algorithms based on dynamic programming, such as BLAST and PSI-BLAST [8]. In addition to BLAST, other well-known sequence alignment tools now include PROSITE [18, 143] and PFAM [327, 328].

The comparison of two sequences aims to determine an evolutionary relationship and infer whether the sequences under comparison share a common ancestor; that is, if they are homologs. One cannot infer shared ancestry, thus homology, based on sequence similarity alone. For instance, high sequence similarity might occur because of convergent evolution; when considering shorter sequences, high similarity may occur because of chance. Two sequences can be similar but not homologous. Therefore, function cannot be realizably transferred even if sequence similarity is high.

While early bioinformatics efforts (some of which are summarized below) indeed transferred function between highly similar sequences, later efforts, aware of convergent evolution, focused on integrating additional data beyond sequence. It is worth noting that early efforts focused on understanding when sequence similarity is high enough, in the absence of convergent evolution, to infer function similarity. By now it has been observed that prediction accuracy suffers when the threshold is set to anything less than 30 % sequence similarity [159]; however, a more comprehensive understanding has emerged that shows that, as long as homology is established, even 10 % sequence identity can allow extracting information on function (for instance, remote homologs can have very low sequence identities due to early branching points in evolution). On the other hand, even 30 % sequence identity between two non-homologous sequences can be misleading [184].

Sequence-based methods have been shown to have limited applicability for some of the reasons listed above. In response, more sophisticated sequence-based methods have been developed. These methods either enhance the basic framework, where the functional unit is still the comparison of two protein sequences in their entirety, or pursue complementary frameworks of comparing subsequences or physico-chemical properties extracted from two given protein sequences. Based on this distinction, sequence-based methods can be organized in three categories:

- **Sequence-based methods**: methods in this category rely on the comparison of a query protein sequence to functionally annotated sequences in a database. More sophisticated methods pursue a probabilistic setting, incorporate data from various sources, and even pursue unsupervised learning frameworks to improve the accuracy and confidence of annotations.
- **Subsequence-based methods**: methods in this category realize that only a subset of the amino acids in a protein comprise the site onto which molecular partners bind. Methods in this category mainly differ in what subsequences are considered, domains or shorter subsequences known as motifs.
- **Feature-based methods**: these methods aim to extract more information from a given protein sequence than what is directly available in the identity of amino acids. By additionally encoding physico-chemical properties of amino acids, these methods construct features for all or a subset of a sequence and pursue functional annotation in a machine learning setting.
- **Ensemble-based methods**: these methods combine the above three approaches via the concept of ensemble classifiers in machine learning.

## *2.1 Sequence-Based Methods for Functional Annotation*

Several directions have been investigated to improve the performance and/or extend the applicability of sequence-based methods. We review representative methods below.

### 2.1.1 Sequence Alignment-Based Methods

The accuracy of transferring functional annotations was investigated in [133], where it was shown on annotation transfer among enzymes that transfer was only reliable when sequence similarity was high. Many subsequent studies trying to determine the sequence identity threshold below which functional annotation transfer was not reliable resulted in the recognition of a twilight zone 25–30 % sequence identity [299]. While basic sequence-based methods are not reliable when sequence identities are 25 % or less, sophisticated methods can handle the twilight zone. The ConFunc method proposed in [368] operates in this range. PSI-BLAST is used to align a query sequence with annotated sequences. The sequences returned by PSI-BLAST are then split into sub-alignments according to the sequences' GO annotations. Conserved residues are then identified within each GO-term sub-alignment, and a position-specific scoring matrices (PSSM) profile is constructed for each sub-alignment. The query sequence is scored against the PSSMs of all sub-alignments, and these scores are then used to calculate expectation values for the GO annotations corresponding to each sub-alignment. Prediction is made based on careful filtering of the different GO annotations based on their expectation values. ConFunc is shown to outperform both BLAST and PSI-BLAST. On a large testing set of query sequences with known homologs with sequence identities in the twilight zone, ConFunc's recall is six times greater than BLAST.

A related method, GoFDR, is proposed in [116], which processes the PSI-BLAST query-sequence based on multiple sequence alignment (MSA). For each GO term of the homologs in the MSA, GFDR identifies functionally discriminating residues (FDR) specific to the GO term. The query sequence is then scored using a position-specific scoring matrix constructed for the FDRs alone. The raw score is converted into a probability based on a score-probability table prepared over training sequences. GoFDR outperforms three sequence-based methods for predicting GO terms, PFP [129], GOtcha [236], and ConFunc [368], and is ranked as the top method in the preliminary evaluation report in CAFA2.

### 2.1.2 Probabilistic Whole-Sequence Annotation Transfer

One direction pursues sequence-based functional annotation in a probabilistic setting. The approach proposed in [207] assumes that a protein can only belong to a functional class if its BLAST score distribution with members of the class is the same as that of these members with one another. A univariate and multivariate probabilistic scheme are investigated. The univariate scheme makes predictions based on the total score of the query protein, by assigning to it a probability of belonging to each functional class. This can lead to ambiguous results, as the query can have similar scores with different functional classes. For this reason, the univariate scheme is extended to a multivariate one by constructing a vector of BLAST scores of the query with all classes. The vector is compared to the distribution of each class. Evaluation shows that the approach reaches an accuracy above 90 % [207]. However, the evaluation is performed on enzymes, where sequences are more strongly correlated with function than on other proteins. In addition, this probabilistic approach performs well on the most specific GO level due to the ambiguity with comparisons to the less specific GO levels (such as cellular location).

### 2.1.3 Integration of Data Sources in Whole-Sequence Annotation Transfer

Another way to improve whole-sequence methods is to integrate additional data sources. The GOtcha method in [236], for instance, organizes the annotations of sequences similar to a query into a set of GO-like directed acyclic graphs. A P-score is calculated based on the frequency of occurrence of respective annotations and BLAST E-values of the corresponding matches. The P-score estimates the confidence attached to the annotation of the query sequence with that term, and a threshold value for the P-score allows extracting a final set of annotations. Evaluation of this approach on the *Drosophila melanogaster* genome showed that the results were more sensitive and specific than those obtained with the baseline approach.

### 2.1.4 Unsupervised Learning in Whole-Sequence Annotation Transfer

Work in [1, 380] pursues an unsupervised learning approach. Sequences similar to a query sequence are identified via BLAST. All pairwise sequence similarities in the set, including the query, are stored in a similarity matrix. The latter is employed for clustering the set of sequences. The annotation of the query sequence is then based not on individual high-similarity sequences but on the cluster of sequences to which the query sequence belongs. In [380], progressive single-linkage clustering and text information analysis are employed to assign GO terms to the query sequence. In [1], the sequence similarity space is encoded in a graph, and the normalized cut clustering algorithm is used to identify groups of sequences that are closely related to the query sequence. In [289], the space of annotated sequences is first organized via hierarchical clustering according to functional and evolutionary relationships. The function of the query sequence is then predicted based on the position of the query in the tree. The approach employed in these methods uses whole-sequence comparisons only as an intermediate step and maps a query protein to a cluster or a level of a hierarchy. This approach is shown to perform well and be more robust to errors in individual entries [1, 289, 380].

A specific subgroup of methods that fall in the same category do not directly address function prediction but rather construct an informative organization of protein sequences into functional groups. The objective is to extract from the groups rules and features that can then be utilized by other methods in a machine learning setting. Specifically, protein sequences are clustered into functional groups based on their evolutionary relationships, structural properties (these can be structural classes based on secondary structure content, folds, or even structural motifs), or subsequences. Manual curators can be employed to provide such clustering, but our focus here is on computational methods. Iterative clustering is proposed in [314, 382]. The method in [94] encodes proteins into a graph with edges encoding pairwise similarity, and then applies Markov Clustering to group known homologs from different species (orthologs) [211]. In [385], clusters of related proteins are identified by analyzing strongly connected sets of vertices in the graph. In [242], pairwise sequence comparisons are used to organize proteins into pre-families, which are then further divided into homogeneous clusters based on the topology of the similarity graph. Spectral clustering is employed in [257] to infer protein families. A heuristic approach is proposed in [10] to improve the performance of these graph-based clustering methods related to the choice of the similarity threshold determining whether two proteins should be connected by an edge or not. Others use hierarchical clustering [56, 226, 305, 342, 350]. Partitioning clustering algorithms have been shown recently to outperform all these other clustering methods [99].

### 2.1.5 Supervised Learning and Generative Models in Whole-Sequence Annotation Transfer

The most successful methods for function prediction rely on supervised learning or generative models. Methods proposed in [66, 83, 150, 213] address the detection of remote homologs by employing sequence alignment profiles to train hidden Markov models (HMMs) or pairwise sequence similarities to train support vector machines (SVMs) or neural networks (NNs). Specifically, the FANN-GO method proposed in [66] aligns a query sequence to a database of annotated sequences to calculate the i-score proposed by the GOtcha method [236] that the query is associated with a specific functional term. The scores are then fed to an ensemble of multi-output neural networks trained to predict the probability of a sequence associated with each function term. The FANN-GO method is shown able to model dependencies between functional terms and outperforms the GOtcha method [236].

## *2.2 Subsequence-Based Methods for Functional Annotation*

Subsequence-based methods are motivated by a deeper understanding of the role of protein sequence in recognition events. In particular, only a subset of the amino acids that comprise a protein chain assist in the sticky interactions with other molecules. This subset is typically comprised of few amino acids. In addition, contiguous, long segments of a protein chain may fold independently to form a domain. Multi-domain proteins employ different domains to interact with different molecules and thus enrich their molecular functions [314]. Subsequence-based methods are organized into two categories:

- **Domain-based methods**: It has long been recognized that a multi-domain protein's array of functions is due to different molecular functions of its

(continued)

domains [314]. Domain-based methods seek to identify all the domains in a protein in order to compile its array of molecular functions.

- **Motif-based methods**: Functional sites in a protein that are employed to recognize and bind ligands, DNA, RNA, and other proteins are comprised of only a subset of the amino acids comprising a protein's polypeptide chain. Since functional sites are under higher evolutionary pressure to be conserved, a way of identifying functional sites on proteins is through detection of evolutionary-conserved (sequence) subsequences. Subsequences that are conserved among protein sequences belonging to a family are referred to as *motifs* [35]. These methods detect motifs in a protein sequence and use these motifs as signatures of specific functional classes [141].

Motif-based methods are rich in machine learning techniques. In contrast, domain detection methods focus more on integrating biological insight. We review domain-based methods first, and then devote the rest of the description of subsequence-based methods to the machine learning strategies employed to detect motifs for functional annotation.

### 2.2.1 Domain-Based Methods

A protein domain is an independent evolutionary and functional unit of a protein that folds independently of the rest of the protein where it is contained. A transcribed exon is referred to as a module, and a domain may be comprised of several modules [346]. More than 80 % of known domains are about 50–150 amino acids long, but exceptionally long domains of more than 800 can be found, as well. Several small domains of 30 amino acids or less are also reported. At least two-thirds of mammalian proteins have more than one domain. Only multi-cellular eukaryotic organisms have a significant proportion of proteins with repeating domains [344].

A domain is an independently folding unit of a protein. So, a domain is a structural unit that can be found in multiple protein contexts. Biologists usually break up large proteins into domains based on a process that involves analysis of sequence, structure, and domain-specific expertise. Protein domains can be found in several databases, such as ProDom [314] and the Conserved Domain database (CDD) [230]. These two databases describe domains at the sequence level. A query protein sequence is aligned to the deposited domains, and BLAST E-values are employed to determine the domains present in the query. In particular, CDD is the protein classification component of NCBI's Entrez query and retrieval system by which one can identify conserved domains in query protein sequences. An illustration of the results returned by the CD-Search tool (which stands for conserved domain search) in CDD is shown in Fig. 1.

**Fig. 1** Output obtained by the CD-Search tool on the human methionine aminopeptidase 2 protein. Domain hits are provided visually and ranked from non-specific (high E-values) to specific (low E-values). Details on each of the domain hits are listed in the *bottom panel*

Other databases also store domains and provide more than sequence and functional information. The CATH [266, 277] and SCOP [251] databases contain structural information, as well. These databases provide hierarchical classifications of protein structures. For instance, CATH breaks down a query protein structure into Class (C), then Architecture (A), followed by Topology (T), and then Homology(H). SCOP breaks down a query protein structure into fold, then superfamily, then family. Both CATH and SCOP are discussed in greater detail in Sect. 5, which discusses structure-based methods for functional annotation. It is worth noting, however, that due to different working definitions of domains and different techniques used to determine what makes a domain, these two databases contain a different number of domains. CATH contains at the moment about 65,000 domains, whereas SCOP contains about 110,800 domains. SCOP defines a domain as an evolutionary unit.

Manual curation of domains provides at the very least sequence and functional information that has been exploited early on to functionally annotate new proteins. In [311], a query protein sequence is aligned to sequences of domains extracted from ProDom and CDD, and rules for function assignment are based on BLAST E-values. Application to a set of 4357 manually curated human proteins results in a recall of 81 % and a precision of 74 %; on data sets from other organisms, precision and recall are decidedly lower, at about 50 %.

A machine learning approach is proposed in [48], where domains extracted from the SBASE library of protein domains [359] are employed as attributes. Sequence alignment is used to determine whether a domain is present or not in a given protein sequence, allowing a protein sequence to be represented as a binary vector recording domain membership. k-nearest neighbor (kNN) and SVM were then trained on functionally annotated protein sequences. kNN is reported to outperform SVM on 13 functional classes obtained from the MIPS database [48].

Work in [281] follows a slightly different approach. Instead of using domains for function annotation, constant-length statistically significant sequence patterns known as promotifs [343] are used instead. Correlations found between SWISS-PROT keywords assigned to the sequences and positions of promotifs are then used to establish rules for function annotation. Precision on a set of PROSITE [144] protein sequences is reported similar to that obtained from work in [311]. A low sensitivity of 50 % is reported. A similarly-low sensitivity is obtained by application of decision trees classifiers to recognize PFAM domains [329] in [131]

Work on domain detection is very rich in computational biology, and methods often include more than sequence information to reliably identify domains in a given protein. One reason for domain-based functional prediction is that relying upon sequence to detect domains falls short. Sequence information is found to be insufficient for identifying structural domains in a protein, because the same structure can be assumed from highly divergent sequences of less than 30 % sequence identity. Structural-based domain detection is more reliable to identify the domains in a protein, but this requires knowledge of a protein's folded structure, which is not readily available for many unannotated protein sequences. A comprehensive review of structural-based domain detection methods can be found in [356].

Independent of the technique used to identify domains present in a given protein, the problem of how to use this information to assign function(s) to an unannotated protein remains open. This is due to the fact that a unit designated to be a domain may not meet the definition of a domain as an independent evolutionary and functional unit of a protein that folds independently of the rest of the protein. Hence, when the criteria for what makes a unit a domain are indeed relaxed, the relationship between a so-called domain and its function is not clear. In addition, domain domain interactions need to be detected, as they may give rise to more complex molecular functions. A review of the current understanding of the domain function relationship can be found in [346, 361]. As such understanding improves, sophisticated machine learning methods that take into account context rather than membership and domain-domain interactions are expected to improve the state of domain-based functional annotation.

### 2.2.2 Motif-Based Methods

The Motif Alignment and Search Tool (MAST) [17] was one of the first to be used for motif-based function annotation. MAST employs the MEME algorithm [16] to construct groups of probabilistic motifs. MAST combines these motifs to construct protein profiles for protein families of interest. MAST successively estimates the significance of the match of a query protein sequence to a particular family model as the product of the p-values of each motif match score. This measure is then used to select the family of the query sequence. MAST's average classification accuracy (ROC50) over 72 distinct queries is shown to be above 0.95.

While MAST is a statistical approach, other methods based on unsupervised and supervised learning approaches have been proposed for motif-based functional annotation, as well. The algorithm in [217] implements unsupervised learning, but relies on SPLASH rather than MEME to detect motifs. Other algorithms, some of which implement supervised learning, employ motifs extracted from motif libraries and databases.

The SPLASH algorithm proposed in [49] is the first to define motifs as sparse amino acid patterns that match repeatedly in a set of protein sequences. This definition is amenable to computational approaches for motif detection. The SPLASH algorithm is a deterministic pattern discovery algorithm that is extremely efficient and can be used in a parallel setting to systematically and exhaustively identify conserved subsequences in protein family sets. The algorithm is employed in the first motif-based method for function annotation [217] to extract motifs. The resulting set of motifs are enriched via a HMM. Proteins are then clustered based on motif membership in a top-down clustering algorithm reminiscent of the construction of a decision tree. The levels of the tree correspond to the motifs, ranked from most to least significant, and proteins are divided at each level/node of the tree based on whether they contain the corresponding motif or not. In this manner, the leaves contain sets of proteins that contain all motifs that together define the signature of a particular functional family. The method performs well, achieving a classification rate between 57 % and 72 % on the exceptionally challenging and sequence-diverse G Protein-Coupled Receptor (GPCR) superfamily.

Instead of the unsupervised method in [217], work in [365] pursues a supervised learning setting to classify proteins based on motif membership. The latter is used to represent each protein sequence as a binary vector, with each entry indicating the presence or absence of a particular motif; motifs are extracted from the PROSITE database [144]. A decision tree is then learned on manually curated protein families extracted from the MEROPS database [290]. Classification accuracy is reported to be significantly better than the clustering approach in [49].

An NN-based method is proposed in [33], using two different types of motifs, class-independent and class-dependent motifs. The former refer to motifs extracted from the training dataset, and the latter refer to motifs extracted for each known functional class. About 30 motifs from each class are extracted with the MEME algorithm [16]. Class-dependent motifs are found to confer the best classification performance to the NN. Application of the NN on the GPCR superfamily yields better classification performance than MAST [17]. However, on this challenging superfamily, a naive Bayes classifier with a $\chi^2$ feature selection algorithm is shown to obtain the best performance [60]. It is worth noting that the special focus on GPCRs is due to their central role in drug design; about 60 % of the approved drugs target some member of the GPCR family [107].

Work in [27] conducts a comprehensive machine learning study for motif-based function annotation. A motif kernel uses the occurrence count of each motif in a protein sequence as a similarity measure between the motif and the sequence. Classification is carried out via an SVM. Issues such as feature selection and multi-class classification are also investigated. An evaluation of different strategies for feature selection and classification schemes is conducted on a set of enzymes. The best performance is obtained when using the RFE method [123] for feature selection, many one-against-the rest classifiers [295], and counting the multiple classes of a protein as a single class. In addition, SVM is found to perform better than kNN. This result is further confirmed in [191], where motif-based SVMs perform best on the classification of enzymes.

## 2.3 *Feature-Based Methods for Functional Annotation*

Many of the methods already described can be viewed as feature-based methods, since they employ term identify or term frequency representations of protein sequences in unsupervised or supervised learning settings, with terms being domains, motifs, modules, promotifs, or even genomes. However, here we describe a category of sequence-based methods that expand their treatment of a protein sequence beyond the identity of amino acids to physico-chemical properties of amino acids. At their core, these methods expand a protein sequence into a vector of physico-chemical properties or attributes of amino acids in the sequence, otherwise referred to as a feature vector. Once such a transformation is made, standard classification techniques can be used. The most popular ones for feature-based functional annotations of protein sequences are the SVM, NN, kNN, and naive Bayes classifiers.

Supervised learning has also been employed to improve and extend the applicability of whole-sequence annotation transfer in the twilight zone of low sequence similarity. The main approach is to organize BLAST or PSI-BLAST results into a positive set, which contains sequences with high similarity to the query sequence, and the negative set, which contains all other sequences. The sets are then used to train a classifier to discriminate sequences in the twilight zone.

One of the earliest methods, PROCANS [376], employs a three-layered NN and represented protein sequences as binary vectors. The attribute for each amino acid in a sequence records the conservation of the amino acid at that position in an MSA and valued 1 when at a position with 50–100 % identity and 0 otherwise. Later improvements to this algorithm employ n-grams as features. In [374], n-gram counts are used to construct feature vectors. In [373, 375], the order of n-grams in a sequence is found to be important, and positions of n-grams in a sequence are added to the feature vector. This collection of works, however, is limited to enzymes, where there are stronger correlations between sequence and function.

The PRED-CLASS method in [274] also employs an NN but restricts its focus to three specific protein classes. The NN contains three levels, classifying transmembrane proteins at the first level, fibrous proteins at the second level, and globular proteins at the third level. Different features are employed at different levels. Compositional features are employed at the first level, as transmembrane proteins have specific subsequence signature; 30 features record the composition of a sequence in all 20 amino acids and 10 different groupings of residues with common structural and physico-chemical properties. 30 features are employed at the second and third level, but these features correspond to the 30 highest intensity for periodicities detected for each residue or each of the 10 considered groups of residues. PRED-CLASS is reported to correctly classify 96 % of 387 proteins.

Work in [172] expands the applicability of feature-based methods. Propositional data mining and inductive logic programming are employed in [172] on binary feature vectors constructed from protein sequences. The features correspond to sets of characteristics found in a significant fraction of the protein sequences. A C4.5 decision tree classifier is trained to predict rules from these features. An encouraging prediction accuracy of 65 % is obtained on ORFs of *Mycobacterium tuberculosis* and *Escherichia coli*.

An extension of the above method is described in [173], where a comparative study is conducted to determine representations of protein sequences that improve function prediction. Three types of representations are investigated, sequence-based attributes, phylogeny-based attributes, and structure-based attributes. Sequence-based attributes (SEQ) are based on a sequence's composition of single and pairs of amino acids. Phylogeny-based attributes (SIM) are computed from the sequences returned from a PSI-BLAST search of a given ORF sequence. The attributes capture, via a first-order language such as Datalog, the distribution of sequences, their evolutionary distance from the ORF, the phylogenetic relationship, as well as keywords describing the sequences; the latter could be easily computed from a sequence, such as the presence of a membrane/trans-membrane binding sequence. Structure-based attributes (STR) are computed from secondary structure segments predicted for a sequence from the Prof program [270]. Pair combinations of these representations, as well as a combination of all of them, are also investigated. Evaluation on ORFs in *E. coli* demonstrates that SIMs confer the highest performance. This important result was one of the first indications that evolutionary information is powerful and perhaps more informative than sequence. Section 4 describes methods that exploit evolutionary history for function prediction.

Work in [153] proposes the now popular ProtFun method, which trains a set of NNs on carefully constructed attribute-value pairs from a training data set. The attributes are compiled via different tools that provide information on post-translational modifications, such as N- and O-glycosylation, phosphorylation, cleavage of N-terminal signal peptides, and other modifications and sorting events that a protein is subjected to before performing its function. An extensive evaluation of ProtFun on 5500 human proteins from the TrEMBL database, with annotations assigned based on SWISS-PROT keywords via the EUCLID system, demonstrates that ProtFun has a high sensitivity of 90 % and a low false positive rate of 10 % on some functional categories. Similar high performance is reported in [154], where ProtFun is applied to predict GO categories for human proteins.

SVMs and kNNs have recently shown to be superior for feature-based function classification. The SVM-Prot method proposed in [47] employs physico-chemical properties of amino acids to represent a protein sequence as a 5-entry feature vector. Five attributes are considered for each amino acid, normalized van der Waals volume, polarity, charge, and surface tension. The attributes are averaged over all amino acids in a sequence, resulting in a five-dimensional feature vector. Binary classification is carried out for each protein family, with proteins in the family comprising the positive data set and all other proteins the negative data set. Evaluation is carried out on annotated proteins extracted from several databases, and reported classification accuracies are in 69.1–99.6 % range. In [125], SVM-Prot is reported to obtain an accuracy of 71.4 % on a testing data set of 49 plant proteins.

Work in [86] attempts to address some of the issues with predicting GO categories for protein sequences. The Classification in a Hierarchy Under Gene Ontology (CHUGO) system is proposed, which recognizes that assignment of a protein to a particular GO node, immediately assigns the protein to all ancestors of the node in the GO hierarchy. Therefore, labels in CHUGO are not specific GO categories, but GO subgraphs in the GO hierarchy. Since CHUGO trains a separate binary classifier for each GO node, effectively an ensemble of classifiers are used for a protein, as a protein can belong to multiple classes at a particular level in the GO hierarchy. While these ideas are merit-worthy, the performance of CHUGO is not higher than sequence-based methods for function prediction.

Work in [240] uses the HMM proposed in [164] to construct profiles for protein families and classify novel proteins via profile comparisons. Families are identified via a single-link hierarchical clustering algorithm of 256,413 manually annotated proteins. The families are available via the PANTHER/Lib library provided as part of this work, and families are carefully indexed according to their GO-based ontology terms documented in the PANTHER/X to permit fast GO annotation transfer on a query sequence. PANTHER v.8.0 now has 82 complete genomes organized into gene families and subfamilies and has evolved to providing not only gene function, ontology, but also pathways and statistical analysis tools. The PANTHER system has emerged as a highly popular tool that enables browsing and query of gene functions, and a large-scale gene functional analysis has recently been reported via the PANTHER classification system [239].

Work on effective feature representations that encode the low-level constraints that function places on sequence is expected to continue. In fact, feature engineering is considered the most promising direction in sequence-based methods for function prediction. Recent work in [262] focuses on this direction and provides the community with hundreds of features of high biological interpretability and are shown promising in predicting subcellular localization, structural classes, and unique functional properties (such as thermophilic and nucleic acid binding).

## 2.4 *Ensemble-Based Methods*

State-of-the-art sequence-based approaches typically employ ensemble techniques that combine the three categories of sequence-based methods described above. For instance, work in [170] proposes two ensemble methods, the consensus (CONS) method and the frequent pattern mining (FPM) method. Each method combines GO predictions from PFP [129, 130], ESG [63], PSI-BLAST [8], PFAM [329], FFPred [225], and HHblits [293]. Each of the ensemble methods in [170] improved performance over the individual methods in the CAFA1 and CAFA2 categories of the CAFA competition.

The GOPred method proposed in [304] combines heterogeneous classifiers that cover the three main sequence-, subsequence-, and feature-based approaches to improve GO annotations. Positive and negative training data sets prepared for each of the 300 GO molecular function terms are subjected to three classification methods, each representative of the three main approaches: BLAST k-nearest neighbor (BLAST kNN), the subsequence profile map (SPMap), and the peptide statistics with SVMs (Pepstats-SVM) method. Four classifier combination techniques are investigated: majority voting, mean, weighted mean, and addition. Evaluation in [304] demonstrates that the weighted mean classifier combination technique, which assigns different weights to the classifiers depending on their discriminative power for a specific functional term, achieves the best performance in 279 of 300 classifiers.

## 2.5 *State of Sequence-Based Function Prediction*

An interesting observation based on evaluation of different methods for function prediction at CAFA has been that sequence-based methods set the bar high for function prediction [124]. While any current function prediction method ought to outperform sequence-based methods, in practice, improvements are small, as sequence-based methods, when implemented correctly, do very well at CAFA. Such are the conclusions of a survey of such methods in 2013 [124]. In particular, it is observed that sequence-based methods can perform at the top-20 provided precise details of the implementation are followed. For instance, score normalization across targets and poor choices for values of free parameters can lead to lower performance. Careful implementations of sequence-based methods can lead to high performance on the CAFA top-20 and threshold measures.

Many challenges remain regarding sequence-based function prediction. One particular challenge regards the detection of remote homologs, which are homologous sequences with less than 25 % sequence identity thought to make up about 25 % of all sequenced proteins. It is worth noting that a particular group of machine learning methods address the problem of remote homology detection. These methods are both sequence-, subsequence-based, and profile-based and predominantly employ SVMs [26, 189, 288] or HMMs [151, 164, 218–220, 249, 323]. A comparative review of some of these methods can be found in [104]. These methods are typically evaluated not directly on function prediction but instead on reproducing the SCOP superfamily classification [9], which is considered the gold standard due to its manual curation [296] (though recently a different picture is emerging of errors in SCOP and increasing reliability of CATH classification due to improvements of machine learning methods for automated structural classification). It remains to be seen whether these methods, by improving the detection of remote homologs can further improve automated functional annotation of proteins.

## 3 Genomic Context Methods for Function Prediction

Genomic context-based methods can be the only viable approach in cases of query proteins with novel sequences, and for which interacting molecular partners have yet to be discovered. These methods are predicated on the knowledge that location of the gene encoding a query protein is important information that can be exploited for function prediction. These methods fall into two main categories:

- **Gene neighborhood- or gene-order based methods**: these methods operate under the hypothesis that two proteins with corresponding genes located in proximity of each other in multiple genomes are expected to interact functionally.
- **Gene fusion-based methods**: these methods operate under the hypothesis that pairs or sets of genes identified in a genome that are merged into a single gene in another genome are functionally related.

### 3.1 Gene Neighborhood- and Gene Order-Based Methods

The hypothesis that gene proximity in a genome implies functional interactions between the proteins they encode is supported by the concept of an operon, which contains one or more genes that are transcribed as a unit in mRNA. The concept of the neighborhood was originally exploited in [71], which found that 75 % of neighboring genes were known to interact physically, with the rest representing potentially novel interactions. Work in [271, 272] inferred functional coupling between genes in 24 genomes, which additionally employed the concept of a pair of close bidirectional best hits (PCBBH); neighboring genes in a genome G1 with neighboring orthologs in a genome G2. PCBBH entries were scored based on the evolutionary distance between two genomes, and the scores were employed to report as predictions those PCBBH entries with scores above a pre-defined threshold. Work in [185] improves upon the idea of PCBBHs by addressing the issue that gene proximity is not sufficient in itself to infer functional coupling. Additional constraints beyond neighboring orthologs, such as proximity of transcription start sites and opposite direction of transcription are enforced in [185] to infer functional coupling.

The SNAPper method in [179, 180] relies on the construction of a similarity-neighborhood graph (SN-graph). Vertices in the graph are the genes in a given set of genomes. Edges connect vertices corresponding to orthologs or neighboring genes. The notion of an SN-cycle is employed, which is hypothesized to preferentially join functionally related gene products that participate in the same biochemical or regulatory process. Evaluation demonstrates that SNAPper is more effective at reconstructing metabolic pathways than directly predicting functional annotations.

### 3.2 Gene Fusion Methods

Gene fusion was first proposed in [233], which hypothesized that if two genes are separate in one genome but are merged or fused into a single gene in another genome, then these genes are expected to be functionally related. There is strong biological reasoning to support this hypothesis. Gene fusion reduces entropy of disassociation, indicating that genes that encode for two domains of one protein in an earlier organism evolve into independent genes in a descendant organism [95]. The hypothesis is also supported at a structural level, since it has been observed that protein–protein interfaces are highly similar to domain–domain interfaces in a multi-domain protein [349].

The effectiveness of the basic idea of gene fusion was demonstrated in [233] by analyzing 6809 pairs of non-homologous genes in the *E. coli* genome. A significant fraction of these pairs were found to have been reported physically or functionally. The basic premise of gene fusion was validated at a large scale in [383], which applies the method to 30 microbial genomes and reports an average sensitivity of

72 % and an average specificity of 90 % on predicted functional links. Another study on 24 genomes reports similarly high performance [93]. The method in [232] expands upon the basic gene fusion approach by replacing orthology for the broader concept of homology. An association scoring function based on the hypergeometric distribution measures the probability of the chance occurrence of a given number of fusion events between a pair of genes. The log of this association score is found to have a linear correlation with functional similarity.

Gene neighborhood- and gene fusion-based methods have been shown in various large-scale studies [147] to yield interactions that are functionally meaningful, such as direct physical interactions, co-membership in a protein complex, co-presence in metabolic or non-metabolic pathways, or other biological processes. Many databases, such as Phydbac [91] and Phybac [89], now store functional associations detected by gene neighborhood- and gene fusion-based methods, as well as phylogenetic profiles. In general, however, methods based on the notion of gene neighborhoods are more accurate at finding functional links than methods based on gene fusion and phylogenetic profiles (described in Sect. 4).

Several systematic studies of genome context methods are now available. For instance, work in [101] carries out a thorough comparison of many different methods on data from several organisms. Several conclusions are drawn. For instance, the study finds gene fusion methods to generally perform the worst, and gene neighbor methods to outperform phylogenetic profile methods by as much as 40 % in sensitivity on most organisms.

## 4 Phylogenomics-Based Methods for Function Prediction

Phylogenomics-based methods expand genomic context and exploit evolutionary relationships between organisms to detect functional similarities between genes. These methods fall into three categories:

- **Phylogenetic profile-based methods**: these methods encode the presence or absence of a gene across genomes in a binary vector referred to as the phylogenetic profile. The underlying hypothesis is that two genes with similar phylogeny profiles will also be functionally similar.
- **Phylogenetic tree-based methods**: these methods exploit the concept of a phylogeny tree, which encodes evolutionary relationships and distances between organisms in a tree. The pattern of evolution of a set of proteins, as present in phylogeny trees, can be exploited in a machine learning setting to detect functional similarity.

(continued)

- **Phylogeny hybrid methods**: Recent machine learning methods combine the information present in phylogenetic profiles and trees.

## 4.1 Phylogenetic Profile-Based Methods

The operating hypothesis for phylogenetic profile-based methods for function prediction is that proteins that participate in the same pathway or molecular complex in the cell are under pressure to evolve together to preserve their role in the cell. Hence, the comparison of phylogenetic profiles, which store in a binary vector the presence or absence of a particular gene in a genome is expected to be valuable for function comparison.

This hypothesis is tested in [279]. Phylogenetic profiles are constructed from 16 genomes of different organisms. Three *E. coli* proteins are used as the test data set to verify that indeed proteins with profiles different at most one position are functionally related. SWISS-PROT annotations are employed for the verification. Similar results are derived from the EcoCyc database of metabolic pathways [168]. In addition, a comparative study in [214] demonstrates that comparison of phylogenetic profiles is more accurate in predicting function than comparison of whole protein sequences. The study additionally shows that function prediction accuracy improves if more genomes are included to construct phylogenetic profiles.

An important extension of phylogenetic profile-based methods is also proposed in [214] regarding the scenario of redundant genes in an organism that are eventually lost. Such genes can be detected by allowing for complementary phylogenetic profiles, as shown in [214] for DNA-directed DNA polymerases, DNA repair proteins, and isomerases. The PhylProM database computes these results and other phylogenetic data [345].

Work on phylogenetic profile-based methods has mainly proceeded in three directions.

The first direction concerns investigating distance functions for comparing two phylogenetic profiles. In [377], Hamming Distance, Pearson's correlation coefficient, and Mutual Information (MI) are compared on the ability to determine co-membership in a metabolic pathways in KEGG [162], and MI is found to confer the best performance.

The second direction concerns investigating different representations of phylogenetic profiles beyond the binary vector. Work in [88] proposes real-valued profiles, where entries record not just membership in a genome but instead record the normalized BLAST score denoting the best match for a protein in a genome. This is effectively a relaxation of the phylogenetic profile to cases where an exact match for a gene cannot be found in a genome. Cosine similarity is used to identify the neighborhood of a profile, and annotation is carried out by finding the statistically dominant class of the MultiFun database [313]. Comparison with the original profile-based method in [279] shows that the relaxation idea provides better performance. It is worth noting that the Phydbac database [89] includes profile-based annotations obtained from real-valued profiles. The Phydbac2 database [90] by the same team of researchers strengthens the annotation procedure by combining predictions based on the genomic context methods described in Sect. 3.

Work in [29] pursues both above directions of what information to encode in profiles and how to compare phylogenetic profiles. Two modifications are proposed. Based on prior studies showing that including more genomes improves accuracy, partially complete genomes are proposed to be included in constructing the phylogenetic profile of a gene. Second, the distance function is proposed to take into consideration both the number of genomes and the evolutionary history of the unannotated gene. Comparing two profiles constructed from a large number of genomes is assigned greater significance than when comparing profiles constructed from fewer genomes. The farther a genome where the query gene is found is in evolutionary history from the genome containing a gene, the higher the weight of the corresponding entry in distance calculations. Evaluation of these ideas on detecting co-occurrence in the KEGG database for two distinct test sets does not show significant performance improvements over earlier work; however, performance is impacted more by the number of genomes than differently weighting profile entries in the distance function [29].

The third direction of research on phylogenetic profile-based methods pursues the combination of information extracted from phylogenetic profiles with that extracted from gene neighborhoods and gene fusion from genomic context-based methods. For instance, the PLEX method in [73] proceeds in iterations. In the first, genes with similar phylogenetic profiles to a query gene are first identified. These genes are then used as queries to identify possibly more genes with similar profiles in another iteration, and this process continues until no new genes are identified. The predicted functional links are then combined with those obtained from gene neighbor and gene fusion links. This approach has been shown capable of reconstructing the important urease enzyme complex and the isoprenoid biosynthesis pathway in *M. tuberculosis*. Another method employs similar ideas [392] but constructs a single, tandem phylogenetic profile for pairs of neighboring genes in a genome. The profile records whether a pair of neighboring genes in a genome are also neighboring in other genomes. This is in itself an interesting genomic context-based extension of the concept of phylogenetic profiles. Pairs with similar profiles are collected, and the functional coherence is tested with respect to functional categories in the Clusters of Orthologuous Groups (COG) proteins database [340]. Comparison of purity and the Jaccard coefficient show that pair profiles confer better performance than single profiles. Comparison of different distance functions shows that MI is more powerful.

Due to the incredibly intuitive relationship between phylogenetic profiles and functional annotations, methods exploiting phylogenetic profiles have not had to investigate sophisticated techniques from machine learning beyond a comparison of distance functions. However, several directions can be investigated. For instance, other similarity measures are known to be more powerful than MI in machine learning, such as odd ratio, Yule's Q and Yule's Y, as well as Piatetsky-Shapiro and collective strength [335]. Association rule mining based on market basket analysis [2] has also not been investigated, though it may identify meaningful frequent patterns from a matrix of phylogenetic profiles. These lines of investigation may further improve the performance of phylogenetic profile-based methods for function prediction.

## 4.2 *Phylogenetic Tree-Based Methods*

Phylogenetic tree-based methods are also prime for investigating different ideas and techniques from machine learning; however, current research on phylogenetic tree-based functional annotation is scarce, primarily for two reasons. First, it is decidedly more difficult and intricate to compare trees. Second, phylogenetic trees are constructed via algorithms themselves and so are approximations of the actual, unknown evolutionary tree for the organisms under consideration.

Work in [85] demonstrated in 1998 how phylogenetic trees could be used for predicting function. Specifically, the homologs of a query protein can be identified via homology-based methods. The query protein and its homologs can then be embedded in a phylogenetic tree via tree reconstruction algorithms, such as PHYLIP [100]. Gene duplication and speciation events can then be identified in the tree and be used to assign function to the query protein. Work in [85] shows that this approach can be more reliable than homology-based methods when there are variations to the rate of functional change and gene duplication events and changes to the function of homologs during evolution. The promise of phylogenetic trees to improve function prediction over homology-based methods is also demonstrated in [80].

Based on this foundational work, later methods have focused on exploiting phylogenetic trees for identifying domain–domain and protein–protein interactions [276], and for training generative and discriminative models of molecular function [92, 285, 321]. In [276], the similarity between two phylogenetic trees is interpreted as an indication of coordinated evolution and similar evolutionary pressure to members of a given molecular complex. Rather than comparing phylogenetic trees, the method in [276] compares the distance matrices that are used to build phylogenetic trees. Earlier work in [112] introduces such a similarity measure by measuring the linear correlation coefficient between all sets of pairwise distances in the tree.

In [285], a HMM is constructed at each parent node of the phylogenetic tree by using the multiple sequence alignment of the reconstructed sequences of the child nodes. A score is associated with each node, and the query protein is assigned the class whose tree scores the highest. A very high accuracy of 99 % is achieved for a set of 1749 GPCRs. In [321], a multi-step strategy is proposed and disseminated via the GTREE software. The strategy begins with identifying homologs of the query, constructing a multiple sequence alignment of the query and its homologs, using the alignment to eventually construct a phylogenetic tree, identifying the high support subtrees, integrating additional experimental data, differentiation of orthologs and paralogs to infer molecular function.

Work in [92] proposes the SIFTER method, which is based on probabilistic graphical models. The seminal idea is that a reconciled phylogenetic tree can be considered a probabilistic graphical model if transition probabilities are associated with its edges. So, a transition probability function is assigned for the transfer of molecular function from a parent to a node. Then, standard propagation algorithms are used to compute the posterior probability of a node being assigned a certain molecular function. An extensive comparative evaluation shows that the method is superior to sequence-based methods and other phylogenetic tree-based methods. A very high precision is reported with complete coverage on 100 Pfam families supplemented with GO annotations. In 2005, SIFTER beat other methods, such as BLAST, GeneQuiz, GOtcha, GOtcha-exp, and Orthostrapper. Specifically, SIFTER achieves 96 % prediction accuracy against a gold standard dataset of 28 manually annotated proteins in the AMP/denosine deaminase Pfam family. SIFTER performs better than BLAST, GeneQuiz, GOtcha, GOtcha-exp (GOtcha transferring only experimental GO annotations), and Orthostrapper, which achieve 75 %, 64 %, 89 %, 79 %, and 11 % prediction accuracy, respectively.

## *4.3 Hybrid Phylogenetic Profile and Tree Methods*

Since the phylogenetic profile of a protein sequence and phylogenetic trees encode different evolutionary knowledge, they can be combined. The method in [358] uses SVMs to learn protein function from phylogenetic profiles. The phylogenetic tree is used to define a tree kernel that calculates profile similarity. Performance comparison between the SVM with the tree kernel vs. the SVM with a linear kernel on the genes of *Saccharomyces cerevisiae* shows that the tree kernel confers better classification performance. The method in [358] is adapted in [256] to use real-valued rather than binary phylogenetic profiles. The elements of a phylogenetic profile are obtained by a post-order traversal of the phylogenetic tree; the internal nodes of the tree are assigned scores that are averages of the children scores. An SVM with a polynomial kernel is trained on the resulting real-valued profiles and shown to outperform the original method in [358].

# 5 Structure-Based Methods for Function Prediction

Despite increasingly sophisticated sequence-based methods for function prediction, research has shown that sequence is not under a strong selective pressure to preserve function. A prime example of this is the presence of remote homologs, which were identified as early as the 1960s, when Perutz and colleagues showed through structural alignment that myoglobin and hemoglobin had similar structures but indeed different sequences [282]. By now many studies have shown that the correlation between sequence and structure is stronger than that between sequence and function [79, 138, 369]. In particular, structure is under more evolutionary pressure than sequence, and structure-based methods effectively cast a wider net at detecting functional similarity based on structural similarity. Some studies suggest that utilizing both sequence-to-structure and structure-to-function relationships may allow for more accurate function prediction [102].

The goal in structure-based methods for function prediction is to detect a level of similarity between two given protein structures, which in terms allows for the transfer of functional annotations from one protein to another. Similarity can be detected by comparing the two structures in their entirety or only in part. This distinction allows organizing structure-based methods into the following categories:

- **Whole Structure-based methods**: these methods identify similar protein structure utilizing a distance metric to transfer functional annotations from a query structure. Most methods rely on a structural alignment which is prohibitive at a large scale, when comparing a query structure to structures in a database. The high computational cost can be addressed by filter-based methods, which attempt to reduce the number of structures of relevance for comparison to the query.
- **Substructure-based methods**: Similar to the subsequence-based methods discussed earlier, substructure-based methods consider that only a portion of the structure may contain the binding sites critical for binding with molecular partners. A query protein is effectively searched for substructures known to exist between functionally related proteins.

## 5.1 *Whole Structure-Based Methods*

The growth in the size of structural databases such as the Protein Data Bank (PDB) [28] is enabling the transfer of functional annotations between structurally similar proteins. Similar to sequence comparisons, structural alignments are used to enable the comparison of proteins of unequal size. Alignment methods strive to maximize the number of residues in the alignment while minimizing some distance or similarity measure. While many structural alignments methods have been proposed, it remains unclear which method provides the most biologically significant alignment [320, 331]. For a given alignment, an optimal superimposition of two proteins is commonly performed using the Kabsch method [160], which finds the rotations require to minimize the root-mean-squared-deviation (RMSD), between the atoms identified in the alignment. RMSD is a popular distance measure for protein structures. Another popular distance measure is GDT_TS [387], which measures the maximum percentage of $C_\alpha$ atoms that can be aligned within a set of difference tolerances (typically 1Å, 2Å, 4Å, and 8Å).

Below, we discuss two main subclasses of whole sequence-based methods. The first uses superimposition and alignments. While these methods are accurate, they are also computationally demanding. The second subclass of methods lower computational demands by reducing or eliminating the need for alignments and superimpositions.

### 5.1.1 Alignment-Based Whole Structure-Based Methods

Some of the most popular alignment-based structure comparison methods are SALIGN [36], SSM [175], MAMMOTH [268], CE [319], STRUCTAL/LSQMAN [176, 206, 332] SSAP [267], VAST [108], SARF2 [6], and DALI [136]. These methods have to address three main issues: how to represent a protein structure so as to facilitate comparison, how to efficiently explore the space of possible alignments in search of the optimal alignment or near-optimal ones, and how to score a given alignment and determine its statistical significance.

Structural alignment is typically cast as a multi-objective optimization problem, where the best alignment is obtained from maximizing the number of amino acids included in the alignment and in tandem improving a structural alignment score measuring structural similarity. The choice of the alignment technique and the particular score employed are independent of each other, but many methods couple them together.

Computing the alignment is computationally demanding. Some methods compute alignments that are optimal with respect to a similarity score, while others compute approximate alignments in return for computational expediency. The secondary-structure mapping (SSM) method presented in [175] aligns two given proteins using their secondary structure units and employs a scoring function similar to VAST [108]. VAST normalizes aligned structures by incorporating the chances of randomly finding these structures in the PDB (much like term frequency inverse document frequency (TF-IDF) in text mining). VAST is now part of the NCBI's structure computational services.

DALI [136] utilizes internal distance matrix representations of protein structures and a sequence score between matched residues to perform the alignment. An initial alignment is generated and a Monte Carlo scheme is used to find better alignments. An integer linear programming approach utilizing the DALI scoring function is proposed in [370] to compute the optimal alignment. The contact map overlap (CMO) scoring method is proposed by Godzik and Skolnick in [111], which also uses internal distances to maximize the number of contacts in the alignment. The TM-Align method and the TM-Score are proposed in [390], where coordinate superimposition and dynamic programming are coupled with an optimized scoring matrix.

Many of the alignment-based methods are available to the community via web servers. Web servers that allow users to employ and compare many of these methods are also available. For instance, the CSA web server for comprehensive structural alignments [372] provides the community with access to both exact and heuristic alignment methods, as well as many different scoring functions.

Alignment-based methods have been the focus of periodic reviews [177, 182, 234]. In particular, in [177, 182], many of these alignment-based methods are tested on a benchmark set of almost 3000 structures and found to perform well but to carry a large computational cost that becomes prohibitive when the goal is to find the structural neighbor of a protein in a database of 50,000 or more structures.

Recently, alternative approaches have been proposed that do not rely on alignment. For instance, work in [371] utilizes a metric based on contact map overlaps. Another set of methods avoid the alignment and superimposition requirement altogether by employing a filtering approach and exploiting lightweight representations of protein structures. We review these methods next.

### 5.1.2 Filtering-Based Whole Structure-Based Methods

The goal of filtering methods is to rapidly eliminate structures that are not likely to share structural features with a query structure, thus making it practical to employ alignment-based methods on remaining structures. How to represent a protein structure is key to the performance of filtering-based methods, as the choice of representation directly dictates what distance metrics can then be employed to accurately and efficiently score the similarity between two structures [13, 45, 51, 52, 138, 174, 216, 235, 245, 391].

For instance, work in [298], inspired by Vassiliev knot invariants, describes the topology of a protein structure via 30 real-valued features. The scaled Guass metric (SGM) is then employed to compare two such 30-element vectors corresponding to two protein structures. A simple classification procedure employing this metric is shown to correctly identify the fold for 95 % of the structures in CATH2.4 [298]. Another method, PRIDE, compares protein structures via a fast algorithm based on the distribution of inter-atomic distances [53]. PRIDE has been shown to accurately assign query proteins to their CATH superfamily 98.8 % of the time. PRIDE and other structure analysis methods are available for the community via a web server [360]. Recently, a group of methods have exploited the fragbag representation, which is a bag-of-word (BOW) representation of a protein structure.

Utilizing techniques from machine learning and data mining, fragment libraries have been utilized to form a dictionary of "words." Utilizing libraries created

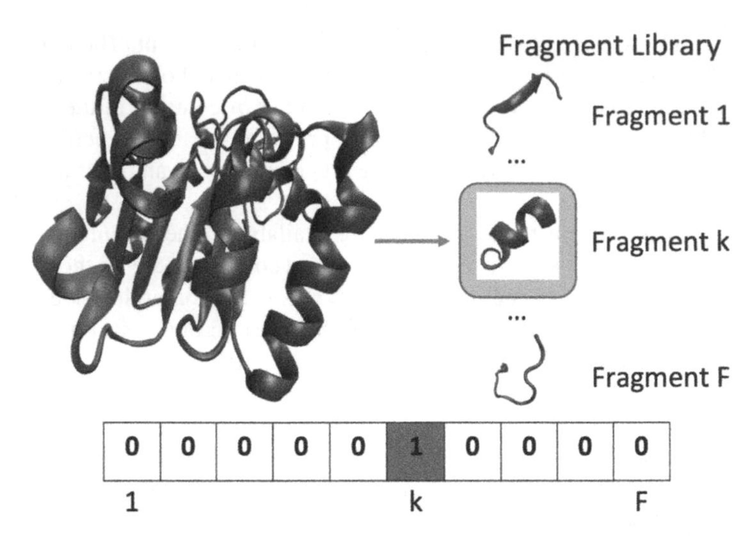


**Fig. 2** Fragbag creation process. A fragment library is shown *top the right* (containing F fragments). The protein structure on the *left* (rendered with VMD [145]) is scanned one fragment at a time (first fragment highlighted in *red*) using a sliding window the same size as the fragments in the library. The library fragment that approximates the current fragment in the protein is located and a term frequency vector is constructed, with each position representing the number of times that library fragment was used in describing the protein. Fragbags are described in [45] and this figure is taken from [245]

in [181], a protein is represented as a term frequency count of the number of times each fragment is used to approximate a segment of the protein's backbone. Figure 2 from work in [245] illustrates the fragbag creation process. Fragbags when combined with a cosine similarity metric were shown to allow the fast detection of remote homologs in [45] and to provide some insight into the relationship between sequence, structure, and function [165, 269].

Work in [245] uses fragbags to construct low-dimensional categorization of the protein structure space and utilize these representations to automatically assign protein function by identifying the SCOP superfamily to which a query protein resides. This categorization is constructed using the Latent Dirichlet Allocation (LDA) model [32] made popular in topic modeling, which is an unsupervised learning approach allows further reducing the fragbag representation to a representation of 10 topics. The topic-based representation introduced in [245] is employed to identify remote homologs of a query protein structure as well as map the query to its SCOP superfamily. The latter is done via SVM, which allows reaching a classification accuracy of over 80 %.

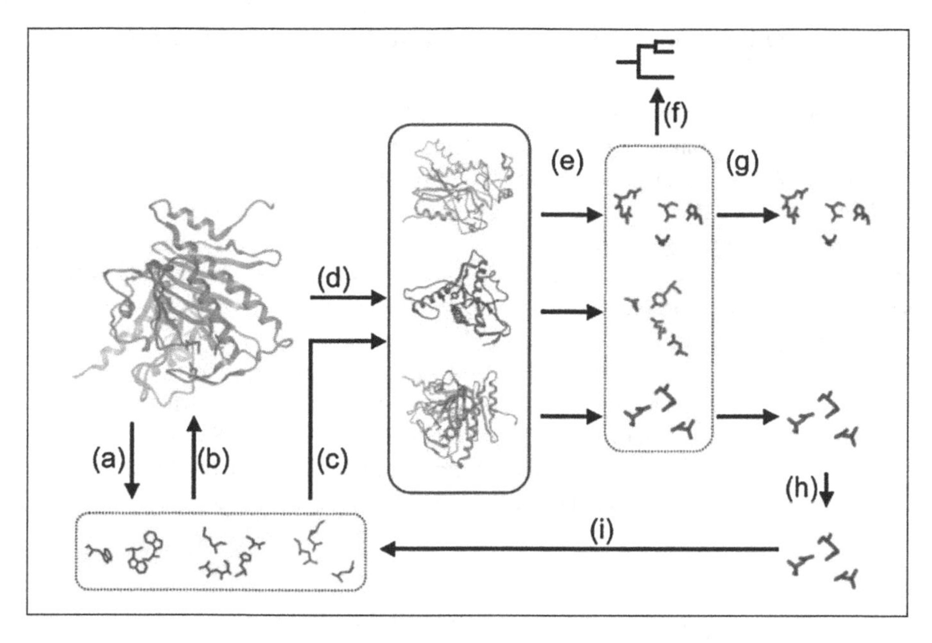


**Fig. 3** Methods to detect common local substructures as shown in [254]. Template-based methods first (**a**) extract known functional motifs from an experimental structure and then (**b**) use this information to search for other structures containing these motifs. (**c**) Illustrates that expert-generated templates can also be used for search queries. The processes for pairwise methods are shown in (**d**)–(**i**). (**d**) Represents a database of protein structures which are processed to reveal local structures shown in (**e**). These structures are then clustered in (**f**) to reveal highly populated clusters extracted in (**g**) and undergo statistical analysis (**h**). The identified motifs can then be fed to (**i**) template-based methods

## 5.2 *Substructure-Based Methods*

Similar to sequence motifs, structural motifs are used to identify common structural components amongst sets of functionally related proteins. The main premise of these methods is to tabulate motifs that act as structure-function signatures and then search a given protein structure for the presence of function-known motifs.

Substructure-based methods can be categorized into two types: template-based methods and pairwise methods. The main difference between these methods is that the motifs may be pre-compiled from experimental structures or identified as shared substructures of two structures under comparison. These two groups of methods are summarized in Fig. 3. Work in [254] summarizes substructure-based methods. Many databases are now available that store structural motifs, including the Database of Structural Motifs in Proteins (DSMP) [121] and the Structural Motifs of Superfamilies (SMoS) [54].

### *5.3 Structure Feature-Based Supervised Learning Methods*

While some of the methods already discussed use supervised learning methods, such as kNN to transfer functional annotations from neighbors to the query, others use SVMs, decision trees, and NNs. For instance, work in [139] introduces SVM-I, a method that uses structural features in lieu of sequence features to detect remote homologs, obtaining an almost four-fold improvement in runtime and improved accuracy over other SVM-based methods that utilize only protein sequence. Machine learning methods have also been applied to predict functionally important sites within a protein. In [381], Yahalom et al. train an SVM using structural features to identify catalytic residues. They utilize that these residues have specific spatial proximities and are deeply embedded in the structure. In this work they investigate other classification methods, but find that SVMs provide the best performance. Many feature-based methods that employ structural features also integrate other sources of data and are the subject of our review in Sect. 8.

### *5.4 State of Structure-Based Function Prediction*

As the number of structures and number of functional annotations increase in structural databases, function prediction methods that utilize structure will gain even more discriminative power over methods that employ only sequence. Work in [102] provides some early evidence to this effect, by showing that structure prediction, coupled with biochemically relevant structural motifs, can outperform sequence-based methods and provide more detailed, robust function annotation of genome sequences. Computational efficiency concerns will remain a challenge. However, techniques that have originated in the data mining community are showing good potential at addressing both computational efficiency and accuracy concerns via smart representations and representation-aware distance metrics.

## 6 Interactions-Based Methods for Function Prediction

Direct binding can be tested at a high throughput scale in the wet laboratory via the yeast two-hybrid (screening) system (Y2H) or affinity purification coupled to mass spectrometry. These two technologies have allowed researchers to amass hundreds of thousands of protein–protein interaction (PPI) data [202]. This data can be found in databases, such as the the Database of Interacting Proteins (DIP) [379], the Biomolecular Interaction Network Database (BIND) [15], the Biological General Repository for Interaction Datasets (BioGRID) [55], the Human Protein Reference Database (HPRD) [169], the IntAct Molecular Interaction Database and the Molecular Interactions Database (MINT) [265], the MIPS Protein Interaction Resource on

Yeast (MIPS-MPact) [119], and the MIPS Mammalian Protein-Protein Interaction Database (MIPS-MPPI) [273]. These databases can be integrated, as in the Agile Protein Interaction DataAnalyzer (APID) [284], the Microbial Protein Interaction Database (MPIDB) [114], and the Protein Interaction Network Analysis (PINA) platform [69].

Other databases combine PPI data obtained via Y2H and/or affinity purification with data predicted in silico. Examples include the Michigan Molecular Interactions (MiMI) [339], Human Protein-Protein Interaction Prediction Database (PIPs) [238], Online Predicted Human Interaction Database (OPHID) [40], the online database of comprehensive Human Annotated and Predicted Protein Interactions (HAPPI) [59], Known and Predicted Protein-Protein Interactions (STRING) [334], and the Unified Human Interactome (UniHI) [161].

PPI data presents a unique opportunity for function prediction methods, particularly when the data is encoded in protein–protein interaction (PPI) networks, where vertices represent proteins and edges represent direct binding. Interactions can be used to infer functional relationships. This principle is known as "guilt by association" (GBA) [263]. Graph-theoretic concepts and algorithms can be employed on PPI networks to predict the function of a query protein. Methods that do so can be organized in four main categories:

- **Neighborhood-based methods**: these methods exploit the most dominant annotations among neighbors of a query protein in a PPI network.
- **Module-assisted methods**: these methods exploit the local topological structure of a PPI network to identify functional modules from which to infer functional annotations of unannotated proteins. Two subgroups of methods can be found in this category, clustering-based methods, which seek dense subgraphs in a PPI network, and non-clustering methods, which seek dense subgraphs via graph-theoretic concepts.
- **Global optimization-based methods**: these methods go beyond neighborhood information and consider the structure of the entire network. Optimization of an objective function allows exploiting annotations of other proteins indirectly connected to the query.
- **Association-based methods**: these methods take a complementary approach to detection of dense subgraphs, employing association rule mining to detect frequently-occurring sets of interactions.

Methods that are not module-assisted are also referred to as direct methods, as they essentially propagate functional information through the network. A detailed review of these methods for function prediction is provided in [315]. In the following, provide an up-to-date summary of these methods.

## 6.1 Neighborhood-Based Methods

Neighborhood-based methods transfer to a query protein the most dominant annotations among neighbors of the query in a PPI [65, 109, 137, 208, 247, 283, 312, 363]. Work in [312] was one of the earliest to demonstrate that 63 % of the interacting proteins in a yeast PPI of 2709 interactions had a common functional assignment, and 76 % were found in the same subcellular compartment. These simple statistics laid the foundation for exploiting the connectivity information in a PPI network to infer function. Functional assignment based on the majority (voting) annotation shared by direct neighbors was shown to be a viable approach [312]. This approach, deemed Majority Voting in [312], later came to be known as the nearest-neighbor voting, or basic GBA (BGBA), and is persistently shown to perform well [283] in the Critical Assessment of Function Annotation (CAFA) challenge. An earlier precursor of nearest-neighbor voting or neighbor counting methods is the chi-square method [135], which considers neighbors indirectly and assigns to a protein $k$ functions with the $k$ largest chi-square scores; the chi-square score for a function $j$ and a protein $P_i$ is defined as $S_i(j) = \frac{|n_i(j)-e_i(j)|^2}{e_i(j)}$, where $n_i(j)$ is the number of direct neighbors of protein $P_i$ that have function $j$ and $e_i(j) = n_i(j)xp_j$ is the expected number of neighbors with function $j$, with $p_j$ denoting the fraction of proteins having function $j$ among all proteins in the PPI network. Work in [75] shows that BGBA-based methods perform comparably to the chi-square method.

Work in [65] extends BGBA by including indirect, level-2 neighbors. The FSWeighted algorithm is proposed. The local topology of the query is compared with that of its direct and indirect neighbors to estimate functional similarity between a query and its neighbors. The experimental reliability of interactions is combined with the functional similarity to associate a weight with each neighbors. The query is then assigned the various GO terms of its neighbors, scoring each term by its weighted frequency among the neighbors. Leave-one-out cross validation shows that the performance of this method is comparable with other neighbor- and similarity-based methods, as well as Markov random field-based global optimization methods summarized below.

There are several issues with neighbor-based methods that are addressed at various levels in existing literature. For instance, the query needs to have a sufficient number of annotated neighbors in the network for a reliable prediction to be made. While early methods did not consider distances between the query and its neighbors, later methods do so, as summarized above. While GBA focuses on annotated neighbors, there is information to be exploited in interacting unannotated neighbors. Recent methods, such as those proposed in [247, 363], have begun to exploit unannotated interacting pairs in PPI networks.

Another issue concerns which neighbors to consider, and whether to restrict the GBA approach to direct neighbors, neighbors within a radius, or extend it to indirect neighbors. Work in [109] shows that indirect connections improve gene function prediction and proposes a new method based on the concept of extended GBA, where networks are extended by self-multiplication. The multiplication allows estimating the number of paths of a certain length connecting a given pair of nodes in the network, and the method in [109] weights the paths in an extended network before conducting GBA as on the original network. This network extension approach sits at the boundary of neighbor-based and global optimization methods. The approach is applied not only to PPI networks but to co-expression networks, as well, showcasing its generality for processing biological data encoded in graphs.

Yet another issue concerns the existence of mutual dependencies among neighbors of the query. If two or more neighbors have similar function, their contribution is likely to accumulate in existing neighbor-based methods. This should not be the case. Instead, dissimilar neighbors should be more important for annotation of a query. Work [137] proposes a way to take into account correlations among neighbors. In particular, the Choquet-Integral for fuzzy theory is employed to aggregate functional correlations among neighbors. The functional aggregation measures the impact of each relevant function on the final prediction and reduces the impact of repeated functional information on the prediction. The functional aggregation is employed in a new protein similarity and a new iterative prediction algorithm proposed in [137]. Evaluation of this approach shows that removing neighbor correlations results in improved performance over neighboring methods based on majority voting and sophisticated distance metrics such as the functional similarity metric proposed in [65].

Finally, the majority of neighbor-based methods ignore the scale-free property found in many biological networks, including PPI networks [3, 21]. In [209], neighbor sharing is assumed to be constrained by preferential attachment, and the Preferential Attachment based common Neighbor Distribution (PAND) method is proposed to calculate the probability of a neighbor-sharing event between any two nodes in a network. This probability distribution was shown to match very well the observed probability in simulations of scale-free networks. PAND was applied to a PPI network in [209] and shown to reveal smaller probabilities correlating with closer functional linkages between proteins. PAND-derived linkages were used to construct new networks with more functionally reliable links than links in PPI networks. Simple annotation schemes on the new networks were found to be more accurate [209].

## 6.2 *Module-Assisted Methods*

The local topological structures and properties of PPI networks are subject to theoretical investigation and empirical exploration via ideas from network science. Module-assisted methods seek to identify local topological structures that can represent functional modules in a PPI network. Clustering-based methods rely on clustering to identify dense regions with a large number of connections in PPI networks, whereas non-clustering methods employ graph-theoretic concepts and algorithms to identify local topological structures. We review each next.

### 6.2.1 Clustering-Based Methods

Clustering methods focus on finding dense regions with a large number of connections as a way of identifying functional modules representing protein–protein complexes/assemblies [23, 330]. MCODE [23] uses a vertex weighting scheme based on the clustering coefficient to measure the cliquishness of the neighborhood of a node. Work in [330] proposes two clustering algorithms. The super paramagnetic clustering algorithm [30] is a physics-inspired hierarchical clustering algorithm. The Monte Carlo algorithm instead maximizes the density of predicted clusters. The Markov clustering (MCL) algorithm is proposed in [94, 280], whereas the highly connected subgraph (HCS) algorithm is proposed in [128]. HCS is a graph-theoretic algorithm that separates a graph into several subgraphs using minimum cuts. A cost-based local search based on tabu search metaheuristic is proposed in [171].

Other methods employ classic clustering algorithms after defining a similarity measure that takes into account the interactions of a protein in the network [11, 42, 113, 228, 302]. The SL method in [302] employs the number of common neighbors to define the similarity between two proteins and then uses k-means to partition the nodes in a PPI network into different groups/assemblies. Following work in [228] modifies the similarity measure via a weighted form of the mutual clustering coefficient approach [113]. Work in [11, 42] uses hierarchical clustering. In [11], the shortest path distance between two proteins is used to estimate their similarity, while work in [42] uses the Czekanovski-Dice (CD) metric.

The CD distance between two proteins $u$ and $v$ is based on the number of neighbors they share and is measured as: $\frac{|N_u \delta N_v|}{|N_u \cup N_v| + |N_u \cap N_v|}$, where $N_*$ refers to the direct neighbors of a vertex, and $\delta$ refers to the symmetric difference between two sets. Work in [42] proposes the PRODISTIN method, which employs the BIONJ algorithm (an improved version of the popular neighbor-joining algorithm) [106] to cluster proteins in a PPI based on their CD distance. The BIONJ algorithm produces a hierarchical classification tree. A PRODISTIN functional component is the largest subtree that contains at least three proteins with the same function and has at least 50 % of its annotated members sharing that function. This function is transferred to the unannotated proteins in the functional component. It is worth noting that the neighbor-based method in [65] proposes a new CD-based functional similarity that penalizes the similarity weights between protein pairs when any of them have few direct neighbors.

Work in [364] shows that clustering-based methods can produce many false positives. To remedy this issue, the EVDENSE method is proposed in [364] to efficiently mine frequent dense subgraphs in a PPI network. EVDENSE produces frequent dense patterns by extending vertices and by using relative support. Improved performance is reported over other clustering-based methods.

A detailed analysis of clustering-based methods is conducted in [325] to evaluate the hypothesis that dense clusters correspond to functional modules. Six different clustering algorithms are applied to a yeast PPI network, and evaluation shows that the performance of these algorithms is dependent on the topological characteristics of the network. For the specific task of function prediction, a non-clustering, BGBA approach outperforms the clustering algorithms. Guidelines are provided in [325] to evaluate and justify novel clustering methods for biological networks.

One issue of clustering-based methods regarding the dynamic process in clustering is addressed in the PClustering method proposed in [301]. Saini and Hou track function appearance across all relevant clusters rather than just the cluster to where the query is mapped by a specific clustering algorithm. A recursive clustering algorithm reclusters until a cluster is obtained where the query gets separated from the rest of the other proteins. The recursion tree tracks how the clusters are split down to the leaf where the query is separated from all other proteins. The particular path from the root to this leaf is then inspected to accumulate all functions of all proteins in the path. This set is the list of relevant functions that can be assigned to the query. The prediction for the query is then made based on how the proteins in the path, with relevant functions, are split during the recursive clustering process. A score is proposed to select functions that are stable in terms of their frequency across the clusters in the path. PClustering is compared to PRODISTIN [42], MCL [94], SL [302], Chi-square [135], Majority Voting [312], and the FSWeighted method [65] on the yeast PPI dataset and shown to outperform many of these methods, particularly on being able to accurately predict functions of more unannotated proteins.

### 6.2.2 Non-clustering, Graph-Theoretic Methods

Clustering-based methods essentially aim to uncover communities in a network. A different group of module assisted methods circumvent clustering and instead employ concepts and algorithms from network science for community detection. A community is a more general concept than a cluster, and various methods exist on community detection in networks, a review of which is beyond the scope of this paper. Here we summarize recent methods that uncover functional modules by exploiting the concept of communities and then employ such modules for function annotation.

Work in [222] introduces the concept of k-partite "protein" cliques as functionally coherent but not necessarily dense subgraphs. This concept is more suitable for PPI networks, which are known to be non-uniform in subgraph density for reasons that are often artifacts of wet-laboratory studies. Briefly, a k-partite protein clique is a maximal k-partite clique comprising two or more non-overlapping subsets between any two of which full interactions are exhibited. In [222], a PPI network is transformed into induced k-partite graphs, where edges exist only between the partites. A maximal k-partite clique mining (MaCMik) algorithm is proposed to enumerate maximal k -partite cliques on these k-partite graphs. MaCMik is applied to a yeast PPI network, and unusually high functional coherence is observed in the k-partite cliques. This direction of work suggests that graph-theoretic concepts can be more powerful at capturing the concept of functional modules and in turn assist with function prediction.

While work in [222] restricts itself to the proposal of a new concept for a functional module, work in [199] shows that the concept of a community can be exploited for function prediction. In general, a cost function is designed to measure the extent to which a subgraph constitutes a community, and then optimization algorithms partition the graph into subgraphs that optimize the cost function. The modularity (Q) measure is one example of a popular cost function that measures the relative density of intra-community connectivity compared to a randomly re-wired counterpart with the same degree of nodes. The conformational space annealing (CSA) algorithm [200, 201] is employed in [222] to detect maximum-Q subgraphs in the yeast PPI network. Figure 4 showcases the ability of the CSA algorithm to detect more subgraphs than a baseline, popular simulated annealing approach.

After the high-modularity subgraphs are detected, Random Forest (RF) is then employed, representing each protein as a vector of features generated only from the network community (including which communities the neighbors belong to and their functions). The resulting RF-comm-CSA method is compared to MRF-based methods in [77, 163], neighborhood enrichment methods [315, 325], and the Majority Voting method [312]. RF-comm-CSA is reported to achieve the best performance, followed by the MRF-based method in [163].

## 6.3 Global Optimization-Based Methods

To overcome these setbacks, global optimization methods consider the full topology of the network and employ techniques, such as Markov random fields, simulated annealing, and network flow [57, 77, 204, 252, 355]. The Markov random field (MRF) method proposed in [77] computes the probability that a protein has a function given the functions of all other proteins in the interaction dataset. MRFs are particularly suitable for modeling the probability that a query has a certain function by capturing the local dependency of the query on its neighbors in a PPI network. The latter is in essence the GBA principle. The Markov property is valid here, as the function of the query is assumed to be independent of all the other proteins given its neighbors in the PPI network. The MRF method in [77] is shown to be more sensitive at a given specificity than neighbor-based and chi-square methods.

Work in [355] proposes a simulated annealing method, which was later shown in [76] to be a special case of the MRF method [77]. The approach proposed in [204] is identical to the MRF method in [77]. More recent work in [57] proposes a bagging MRF-based method (BMRF). The method follows a maximum a posteriori principle to form a novel network score that considers interactions in a PPI. The score is used by the method to search for subnetworks with maximal scores. A bagging scheme based on bootstrapping samples is also employed to statistically select high-confidence subnetworks. While work in [57] applies BMRF to identify subnetworks associated with breast cancer progression, later work in [317] shares the BMRF-Net software with the community to identify subnetworks of interest in a PPI by the BMRF method.

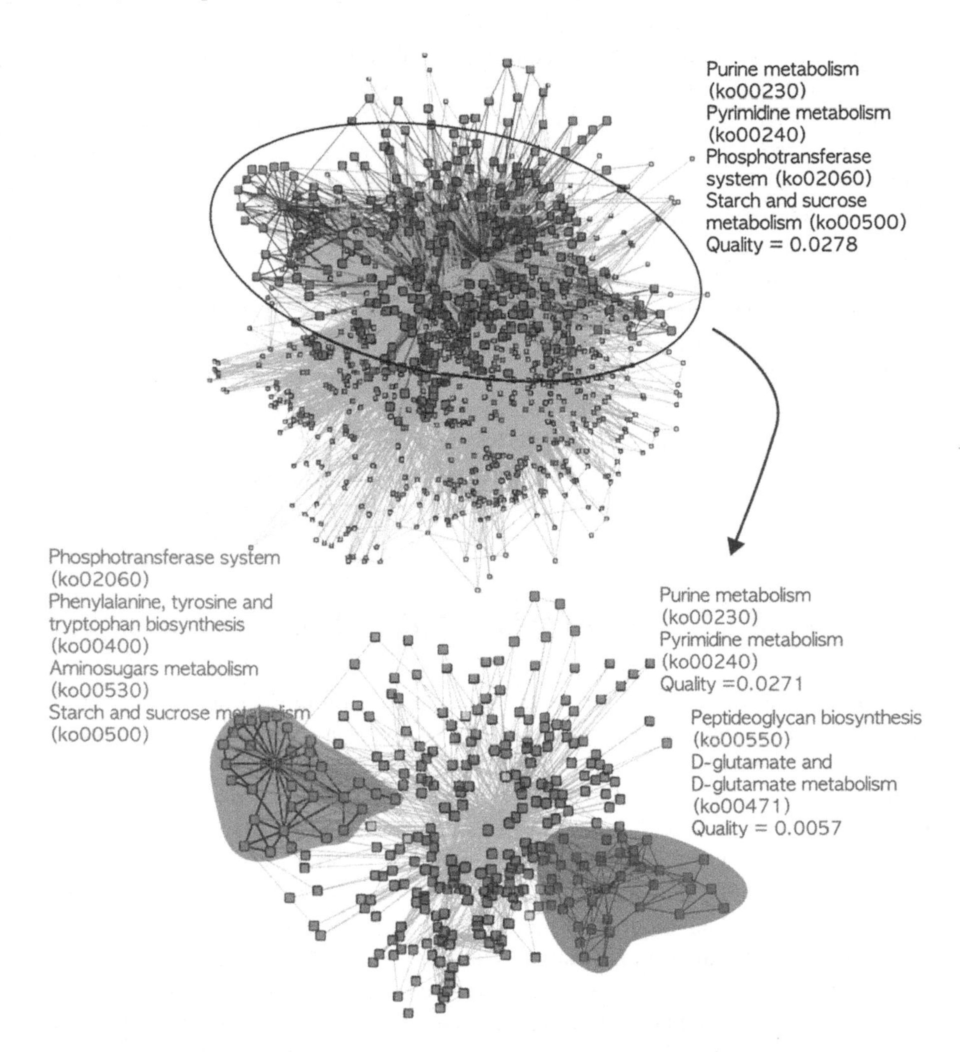


**Fig. 4** In [199], nodes of one community obtained by simulated annealing are split into three communities by the CSA algorithm. Two of these are in *red and blue shaded areas*, and the third is drawn with *large squares*. Within each of these communities, meaningful functional clusters of KEGG pathway annotations with P-value less than $10^{-4}$ are listed

While the MRF-based method in [77] does not consider unannotated proteins when training the regression model for parameter estimation, work in [187] uses the same MRF model but then applies adaptive Markov chain MC (MCMC) to draw samples from the joint posterior. This approach gives an AUC value of 0.8915 on 90 GO terms compared to the 0.7578 reported in [77] and 0.7867 reported in [204]. Work in [76] extends the MRF model by integrating various data sources, such as PPI data, expression profiles, protein complex data, and domain information.

Graphical models such as MRFs continue to solicit interest for predicting protein function from biological networks such as PPI networks. A recent review in [338] focuses on MRFs and conditional random fields (CRFs) and their applications for predicting protein function and protein structure. Interestingly, the review finds that, while CRFs have become popular in protein structure prediction and protein design, they have yet to solicit interest in the function prediction community.

Work in [163, 252] represents a complementary approach based on the concept of flow. Karaoz and co-workers visualize propagation diagrams to illustrate the flow of functional evidence from annotated to unannotated proteins in a PPI network [163]. However, it is Nabieva and co-workers that operationalize on the concept of flow in the Functional Flow method [252]. The method simulates functional flow between proteins. A protein annotated with a specific function is assigned an infinite potential for that function, whereas an unannotated protein is assigned a 0 potential. Functions then flow from proteins with higher potentials to their direct neighbors with lower potential. The amount of flow depends on the reliability of the interactions.

## 6.4 Association-Based and Other Mining Methods

Association-based methods employ association rule mining to detect frequently occurring sets of interactions. The local topology of a query in a PPI, whether restricted to direct and/or indirect neighbors, or extended to the more general concept of a community, can be encoded via features in a feature-based representation of the query and other proteins in the PPI network. Such representations open the way for application of popular supervised learning algorithms for function prediction. For instance, work in [260] employs logistic regression on representations that use only functional annotations of the direct neighbors of a protein. Work in [64] mines a PPI for frequent functional association patterns. The set of functions that an annotated protein performs is assigned to the protein as a label, and a functional association pattern is represented as a labeled subgraph. A frequent labeled subgraph mining algorithm efficiently searches for functional association patterns in a PPI network. The algorithm increases the size of frequent patterns one node at a time by selective joining, while simplifying the network by a priori pruning. The algorithm is reported to identify more than 1400 frequent functional association patterns in the yeast PPI network. Function prediction is carried out by matching the subgraph that contains the query with the frequent patterns analogous to it. Leave-one-out cross validation shows that this function prediction approach outperforms neighbor-based methods [64].

In general, however, these methods do not restrict themselves to features extracted only from the local topology of a node in a PPI network, but integrate various data sources to represent proteins. For instance, work in [50] proposes three probabilistic scores, MIS, SEQ, and NET, to combine protein sequence, function association, protein–protein interactions, and gene–gene interaction networks. The MIS score is generated from homologs found for a query via PSI-BLAST and association rule between GO terms learned by mining the SWISS-PROT database. The SEQ score is based on sequences, and the NET score is generated from PPI and gene–gene interaction networks. The three scores are combined in the Statistical Multiple Integrative Scoring System (SMISS), which is reported to outperform three baseline methods that combine profile-sequence homology search, profile–profile homology search, and domain co-occurrence network [366] on the CAFA1 dataset.

A detailed description of data integration methods is provided in Sect. 8.

# 7 Gene Expression-Based Methods for Function Prediction

A complementary source of data that can be used for function prediction are gene expression data. cDNA microarray technology allows measuring the amount of protein a gene makes at a given time under specific conditions. cDNA chains can be designed to bind complementary mRNA so as to detect the transcription of specific genes. Gene expression data, also referred to as microarray data, can be measured in time, as well. Often, expression data measured across different labs and under different conditions can vary, and standard normalization strategies help eliminate discrepancies [58] and allow the employment of machine learning techniques to mine such data. Gene expression data are typically viewed in matrix format, with genes in rows and gene expressions under different conditions in columns. By now, many such data exist for different organisms. The Stanford MicroArray Database [316], now retired, contained expression data for genes in the human genome. Similar data can now be found in databases, such as the NCBI Gene Expression Omnibus [22, 84] or EMBL-EBI's ArrayExpress [178].

Microarray data are prime for machine learning methods to detect changes in gene expressions indicating the presence of specific types of diseases. Such data can also be used for function prediction, as similarities between expression profiles of genes can indicate functional similarities. This hypothesis was investigated early on in [362], where 40,000 genes were examined for co-expression with five genes known to be associated with prostate cancer. The guilt-by-association principle was employed to identify uncharacterized genes significantly co-expressed with at least one known prostate gene. This approach detected eight novel genes associated with prostate cancer. While technically this study pursued the exploitation of gene expression data to detect gene signatures of disease, it opened the way to more sophisticated machine learning methods for the more detailed problem of function prediction. Such methods can be organized in the following three categories:

- **Unsupervised learning methods**: these methods cluster expression profiles to identify genes with similar profiles that can be hypothesized to share functional annotations.

(continued)

- **Supervised learning methods**: Predicting function from gene expression data is a natural learning problem in the supervised setting, and these methods investigate classifiers, such as SVMs, naive Bayes, NNs, and more.
- **Temporal analysis methods**: these methods exploit the ability to measure gene expression in time during, for instance, a disease. Temporal expression data can also be used by classifiers to predict the function of unannotated genes.

## *7.1 Unsupervised Learning of Gene Expression Data for Function Prediction*

Many different clustering algorithms can be used to organize gene expression profiles, including clustering algorithms specifically designed for gene expression data, such as CAST [25]. The issue, however, is what to do with the clusters and which cluster to employ for function annotation. Measures such as majority [122] have not been effective [395]. Work in [85] demonstrated an effective approach relying on hierarchical average-linkage clustering with a variation of the correlation coefficient as a similarity measure. Analysis showed that genes mapped to the same cluster were for the most part involved in common cellular processes, directly validating the hypothesis that clusters of co-expressed genes are also functionally coherent. This result is often credited as bringing clustering to the forefront of techniques for analysis of biological data.

The results of the relevance of clustering for functional annotation motivated the development of a novel clustering algorithm, the Cluster Affinity Search Technique (CAST) [25]. CAST proceeds in two phases, an add phase where elements with high affinity to the current cluster are added to the cluster, and a remove phase, where elements with low affinity are removed from the cluster. Clusters are constructed one by one, and the algorithm terminates when no changes occur. Evaluation on static and temporal gene expression data showed that CAST is able to preserve functional categories. The ability of the algorithm to extract knowledge about diseases from expression data has also been demonstrated, and has been employed in other works to study gene signatures of disease [24, 333].

Work in [259] focused on removing noisy and redundant dimensions from expression profiles via latent semantic indexing [205]. The retained dimensions are then employed for clustering based on the concept of neighborhood, where intra-cluster similarity needs to be significantly higher than inter-cluster similarity. Unannotated genes in a cluster then are assigned the majority annotation of the characterized genes in the cluster.

The known issues with clustering regarding how to define similarity and how many clusters to extract have prompted many researchers to pursue additional directions. For instance, work in [395] increases confidence in the results via the novel ontology-based pattern identification (OPI) strategy. Briefly, all decisions such as attribute weights, choice of similarity threshold, choice of mean or median, and other decisions that need to be made in clustering algorithms are embedded in a Euclidean space. Then, a hill climbing algorithm is employed to navigate this space in order to find decisions that are optimal with regard to an objective function encoding expected characteristics of optimal clustering. The hill climbing algorithm minimizes this function for all the GO functional categories and in the process identifies the best cluster for each category. As in earlier work, unannotated genes in a cluster are functionally linked to annotated genes in the same cluster. OPI has been shown able to identify more statistically-significant clusters than other related work employing the k-means clustering algorithm [297]. OPI has been additionally validated. About 12 of the 50 genes predicted by OPI to have the antigenic variation have now been verified.

Dealing with cluster overlaps is another way to increase confidence in the clustering results. Work in [378] employs different clustering algorithms and annotates a cluster with the functional class of the least p-value, as calculated from the fractions of the different functional classes in a cluster. An unannotated gene is then assigned the functional class of the cluster to which it is mapped, and the assignment is also associated a confidence value based on the p-value of the cluster. Consensus clustering is proposed in [246] as an analysis approach to assist clustering algorithms. Consensus clustering, together with resampling, is proposed to represent the consensus over multiple runs of clustering algorithm with random restarts, such as K-means, model-based Bayesian clustering, and self-organizing maps (SOM), which are typically sensitive to initial conditions.

The method in [333] fuses consensus clustering with ideas proposed in [378] to carry out both robust and consensus clustering of gene expression data, as well as assign statistical significance to clusters from known gene functions. While the method in [246] perturbs gene expression data for a single algorithm, the method in [333] uses different clustering algorithms. This method proposes a robust clustering algorithm, which seeks maximum agreement across different clustering methods by reporting only the co-clustered genes grouped together by all the different algorithms. To address the issue of robust clustering discarding gene expression vectors if only one clustering method performs badly, consensus clustering is proposed, which seeks a minimum agreement. An objective function is defined to reward clusters with instances of high agreement and penalize clusters with instances of low agreement. The function is minimized via simulated annealing. Consensus clustering is reported to improve upon the performance of individual clustering methods, as measured by the weighted-k measure [7]. Clusters identified for ten functional classes in [333] were also more likely to be annotated with the same classes by consensus clustering than individual clustering methods.

The functional coherence of gene expression clusters is called into question in [394], which investigates a graph-theoretic approach. Genes are encoded as vertices, and edges connect genes with correlated expression profiles. Shortest paths in the graph allow identifying transitively related genes. A simple experiment is conducted, where shortest paths between genes of the same GO category are analyzed to check if the genes in these paths are annotated with the same GO category or a parent or child function in the *S. cerevisiae* Genome Database (SGD) [82]. The analysis shows that high accuracy is obtained for mitochondrial and cytoplasmic genes, but medium accuracy is obtained for nuclear genes. The graph-theoretic approach in [394] additionally provided functional annotations to 146 genes that were weakly correlated to other genes.

Another approach is employed in [229] to relax the functional coherence requirement of a cluster. The latter may be weak if one considers all conditions, as some may act as noise and strong correlation can be obtained upon removal of such entries. Biclustering or coclustering is employed to address this. Briefly, biclustering is a specific type of sub-space clustering and refers to the simultaneous clustering of both rows and columns of a data matrix [127]. Two-way analysis of variance is then used to identify constant valued sub-matrices. While many methods pursued biclustering for gene expression data [62, 384], it was work in [43] that demonstrated a simulated annealing approach to biclustering to perform well on yeast cell cycle data sets. The two largest clusters were found to contain largely members of two different families, the ribosomal proteins and the nucleotide metabolism proteins. The method in [221] further demonstrated the potential of biclustering for functional annotations. The structure of the GO hierarchy is incorporated in the hierarchical biclustering process. Genes are first clustered via hierarchical clustering, and then each node in the hierarchy is annotated with the GO functional class with which it is most enriched.

While much progress has been made in clustering algorithms, many issues inherent to clustering remain. One issue, in particular, pertains to the inability of clustering algorithms to exploit already labeled instances. Supervised learning methods exploit labeled data, and we review such methods for gene expression data next.

## *7.2 Supervised Learning of Gene Expression Data for Function Prediction*

Gene expression data are subjected to a classification setting in [41], where three different classifiers are compared to learn functions from yeast gene expression data: Parzen's window, Fisher's linear discriminant analysis, two decision tree classifiers (C4.5 and MOC), and SVMs with different kernels. Comparative analysis demonstrates SVM with the radial basis kernel to perform best. Work in [192] compares SVM to kNN classifiers on gene expression data. The cosine similarity measure is employed, as it better captures the shape of an expression vector in high-dimensional space rather than its magnitude. kNN is reported to be better at predicting the *m* most appropriate classes for a test gene over SVM.

Multilayer perceptrons are employed in [237] to learn 96 different functional classes from annotated/labeled yeast gene expression data. Better performance is obtained over work in [41], but three sources of errors are identified: class size, class heterogeneity, and Borges effect (termed in [237] to indicate simultaneous membership of a gene in different functional classes). An iterative learning procedure is proposed in [237] to address these three sources of error better than a one-pass learning procedure.

Work in [258] employs multiple expression data sets for learning with SVMs and presented a strategy to select the most informative data set for learning individual classes. The learning cost savings measure introduced in [41] is used to show that blindly combining different data sets is not optimal. A hill-climbing algorithm is proposed instead to incrementally add the data set that provides the maximum learning cost savings until a maximum is reached. Comparison with other classification methods showed this approach to be superior.

A different direction is investigated in [388], namely, that of training on larger mammalian expression data sets (the earlier works above focused on *S. cerevisiae* or *Caenorhabditis elegans*). An SVM is employed to learn each of the 992 GO biological process categories and classify 10,000 unannotated genes. Performance is reported to be mixed, suggesting that more sophisticated techniques may be needed for mammalian gene expression data sets. Specifically, while SVMs are at the moment the state of the art in classification of gene expression data [250], there is room for investigating different classification techniques, such as boosting, active learning [81], and more.

## 7.3 Temporal Gene Expression Data Analysis for Function Prediction

Temporal gene expression data provide dynamic information on the simultaneous expression levels of genes, effectively providing a dynamic picture of what goes on in the cell via expression measurements. Both unsupervised and supervised learning techniques have been applied to temporal expression data. Work in [19, 46, 96, 132, 155, 243] exposes challenges posed by time-series expression data to clustering algorithms, such as similarity measures, co-clustering, short profile lengths, and unevenly sampled genes. The issue of possible time offsets in the expression of different but functionally associated genes is another one that still challenges temporal gene expression analysis. Some progress has been made.

Temporal gene expression data are used in [148, 193] to learn GO biological process annotations for unannotated genes. The gene expression profiles are transformed into attribute-valued vectors. Attributes constructed by calculating the increase or decrease of expression values between two instances separated by an interval of three time points. Three possible values are assigned, high, medium, and low. This mapping is central to the ability to use set-based classifiers, which can only robustly handle nominal attributes. The classifier in [148] achieves a cross-validation AUC score of 0.8. Testing on the human serum response expression data set results in labels being correctly predicted for 211 of the 213 genes.

Inductive logic programming and description logics are proposed in [14] for learning the classification rule set from temporal gene expression patterns. Work in [241] presents another rule-based classification model which outperforms the approach in [148]. HMMs are employed in [78] to model the interdependence between conditions and the dependence of the functional class on them. A dual HMM modeling both expression values and experiment order is shown to perform best on yeast gene expression data. Another statistical approach, Mixture Functional Discriminant Analysis (MFDA), is proposed in [118] to operationalize upon the observation that temporal profiles of genes belonging to the same functional class are highly similar. Each individual class is modeled as a mixture of sub-classes, and the Expectation Maximizaton (EM) algorithm [81] is used to learn the parameters of the model. MFDA is marginally better over other discriminant analysis methods on yeast cell cycle expression data.

Work in [353] injects evolution and evaluates the hypothesis that the conservation of co-expression between pairs of genes that share an evolutionary history can enable more confident prediction of their functional association and pathways in which they are involved. The evaluation is conducted on *S. cerevisiae* and *C. elegans*, using correlation as a measure of co-expression. Two types of co-expression conservation are defined: Paralogous conservation, which refers to two pairs of genes (A, B) and (A', B') in the same organism, where A is homologous to A' and B is homologous to B; Orthologous conservation, where the two pairs (A, B) and (A', B') belong to different organisms. A correlation threshold of 0.6 results in an accuracy of 93 % and 82 % on orthologous and paralogous conservation in *S. cerevisiae*.

Research on temporal gene expression analysis is very active. Work in [244] proposes a clustering algorithm capable of handling unevenly sampled temporal gene expression data. A novel dissimilarity measure is proposed in [72] to assist graph-based clustering methods on temporal gene expression data. Smoothing spline derivatives are combined with hierarchical and partitioning clustering algorithms in [74] to capture the effects of fasting on the mouse liver. Dynamic clustering is shown in [215] to statistically estimate the optimal number of clusters and distinguish significant clusters from noise. A novel sub-space clustering algorithm is proposed in [341]. A detailed review of the state of analysis methods for temporal gene expression data is presented in [20].

# 8 Data Integration Methods for Protein Function Prediction

Integrative methods exploit and integrate heterogeneous data to improve the accuracy of function prediction. This category of methods operates under the umbrella of machine learning and can be organized in mainly four categories:

- **Vector-space integration methods**: these methods combine features extracted from different sources of biological data into one typically long feature vector. The transformation then allows investigating function prediction under the umbrella of machine learning.
- **Classifier integration methods**: these methods do not combine features into one long vector but instead train separate classifiers on separate feature vectors extracted from the different biological data. The results of the classifiers are then combined via the ensemble approach.
- **Kernel integration methods**: these methods employ a special similarity matrix known as the kernel matrix. A kernel matrix records the pairwise similarities between the proteins under investigation. Data sources can be kept separate, with a kernel matrix for each data source. The kernel matrices can also be combined via basic algebraic operations. Standard supervised classifiers, such as SMV and kNN, can be then used.
- **Network integration methods**: these methods rely on encoding pairwise similarities as edges of a graph. Different graphs can be constructed for the different data sources under consideration. The graphs can be then unified, and function prediction can proceed via generative or discriminative machine learning models.

## *8.1 Vector-Space Integration Methods*

Methods already described in this review that combine sequence, physico-chemical, and secondary structure information about a protein into one long feature vector fall in this category. Other methods that combined genomic context, phylogenetic profiles, and phylogenetic trees also belong to this category. Here we describe some recent methods that combine additional data sources to improve the accuracy of function prediction. However, these methods integrate data from essentially the same source. The first employ data that are extracted from the amino-acid sequence, whereas the latter employ data that are extracted from the evolutionary history of a protein. The ProtFun method proposed in [153] and then applied to predicting GO annotations in [154], described as a feature-based method for inferring function from sequence, is indeed a vector-space integration method. We recall, as described in Sect. 2.3, that ProtFun integrates sequence data with post-translational modifications, such as N- and O-glycosalization, phosphorylation, cleavage of N-terminal signal peptides, and other modifications and sorting events that a protein is subjected to before performing its function.

Work in [223] takes a unique approach to function prediction by focusing on proteins with intrinsically disordered regions (IDRs). Some studies estimate that between 30 and 60 % of eukaryotic proteins contain long stretches of IDRs, and work in [223] investigates the extent to which function can be inferred from information hidden in these regions. Specifically, pattern analysis of the distribution of IDRs in human protein sequences shows that the functions of intrinsically disordered proteins are length- and position-dependent. A total of 122 features are extracted from a protein in [223], and the features cover 14 different sources of biological information about a protein. The latter range from sequence-based features, such as sequence length, molecular weight, average hydrophobicity, charge, and more, to transmembrane-based features, such as number of transmembrane residues, percentage of N-terminal and C-terminal residues, and more, to secondary-structure features, Pest region features, phosphorylation features, O- and N- glycosylation features, and peptide features, and disorder-related features. The latter can be easily extracted from sequence via tools, such as disEMBL [149]. Correlations between the 122 features are investigated in [223], and multidimensional scaling (MDS) is applied to see how organization in a three-dimensional embedded space. Visualization of the three-dimensional embedded space obtained by MDS shows not only correlations between features extracted from the same category/source of biological information, but also correlations across features of different categories.

The attributes are valued and recorded in one long feature vector for each protein, and an SVM is trained on 26 GO categories. Improvements in accuracy are observed over a version of the classifier without the disorder features and ProtFun [153] and the method with no disorder. The individual contribution of each feature is also estimated via loss of classification accuracy upon feature removal. Significant improvements are observed for specific functional categories, such as kinases, phosphorylation, growth factors, and helicases.

Work [367] proposes the CombFunc method, which incorporates the ConFunc method proposed by the same authors (ConFunc is a homology-based method described in Sect. 2.1) and other methods that use sequence, gene expression, and protein–protein interaction data. Three categories of features are employed, sequence-based features, protein–protein interactions, and gene co-expression. Sequence-based features include those used by ConFunc, the E-values of the top annotated BLAST and PSI-BLASt hits, the sequence identity between the query and the top hits, and the sequence coverage of the query by the top hits. The i-score proposed in the GOtcha method [236] to take into account the annotations of multiple sequences returned by PSI-BLAST is included in the sequence-based features.

Other sequence-based sources of data include domain information about they query, as obtained with Interpro [146] and structures homologous to the query in the fold library of Phyre2 [167]. The domains and corresponding GO term annotations of the domains identified by InterPro are used to encode additional features. For each of the identified GO terms, the lowest E-value of a domain hit annotated with that term is recorded and added to the feature set. Pfam domain combinations [329] are also used to make predictions as in [103]. In this case, only one feature is added to the feature set, 1 if predicted by the method and 0 otherwise. Features from the Phyre2 fold library employ GO terms present in the top annotated hit and the probability score from the HHsearch [323] between the query and hit and the sequence coverage of the query by the hit.

For the GO terms identified by the interactome analysis, the features added to the feature set are the fraction of direct and indirect neighbor annotated with each term. For GO terms identified from the gene expression data, the features added to the feature set are the fraction of co-expressed genes annotation with the particular term, as well as the minimum, average, and minimum mutual rank and correlation coefficients of the co-expressed genes. In this manner, each protein sequence is transformed into a 30-dimensional feature vector.

Three different SVMs are employed in CombFunc for the different levels of the GO hierarchy under the molecular function and biological process categories. One SVM considers only terms one level below the root (for instance, catalytic activity or binding for the molecular function category). Another SVM considers the terms in the next two levels, and the third SVM considers the rest of the more specific terms. The reason for training three separate classifiers is due to the insight that potentially different subsets of features may be correlated with different levels in the GO hierarchy.

CombFunc is evaluated on predictions of GO molecular function terms on a set of 6686 proteins. UniProt-GOA annotations are extracted for the proteins, but only 5000 of them are used for training, with the rest used for testing. On the testing data set, CombFunc obtains a precision of 0.71 and recall of 0.64. Performance on prediction of GO biological process terms is slightly lower, with a precision of 0.74 and recall of 0.41.

## 8.2 *Classifier Integration Methods*

Integrating data by combining it into essentially a common representation often results in information loss [195, 397]. For this reason, classifier and kernel integration methods are pursued as better alternatives over vector space integration methods.

Integration of different classifiers has been investigated for sequence-based function prediction. We recall that the GoPred method proposed in [304] and described in Sect. 2.4 combines different classifiers and then evaluates the performance of different combination strategies, such as majority voting, mean, weighted mean, and addition.

Work in [292] demonstrates that competitive or superior performance can be obtained on prediction of top-level classes in the FunCat taxonomy [300] by using an ensemble of classifiers for data integration than by vector space integration and kernel fusion-based methods. The data sources considered in [292] are protein sequence, gene expression data, domain information, and protein–protein interactions. In [292], binary SVMs are trained on each data source, and three combination strategies are evaluated, weighted majority voting, naive Bayes, and decision templates [190]. It is worth noting that the naive Bayes and logistic regression combination strategy for integrating the outputs of several SVMs trained with different data sources and kernels has already been proposed in [117] and [261], respectively, to produce probabilistic outputs corresponding to GO terms.

Work in [117] is part of the MouseFunc function prediction project and integrates new data sources not previously considered such as disease, phenotype, and phylogenetic profiles in training of three different SVMs. Three combination strategies are evaluated, bootstrap aggregation, hierarchical Bayesian, and naive Bayes combination. One of the results in [117] is that the naive Bayes combination of the per-dataset SVMs outperforms a single SVM classifier for several GO terms. A comparison of the different combination strategies shows that the naive Bayes performs best, followed by the hierarchical Bayesian over the bootstrap aggregation.

Work in [261] proposes "reconciliation" to address the drawback of making predictions for GO terms independently; the latter often results in assigning to a query protein a set of GO terms that are inconsistent with one another; that is, that do not obey the GO hierarchy. In [261], the different, independent predictions are calibrated and combined to obtain a set of probabilistic predictions consistent with the GO topology. A total of 11 distinct reconciliation techniques are considered to combine predictions for each term obtained from different SVM classifiers with different kernels. The techniques are three heuristic ones, four variants of a Bayesian network, an extension of logistic regression to the structured case, and three novel projection techniques, such as isotonic regression and two variants of a Kullback–Leibler projection technique. Isotonic regression is shown to perform best in being able to use the constraints from the GO topology.

Work in [386] addresses the multi-label setting in GO annotation predictions. A transductive multi-label classifier (TMC) and a transductive multilabel ensemble classifier (TMEC) are proposed to predict multiple GO terms for unannotated proteins. The TMC is based on a bidirected birelational graph with edges connecting protein pairs, function pairs, and protein–function pairs. An interfunction similarity measure is used to encode function–function edges. Protein–protein similarity is specific to a data source. Directionality is added to the graph to avoid issues of annotation change and function label override. The TMC uses network propagation via a nonsymmetric propagation matrix on the resulting directed bidirectional graph by optimizing local and global consistency functions [393]. Three TMCs are trained simultaneously, each one considering a different data source. Sequence, protein–protein interaction, and gene expression data are considered. The TMEC then combines the output of the three classifiers via a weighted majority vote scheme, where a classifier's influence on determining a particular GO term for a protein is proportional to its confidence on that prediction. Evaluation is carried out on predicting biological process GO categories on benchmark yeast, human, and fly protein data sets proposed originally in [248]. Comparisons on Ranking Loss, Coverage, and AUC with related multi-label methods, such as PfunBG [156], GRF [389], SW [248], MKL-Sum [337], and MKL-SA [44], show comparable or superior performance by the TMEC [386].

While work in [117, 261, 386] ignores the hierarchy of the taxonomy and then relies on ensemble techniques to reconcile conflicting predictions, work in [307, 352] either proposes classifiers that obey the GO hierarchy directly, or ensemble techniques that make final decisions based on the hierarchy of the taxonomy. For instance, hierarchical multi-label decision trees are combined via bagging in [307]. Hierarchical multi-label decision trees are intuitive in that they exploit the ability of the decision tree model to obey intrinsic hierarchy in the target taxonomy. Essentially, the query gene can be compared via sequence similarity to all genes annotated with a specific GO term, and the tree proceeds down the GO hierarchy. Bagging is shown in [307] to best combine decision tree classifiers over random forest and boosting. In [352], the topology of the GO hierarchy is not considered in the classifiers, but it is directly integrated in a novel ensemble technique.

The key observation employed in [352] is that an annotation for a class/node in the hierarchy automatically transfers to the ancestors. This is also known as the "true path" rule (TPR), which governs hierarchical taxonomies, such as GO and FunCat. The TPR ensemble technique proposed in [352] is a hierarchical ensemble algorithm that puts together predictions made each node by local base classifiers to realize an ensemble that obeys TPR. As in ensemble methods, the classifiers are trained independently, and they make predictions for their corresponding nodes. The algorithm then combines these predictions via an information propagation mechanism that can be characterized as a two-way asymmetric information flow. The information traverses the graph-structured ensemble. While positive predictions for a node influence in a recursive way its ancestors, negative predictions influence the offsprings. This is related to work in [157], where negative information propagates from a node to its offspring. In [352], in addition, positive information propagates from a node to its ancestors.

Seven biomolecular data sources are integrated in [352], such as sequence, domain, phylogenetic, protein–protein interaction, and gene expression data. SVMs and logistic regression are used as base classifiers. Evaluation of the hierarchical ensemble technique is carried out on *S. cerevisiae*. Best performance is obtained by a weighted version of the TPR algorithm, followed by the TPR, a related hierarchical ensemble technique where information flows only from a node to its offspring, and a non-hierarchical ensemble technique that ignores the hierarchy in the ontology [157]. In another related work [188], a discrete approach is proposed that infers the most probable TPR-consistent assignments. The GO DAG is modeled as a Bayesian network that infers the most probable assignments via global optimization. The differential evolution algorithm is adapted for this purpose.

## 8.3 Kernel Integration Methods

Kernel-based methods [310] encode the pairwise similarity between proteins in a similarity matrix, also called a kernel matrix. This is a positive definite and symmetric matrix $K(x, y)$, where elements record the similarity between proteins $x, y$. Different kernel matrices can be defined to encode protein-pair similarities according to different data sources. For instance, when employing protein sequences, there are various options. BLAST E-values can be used to fill the entries of the kernel matrix. Alternatively, the spectrum/string kernel [203], motif kernel [26], and Pfam kernel can be used [115]. In the string kernel, a protein sequence $x$ is represented by a vector $\phi(x)$ of frequencies of all k-mers, and then the inner product of two vectors $\phi(x), \phi(y)$ corresponding to two proteins $x, y$ is taken to obtain $K(x, y)$.

For structured data sources, such as protein–protein interaction data, the random walk kernel [315] and diffusion kernels [183] are employed. Diffusion kernels encode similarities between the nodes of a network and are variants of $K = e^{-\beta L}$, where $\beta > 0$ is the parameter that quantifies the degree of diffusion, and $L$ is the network Laplacian. A comprehensive evaluation in [227] shows that diffusion kernels give superior performance on function prediction, prioritizing genes related to a phenotype, and identifying false positives and false negatives from RNAi experiments. Work in [227] concludes that diffusion kernels should be the kernel of choice to measure network similarity over other similarity measures, such as direct neighbors and short path distance.

Kernel matrices corresponding to different data sources can be combined by carrying out basic algebraic operations such as addition, multiplication, or exponentiation. When addition is employed, the individual kernels can be weighted by fixed coefficients [196, 275], or by coefficients learned via semi-definite programming [195]. The latter can be computationally demanding, particularly on large and multiple data sets. In response, more efficient combination schemes have been proposed recently based on semi-infinite programming [326]. Whatever the strategy employed to combine individual kernels, the resulting kernel matrix can then be fed to popular classifiers, such as SVM or kNN.

## 8.4 Network Integration Methods

Instead of encoding protein similarities on different data sources via kernels, network methods encode similarities as edges connecting protein pairs in a network. For instance, sequence similarities between proteins can be estimated via BLAST E-value or other means (analogous to spectrum or string kernels) and encoded in a network connecting proteins of similar sequences. Co-expressed genes can also be connected by edges in a gene expression network. Similar ideas can be employed to encode similarities based on phylogenetic profile or phylogenetic tree in networks. These network-based representations of biological data can be as powerful as PPI or gene–gene interaction networks and can be simultaneously exploited to predict function.

Work in [152] focuses on a drawback of many network-based methods that ignore dependencies between interacting pairs and predict them independently of one another. In [152], relational Markov networks are employed to build a unified probabilistic model that allows predicting unobserved interactions concurrently. The model integrates various attributes and models measurement noise. In essence, PPI networks, interaction assay readouts, and other protein attributes are represented as random variables. Variable dependencies are modeled by joint distributions. Since a naive representation of the joint distribution requires a large number of parameters, relational Markov models are used instead. Improved performance is reported over related methods for predicting sub-cellular localization and interaction partners of the mediator complex.

The MAGIC method proposed in [348] integrates yeast PPI data from the General Repository of Interaction Datasets (GRID) [37], pairs of genes that have experimentally determined bindings sites for the same transcription factor, as extracted from The (SCPD) Promoter Database of *S. cerevisiae* [396], and gene expression data. Three separate gene–gene relationship matrices are constructed from each data source, with an entry encoding whether a particular gene pair has a functional relationship or not; 0 indicates lack of relationship, and a numeric value indicates confidence of putative relationship. Different algorithms are used on each data sources to obtain these relationships. For instance, gene expression data are subjected to different clustering algorithms, such as K-means, SOM, and hierarchical clustering, and each of these algorithms are nodes in a Bayesian network constructed for each gene–gene pair. SCPD data provide gene–gene pairs directly. These matrices are provided as input to Bayesian networks, one for each gene-gene pair. A network combine evidence from the different clusters to generate a posterior belief for whether its corresponding gene-gene pair has a functional relationship. MAGIC is reported to improve accuracy of the functional groupings compared with gene expression analysis alone [348].

Work in [255] integrates functional linkage graphs constructed from PPI and gene expression data. Functional linkage graphs are constructed to encode via edges evidence for functional similarity. These graphs are used in concert with categorical data, such as protein motif data, mutant phenotype data, and protein localization data, to make a final prediction. The categorical features of a query protein are used as random variables/nodes in a Bayesian network, together with annotated neighbors of a query protein in the functional linkage graphs. The posterior probability of the query annotated with a particular GO term is then calculated. This approach is employed to predict functions for yeast proteins. A cross validation setting shows that this integrated approach increases recall by 18 %, compared to using PPI data alone at the 50 % precision. The integrated predictor also outperforms each individual predictor. However, improvements in performance are not uniform and depend on the particular functional category predicted.

In [126], information present in metabolic networks and gene co-expression data is indirectly combined. A graph distance function is first defined on metabolic networks, and the function is combined with a correlation-based distance function for gene expression measurements. The resulting distance function is used to jointly cluster genes and network vertices via hierarchical clustering. The resulting clusters are shown to be interpretable in terms of biochemical network and gene expression data. A related, clustering-based method is proposed in [347]. Co-expression and PPI networks are separately evaluated by computing the probability of groups of genes to be correlated in the networks. The groups of correlated genes are found via super-paramagnetic hierarchical clustering.

Different data sources, such as PPI data, gene expression, phenotypic sensitivity, and transcription factor binding are integrated in [336] in a bipartite graph, with genes on one side of the graph, and their properties on the other. Biclustering algorithms based on combinatorial principles are then used to detect statistically significant subgraphs that correspond to functionally related genes. In [318], a clustering method based on learning a probabilistic model, referred to as a hidden modular random field. The relation between hidden variables represents a given gene network. The learning algorithm minimizes an energy function that considers network modularity. The method is shown to be highly sensitive for gene clustering and annotation of gene function.

Recent work in [231] proposes semi-supervised parametric neural models to combine different bio-molecular networks and predict protein functions. The models take into account the unbalance between annotated and unannotated proteins in the construction of the integrated network and in the final prediction of annotations for each functional class. Evaluation on full-genome and ontology-wide experiments on three eukaryotic organisms show that the UNIPred method proposed in [231] compares favorably with state-of-the-art methods, such as SW [248] and MS-kNN [194].

## 8.5 State of Data Integration Methods for Function Prediction

An increasing volume and diversity of biological data presents both opportunities and challenges for data integration methods. One such challenge that is relatively under-explored in data integration methods is how to explore topologies of multiple different networks. The majority of data integration methods for function prediction exploit PPI networks but largely ignore other important network data, such as gene-gene interaction networks and metabolomic interaction networks. Some work exists in this direction via methods that use random walk or diffusion processes to infer knowledge from all networks concurrently, though in the context of predicting disease interactions, disease gene associations, drug target interactions, and drug disease associations [61, 87, 120, 142, 354]. In this context, many issues regarding data integration that are currently under-pursued in function prediction methods are being addressed, such as noise, bias in data collection, concordant and discordant data sets, and scalability. A comprehensive review of data integration methods for disease- and drug-driven problems in molecular biology can be found in [110].

Data integration is identified as a key direction to improve the performance of function prediction methods. A detailed study in [357] pitches two different network-based function prediction approaches against each other, ensemble techniques that combine classifiers versus state-of-the-art classifiers that integrate various datasets. The study reports that a modest benefit of 17 % in the area under the ROC (AUROC) is obtained from ensemble techniques over the baseline classifiers. In contrast, data aggregation results in an 88 % improvement in mean AUROC. The study concludes that substantial evidence supports the view that additional algorithm development has little to offer for gene function prediction as opposed to data aggregation. While a saturation point may have been reached for off-the-shelf machine learning methods, there may be further ground to explore for novel methods capable of efficiently and effectively integrating noisy and non-uniformly dense data.

## 9 Text Mining-Based Methods for Function Prediction

Text mining is a promising machine learning technology for the analysis of biomedical literature in the problem of protein function classification for the fundamental reason of the abundance of literature that links proteins with each other. This offers the hope of increasing the size of labeled data available for training and evaluation.

Following the idea of using query proteins to find their homologous proteins, one of the first applications of text mining was to utilize this notion. In Renner and Aszodi [294] the authors describe a procedure for the prediction of functions of novel products. The last steps of the procedure are based on text mining. First, a protein whose function remains unknown is used as input across multiple databases (e.g., SWISS-PROT, PIR, PROSITE). The result of these searches are annotation documents that can be subjected to text mining procedures. In particular, the documents are compared by checking the terms that occur in them and using those terms to produce clusters. The principle is that if two documents contain terms that belong to the same cluster, then the documents probably describe the same phenomenon. In order to cluster the terms, the authors analyze their co-occurrence in documents, proceeding to build clusters starting at a term and adding terms that often co-occur with it, recursively. The probability of a term belonging to a cluster is then computed as the ratio of the sum of the number of times the term co-occurred with all the other terms in the same cluster, divided by the total number of occurrences of the term. For a given document, the "match" score of a cluster is defined as the maximum probability of belonging to that cluster for all the terms found in the document. Comparing documents is now a matter of computing the normalized sum of differences of their match scores across all the clusters. With such a distance measure, the documents can be clustered. (While this is a valid way to compute distances, it is intriguing to think what the results would have been if a more modern way of clustering documents, such as Topic Modeling [31] had been utilized.) It was observed that for most proteins, all the documents clustered into a single clustering, indicating coherence.

Simple text classification approaches have been used for the prediction of functionality, using the documents with which the proteins are associated. Raychaudhari [291] uses the maximum entropy, naive Bayes, and nearest neighbor classifiers using training abstracts from PubMed. The features for the classifiers are bigrams of two co-occurring words subjected to a $\tilde{\chi}^2$ test of correlation with the class. Results show that the most accurate of classifiers, namely maximum entropy, is capable of finding the proper class 72.8 % of the time.

A kNN classifier is applied by Keck and Wetter [166] using BLAST searches and a variety of databases (GenProtEC, MIPS). Results show a very low recall of 0.4, which can be attributed to a very weak distance metric.

A more elaborate approach of integrating text mining into this problem can be found in the work of Eskin and Agichtein [97]. In that paper, the authors use the SVM classifier, combining a variety of text and sequence kernels. First, a seed set set is created. The set consists of labeled proteins as positive examples and other proteins with different labels as negative. This set is small due to the availability of known cases. A text classifier is then trained over the annotations of the sequences in the set (found in a variety of databases). The feature space employed is the bag-of-words representation of the annotations. Each word in the annotation receives a 1 for the word's dimension; words not present receive a 0. This results in a feature space that is high-dimensional but very sparse. The text kernel is the dot product of the representation vectors. This classifier, after being trained with the original set, is used to predict the function of unknown proteins, using textual information available in databases such as SWISS-PROT. The result of applying the trained classifier is an enriched, larger set of labeled proteins.

Next, a joint classifier that uses sequence information and text annotations is trained with the new set. To this end, a kernel for both sequences and text is defined. The text part of the kernel is, as described above, based on the bag-of-words representation. The sequence part of the kernel represents sequences as substrings of length $k$, or $k$-mers, obtained by segmenting the sequence with a moving window. The feature space contains a dimension for each possible $k$-mer, with a 1 for a $k$-mer that appears in the sequence and 0s for those that do not. The kernel is then the inner product of two such representations. Since matching $k$-mers in practice are very rare, the authors utilize the sparse kernel representation that allows for approximate matching. In it, the kernel is defined as a parameter $\alpha$ raised to the Hamming distance between the sequences being compared. The combination of the text and sequence kernels is achieved by kernel composition. The result is a kernel that adds the two components and a degree two polynomial kernel over the sum of the two original kernels. The rationale for the polynomial term is to include features for all pairs of sequences and words. With such combined kernel, the classifier effectively learns from both sequence and text annotations and the interactions between them.

An additional benefit of this approach is obtained by projecting the classifier onto the original sequences to learn which regions of the protein have a high positive weight with respect to the class and as such, are likely candidates for relevant functional regions. Experiments conducted by leaving 20 % of the set as a test set and using the remaining as the training set indicate that the results obtained by the joint classifier are superior across a variety of functional classes to those obtained by applying any of the two (text or sequence) classifiers independently. The task of identifying relevant regions is shown to perform well when comparing the results to a searchable database, such as NLS [253].

## 10 Discussion and Prospects for Future Research

A pervasive theme of this survey of function prediction methods has been that while significant advances are occurring in each of the five categories of methods for function prediction, significant performance gains are obtained by methods that ingrate data from diverse sources. Two recent studies point to the fact that data integration is expected indeed to be the most promising avenue for improved function prediction performance [231, 357]. Readers with interests beyond computational protein function prediction may find useful information in this survey on how data are integrated in the machine learning methods summarized here. Interesting trends can be observed regarding how different types of features are combined, and how such trends have evolved over time as driven by the need to balance between accuracy and computational efficiency.

Considering all the rapid advancements in novel methodologies for function prediction, it is not easy to keep track of the current state of automated function prediction. Nor is it easy to objectively conclude whether certain methods are better than others from summaries of published works, where performance is evaluated in a controlled setting and on some specific dataset of interest. The CAFA experiment provides just the avenue for objective comparisons. A large-scale evaluation of 54 automated function prediction methods in CAFA is reported in [287]. Two main findings are reported: first, that current methods significantly outperform first-generation ones on all types of query proteins; second, that, although current methods perform well enough to guide experiments, there is significant room for improvement.

Specifically, the top five labs/methods in 2013 CAFA on all targets, one- and multiple-domain proteins are Jones-UCL [70], GOstruct [324], Argot2 [98], ConFunc [368], and PANNZER [186]. Their comparative performance is summarized in Fig. 5. The Jones-UCL team consistently outperformed other methods due to a massive integration of evolutionary analyses and multiple data sources, combining in a probabilistic manner GO term predictions from PSI-BLAST, SWISS-PROT text mining, amino-acid trigram mining, FFPred sequence features [225], orthologous groups, PSSM profile–profile comparisons, and FunctionSpace [224]. A network propagation algorithm based on the GO graph structure combines the various predictions. ConFunc [368], we recall, is another data integration method. GOstruct models the structure of the GO hierarchy in the framework of kernel methods for

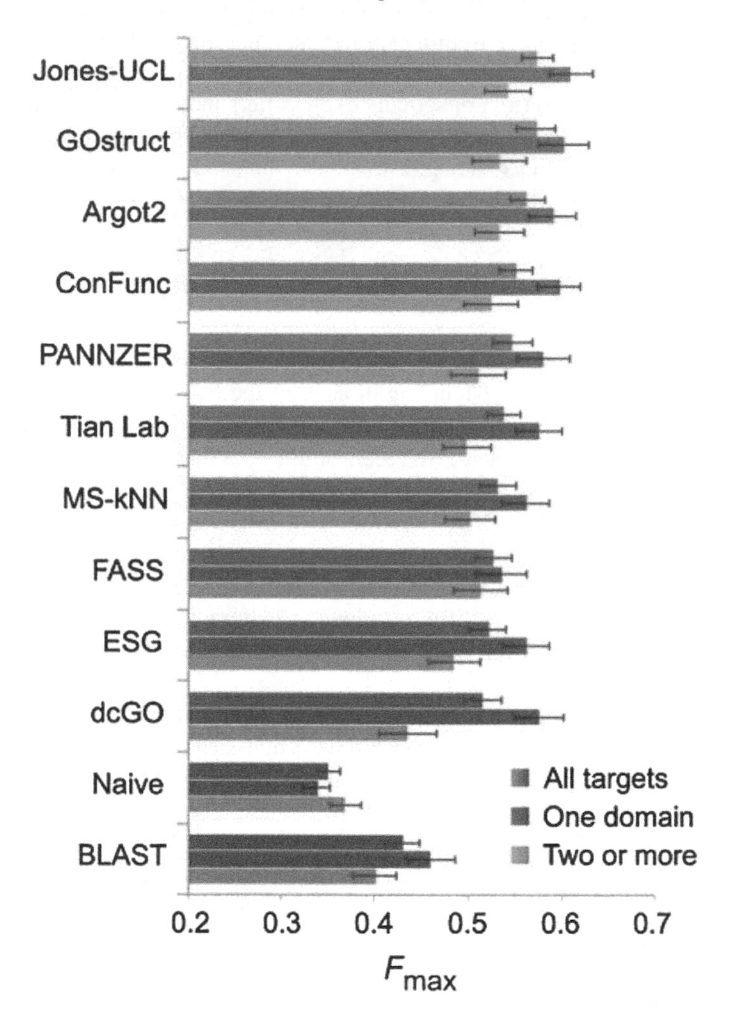


**Fig. 5** The performance of the top ten methods is shown here. The methods achieved higher accuracy on single-domain proteins. Confidence intervals (95 %) were determined using bootstrapping with 10,000 iterations on the target sequences

structured-output spaces. The structured output SVM in [324] does not only well in CAFA 2013, but it also confers high performance to approaches that investigate text-mined features for automated function prediction with GOstruct [105].

In contrast, The Argot2 web server debuted in [98] is mainly a sequence-based method, that employs BLAST and HMMER searches of a query sequence against UnitProtKB and PFAM databases. GO terms are weighted based on E-values returned by the searches, and the weights are processed according to semantic similarity relations between the terms. A recent version of the web server, Argot2.5, is enriched with more features, including a new semantic similarity measure, and shown to improve performance even further over Argot2 [197]. Argot2.5 is also shown to outperform PANNZER [186], which is another sequence-based method. PANNZER relies on a weighted k-nearest neighbor approach with statistical testing, after partitioning sequence-similarity results into clusters according to description similarity. A sophisticated regression model evaluates the support for the candidate cluster.

The list of the top performers is diverse in terms of methodologies. Perhaps not surprisingly, and in line with other studies drawn similar observations [124], sophisticated sequence-based methods can perform comparably to state-of-the-art data integration methods. A pre-print of a recent, 2016 report can also be found [158]. In the report, 126 methods from 56 research groups are compared against one another in a set of 3681 proteins from 18 species. One of the findings in the report is that top-performing methods in CAFA2 outperform top-performing ones in CAFA1. This finding suggests that indeed computational function prediction is improving, possibly due to both an increase in experimental annotations (via high-throughout wet-laboratory techniques) and improvements in methodology.

Carefully drawn case studies in [287] show that there is room for improvement. One challenge in function prediction that is not often mentioned is related to the existence of promiscuous proteins that are multi-functional; indeed, more than 30 % of the proteins in SWISS-PROT have more than one leaf in the Molecular Function ontology; more than 60 % have more than one leaf in the Biological Process ontology. In addition, while data integration is often touted as the most promising direction, the presence of noisy and erroneous experimental data may be an additional source of error that needs to be addressed for robust performance. Finally, machine learning is shown to generally improve performance, and it is expected that there is more performance to be gained by approaches based on principles of statistical learning and inference.

**Acknowledgements** Funding for this work is provided in part by NSF-IIS1144106.

## References


1. Abascal, F., Valencia, A.: Automatic annotation of protein function based on family identification. Proteins **53**(3), 683–692 (2003)
2. Agrawal, R., Imielinski, T., Swami, A.: Mining association rules between sets of items in large databases. In: SIGMOD Intl Conf on Management of Data, pp. 207–216. ACM (1993)
3. Albert, R.: Network inference, analysis, and modeling in systems biology. Plant Cell **19**(11), 3327–3338 (2007)
4. Alberts, B., Johnson, A., Lewis, J., et al.: From RNA to protein. In: Molecular Biology of the Cell, 4 edn. New York: Garland Science (2002)
5. Alberts, B., Johnson, A., Lewis, J., et al.: Studying gene expression and function. In: Molecular Biology of the Cell, 4 edn. New York: Garland Science (2002)
6. Alexandrov, N.N.: SARFing the PDB. Protein Eng **9**(9), 727–732 (1996)
7. Altman, D.G.: Practical Statistics for Medical Research. Chapman and Hall (1997)

8. Altschul, S.F., Madden, T.L., Schaeffer, A.A., Zhang, J., Zhang, Z., Miller, W., Lipman, D.J.: Gapped BLAST and PSI-BLAST: a new generation of protein database search programs. Nucl. Acids Res. **25**, 3389–3402 (1997)
9. Andreeva, A., Howorth, D., Brenner, S.E., Hubbard, T.J., Chothia, C., Murzin, A.G.: Scop database in 2004: refinements integrate structure and sequence family data. Nucleic Acids Res **32**(Database issue), D226–D229 (2004)
10. Apeltsin, L., Morris, J.H., Babbitt, P.C., Ferrin, T.E.: Improving the quality of protein similarity network clustering algorithms using the network edge weight distribution. Bioinformatics **27**(3), 326–333 (2011)
11. Arnau, V., Mars, S., Marin, I.: Iterative cluster analysis of protein interaction data. Bioinformatics **21**(3), 364–378 (2005)
12. Ashburner, M., Ball, C., Blake, K., et al.: The gene ontology consortium. Nature Genetics **25**(1), 25–29 (2000)
13. Aung, Z., Tan, K.L.: Rapid 3D protein structure database searching using information retrieval techniques. Bioinformatics **20**(7), 1045–1052 (2004)
14. Badea, L.: Functional discrimination of gene expression patterns in terms of the gene ontology. In: Pacific Symp Biocomput (PSB), pp. 565–576 (2003)
15. Bader, G.D., Betel, D., Hogue, W.V.: BIND: the biomolecular interaction network database. Nucleic Acids Res **31**(1), 248–250 (2003)
16. Bailey, T.L., Elkan, C.: Fitting a mixture model by expectation maximization to discover motifs in biopolymers. In: Intl Conf Intell Sys Mol Biol (RECOMB), pp. 28–36 (1998)
17. Bailey, T.L., Gribskov, M.: Combining evidence using p-values: application to sequence homology searches. Bioinformatics **14**(1), 48–54 (1998)
18. Bairoch, A., BUcher, P., Hoffmann, K.: The PROSITE database, its status in 1997. Nucl. Acids Res. **25**(1), 217–221 (1997)
19. Bar-Joseph, Z.: Analyzing time series gene expression data. Bioinformatics **20**(16), 2493–2503 (2004)
20. Bar-Joseph, Z., Gitter, A., Simon, I.: Studying and modelling dynamic biological processes using time-series gene expression data. Nat Rev Genet **13**(8), 552–564 (2012)
21. Barabasi, A.L., Oltvai, Z.N.: Network biology: understanding the cell's functional organization. Nature Rev Genet **5**(2), 101–113 (2004)
22. Barrett, et al.: NCBI GEO: archive for functional genomics data sets–update. Nucleic Acids Res **41**(Database issue), D991–D995 (2013)
23. Bder, G., Hogue, C.: An automated method for finding molecular complexes in large protein interaction networks. BMC Bioinf **4**(1), 2 (2003)
24. Bellaachia, A., Portnov, D., Chen, Y., Elkahloun, A.G.: E-CAST: a data mining algorithm for gene expression data. In: Workshop on Data Mining in Bioinformatics (BIOKDD), pp. 49–54 (2002)
25. Ben-Dor, A., Shamir, R., Yakhini, Z.: Clustering gene expression patterns. J Comput Biol **6**(3–4), 281–297 (1999)
26. Ben-Hur, A., Brutlag, D.: Remote homology detection: a motif based approach. Bioinformatics **19**(Suppl 1), i26–i33 (2003)
27. Ben-Hur, A., Brutlag, D.: Sequence motifs: Highly predictive features of protein function. In: I. Guyon, S. Gunn, M. Nikravesh, L. Zadeh (eds.) Feature extraction and foundations and applications. Springer Verlag (2005)
28. Berman, H.M., Westbrook, J., Feng, Z., Gilliland, G., Bhat, T.N., Weissig, H., Shindyalov, I.N., , Bourne, P.E.: The protein data bank. Nucl. Acids Res. **28**(1), 235–242 (2000)
29. Bilu, Y., Linial, M.P.: Functional consequences in metabolic pathways from phylogenetic profiles. In: Intl Workshop on Algorithms in Bioinformatics (WABI), pp. 263–276 (2002)
30. Blatt, M., Wiseman, S., Domany, E.: Superparamagnetic clustering of data. FEBS Lett **76**, 3251–3254 (1996)
31. Blei, D.: Probabilistic topic models. Communications of the ACM **55**(4), 77–84 (2012)
32. Blei, D.M.: Latent Dirichlet Allocation. J. Mach. Learn. Res. **3**, 993–1022 (2003)

33. Blekas, K., Fotiadis, D.I., Likas, A.: Motif-based protein sequence classification using neural networks. J Comput Biol **12**(1), 64–82 (2005)
34. Boeckmann, B., Bairoch, A., Apweiler, R., Blatter, M.C., Estreicher, A., Gasteiger, E., Martin, M.J., Michoud, K., O'Donovan, C., Phan, I., Pilbout, S., Schneider, M.: The SWISS-PROT protein knowledgebase and its supplement TrEMBL in 2003. Nucleic Acids Res **31**(1), 365–370 (2003)
35. Bork, P., Koonin, E.V.: Protein sequence motifs. Curr Opin Struct Biol **6**(3), 366–376 (1996)
36. Braberg, H., Webb, B.M., Tjioe, E., Pieper, U., Sali, A., Madhusudhan, M.S.: SALIGN: a web server for alignment of multiple protein sequences and structures. Bioinformatics **15**(28), 2071–2073 (2012)
37. Breitkreutz, B., Stark, C., Tyers, M.: The GRID: The general repository for interaction datasets. Genome Biol **4**(3), R3 (2003)
38. Brenner, S.E.: Errors in genome annotation. Trends Genet **15**(4), 132–133 (1999)
39. Brenner, S.E., Levitt, M.: Expectations from structural genomics. Protein Sci. **9**(1), 197–200 (2000)
40. Brown, K.R., Jurisica, I.: Online predicted human interaction database. Bioinformatics **21**(9), 2076–2082 (2005)
41. Brown, M.P., et al.: Knowledge based analysis of microarray gene expression data by using support vector machines. Proc Natl Acad Sci USA **97**(1), 262–267 (2000)
42. Brun, C., Chevenet, F., Martin, D., Wojcik, J., Guénoche, A., Jacq, B.: Functional classification of proteins for the prediction of cellular function from a protein-protein interaction network. Genome Biol **5**(1), R6 (2003)
43. Bryan, K., Cunningham, P., Bolshakova, N.: Biclustering of expression data using simulated annealing. In: IEEE Symp Computer-based Medical Systems (CBMS), pp. 383–388 (2005)
44. Bucak, S., Jin, R., Jain, A.: Multi-label multiple kernel learning by stochastic approximation: Application to visual object recognition. In: Advances Neural Inform Processing Systems (NIPS), pp. 1145–1154 (2010)
45. Budowski-Tal, I., , Nov, Y., Kolodny, R.: Fragbag, an accurate representation of protein structure, retrieves structural neighbors from the entire PDB quickly and accurately. Proc. Natl. Acad. Sci. USA **107**, 3481–3486 (2010)
46. Butte, A.J., Bao, L., Reis, B.Y., Watkins, T.W., Kohane, I.S.: Comparing the similarity of time-series gene expression using signal processing metrics. J Biomed Bioinf **34**(6), 396–405 (2001)
47. Cai, C.Z., Han, L.Y., Ji, Z.L., Chen, X., Chen, Y.Z.: SVM-Prot: Web-based support vector machine software for functional classification of a protein from its primary sequence. Nucleic Acids Res **31**(13) (2003)
48. Cai, Y.D., Doig, A.J.: Prediction of saccharomyces cerevisiae protein functional class from functional domain composition. Bioinformatics **20**(8), 1292–1300 (2004)
49. Califano, A.: SPLASH: structural pattern localization analysis by sequential histograms. Bioinformatics **16**(4), 341–357 (2000)
50. Cao, R., Cheng, J.: Integrated protein function prediction by mining function associations, sequences, and protein-protein and gene-gene interaction networks. Methods **93**, 84–99 (2016)
51. Carpentier, M., Brouillet, S., Pothier, J.: YAKUSA: a fast structural database scanning method. Proteins: Struct. Funct. Bioinf. **61**(1), 137–151 (2005)
52. Carugo, O.: Rapid methds for comparing protein structures and scanning structure databases. Current Bioinformatics **1**, 75–83 (2006)
53. Carugo, O., Pongor, S.: Protein fold similarity estimated by a probabilistic approach based on c(alpha)-c(alpha) distance comparison. J Mol Biol **315**(4), 887–898 (2002)
54. Chakrabarti, S., Venkatramanan, K., Sowdhamini, R.: SMoS: a database of structural motifs of protein superfamilies. Protein Eng **16**(11), 791–793 (2003)
55. Chatr-Aryamontri, A., et al.: The BioGRID interaction database: 2015 update. Nucleic Acids Res **43**(Database Issue), D470–D478 (2015)
56. Chen, C., Chung, W., Su, C.: Exploiting homogeneity in protein sequence clusters for construction of protein family hierarchies. Pattern Recognition **39**(12), 2356–2369 (2006)

57. Chen, L., Xuan, J., Riggins, R.B., Wang, Y., Clarke, R.: Identifying protein interaction subnetworks by a bagging markov random field-based method. Nucleic Acd Res **41**(2), e42 (2013)
58. Chen, Y.J., Kodell, R., Sistare, F., Thompson, K.L., Moris, S., Chen, J.J.: Studying and modelling dynamic biological processes using time-series gene expression data. J Biopharm Stat **13**(1), 57–74 (2003)
59. Chen, Y.J., Mamidipalli, S., Huan, T.: HAPPI: an online database of comprehensive human annotated and predicted protein interactions. BMC Genomics **10**(Suppl 1), S16 (2009)
60. Cheng, B.Y., Carbonell, J.G., Klein-Seetharaman, J.: Protein classification based on text document classification techniques. Proteins **58**(4), 955–970 (2005)
61. Cheng, F., et al.: Prediction of drug-target interactions and drug repositioning via network-based inference. PLoS Comput Biol **8**(5), e1002,503 (2012)
62. Cheng, Y., Church, G.M.: Biclustering of expression data. In: Intl Conf Intell Sys Mol Biol (RECOMB), pp. 93–103 (2000)
63. Chitale, M., Hawkins, T., Park, C., Kihara, D.: ESG: extended similarity group method for automated protein function prediction. Bioinformatics **25**(14), 1739–1745 (2009)
64. Cho, Y., Zhang, A.: Predicting protein function by frequent functional association pattern mining in protein interaction networks. IEEE Trans Info Technol Biomed **14**(1), 30–36 (2009)
65. Chua, H.N., Sung, W.K., Wong, L.: Exploiting indirect neighbours and topological weight to predict protein function from protein-protein interactions. Bioinformatics **22**(13), 1623–1630 (2006)
66. Clark, W.T., Radivojac, P.: Analysis of protein function and its prediction from amino acid sequence. Proteins: Struct Funct Bioinf **79**(7), 2086–2096 (2011)
67. Cong, L., Ran, F.A., Cox, D., Lin, S., Barretto, R., Habib, N., Hsu, P.D., Wu, X., Jiang, W., Marrafini, L.A., Zhang, F.: Multiplex genome engineering using CRISPR/Cas systems. Science **339**(6121), 819–823 (2013)
68. Consortium, T.U.: Ongoing and future developments at the universal protein resource. Nucleic Acids Res **39**(Database issue), D214–D219 (2011)
69. Cowley, M.J., Pinese, M., Kassahn, K.S., Waddell, N., Pearson, J.V., Grimmond, S.M., Biankin, A.V., Hautaniemi, S., Wu, J.: PINA v2.0: mining interactome modules. Nucleic Acids Res **40**(Database issue), D862–D865 (2012)
70. Cozzetto, D., Buchan, D.W.A., Jones, D.T.: Protein function prediction by massive integration of evolutionary analyses and multiple data sources. BMC Bioinf **14**(Suppl 1), S1 (2013)
71. Dandekar, T., Snel, B., Huynen, M., Bork, P.: Conservation of gene order: a fingerprint of proteins that physically interact. Trends Biochem Sci **23**(9), 324–328 (1998)
72. Das, R., Kalita, J., Bhattacharyya, D.K.: A new approach for clustering gene expression time series data. Intl J Bioinform Res Appl **5**(3), 310–328 (2009)
73. Date, S.V., Marcotte, E.M.: Protein function prediction using the Protein Link EXplorer (PLEX). Bioinformatics **21**(10), 2558–2559 (2005)
74. Déjean, S., Martin, P.G.P., Besse, P.: Clustering time-series gene expression data using smoothing spline derivatives. EURASIP J Bioinf Sys Biol **2007**(1), 70,561 (2007)
75. Deng, M., Sun, T., Chen, T.: Assessment of the reliability of protein-protein interactions and protein function prediction. In: Pacific Symp Biocomput (PSB), vol. 8, pp. 140–151 (2003)
76. Deng, M., Tu, Z., Sun, F., Chen, T.: Mapping gene ontology to proteins based on protein-protein interaction data. Bioinformatics **20**(6), 895–902 (2004)
77. Deng, M., Zhang, K., Mehta, S., Chen, T., Sun, F.: Prediction of protein function using protein-protein interaction data. J Comput Biol **10**(6), 947–960 (2003)
78. Deng, X., Ali, H.H.: A hidden markov model for gene function prediction from sequential expression data. In: IEEE Comput Sys Bioinf Conf (CSB), pp. 670–671 (2004)
79. Devos, D., Valencia, A.: Practical limits of function prediction. Proteins: Struct Funct Bioinf **41**(1), 98–107 (2000)
80. Doerks, T., Bairoch, A., Bork, P.: Protein annotation: detective work for function prediction. Trends Genet **14**(6), 248–250 (1998)
81. Duda, R.O., Hart, P.E., Stork, D.G.: Pattern Classification, 2 edn. Wiley-Interscience (2000)

82. Dwight, S.S., et al.: Saccharomyces genome database (SGD) provides secondary gene annotation using the gene ontology (GO). Nucleic Acids Res **30**(1), 69–72 (2002)
83. Eddy, S.R.: Profile hidden Markov models. Bioinformatics **14**(9), 755–763 (1998)
84. Edgar, R., Domrachev, M., Lash, A.E.: Gene expression omnibus: NCBI gene expression and hybridization array data repository. Nucleic Acids Res **30**(1), 207–210 (2003)
85. Eisen, J.A.: Phylogenomics: improving functional predictions for uncharacterized genes by evolutionary analysis. Genome Res **8**(3), 163–167 (1998)
86. Eisner, R., , Poulin, B., Szafron, D., Lu, P., Greiner, R.: Improving protein function prediction using the hierarchical structure of the gene ontology. In: IEEE Comput Intell Bioinf Comput Biol (CIBCB), pp. 1–8 (2005)
87. Emig, D., Ivliev, A., Pustovalova, O., Lancashire, L., Bureeva, S., Nikolsky, Y., Bessarabova, M.: Drug target prediction and repositioning using an integrated network-based approach. PLoS One **8**(4), e60,618 (2013)
88. Enault, F., Suhre, K., Abergel, C., Poirot, O., Claverie, J.: Annotation of bacterial genomes using improved phylogenomic profiles. Bioinformatics **19**(Suppl 1), i105–i107 (2003)
89. Enault, F., Suhre, K., Abergel, C., Poirot, O., Claverie, J.: Phydbac (phylogenomic display of bacterial genes): An interactive resource for the annotation of bacterial genomes. Nucleic Acids Res **31**(13), 3720–3722 (2003)
90. Enault, F., Suhre, K., Abergel, C., Poirot, O., Claverie, J.: Phydbac2: improved inference of gene function using interactive phylogenomic profile and chromosomal location analysis. Nucleic Acids Res **32**(Web Server Issue), W336–W339 (2004)
91. Enault, F., Suhre, K., Claverie, J.: Phydbac "gene function predictor": a gene annotation tool based on genomic context analysis. BMC Bioinf **6**(247) (2005)
92. Engelhardt, B.E., Jordan, M.I., Muratore, K.E., Brenner, S.E.: Protein molecular function prediction by bayesian phylogenomics. PLoS Comput Biol **1**(5), e45 (2005)
93. Enright, A.J., Ouzounis, C.A.: Functional associations of proteins in entire genomes by means of exhaustive detection of gene fusions. Genome Biol **2**(9), RESEARCH0034 (2001)
94. Enright, A.J., Van Dongen, S., Ouzounis, C.A.: An efficient algorithm for large-scale detection of protein families. Nucleic Acids Res **30**(7), 1575–1584 (2002)
95. Erickson, H.P.: Cooperativity in protein-protein association: the structure and stability of the actin filament. J Mol Biol **206**(3), 465–474 (1989)
96. Ernst, J., Nau, G.J., Bar-Joseph, Z.: Clustering short time series gene expression data. Bioinformatics **21**(Suppl 1), i159–i168 (2005)
97. Eskin, E., Agichtein, E.: Combining text mining and sequence analysis to discover protein functional regions. In: Pac. Symp. Biocomputing, pp. 288–299 (2004)
98. Falda, M., et al.: Argot2: a large scale function prediction tool relying on semantic similarity of weighted gene ontology terms. BMC Bioinf **28**(Suppl 4), S14 (2012)
99. Fayech, S., Essoussi, N., Limam, M.: Partitioning clustering algorithms for protein sequence data sets. BioData Mining **2**(1), 3 (2009)
100. Felsenstein, J.: PHYLIP - phylogeny inference package (version 3.2). Cladistics **5**, 164–166 (1989)
101. Ferrer, L., Dale, J.M., Karp, P.D.: A systematic study of genome context methods: calibration, normalization and combination. BMC Genomics **11**(1), 1–24 (2010)
102. Fetrow, J.S., Siew, N., Di Gennaro, J.A., Martinez-Yamout, M., Dyson, H.J., Skolnick, J.: Genomic-scale comparison of sequence- and structure-based methods of function prediction: Does structure provide additional insight? Protein Science : A Publication of the Protein Society **10**(5), 1005–1014 (2001)
103. Forslund, K., Sonnhammer, E.L.: Predicting protein function from doma in content. Bioinformatics **24**(15), 1681–1687 (2008)
104. French, L.: Fast protein superfamily classification using principal component null space analysis. appendix a: A survey on remote homology detection and protein superfamily classification. Master's thesis, University of Windsor, Ontario, Canada (2005)
105. Funk, C.S., Kahanda, I., Ben-Hur, A., Verspoor, K.M.: Evaluating a variety of text-mined features for automatic protein function prediction with GOstruct. J Biomed Semantics **18**(6), 9 (2015)

106. Gascuel, O.: BIONJ: an improved version of the nj algorithm based on a simple model of sequence data. Mol Biol Evol **14**(7), 685–695 (1997)
107. Gether, U.: Uncovering molecular mechanisms involved in activation of g protein-coupled receptors. Endocr Rev **21**(1), 90–113 (2000)
108. Gibrat, J.F., Madej, T., Bryant, S.H.: Surprising similarities in structure comparison. Curr. Opinion Struct. Biol. **6**(3), 377–385 (1996)
109. Gillis, J., Pavlidis, P.: The role of indirect connections in gene networks in predicting function. Bioinformatics **27**(13), 1860–1866 (2011)
110. Gligorijevic, V., Przulj, N.: Methods for biological data integration: perspectives and challenges. Roy Soc Interface **12**(112), 20150,571 (2015)
111. Godzik, A., Skolnick, J.: Flexible algorithm for direct multiple alignment of protein structures and sequences. Comput Appl Biosci **10**(6), 587–596 (1994)
112. Goh, C., Bogan, A.A., Joachimiak, M., Walther, D., Cohen, F.E.: Co-evolution of proteins with their interaction partners. J Mol Biol **299**(2), 283–293 (2000)
113. Goldberg, D.S., Roth, F.P.: Assessing experimentally derived interactions in a small world. Proc Natl Acad Sci USA **100**(8), 4372–4376 (2003)
114. Goll, J., Rajagopala, S.V., Shiau, S.C., Wu, H., Lamb, B.T., Uetz, P.: MPIDB: the microbial protein interaction database. Bioinformatics **24**(15), 1743–1744 (2008)
115. Gomez, S.M., Noble, W.S., Rzhetsky, A.: Learning to predict protein-protein interactions from protein sequences. Bioinformatics **19**(15), 1875–1881 (2003)
116. Gong, Q., Ning, W., Tian, W.: GoFDR: A sequence alignment based method for predicting protein functions. Methods **S1046–2023**(15), 30,048–7 (2015)
117. Guan, Y., Myers, C.L., Hess, D.C., Barutcuoglu, Z., Caudy, A.A., Troyanskaya, O.G.: Predicting gene function in a hierarchical context with an ensemble of classifiers. Genome Biol **9**(Suppl 1), S3 (2008)
118. Gui, J., Li, H.: Mixture functional discriminant analysis for gene function classification based on time course gene expression data. In: Joint Statistical Meeting: Biometrics Section (2003)
119. Gúldener, U., Muensterkoetter, M., Oesterheld, M., Pagel, P., Ruepp, A., Mewes, H.W., Stúmpflen, V.: MPact: the MIPS protein interaction resource on yeast. Nucleic Acids Res **34**(Database issue), D436–D441 (2006)
120. Guo, X., Gao, L., Wei, C., Yang, X., Zhao, Y., Dong, A.: A computational method based on the integration of heterogeneous networks for predicting disease-gene associations. PLoS One **6**(e24171) (2011)
121. Guruprasad, K., Prasad, M.S., Kumar, G.R.: Database of structural motifs in proteins. Bioinformatics **16**(4), 372–375 (2000)
122. Guthke, R., Schmidt-Heck, W., Hahn, D., Pfaff, M.: Gene expression data mining for functional genomics. In: European Symp Intelligent Techniques, pp. 170–1777 (2000)
123. Guyon, I., Weston, J., Barnhill, S., Vapnik, V.: Gene selection for cancer classification using support vector machines. Mach Learn **46**(1–3), 389–422 (2002)
124. Hamp, T., et al.: Homology-based inference sets the bar high for protein function prediction. BMC Bioinf **14**(Suppl 1), S7 (2013)
125. Han, L.Y., Zheng, C.J., Lin, H.H., Cui, J., Li, H., Zhang, H.L., Tang, Z.Q., Chen, Y.Z.: Prediction of functional class of novel plant proteins by a statistical learning method. New Phytol **168**(1), 109–121 (2005)
126. Hanisch, D., Zien, A., Zimmer, R., Lengauer, T.: Co-clustering of biological networks and gene expression data. Bioinformatics **18**(Suppl 1), S145–S154 (2002)
127. Hartigan, J.A.: Direct clustering of a data matrix. J Amer Stat Assoc **67**(337), 123–129 (1972)
128. Hartuv, E., Shamir, R.: A clustering algorithm based on graph connectivity. Information Processing Letters **76**(4–6), 175–181 (2000)
129. Hawkins, T., Chitale, M., Luban, S., Kihara, D.: PFP: Automated prediction of gene ontology functional annotations with confidence scores using protein sequence data. Proteins: Struct Funct Bioinf **74**(3), 566–582 (2009)
130. Hawkins, T., Luban, S., Kihara, D.: Enhanced automated function prediction using distantly related sequences and contextual association by PFP. Protein Sci **15**(6), 1550–1556 (2006)

131. Hayete, B., Bienkowska, J.R.: GOTrees: Predicting go associations from protein domain composition using decision trees. In: Pacific Symp Biocomput (PSB), pp. 140–151 (2005)
132. Heard, N., Holmes, C.C., Stephens, D.A., Hand, D.J., Dimopoulos, G.: Bayesian coclustering of anopheles gene expression time series: Study of immune defense response to multiple experimental challenges. Proc Natl Acad Sci USA **102**(47), 16,939–16,944 (2005)
133. Hegyi, H., Gerstein, M.: The relationship between protein structure and function: a comprehensive survey with application to the yeast genome. J Mol Biol **288**(1), 147–164 (1999)
134. Hinson, J.T., Chopra, A., Nafissi, N., Polacheck, W.J., Benson, C.C., Swist, S., Gorham, J., Yang, L., Schafer, S., Sheng, C.C., Haghighi, A., Homsy, J., Hubner, N., Church, G., Cook, S.A., Linke, W.A., Chen, C.S., Seidman, J.G., Seidman, C.E.: Heart disease. titin mutations in iPS cells define sarcomere insufficiency as a cause of dilated cardiomyopathy. Science **349**(6251), 892–986 (2015)
135. Hishigaki, H., Nakai, K., Ono, T., Tanigami, A., Takagi, T.: Assessment of prediction accuracy of protein function from protein-protein interaction data. Yeast **18**(6), 523–531 (2001)
136. Holm, L., Sander, C.: Protein structure comparison by alignment of distance matrices. jmb **233**(1), 123–138 (1993)
137. Hou, J., Chi, X.: Predicting protein functions from PPI networks using functional aggregation. Mathematical Biosciences **240**(1), 63–69 (2012)
138. Hou, J., S.-R., J., Zhang, C., Kim, S.: Global mapping of the protein structure space and application in structure-based inference of protein function. Proc. Natl. Acad. Sci. USA **102**, 3651–3656 (2005)
139. Hou, Y., Hsu, W., Lee, M.L., Bystroff, C.: Efficient remote homology detection using local structure. Bioinformatics **19**(17), 2294–2301 (2003)
140. Hsu, P.D., Lander, E.S., Zhang, F.: Development and applications of CRISPR-Cas9 for genome engineering. Cell **157**(6), 1262–1278 (2014)
141. Huang, J.Y., Brutlag, D.L.: The EMOTIF database. Nucleic Acids Res **29**(1), 202–204 (2001)
142. Huang, Y., Yeh, H., Soo, V.: Inferring drug-disease associations from integration of chemical, genomic and phenotype data using network propagation. BMC Med Genomics **6**(3), S4 (2013)
143. Hulo, N., Sigrist, C.J., Le Saux, V., Langendijk-Genevaux, P.S., Bordoli, L., Gattiker, A., De Castro, E., Bucher, P., Bairoch, A.: Recent improvements to the PROSITE database. Nucl. Acids Res. **32**(1), D134–D137 (2003)
144. Hulo, N., et al.: The PROSITE database. Nucleic Acids Res **34**(Database issue), D227–D230 (2006)
145. Humphrey, W., Dalke, A., Schulten, K.: VMD - Visual Molecular Dynamics. J. Mol. Graph. Model. **14**(1), 33–38 (1996). http://www.ks.uiuc.edu/Research/vmd/
146. Hunter, S., et al.: InterPro in 2011: new developments in the family and domain prediction database. Nucleic Acids Res **40**(Database issue), 306–312 (2012)
147. Huynen, M., Snel, B., Lathe, W., Bork, P.: Predicting protein function by genomic context: quantitative evaluation and qualitative inferences. Genome Res **10**(8), 1204–1210 (2000)
148. Hvidsten, T., Komorowski, J., Sandvik, A., Laegreid, A.: Predicting gene function from gene expressions and ontologies. In: Pacific Symp Biocomput (PSB), pp. 299–310 (2001)
149. Iakoucheva, L.M., Dunker, A.K.: Order, disorder, and flexibility: Prediction from protein sequence. Structure **11**(11), 1316–1317 (2003)
150. Jaakkola, T., Diekhans, M., Haussler, D.: Using the fisher kernel method to detect remote protein homologies. In: T. Lengauer, R. Schneider, P. Bork, D. Brutlag, J. Glasgow, H.W. Mewes, R. Zimmer (eds.) Int Conf Intell Sys Mol Biol (ISMB), pp. 149–159. AAAI Press, Menlo Park, CA (1999)
151. Jaakkola, T., Diekhans, M., Haussler, D.: A discriminative framework for detecting remote protein homologies. J Comput Biol **7**(1–2), 95–114 (2000)
152. Jaimovich, A., Elidan, G., Margalit, H., Friedman, N.: Towards an integrated protein-protein interaction network: A relational markov network approach. J Comput Biol **13**(2), 145–164 (2006)
153. Jensen, L., et al.: Prediction of human protein function from post-translational modifications and localization features. J Mol Biol **319**(5), 1257–1265 (2002)

154. Jensen, L.J., Gupta, R., Staerfeldt, H., Brunak, S.: Prediction of human protein function according to gene ontology categories. Bioinformatics **19**(5), 635–642 (2003)
155. Jiang, D., Pei, J., Ramanathan, M., Tang, C., Zhang, A.: Mining coherent gene clusters from gene-sample-time microarray data. In: ACM Intl Conf Knowledge Discovery Data Mining (SIGKDD), pp. 430–439 (2004)
156. Jiang, J.Q.: Learning protein functions from bi-relational graph of proteins and function annotations. In: Algorithms in Bioinformatics, *Lecture Notes in Computer Science*, vol. 6833, pp. 128–138. Springer Verlag (2011)
157. Jiang, X., Nariai, N., Steffen, M., Kasif, S., Kolaczyk, E.: Integration of relational and hierarchical network information for protein function prediction. BMC Bioinf **9**, 350 (2008)
158. Jiang, X., et al.: An expanded evaluation of protein function prediction methods shows an improvement in accuracy. Quantitative Methods arXiv pp. 1–70 (2016)
159. Joshi, T., Xu, D.: Quantitative assessment of relationship between sequence similarity and function similarity. BMC Genomics **8**(1), 1–10 (2007)
160. Kabsch, W.: Efficient remote homology detection using local structure. Acta. Crystallog. sect. A **34**, 827–828 (1978)
161. Kalathur, R.K., Pinto, J.P., Hernández-Prieto, M.A., Machado, R.S., Almeida, D., Chaurasia, G., Futschik, M.E.: UniHI 7: an enhanced database for retrieval and interactive analysis of human molecular interaction networks. Nucleic Acids Res **42**(Database issue), D408–D414 (2014)
162. Kanehisa, M., Goto, S., Kawashima, S., Okuno, Y., Hattori, M.: The KEGG resource for deciphering the genome. Nucleic Acids Res **32**(Database Issue), D277–D280 (2004)
163. Karaoz, U., Murali, T.M., Letovsky, S., Zheng, Y., Ding, C., Cantor, C.R., Kasif, S.: Whole-genome annotation by using evidence integration in functional-linkage networks. Proc Natl Acad Sci USA **101**(9), 2888–2893 (2004)
164. Karplus, K., Barret, C., Hughey, R.: Hidden markov models for detecting remote protein homologies. Bionformatics **14**(10), 846–856 (1998)
165. Keasar, C., Kolodny, R.: Using protein fragments for searching and data-mining protein databases. In: AAAI Workshop, pp. 1–6 (2013)
166. Keck, H., Wetter, T.: Functional classification of proteins using a nearest neighbor algorithm. In Silico Biology **3**(3), 265–275 (2003)
167. Kelley, L.A., Sternberg, M.J.: rotein structure prediction on the web: a case study using the phyre server. Nat Protocols **4**(3), 363–371 (2009)
168. Keseler, I.M., Collado-Vides, J., Gama-Castro, S., Ingraham, J., Paley, S., Paulsen, I.T., Peralta-Gil, M., D., K.P.: EcoCyc: a comprehensive database resource for escherichia coli. Nucleic Acids Res **33**(Database Issue), D334–D337 (2005)
169. Keshava, P., et al.: Human protein reference database–2009 update. Nucleic Acids Res **37**(Database issue), D767–D772 (2009)
170. Khan, I., Wei, Q., Chapman, S., Dukka, B.K., Kihara, D.: The PFP and ESG protein function prediction methods in 2014: effect of database updates and ensemble approaches. GigaScience **4**, 43 (2015)
171. King, A., Przulj, N., Jurisica, I.: Protein complex prediction via cost-based clustering. Bioinformatics **20**(17), 3013–3020 (2004)
172. King, R.D., Karwath, A., Clare, A., Dehaspe, L.: Accurate prediction of protein functional class from sequence in the mycobacterium tuberculosis and escherichia coli genomes using data mining. Yeast **17**(4), 283–293 (2000)
173. King, R.D., Karwath, A., Clare, A., Dehaspe, L.: The utility of different representations of protein sequence for predicting functional class. Bioinformatics **17**(5), 445–454 (2001)
174. Kirilova, S., Carugo, O.: Progress in the PRIDE technique for rapidly comparing protein three-dimensional structures. BMC Research Notes **1**, 44 (2008)
175. Kissinel, E., Henrick, K.: Secondary-structure matching (SSM), a new tool for fast protein structure alignment in three dimensions. Acta Crystallographica D Bio Crystallogr **60**(12.1), 2256–2268 (2004)
176. Kleywegt, G.J.: Use of noncrystallographic symmetry in protein structure refinement. Acta Crystallogr D. **52**(Pt. 4), 842–857 (1996)
177. Koehl, P.: Protein structure similarities. Curr. Opinion Struct. Biol. **11**, 348–353 (2001)

178. Kolesnikov, N., et al.: Arrayexpress update–simplifying data submissions. Nucleic Acids Res **43**(Database issue), D1113–D1116 (2015)
179. Kolesov, G., Mewes, H.W., Frishman, D.: Snapping up functionally related genes based on context information: a colinearity-free approach. J Mol Biol **311**(4), 639–656 (2001)
180. Kolesov, G., Mewes, H.W., Frishman, D.: Snapper: gene order predicts gene function. Bioinformatics **18**(7), 1017–1019 (2002)
181. Kolodny, R., Koehl, P., Guibas, L., Levitt, M.: Small libraries of protein fragments model native protein structures accurately. J. Mol. Biol. **323**, 297–307 (2002)
182. Kolodny, R., Koehl, P., Levitt, M.: Comprehensive evaluation of protein structure alignment methods: Scoring by geometric measures. J. Mol. Biol. **346**, 1173–1188 (2005)
183. Kondor, R.I., Lafferty, J.: Diffusion kernels on graphs and other discrete structures. In: Int Conf Mach Learn (ICML), pp. 315–322 (2002)
184. Koonin, E.V., Galperin, M.Y.: Sequence - evolution - function: Computational approaches in comparative genomics. In: Evolutionary Concept in Genetics and Genomics, 1 edn., chap. 2. Kluwer Academic, Boston, MA (2003)
185. Korbel, J.O., Jensen, L.J., von Mering, C., Bork, P.: Analysis of genomic context: prediction of functional associations from conserved bidirectionally transcribed gene pairs. Nature Biotechnol **22**(7), 911–917 (2004)
186. Koskinen, P., Törönen, P., Nokso-Koivisto, J., Holm, L.: PANNZER: high-throughput functional annotation of uncharacterized proteins in an error-prone environment. Bioinformatics **31**(10), 1544–1552 (2015)
187. Kourmpetis, Y.A., van Dijk, A.D., Bink, M.C., van Ham, R.C., ter Braak, C.J.: Bayesian markov random field analysis for protein function prediction based on network data. PLoS One **5**(2), e9293 (2010)
188. Kourmpetis, Y.A., van Dijk, A.D., ter Braak, C.J.: Gene ontology consistent protein function prediction: the falcon algorithm applied to six eukaryotic genomes. Algorithms Mol Biol **8**(1), 10 (2013)
189. Kuang, R., Ie, E., Wang, K., Wang, K., Siddiqi, M., Freund, Y., Leslie, C.: Profile-based string kernels for remote homology detection and motif extraction. J Bioinf Comput Biol **3**(3), 527–550 (2005)
190. Kuncheva, L.I., Bezdek, J.C., Duin, R.P.W.: Simple ensemble methods are competitive with state-of-the-art data integration methods for gene function prediction. Pattern Recognition **34**(2), 299–314 (2011)
191. Kunik, V., Solan, Z., Edelman, S., Ruppin, E., Horn, D.: Motif extraction and protein classification. In: Pacific Symp Biocomput (PSB), pp. 80–85 (2005)
192. Kuramochi, M., Karypis, G.: Gene classification using expression profiles. In: IEEE Symp Bioinf Bioeng (BIBE), pp. 191–200 (2001)
193. Lagreid, A., Hvidsten, T.R., Midelfart, H., Komorowski, J., Sandvik, A.K.: Predicting gene ontology biological process from temporal gene expression patterns. Genome Res **13**(5), 965–979 (2003)
194. Lan, L., et al.: Ms-knn: Protein function prediction by integrating multiple data sources. BMC Bioinform **14**(Suppl 1), S8 (2013)
195. Lanckriet, G.R.G., De Bie, T., Cristianini, N., Jordan, M.I., Noble, W.S.: A statistical framework for genomic data fusion. Bioinformatics **20**(16), 2626–2635 (2004)
196. Lanckriet, G.R.G., Deng, M., Cristianini, N., Jordan, M.I., Noble, W.S.: Kernel-based data fusion and its application to protein function prediction in yeast. In: Pacific Symp Biocomput (PSB), pp. 300–311 (2004)
197. Lavezzo, E., Falda, M., Fontana, P., Bianco, L., Toppo, S.: Enhancing protein function prediction with taxonomic constraints - the Argot2.5 web server. Methods **93**, 15–23 (2016)
198. Lee, D., Redfern, O., Orengo, C.: Predicting protein function from sequence and structure. Nat. Rev. Mol. Cell Biol. **8**, 995–1005 (2007)
199. Lee, J., Gross, S.P., Lee, J.: Improved network community structure improves function prediction. Scientific Reports **3**, 2197 (2013)
200. Lee, J., Lee, I., Lee, J.: Unbiased global optimization of Lennard-Jones clusters for $n \leq 201$ using the conformational space annealing method. Phys Rev Lett **91**(8), 080,201 (2003)

201. Lee, J., Scheraga, H.A., Rackovsky, S.: New optimization method for conformational energy calculations on polypeptides: conformational space annealing. J Comput Chem **18**(9), 1222–1232 (1997)
202. Legrain, P., Wojcik, J., Gauthier, J.M.: Protein–protein interaction maps: a lead towards cellular functions. Trends Genet **17**(6), 346–352 (2001)
203. Leslie, C.S., Eskin, E., Cohen, A., Weston, J., Noble, W.S.: Mismatch string kernels for discriminative protein classification. Bioinformatics **20**(4), 467–476 (2003)
204. Letovsky, S., Kasif, S.: Predicting protein function from protein/protein interaction data: a probabilistic approach. Bioinformatics **19**(Suppl 1), i197–i204 (2003)
205. Letsche, T.A., Berry, M.W.: Large-scale information retrieval with latent semantic indexing. Inf Sci **100**(1–4), 105–137 (1997)
206. Levitt, M., Gerstein, M.: A unified statistical framework for sequence comparison and structure comparison. Proc. Natl. Acad. Sci. USA **95**(11), 5913–5920 (1998)
207. Levy, E., Ouzounis, C.A., Gilks, W.R., Audit, B.: Probabilistic annotation of protein sequences based on functional classifications. BMC Bioinf **6**, 302 (2005)
208. Li, H., Liang, S.: Local network topology in human protein interaction data predicts functional association. PLoS One **4**(7), e6410 (2009)
209. Li, H., Tong, P., Gallegos, J., Dimmer, E., Cai, G., Molldrem, J.J., Liang, S.: PAND: A distribution to identify functional linkage from networks with preferential attachment property. PLoS One **10**(7), e0127,968 (15)
210. Li, H.L., Fujimoto, N., Sasakawa, N., Shirai, S., Ohkame, T., Sakuma, T., Tanaka, M., Amano, N., Watanabe, A., Sakurai, H., Yamamoto, T., Yamanaka, S., Hotta, A.: Precise correction of the dystrophin gene in duchenne muscular dystrophy patient induced pluripotent stem cells by TALEN and CRISPR-Cas9. Stem Cell Reports **4**(1), 143–154 (2015)
211. Li, L., Stoeckert, C.J., Roos, D.S.: OrthoMCL: Identification of ortholog groups for eukaryotic genomes. Genome Res **13**(9), 2178–2189 (2003)
212. Li, Y., L., C.: Big biologica data: Challenges and opportunities. Genomics, Proteomics, and Bioinformatics **12**(5), 187–189 (2014)
213. Liao, L., Noble, W.S.: Combining pairwise sequence similarity and support vector machines for detecting remote protein evolutionary and structural relationships. J. Comp. Biol. **10**(6), 857–868 (2002)
214. Liberles, D.A., Thorn, A., von Heijne G. AN Elofsson, A.: The use of phylogenetic profiles for gene predictions. Current Genomics **3**(3), 131–137 (2002)
215. Lingling, A., Doerge, R.W.: Dynamic clustering of gene expression. ISRN Bioinformatics **2012**(537217), 1–12 (2012)
216. Lisewski, A.M., Lichtarge, O.: Rapid detection of similarity in protein structure and function through contact metric distances. Nucl. Acids Res. **34**(22), e152 (2006)
217. Liu, A.H., Califano, A.: Functional classification of proteins by pattern discovery and top-down clustering of primary sequences. IBM Systems J **40**(2), 379–393 (2001)
218. Liu, B., Wang, X., Chen, Q., Dong, Q., Lan, X.: Using amino acid physicochemical distance transformation for fast protein remote homology detection. PLoS One **7**(9), e46,633 (2012)
219. Liu, B., Wang, X., Lin, L., Dong, Q., Wang, X.: A discriminative method for protein remote homology detection and fold recognition combining top-n-grams and latent semantic analysis. BMC Bioinf **9**(510) (2008)
220. Liu, B., et al.: Combining evolutionary information extracted from frequency profiles with sequence-based kernels for protein remote homology detection. Bioinformatics **30**(4), 472–479 (2014)
221. Liu, J., Wang, W., Yang, J.: Gene ontology friendly biclustering of expression profiles. In: IEEE Comput Sys Bioinf Conf (CSB), pp. 436–447 (2004)
222. Liu, Q., Chen, Y.P., Li, J.: k-partite cliques of protein interactions: A novel subgraph topology for functional coherence analysis on PPI networks. J Theoretical Biol **340**(7), 146–154 (2014)
223. Lobley, A., Swindells, M.B., Orengo, C.A., Jones, D.T.: Inferring function using patterns of native disorder in proteins. PLoS Comput Biol **3**(8), e162 (2007)
224. Lobley, A.E.: Human protein function prediction: application of machine learning for integration of heterogeneous data sources. Ph.D. thesis, University College London (2010)

225. Lobley, A.E., Nugent, T., Orengo, C.A., Jones, D.T.: FFPred: an integrated feature-based function prediction server for vertebrate proteomes. Nucleic Acids Res **36**(Web server issue), W297–W302 (2008)
226. Ma, Q., Chirn, G.W., Cai, R., Szustakowski, J., Nirmala, N.C.: Clustering protein sequences with a novel metric transformed from sequence similarity scores and sequence alignments with neural networks. BMC Bioinf **6**(1), 242 (2005)
227. Ma, X., Chen, T., Sun, F.: Integrative approaches for predicting protein function and prioritizing genes for complex phenotypes using protein interaction networks. Briefings in Bioinformatics **15**(5), 685–698 (2013)
228. Maciag, K., et al.: Systems-level analyses identify extensive coupling among gene expression machines. Mol Syst Biol **2**(1), 0003 (2006)
229. Madeira, S.C., Oliveira, A.L.: Biclustering algorithms for biological data analysis: A survey. IEEE Trans Comput Biol Bioinf **1**(1), 24–45 (2004)
230. Marchler-Bauer, A., et al.: CDD: a conserved domain database for protein classification. Nucleic Acids Res **33**(Database issue), D192–D196 (2005)
231. Marco, F., Alberto, B., Valentini, G.: UNIPred: Unbalance-aware network integration and prediction of protein functions. J Comput Biol **22**(12), 1057–1074 (2015)
232. Marcotte, C.J.V., Marcotte, E.M.: Predicting functional linkages from gene fusions with confidence. Applied Bioinf **1**(2), 93–100 (2002)
233. Marcotte, E.M., Pellegrini, M., Ng, H., Rice, D.W., Yeates, T.O., Eisenberg, D.: Detecting protein function and protein-protein interactions from genome sequences. Science **285**(5428), 751–753 (1999)
234. Marti-Renom, M.A., Capriotti, E., Shindyalov, I.N., Bourne, P.E.: Structure comparison and alignment. In: J. Gu, P.E. Bourne (eds.) Structural Bioinformatics, 2 edn., chap. 16. John Wiley & Sons (2009)
235. Martin, A.C.: The ups and downs of protein topology; rapid comparison of protein structure. Protein Eng. **13**(12), 829–837 (2000)
236. Martin, D.M., Berriman, M., Barton, G.J.: GOtcha: a new method for prediction of protein function assessed by the annotation of seven genomes. BMC Bioinf **5**(178) (2004)
237. Mateos, A., Dopazo, J., Jansen, R., Tu, Y., Gerstein, M., Stolovitzky, G.: Systematic learning of gene functional classes from dna array expression data by using multilayer perceptrons. Genome Res **12**(11), 1703–1715 (2002)
238. McDowall, M.D., Scott, M.S., Barton, G.J.: PIPs: human protein-protein interaction prediction database. Nucleic Acids Res **37**(Database issue), D651–D656 (2009)
239. Mi, H., Muruganujan, A., Casagrande, J.T., Thomas, P.T.: Large-scale gene function analysis with the PANTHER classification system. Nat Protocol **8**(8), 1551–1566 (2013)
240. Mi, H., et al.: The PANTHER database of protein families and subfamilies and functions and pathways. Nucleic Acids Res **33**(Database issue), D284–D288 (2005)
241. Midelfart, H., Laegreid, A., Komorowski, J.: Classification of gene expression data in an ontology. In: Medical Data Analysis, *Lecture Notes in Computer Science*, vol. 2199, pp. 186–194. Springer (2001)
242. Miele, V., Penel, S., Daubin, V., Picard, F., Kahn, D., Duret, L.: High-quality sequence clustering guided by network topology and multiple alignment likelihood. Bioinformatics **28**(8), 1078–1085 (2012)
243. Möller-Levet, C.S., Cho, K., Yin, H., Wolkenhauer, O.: Clustering of gene expression time-series data. Tech. rep., University of Rostock, Germany (2003)
244. Möller-Levett, C.S., Klawonn, F., Cho, K.: Clustering of unevenly sampled gene expression time-series data. Science **152**(1), 49–66 (2005)
245. Molloy, K., Min, J.V., Barbara, D., Shehu, A.: Exploring representations of protein structure for automated remote homology detection and mapping of protein structure space. BMC Bioinf **15**(Suppl 8), S4 (2014)
246. Monti, S., Tamayo, P., Mesirov, J., Golub, T.: Consensus clustering: a resampling-based method for class discovery and visualization of gene expression microarray data. Mach Learn **52**(1), 91–118 (2003)

247. Moosavi, S., Rahgozar, M., Rahimi, A.: Protein function prediction using neighbor relativity in protein-protein interaction network. Comput Biol Chem **43**, 11–16 (2013)
248. Mostfavi, S., Morris, Q.: Fast integration of heterogeneous data sources for predicting gene function with limited annotation. Bioinformatics **26**(14), 1759–1765 (2010)
249. Muda, H.M., Saad, P., Othman, R.M.: Remote protein homology detection and fold recognition using two-layer support vector machine classifiers. Comput Biol Med **41**(8), 687–699 (2011)
250. Mukherjee, S.: Classifying microarray data using support vector machines. In: D.P. Berrar, W. Dubitzky, M. Granzow (eds.) A Practical Approach to Microarray Data Analysis, chap. 9. Kluwer Academic Publishers (2003)
251. Murzin, A.G., Brenner, S.E., Hubbard, T., Chothia, C.: SCOP: a structural classification of proteins database for the investigation of sequences and structures. J. Mol. Biol. **247**, 536–540 (1995)
252. Nabieva, E., Jim, K., Agarwal, A., Chazelle, B., Singh, M.: Whole-proteome prediction of protein function via graph-theoretic analysis of interaction maps. Bioinformatics **21**(Suppl 1), i302–i310 (2005)
253. Nair, R., Carter, P., Rost, B.: Nlsdb: database of nuclear localization signals. Nucleic Acid Research **31**(1), 397–399 (2003)
254. Najmanovich, R.J., Torrance, W., Thornton, J.M.: Prediction of protein function from structure: Insights from methods for the detection of local structural similarities. Bio Techniques **38**(6), 847–851 (2005)
255. Nariai, N., Kolaczyk, E.D., Kasif, S.: Probabilistic protein function prediction from heterogeneous genome-wide data. PLoS One **2**(3), e337 (2007)
256. Narra, K., Liao, L.: Use of extended phylogenetic profiles with E-values and support vector machines for protein family classification. Intl J Computer Info Sci **6**(1) (2005)
257. Nepusz, T., Sasidharan, R., Paccanaro, A.: SCPS: a fast implementation of a spectral method for detecting protein families on a genome-wide scale. BMC Bioinf **11**(1), 120 (2010)
258. Ng, S., Tan, S., Sundararajan, V.: On combining multiple microarray studies for improved functional classification by whole-dataset feature selection. Genome Informatics **14**, 44–53 (2003)
259. Ng, S., Zhu, Z., Ong, Y.: Whole-genome functional classification of genes by latent semantic analysis on microarray data. In: Asia-Pacific Conf on Bioinformatics, pp. 123–129 (2004)
260. Ni, Q., Wang, Z., Han, Q., Li, G.: Using logistic regression method to predict protein function from protein-protein interaction data. In: IEEE Intl Conf Bioinf Biomed Eng (ICBBE), pp. 1–4 (2009)
261. Obozinski, G., Lanckriet, G., Grant, C., Jordan, M., Noble, W.S.: Consistent probabilistic output for protein function prediction. Genome Biol **9**(Suppl 1), S6 (2008)
262. Ofer, D., Linial, M.: ProFET: Feature engineering captures high-level protein functions. Bioinformatics **31**(21), 3429–3436 (2015)
263. Oliver, S.: Guilt-by-association goes global. Nature **403**(6770), 601–603 (2000)
264. Oliver, S.G.: From DNA sequence to biological function. Nature **379**(6566), 597–600 (1996)
265. Orchard, S., et al.: The MIntAct project–IntAct as a common curation platform for 11 molecular interaction databases. Nucleic Acids Res **42**(Database issue), D358–D363 (2014)
266. Orengo, C.A., Michie, A.D., Jones, S., Jones, D.T., Swindells, M.B., Thornton, J.M.: CATH database: A hierarchic classification of protein domain structures. Structure **5**(8), 1093–1108 (1997)
267. Orengo, C.A., Taylor, W.R.: SSAP: sequential structure alignment program for protein structure comparison. Methods Enzymol **266**, 617–635 (1996)
268. Ortiz, A.R., Strauss, C.E., Olmea, O.: MAMMOTH (matching molecular models obtained from theory): an automated method for model comparison. Protein Sci **11**(11), 2606–2621 (2002)
269. Osadchy, M., Kolodny, R.: Maps of protein structure space reveal a fundamental relationship between protein structure and function. Proc. Natl. Acad. Sci. USA **108**, 12,301–12,306 (2011)

270. Ouali, M., King, R.D.: Cascaded multiple classifiers for secondary structure prediction. Protein Science **9**(6), 1162–1176 (2000)
271. Overbeek, R., Fonstein, M., D'Souza, M., Pusch, G.D., Matlsev, N.: Use of contiguity on the chromosome to predict functional coupling. In Silico Biol **1**(2), 93–108 (1999)
272. Overbeek, R., Fonstein, M., D'Souza, M., Pusch, G.D., Matlsev, N.: The use of gene clusters to infer functional coupling. Proc Natl Acad Sci USA **96**(6), 2896–2901 (1999)
273. Pagel, P., et al.: The MIPS mammalian protein-protein interaction database. Bioinformatics **21**(6), 832–834 (2005)
274. Pasquier, C., Promponas, V., Hamodrakas, S.J.: PRED-CLASS: cascading neural networks for generalized protein classification and genome-wide application. Proteins **44**(3), 361–369 (2000)
275. Pavlidis, P., Cai, J., Weston, J., Noble, W.S.: Learning gene functional classifications from multiple data types. J Comput Biol **9**(2), 401–411 (2002)
276. Pazos, F., Valencia, A.: Similarity of phylogenetic trees as indicator of protein-protein interaction. Protein Eng **14**(9), 609–614 (2001)
277. Pearl, F.M., Bennett, C.F., Bray, J.E., al., e.: The CATH database: an extended protein family resource for structural and functional genomics. Nucl. Acids Res. **31**, 452–455 (2003)
278. Pearson, W.R., Lipman, D.J.: Improved tools for biological sequence comparison. Proc Natl Aca Sci USA **85**(8), 2444–2448 (1988)
279. Pellegrini, M., Marcotte, E.M., Thompson, M.J., Eisenberg, D., Yeates, T.O.: The underlying hypothesis is that two genes with similar phylogeny profiles will also be functionally similar. Proc Natl Acad Sci USA **96**(8), 4285–4288 (1999)
280. Pereira-Leal, J.B., Enright, A.J., Ouzounis, C.A.: Detection of functional modules from protein interaction networks. Proteins: Struct Funct Bioinf **54**(1), 49–57 (2004)
281. Pérez, A.J., Rodriguez, A., Trelles, O., Thode, G.: A computational strategy for protein function assignment which addresses the multidomain problem. Comp Funct Genomics **3**(5), 423–440 (2002)
282. Perutz, M.F., Rossmann, M.G., Cullis, A.F., Muirhead, H., Will, G., North, A.C.T.: Structure of myoglobin: a three-dimensional fourier synthesis at 5.5 angstrom resolution. Nature **185**, 416–422 (1960)
283. Piovesan, D., Giollo, M., Ferrari, C., Tossato, S.C.E.: Protein function prediction using guilty by association from interaction networks. Amino Acids **47**(12), 2583–2592 (2015)
284. Prieto, C., De Las Rivas, J.: APID: Agile protein interaction dataanalyzer. Nucleic Acids Res **34**(Web Server issue), W298–W302 (2006)
285. Qian, B., Goldstein, R.A.: Detecting distant homologs using phylogenetic tree-based HMMs. Proteins **52**(3), 446–453 (2003)
286. Qin, W., Dion, S.L., Kutny, P.M., Zhang, Y., Cheng, A.W., Jillete, N.L., Malhotra, A., Geurts, A.M., Chen, Y.G., Wang, J.: Efficient CRISPR/Cas9-Mediated genome editing in mice by zygote electroporation of nuclease. Genetics **200**(2), 423–430 (2015)
287. Radivojac, P., et al.: A large-scale evaluation of computational protein function prediction methods. Nat Methods **10**(3), 221–227 (2013)
288. Rangwala, H., Karypis, G.: Profile-based direct kernels for remote homology detection and fold recognition. Bioinformatics **21**(23), 4239–4247 (2005)
289. Rappoport, N., Karsenty, S., Stern, A., Linial, N., Linial, M.P.: ProtoNet 6.0: organizing 10 million protein sequences in a compact hierarchical family tree. Nucleic Acids Res **40**(Database Issue), D313–D320 (2012)
290. Rawlings, N.D., Barrett, A.J.: MEROPS: the peptidase database. Nucleic Acids Res **27**(1), 325–331 (1999)
291. Raychaudari, S., Chang, J., Sutphin, P., Altman, R.: Associating genes with gene ontology codes using a maximum entropy analysis of biomedical literature. Genome Research **12**(1), 203–214 (2002)
292. Re, M., Valentini, G.: Simple ensemble methods are competitive with state-of-the-art data integration methods for gene function prediction. J Mach Learn Res **8**, 98–111 (2010)

293. Remmert, M., Biegert, A., Hauser, A., Söding, J.: HHblits: lightning-fast iterative protein sequence searching by hmm-hmm alignment. Nat Methods **9**(2), 173–175 (2011)
294. Renner, A., Aszodi, A.: High-throughput functional annotation of novel gene products using document clustering. In: Proc. Symp. Biocomputing (PSB), pp. 54–68 (2000)
295. Rifkin, R., Klautau, A.: In defense of one-vs-all classification. J Mach Learn **5**, 101–141 (2004)
296. Riley, M.: Systems for categorizing functions of gene products. Curr Opin Struct Biol **8**(3), 388–392 (1998)
297. Roch, K.G.L., et al.: Discovery of gene function by expression profiling of the malaria parasite life cycle. Science **301**(5639), 1503–1508 (2003)
298. Rogen, P., Fain, B.: Automatic classification of protein structure by using gauss integrals. Proc. Natl. Acad. Sci. USA **100**(1), 119–124 (2003)
299. Rost, B.: Enzyme function less conserved than anticipated. J Mol Biol **318**, 595–608 (1999)
300. Ruepp, A., et al.: The FunCat, a functional annotation scheme for systematic classification of proteins from whole genomes. Nucleic Acids Res **32**(18), 5539–5545 (2004)
301. Saini, A., Hou, J.: Progressive clustering based method for protein function prediction. Bulletin Math Biol **75**(2), 331–350 (2013)
302. Samanta, M.P., Liang, S.: Predicting protein functions from redundancies in large-scale protein interaction networks. Proc Natl Acad Sci USA **100**(22), 12,579–12,583 (2003)
303. Sander, J.D., Joung, J.K.: CRISPR-Cas systems for editing, regulating and targeting genomes. Nature Biotechnology **32**(4), 347–355 (2014)
304. Sarac, O.S., Atalay, V., Cetin-Atalay, R.: GOPred: GO molecular function prediction by combined classifiers. PLoS One **5**(8), e12,382 (2010)
305. Sasson, O., Linial, N., Linial, M.P.: The metric space of proteins-comparative study of clustering algorithms. Bioinformatics **18**(Suppl 1), S14–S21 (2002)
306. Sboner, A., Mu, X.J., Greenbaum, D., Auerbach, R.K., Gerstein, M.B.: The real cost of sequencing: higher than you think! Genome Biol **12**(8), 125–134 (2011)
307. Schietgat, L., Vens, C., Struyf, J., Blockeel, H., Kocev, D., Dzeroski, S.: Predicting gene function using hierarchical multi-label decision tree ensembles. BMC Bioinf **11**(1), 2 (2010)
308. Schnoes, A.M., Brown, S.D., Dodevski, I., Babbitt, P.C.: Annotation error in public databases: misannotation of molecular function in enzyme superfamilies. PLoS Comput Biol **5**(12), e1000,605 (2009)
309. Schnoes, A.M., Ream, D.C., Thorman, A.W., Babbitt, P.C., Friedberg, I.: Biases in the experimental annotations of protein function and their effect on our understanding of protein function space. PLoS Comput Biol **9**(5), e1003,063 (2013)
310. Scholkopf, B., Smola, A.J.: Learning with Kernels: Support Vector Machines, Regularization, Optimization, and Beyond. MIT Press (2002)
311. Schug, J.: Predicting gene ontology functions from ProDom and CDD protein domains. Genome Res **12**(4), 648–655 (2002)
312. Schwikowski, B., Uetz, P., Fields, S.: A network of protein-protein interactions in yeast. Nat Biotechnol **18**(12), 1257–1261 (2000)
313. Serres, M.H., Riley, M.: MultiFun, a multifunctional classification scheme for Escherichia coli K-12 gene products. Microb Comp Genomics **5**(4), 205–222 (2000)
314. Servant, F., Bru, C., Carrere, S., et al.: ProDom: Automated clustering of homologous domains. Briefings in Bioinformatics **3**(3), 246–251 (2002)
315. Sharan, R., Ulitsky, I., Shamir, R.: Network-based prediction of protein function. Mol Sys Biol **3**(1), 88 (2007)
316. Sherlock, G., et al.: The stanford microarray database. Nucleic Acid Res **29**(1), 152–155 (2001)
317. Shi, X., et al.: BMRF-Net: a software tool for identification of protein interaction subnetworks by a bagging markov random field-based method. Bioinformatics **31**(14), 2412–2414 (2015)
318. Shiga, M., Takigawa, I., Mamitsuka, H.: Annotating gene function by combining expression data with a modular gene network. Bioinformatics **23**(13), i468–i478 (2007)

319. Shindyalov, I.N., Bourne, P.E.: Protein structure alignment by incremental combinatorial extension (CE) of the optimal path. Protein Eng. **11**(9), 739–747 (1998)
320. Sierk, M.L., Pearson, W.R.: Sensitivity and selectivity in protein structure comparison. Protein Sci. **13**(3), 773–785 (2004)
321. Sjolanderk, K.: Phylogenomic inference of protein molecular function: advances and challenges. Bioinformatics **20**(2), 170–179 (2004)
322. Sliwoski, G., Kothiwale, S., Meiler, J., Lowe, E.W.: Computational method in drug discovery. Pharmacol Rev **66**(1), 334–395 (2014)
323. Soding, J.: Protein homology detection by HMM-HMM comparison. Bioinformatics **21**(7), 951–960 (2005)
324. Sokolov, A., Ben-Hur, A.: Hierarchical classification of gene ontology terms using the GOstruct method. J Bioinform Comput Biol **8**(2), 357–376 (2010)
325. Song, J., Singh, M.: How and when should interactome-derived clusters be used to predict functional modules and protein function? Bioinformatics **25**(23), 3143–3150 (2009)
326. Sonnenburg, S., Ratsch, G., Schafer, C., Scholkopf, B.: Large scale multiple kernel learning. journal of machine learning research. J Mach Learn Res **7**, 1531–1565 (2006)
327. Sonnhammer, E.L., Eddy, S.R., Birney, E., Bateman, A., Durbin, R.: Pfam: Multiple sequence alignments and HMM-profiles of protein domains. Nucl. Acids Res. **26**(1), 320–322 (1998)
328. Sonnhammer, E.L., Eddy, S.R., Durbin, R.: Pfam: a comprehensive database of protein domain families based on seed alignments. Proteins: Struct. Funct. Bioinf. **28**(3), 405–420 (1997)
329. Sonnhammer, E.L., Eddy, S.R., Durbin, R.: Pfam: A comprehensive database of protein domain families based on seed alignments. Proteins **28**(3), 405–420 (1997)
330. Spirin, V., Mirny, L.A.: Protein complexes and functional modules in molecular networks. Proc Natl Acad Sci USA **100**(21), 12,123–12,128 (2003)
331. Stark, A., Sunyaev, S., Russell, R.B.: A model for statistical significance of local similarities in structure. J. Mol. Biol. **326**(5), 1307–1316 (2003)
332. Subbiah, S., Laurents, D.V., Levitt, M.: Secondary-structure matching (SSM), a new tool for fast protein structure alignment in three dimensions. Curr Biol **3**(3), 141–148 (1993)
333. Swift, S., Tucker, A., Vinciotti, V., Martin, N., Orengo, C., Liu, X., Kellam, P.: Consensus clustering and functional interpretation of gene-expression data. Genome Biol **5**(11), R94 (2004)
334. Szklarczyk, D., et al.: STRING v10: protein-protein interaction networks, integrated over the tree of life. Nucleic Acids Res **43**(Database Issue), D447–D552 (2015)
335. Tan, P., Kumar, V., Srivastava, J.: Selecting the right objective measure for association analysis. Information Systems **29**, 293–313 (2004)
336. Tanay, A., Sharan, R., Kupiec, M., Shamir, R.: Revealing modularity and organization in the yeast molecular network by integrated analysis of highly heterogeneous genomewide data. Proc Natl Acad Sci USA **101**(9), 2981–2986 (2004)
337. Tang, L., Chen, J., Ye, J.: On multiple kernel learning with multiple labels. In: Intl Joint Conf Artif Intell (IJCAI), pp. 1255–1260 (2009)
338. Tang, M., et al.: Graphical models for protein function and structure prediction. In: M. Elloumi, A.Y. Zomaya (eds.) Biological Knowledge Discovery Handbook: Preprocessing, Mining, and Postprocessing of Biological Data, Wiley series on Bioinformatics: Computational Techniques nd Engineering, chap. 9, pp. 191–222. Wiley (2013)
339. Tarcea, V.G., et al.: Michigan molecular interactions r2: from interacting proteins to pathways. Nucleic Acids Res **37**(Database issue), D642–D646 (2009)
340. Tatusov, R.L., Fedorova, N.D., Jackson, J.D., et al.: The COG database: an updated version includes eukaryotes. BMC Bioinf **4**, 41 (2003)
341. Tchagang, A.B., et al.: Mining biological information from 3D short time-series gene expression data: the OPTricluster algorithm. BMC Bioinf **13**(54), 2105–2154 (2012)
342. Tetko, I., Facius, A., Ruepp, A., Mewes, H.W.: Super paramagnetic clustering of protein sequences. BMC Bioinf **6**(1), 82 (2005)
343. Thode, G., Garcia-Ranea, J.A., Jimenez, J.: Search for ancient patterns in protein sequences. J Mol Evol **42**(2), 224–233 (1996)
344. Thomas, T.: Multidomain proteins. eLS pp. 1–8 (2014)

345. Thoren, A.: The PhylProm database - extending the use of phylogenetic profiles and their applications for membrane proteins. Master's thesis, Stockholm University, Sweden (2000)
346. Tordai, H., Nagy, A., Farkas, K., Bányai, L., Patthy, L.: Modules, multidomain proteins and organismic complexity. FEBS J **272**(19), 5064–5078 (2005)
347. Tornow, S., Mewes, H.W.: Functional modules by relating protein interaction networks and gene expression. Nucleic Acids Res **31**(21), 6283–6289 (2003)
348. Troyanskaya, O.G., Dolinski, K., Owen, A.B., Altman, R.B., Botstein, D.: A bayesian framework for combining heterogeneous data sources for gene function prediction (in saccharomyces cerevisiae. Proc Natl Acad Sci USA **100**(4), 8348–8353 (2003)
349. Tsai, C.J., Nussinov, R.: Hydrophobic folding units at protein-protein interfaces: implications to protein folding and to protein-protein association. Protein Sci **6**(7), 1426–1437 (1996)
350. Uchiyama, I.: Hierarchical clustering algorithm for comprehensive orthologous-domain classification in multiple genomes. Nucleic Acids Res **34**(2), 647–658 (2006)
351. Valastyan, J.S., Lindquist, S.: Mechanisms of protein-folding diseases at a glance. Disease Models and Mechanisms **7**(1), 9–14 (2014)
352. Valentini, G.: True path hierarchical ensembles for genome-wide gene function prediction. IEEE Trans Comput Biol Bioinform **8**(3), 832–847 (2011)
353. van Noort, V., Snel, B., Huynen, M.A.: Predicting gene function by conserved co-expression. Trends Genet **19**(5), 238–242 (2003)
354. Vanunu, O., Magger, O., Ruppin, E., Shlomi, T., Sharan, R.: Associating genes and protein complexes with disease via network propagation. PLoS Comput Biol **6**(1), e1000,641 (2010)
355. Vazquez, A., Flammini, A., Maritan, A., Vespignani, A.: Global protein function prediction from protein-protein interaction networks. Nature Biotechnol **21**(6), 697–700 (2003)
356. Veretnik, S., Gu, J., Wodak, S.: Identifying structural domains in proteins. In: J. Gu, P. Bourne (eds.) Structural Bioinformatics, 2 edn., chap. 20, pp. 487–515. John Wiley & Sons (2009)
357. Verleyen, W., Ballouz, S., Gillis, J.: Measuring the wisdom of the crowds in network-based gene function inference. Bioinformatics **31**(5), 745–752 (2015)
358. Vert, J.: A tree kernel to analyze phylogenetic profiles. Bioinformatics **18**(Suppl 1), S276–S284 (2002)
359. Vlahovicek, K., Murvai, J., Barta, E., Pongor, S.: The SBASE protein domain library and release 9.0: an online resource for protein domain identification. Nucleic Acids Res **30**(1), 273–275 (2002)
360. Vlahovicek, K., Pintar, A., Parthasarathi, L., Carugo, O., Pongor, S.: CX, DPX and PRIDE: WWW servers for the analysis and comparison of protein 3d structures. Nucleic Acids Res **33**(Web Server issue), W252–W254 (2005)
361. Vogel, C., Bashton, M., Kerrison, N.D., Chothia, C., Teichmann, S.A.: Structure, function and evolution of multidomain proteins. Curr Opin Struct Biol **14**(2), 208–216 (2004)
362. Walker, M.G., Volkmuth, W., Sprinzak, E., Hodgson, D., Klingler, T.: Prediction of gene function by genome-scale expression analysis: prostate cancer-associated genes. Genome Res **9**(12), 1198–1203 (1999)
363. Wang, D., Hou, J.: Explore the hidden treasure in protein-protein interaction networks - an iterative model for predicting protein functions. J Bioinf and Comput Biol **13**(1550026), 22 (2015)
364. Wang, M., Shang, X., Xie, D., Li, Z.: Mining frequent dense subgraphs based on extending vertices from unbalanced PPI networks. In: IEEE Intl Conf Bioinf Biomed Eng (ICBBE), pp. 1–7 (2009)
365. Wang, X., Schroeder, D., Dobbs, D., Honavar, V.: Automated data-driven discovery of motif-based protein function classifiers. Inf Sci **155**(1–2), 1–18 (2003)
366. Wang, Z., Cao, R., Cheng, J.: Three-level prediction of protein function by combining profile-sequence search, profile-profile search, and domain co-occurrence networks. BMC Bioinf **14**(3), S3 (2013)
367. Wass, M.N., Barton, G., Sternberg, M.J.E.: Combfunc: predicting protein function using heterogeneous data sources. Nucleic Acids Res **40**(Web server issue), W466–W470 (2012)

368. Wass, M.N., Sternberg, M.J.: ConFunc-functional annotation in the twilight zone. Bioinformatics **24**(6), 798–806 (2007)
369. Whisstock, J.C., Lesk, A.M.: Prediction of protein function from protein sequence and structure. Q Rev Biophys **36**(3), 307–340 (2003)
370. Wohlers, I., Andonov, R., Klau, G.W.: Algorithm engineering for optimal alignment of protein structure distance matrices. Optimization Letters (2011). DOI 10.1007/s11590-011-0313-3. URL https://hal.inria.fr/inria-00586067
371. Wohlers, I., Le Boudic-Jamin, M., Djidjev, H., Klau, G.W., Andonov, R.: Exact Protein Structure Classification Using the Maximum Contact Map Overlap Metric. In: 1st International Conference on Algorithms for Computational Biology, AlCoB 2014, pp. 262–273. Tarragona, Spain (2014). DOI 10.1007/978-3-319-07953-0_21. URL https://hal.inria.fr/hal-01093803
372. Wohlers, I., Malod-Dognin, N., Andonov, R., Klau, G.W.: CSA: Comprehensive comparison of pairwise protein structure alignments. Nucleic Acids Research pp. 303–309 (2012). URL https://hal.inria.fr/hal-00667920. Preprint, submitted to Nucleic Acids Research
373. Wu, C., Berry, M., Shivakumar, S., McLarty, J.: Neural networks for full-scale protein sequence classification: Sequence encoding with singular value decomposition. Mach Learn **21**(1), 177–193 (1992)
374. Wu, C., Ermongkonchai, A., Chang, T.C.: Protein classification using a neural network proein database (nnpdb) system. In: Anal Neural Net Appl Conf, pp. 29–41 (1991)
375. Wu, C., Whitson, G., McLarty, J., Ermongkonchai, A., Chang, T.C.: Protein classification artificial neural system. Protein Sci **1**(5), 667–677 (1995)
376. Wu, C.H., Whitson, G.M., Montllor, G.J.: PROCANS: a protein classification system using a neural network. Neural Networks **2**, 91–96 (1990)
377. Wu, J., Kasif, S., DeLisi, C.: Identification of functional links between genes using phylogenetic profiles. Bioinformatics **19**(12), 1524–1530 (2003)
378. Wu, L.F., Hughes, T.R., Davierwala, A.P., Robinson, M.D., Stoughton, R., Altschuler, S.J.: Large-scale prediction of saccharomyces cerevisiae gene function using overlapping transcriptional clusters. Nat Genet **31**(3), 255–265 (2002)
379. Xenarios, I., Rice, D.W., Salwinski, L., Baron, M.K., Marcotte, E.M., Eisenberg, D.: Dip: the database of interacting proteins. Nucleic Acids Res **28**(1), 289–291 (2000)
380. Xie, H., Wasserman, A., Levine, Z., Novik, A., Grebinskiy, V., Shoshan, A., Mintz, L.: Large-scale protein annotation through gene ontology. Genome Res **12**(5), 785–794 (2002)
381. Yahalom, R., Reshef, D., Wiener, A., Frankel, S., Kalisman, N., Lerner, B., Keasar, C.: Structure-based identification of catalytic residues. Proteins **79**(6), 1952–1963 (2011)
382. Yan, Y., J., M.: Protein family clustering for structural genomics. J Mol Biol **353**(3), 744–759 (2005)
383. Yanai, I., Derti, A., DeLisi, C.: Genes linked by fusion events are generally of the same functional category: a systematic analysis of 30 microbial genomes. Proc Natl Acad Sci USA **98**(14), 7940–7945 (2001)
384. Yang, J., Wang, H., Wang, W., Yu, P.: Enhanced biclustering on expression data. In: IEEE Symp Bioinf Bioeng (BIBE), pp. 321–327 (2003)
385. Yona, G., Linial, N., Linial, M.P.: ProtoMap: automatic classification of protein sequences and hierarchy of protein families. Nucleic Acids Res **28**(1), 49–55 (2000)
386. Yu, G., Rangwala, H., Domeniconi, C., Zhang, G., Yu, Z.: Protein function prediction using multi-label ensemble classification. IEEE/ACM Trans Comput Biol Bioinform **10**(4), 1045–1057 (2013)
387. Zemla, A.: LGA: a method for finding 3D similarities in protein structures. Nucl. Acids Res. **31**(13), 3370–3374 (2003)
388. Zhang, W., et al.: The functional landscape of mouse gene expression. J Biol **3**(5), 21 (2004)
389. Zhang, X., Dai, D.: A framework for incorporating functional interrelationships into protein function prediction algorithms. IEEE/ACM Trans Comput Biol Bioinform **9**(3), 740–753 (2012)
390. Zhang, Y., Skolnick, J.: TM-align: a protein structure alignment algorithm based on the TM-score. Nucl. Acids Res. **33**(7), 2302–2309 (2005)

391. Zhang, Z.H., Hwee, K.L., Mihalek, I.: Reduced representation of protein structure: implications on efficiency and scope of detection of structural similarity. BMC Bioinformatics **11**, 155 (2010)
392. Zheng, Y., Roberts, R.J., Kasif, S.: Genomic functional annotation using co-evolution profiles of gene clusters. Genome Biol **3**(11), research0060.1–0060.9 (2002)
393. Zhou, D., Bousquet, O., Lal, T., Weston, J., Schlkopf, B.: Learning with local and global consistency. In: Advances Neural Inform Processing Systems (NIPS), pp. 321–328 (2004)
394. Zhou, X., Kao, M.C., Wong, W.: Transitive functional annotation by shortest-path analysis of gene expression data. Proc Natl Acad Sci USA **99**(20), 12,783–12,788 (2002)
395. Zhou, Y., Young, J.A., Santrosyan, A., Chen, K., Yan, S.F., Winzeler, E.A.: In silico gene function prediction using ontology-based pattern identification. Bioinformatics **21**(7), 1237–1245 (2005)
396. Zhu, J., Zhang, M.Q.: SCPD: a promoter database of the yeast saccharomyces cerevisiae. Bionformatics **15**(7), 607–611 (1999)
397. Zitnik, M., Zupan, B.: Data fusion by matrix factorization. IEEE Trans Pattern Anal Mach Intell **37**(1), 41–53 (2015)

# Genome-Wide Mapping of Nucleosome Position and Histone Code Polymorphisms in Yeast

**Muniyandi Nagarajan and Vandana R. Prabhu**

**Abstract** Nucleosomes are the building blocks of chromatin and control the physical access of regulatory proteins to DNA either directly or through epigenetic changes. Its positioning across the genome leaves a significant impact on the DNA dependent processes, particularly on gene regulation. Though they form structural repeating units of chromatin they differ from each other by DNA/histone covalent modifications establishing diversity in natural populations. Such differences include DNA methylation and histone post translational modifications occurring naturally or by the influence of environment. DNA methylation and histone post translational modifications interact with DNA resulting in gene expression level changes without altering the DNA sequences and show high degree of variation among individuals. Therefore, precise mapping of nucleosome positioning across the genome is essential to understand the genome regulation. Nucleosome positions and histone borne polymorphism are usually detected by MNase-Seq and ChIP-CHIP/ChIP-Seq techniques, respectively. Various computational software are put forth to analyze the data and create high resolution maps, which would offer precise knowledge about nucleosome positioning and genomic locations associated with histone tail modifications. This chapter describes genome level mapping of nucleosome positions and histone code polymorphisms in yeast *Saccharomyces cerevisiae*.

**Keywords** Acetylation • ChIP • Epigenomics • Histone • Nucleosome • Yeast

M. Nagarajan (✉) • V.R. Prabhu
Department of Genomic Science, School of Biological Sciences, Central University of Kerala, Kasargod 671314, Kerala, India
e-mail: nagarajan@cukerala.ac.in



K.-C. Wong (ed.), *Big Data Analytics in Genomics*,
DOI 10.1007/978-3-319-41279-5_8

# 1 Introduction

The size of the eukaryotic DNA is very large and therefore, it must undergo higher levels of compaction to fit well inside the tiny nucleus and concurrently must be able to perform its duty. To solve this issue, negatively charged DNA wraps around positively charged histone proteins and by neutralizing the charge they assemble as a DNA–protein complex known as chromatin. Nucleosomes are the basic repeating structural elements of chromatin which function in a paradoxical way by safeguarding and stabilizing the genetic material by compaction on one hand and on the other hand allowing the DNA to be accessible for various cellular processes. The positions of nucleosomes along a DNA sequence impact gene regulation and various other DNA dependent processes to a great extent. Nucleosome positions are non-random and conserved among similar cell types. Studies have succeeded in identifying factors like chromatin remodeling complexes and the underlying DNA sequences that provide substantial importance in nucleosome studies. But the dynamic and complex nature of these factors hampers the prediction of nucleosome positions in a genome leaving the task more challenging ones.

Besides the nucleosome positioning, polypeptide chain of histone tail that undergoes covalent modifications such as acetylation or methylation also play a major role in gene activity. Altogether, the positions of nucleosomes as well as the post translational modifications of histones foster a new realm of research called "*epigenomics*" which analyzes global heritable epigenetic changes (i.e., heritable change in gene expression without altering the DNA sequences) across the whole genome. The precise detection and determination of histone post translational modifications and nucleosome positions are crucial for unveiling the complex nature of chromatin and its control over gene regulation. Various techniques have been employed to understand the interaction between gene and histone post translational modifications. With the advancement in technology, new techniques such as ChIP-CHIP (Chromatin Immunoprecipitation-CHIP) and ChIP-Seq (Chromatin Immunoprecipitation- Sequencing) have come up with the potential to accurately determine the genomic binding sites of DNA associated protein of interest by chromatin precipitation followed by microarray or NGS and subsequent analysis using bioinformatics tools.

The baker's yeast, *Saccharomyces cerevisiae* is an extensively used model for epigenetic research as the chromatin was mapped first in this single cell system using MNase-Sequencing [28]. In addition, smallest genome with only 12,495,682 bp, easy availability, and short generation time makes yeast a perfect model for studying intra species epigenetic variability. In this chapter we present a case study of genome-wide mapping of nucleosome positions in *S. cerevisiae*. It will also confer about Single Nucleosome Epi-Polymorphism (SNEP), its reversible nature and complex pattern in *S. cerevisiae*.

## 2 A Look at the Nucleosome Positioning

Emerging studies on nucleosome positions, histone post translational modifications, and their influence on gene regulation have accelerated our understanding of various cellular processes and disease development. In the last decade, mapping of nucleosomes to find out the position as well as occupancy of nucleosomes has attracted much attention among researchers. With the advent of a number of relevant techniques and computational tools, nucleosome maps can be constructed at genome level and have been succeeded in widening up our apprehension of chromatin structure and functions. Genome-wide nucleosome maps have been constructed for several commonly used biological model systems such as yeast, fruit fly, round worm, and mouse.

Lee et al. [17] have come up with a more and more clear picture of chromatin structure and have constructed well-positioned nucleosome map for single cellular eukaryote, *S. cerevisiae*, which provides information to find out the mechanism behind nucleosome positioning and occupancy. Kaplan et al. [15] studied the role of DNA sequence in determining the histone octamer assembly in yeast through a comparison of in vivo and in vitro nucleosome maps. Nucleosome organization was found similar in both the cases providing evidence that nucleosome organization is characterized by its DNA sequence pattern. Nucleosome landscapes depend on cell type and stage of an organism's development and therefore multicellular eukaryotes show different patterns of arrangement of nucleosome positions and their prediction becomes a challenging task. In order to understand the complexity of nucleosomal positioning in multicellular organisms, a high resolution nucleosome map was established for *Caenorhabditis elegans* using SOLiD NGS technique [32]. Over 44 million putative nucleosome cores were positioned on the *C. elegans* genome, which provided an explicit idea of nucleosome positioning in multicellular eukaryotes. Mavrich et al. [20] constructed a high resolution nucleosome map for fruit fly, *Drosophila melanogaster* to understand the co-evolution of chromatin organization and transcription machinery between drosophila and yeast which provided evidences for the presence of evolutionarily conserved nucleosomes in drosophila and yeast. Human nucleosome map of $CD4^{+}$ T cells provided profound insights into the nucleosome topography and its regulation in human genome [30].

Recent studies focus on to discern the positions of nucleosomes and nucleosome free regions in a specific gene [26] and attempts to identify how the environmental cues directly affect the genomic properties of chromatin [14]. Lieleg et al. [18] reported the clamping activity of ISWI and CHD1 remodelers in the orderly spacing of nucleosome array. The study further suggested that the remodelers, besides mediating the mechanism of nucleosome sliding contribute to the nucleosome clamp.

## 3 Nucleosome Purification and Sequencing

Genome-wide nucleosome map construction requires isolation of mononucleosomes, which is an important aspect towards the prediction of nucleosome position at high resolution. Several protocols have been established for isolating mononucleosomes. Chemical methods use cleavage properties of hydroxyl radicals and methidiumpropyl-EDTA to extract mononucleosome [25]. Nevertheless, the recent approaches provide an effective way to digest the chromatin and to release mononucleosomes by employing various enzymes like micrococcal nuclease [33] alone or in combination with DNase [4, 12] or transposase [29] or CpG methyltransferase [16]. Efficiency of these enzymes to release the mononucleosomes varies remarkably as they are highly influenced by the structure and mechanical properties of chromatin.

Over the last decade, tiling array was the method of choice for constructing nucleosome maps. But now, it is being replaced by the most promising high throughput sequencing methods, which can produce massive parallel sequencing data with the lowest price and higher throughput. The remarkable contributions of NGS techniques in the field of epigenetics have established a new way to solve many of the unresolved clinical complications with the support of computer science.

## 4 Micrococcal Nuclease Sequencing

Micrococcal nuclease based high throughput sequencing (MNase-Seq) is a commonly used technique to construct high resolution genome-wide nucleosome map, which depicts the positions and occupancy of the nucleosomes. Digestion of exposed DNA using endo-exonuclease activity of MNase, which specifically nick the linker DNA until it reaches nucleosome core, serve as the key feature of the process. Samples under investigation are incubated with micrococcal nuclease at optimal digestion conditions. The DNA wrapped around the histone core is then extracted from the obtained crude mononucleosomes using effective genomic DNA isolation protocols. These nucleosomal DNA fragments are concurrently sequenced using NGS technology.

## 5 Software/Algorithms for Processing Nucleosome Experiments Data

The usage of NGS in nucleosome mapping has made the task easier and precise. Initial steps of MNase-seq experiments apply standard NGS software in order to map the reads and for quality control. The software measures many indistinct, small enriched peaks that correspond to individual nucleosomes, which detect 146 bp with precision of one to several base pairs. The following computational tools are commonly used to analyze and confirm whether a DNA sequence is nucleosomal core sequence and precisely locate its positions onto a reference genome.

*Template filtering* It is used to call nucleosome position, occupancy, and length. The position of nucleosome varies from cell to cell while the MNase digestion varies at the end of a nucleosome due to these variabilities, sequencing reads at each nucleosome end forms distribution with variable size. Thus, the template filtering is based on the expected pattern of forward and reverse strand reads that results from sequencing the nucleosome ends. This model identifies occupancy of forward and reverses read templates and also locates and correlates read distributions with model templates that are specific to nucleosomes [33].

*Improved Nucleosome positioning from sequencing (iNPS)* iNPS classifies the nucleosomes based on their shape and unveils their biological feature. It locates positions of nucleosome in a more accurate manner from sequencing data and is a better version of NPS [6].

*NOrMAL* It is a modified version of Gaussian mixture model, which specifically employed to detect and locate the positions of overlapping nucleosomes. It provides additional inferences about distorted nucleosomes and can also be applied to a paired end sequencing product [23].

*NucleoFinder* It is a novel statistical approach that unambiguously detects nucleosome positions by mitigating the noise level. It attempts to rectify the experimental biases and the heterogeneity in nucleosome positions across the cells. The method infers spacing of nucleosome downstream of active promoters and the abundance or desolation of dinucleotide (GC/AT) at the nucleosome center, which are the typical features of nucleosome organization [3].

*Probabilistic Inference for Nucleosome Positioning (PING)* It exploits nucleosomal short read data gathered from MNase digestion or sonication. By using mixture models the method depicts spatial positioning of nucleosomes and also identifies nucleosomes having low read densities [36].

*NUCwave* It constructs nucleosome occupancy maps for single end reads as well as paired end reads obtained from MNase-Seq, ChIP-Seq or CC-Seq [24].

*Dynamic Analysis of Nucleosome Position and Occupancy by Sequencing (DANPOS)* It has been exclusively used for analyzing dynamic nucleosomes at a high resolution of single base pair. The high resolution analysis, by applying a uniform statistical configuration, determines nucleosome dynamics which includes changes in fuzziness, positional and occupancy variation. It can also detect the functional dynamic nucleosomes locating at the distal regulatory regions of mammalian genome [5].

*FineStr* It is a web server that generates map of nucleosome with single base pair resolution. Sequence data are uploaded in FASTA format to infer the positions of nucleosome on the genomic sequences. It has significantly contributed to construct the genome-wide nucleosome position map of *C. elegans* [10].

*iNuc-PhysChem* It identifies the nucleosomal or linker sequences in the genome based on the physicochemical properties. It has been applied on *Homo sapiens*, *D. melanogaster*, and *C. elegans* genomes by integrating the physicochemical properties into an 884-D vector and the software could identify the positions of most of the nucleosomes successfully [7].

*iNuc-PseKNC* Nucleosome position maps are predicted using pseudo k-tuple nucleotide composition. The predictor has been used to position nucleosomes of *H. sapiens*, *D. melanogaster,* and *C. elegans*. The predictor performs better than iNuc-PhysChem and the success rate in nucleosome prediction is much higher [13].

*Interactive chromatin modeling web server (ICM Web)* It is mainly developed for analyzing the stability of nucleosome and for folding the DNA sequence into 3D chromatin structure. It constructs an energy level diagram of nucleosome and also represents nucleosome free regions of DNA [31].

*nuMap* It offers a platform to unambiguously map the positions of nucleosome computationally. nuMap application uses YR and W/S models to predict the transitional and rotational positioning of nucleosomes. It can predict nearly 80 % of the nucleosome position patterns with a better resolution of 2 bp [2].

*Support vector machine classifier* It has drawn major attention in bioinformatics, which analyze the molecular data in a commendable way and therefore have greater importance in nucleosome mapping. It deciphers the data into a high dimensional feature space and uses a kernel trick to infer the optimal separating hyperplane. SVM based approach has been applied to classify nucleosome occupancy and nucleosome free regions in *S. cerevisiae* [22].

*Viterbi algorithm* It is a well-known algorithm which determines the Viterbi path, the most probable sequence of hidden states. The algorithm is run along with Markov models and thus it is well known for bioinformatics analysis. Its involvement in solving problems exclusively of probability makes them useful in predicting nucleosome positions and generating nucleosome occupancy profiles.

*Hidden Markov Models (HMM)* The applications of HMM are well known in data compression, molecular biology, and in the field of artificial intelligence and pattern recognition. The model is a representation of Bayesian network and is capable of finding out the unobserved latent states and therefore its application in predicting nucleosome positions and linker regions is remarkable.

## 6 Prediction of Nucleosome Position in Yeast

The statistical aspect of predicting nucleosomes is much reliable than random guess but limited when compared with experimental determination of nucleosome positioning. The discovery of HMM and its routine use in ChIP techniques provide deep perceptive into the nucleosome positioning and is the most accurate mathematical model in use for mapping the nucleosome. It also potentially involves in solving a vast variety of well-known biological problems like identification of a gene or protein secondary structure by analyzing large scale data in an exhaustive manner. Various parameters like the advancity and duration of methods, the number of states and allowed transitions used need to be taken into account while choosing the model, which finally brings the more complicated data into a simple manner. Applications of HMM in determining the nucleosome positions and linker boundaries and for mapping the nucleosome at a genome-wide scale are phenomenal.

A considerable amount of studies have been conducted using different versions of HMM to predict the nucleosome positions at genome-wide scale. Yuan et al. [35] generated useful nucleosomal data at an enormous rate to find out the nucleosome positions in *S. cerevisiae*. The ChIP-CHIP technique followed by HMM was used exclusively to detect the hybridization status of the probe with nucleosomal DNA and to locate the distorted nucleosomal region and linker DNA on the genome in order to precisely determine the position of nucleosomes. The model could finally expose the presence of 65–69 % of nucleosomal DNA in the well-positioned nucleosomes. Lee et al. [17] used affymetrix tiling array with 4 bp resolution to construct an atlas of nucleosome occupancy in yeast. The HMM used was in comparison with Yuan et al. [35]. A Viterbi algorithm was run in the HMM to reveal the sequences of all possible hidden states which successfully detected 40,095 well positioned as well as 30,776 fuzzy nucleosomes providing a key to mark their positions on the genome. But, HMM could not gather information regarding the internucleosomal occupancy differences. Yassour et al. [34] showed a coherent, accurate, and a much improved picture of nucleosome positions in yeast using raw microarray data of Yuan et al. [35]. A probabilistic graphical model was used for the easy and precise estimation of data and the predicted inferences were noted in comparison with Lee et al. [17] and other previous studies. However, surprisingly with 20 % more accuracy, the model could trace 13 % more nucleosomes than the original study.

Nagarajan et al. [21] employed a 4 bp resolution tiling array to map the yeast nucleosome positions at genome-wide scale (Fig. 1). Out of the 6,553,600 array probes 2,801,855 and 2,570,638 probes have perfectly matched and hybridized with BY4716 and RM11 yeast strains, respectively (BY4716: MATalpha, laboratory strain, isogenic to S288c; RM11-1a: MATa, derived from wild isolate). The signals generated from perfectly matched probes have been considered for analysis (an average of 34 probes per nucleosome). Nucleosomes are mapped in each strain independently using three replicates per strain. A customized version of HMM, by changing the parameters of HMM used by Yuan et al. [35], has been used to map the positions of nucleosomes. HMM program is run on the entire genome of *S. cerevisiae* without restricting to specific region to remove unpredictable trends in hybridization signal. Therefore, independent run of the HMM in the two strains have been applied successively in window of 1 kb all along the genome. Parameters of HMM and all windows posteriors containing a fixed probe are then averaged and

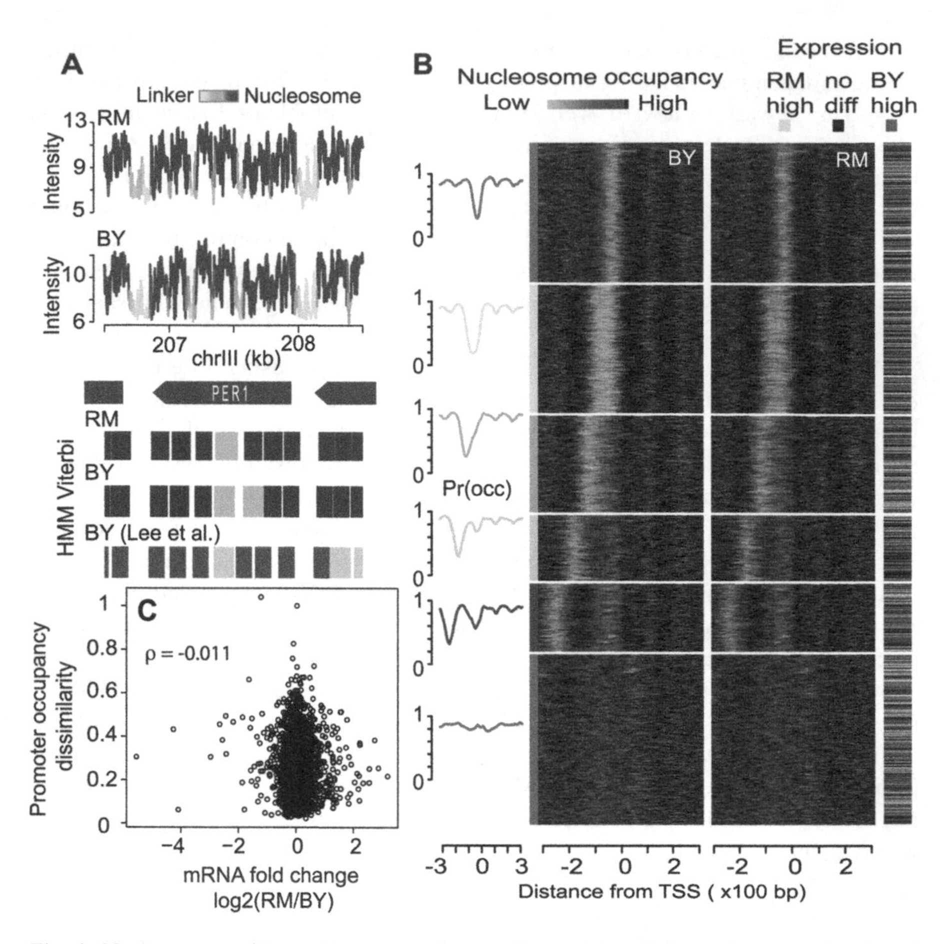


**Fig. 1** Nucleosome position and occupancy in yeast *S. cerevisiae*. (**a**) Raw microarray signals and nucleosome occupancy in the region of *PER1* gene. (**b**) The heatmaps of nucleosome occupancy around transcriptional start site in BY and RM strains and gene expression divergence between the two strains. The *left curve* shows the mean nucleosome occupancy of six main classes of promoters. (**c**) The relationship between promoter occupancy and gene expression divergence. Each *dot* represents one gene. X-axis: inter-strain gene expression difference. Y-axis: inter-strain dissimilarity of promoter occupancy (Adapted from [21])

are used for global computational analysis of both state probabilities and most likely states. HMM also attempts to deal with missing data. The computation of the state probabilities and most-probable states of "missing probes" is done in the same way for observed probes, with the aid of neighboring observed information. It predicted 58,694 nucleosomes which were in accordance with previously published reports.

## 7 An Overview of Epigenetic Modifications

There are epigenetic factors beyond DNA sequences that naturally play a substantial role in an organism's development and cellular differentiation by controlling the gene expression. Despite this fact, epigenetic factors are always under the influence of environment and can alter a gene's function at any stage of an individual's lifespan and are constantly being linked up with various diseases. Their clinical relevance for a variety of cancers has untangled novel methods to diagnose and treat the illness in a better way. Thus apparently, studying the nature and role played by epigenetic marks becomes an inevitable part of current research in the field of biomedical genomics.

The key components responsible for epigenetic changes are DNA methylation and histone post translational modification. The histone post translational modifications such as acetylation, phosphorylation, ubiquitination, methylation, and ADP- ribosylation occur at the N-terminus of histone proteins and regulate most of the DNA dependent processes like replication, repair, and recombination. These covalent modifications alter chromatin architecture resulting in variation at transcription levels. Studies on histone post translational modifications, particularly acetylation and methylation, have unveiled the dynamic nature of chromatin structure and its role in gene expression. The significance of other histone modifications remains poorly understood. Acetylation of histone at lysine has shown both transcriptional activation and repression. However, the genes function either can be up regulated or down regulated depends on which lysine is acetylated and the position of the nucleosome modified. Histone methylation has been well studied on H3K4 and reported that methylation is associated with active transcription in several model organisms, ranging from yeast to mammals. Liu et al. [19] discovered that combinatorial effect of different kinds of histone covalent modifications can result in distinct transcriptional upshots. The study was performed through a 20 bp resolution microarray to analyze 12 distinct histone modifications in individual nucleosomes of *S. cerevisiae*. It showed that these covalent modifications do not occur independently and are enriched especially in the gene or promoter regions.

Earlier studies to map the epigenome were limited to single locus. But advancement in microarray and NGS technologies revolutionized scenario. The technology has been executed to map human epigenome in an unbiased manner and is proven to be cost effective, accurate, and fast [27]. The necessity of studying DNA–protein interaction and chromatin architecture has motivated the development of various ChIP-Seq methods to better understand the interactions. Furey [9] mentioned about modified ChIP-Seq protocols that require only a little amount of cells and detect protein DNA interactions with better resolution. In addition to this, the study also implied that DNase1 hypersensitive site mappings as well as analysis of chromatin interactions mediated by particular proteins provide additional insights into the functions of DNA interacting proteins across the genome of individual. Though ChIP-Seq is considered as a powerful technique to identify the DNA sequence associated with modifications on histones, limitations of antibody development, lack of functional annotation and single modification limit per experiment impel to use additional techniques for better understanding the chromatin dynamics [9]. Similarly, study by Zentner and Henikoff [37] delineated the involvement of recent technologies in generating base pair resolution maps of nucleosomes, suggesting that ChIP-Seq in combination with DNase-Seq and MNase-Seq can be useful in constructing nucleosome map of base pair resolution. Such maps provide detailed information about the epigenetic features at genome-wide level and can be used for epigenetic profiling of specific cell types in multicellular eukaryotes.

## 8 Discovery of Epi-Polymorphisms in Yeast

Recently, there has been a paramount interest to study the existence of histone borne epigenome variations within the natural populations. Similar to DNA polymorphism, epi-polymorphism establishes another basis for diversity among natural populations. Quantitative variations occurring at the levels of DNA methylation/histone post translational modifications can bring about enormous number of polymorphs in a population which is known as "epi-polymorphism." Besides, influence of environment also results in more number of polymorphs. Epi-polymorphisms are known to be involved in gene activity, genome dynamics, and diseases development [21, 11]. Nagarajan et al. [21] explored the histone tail epi polymorphisms in two unrelated yeast strains (BY4716 and RM11-1a). Nucleosomes that differed in the levels of H3 acetylation at Lysine 14 are known as 'Single Nucleosome Epi-Polymorphisms' (SNEP). To detect such nucleosomes, ChIP-CHIP technique, followed by custom algorithms was used (Fig. 2) Only pairs of aligned nucleosomes sharing at least 15 microarray probes (which is the case of 97 % of aligned pairs) were taken into account to detect SNEPs. Subsequently, analysis of variance (ANOVA) was applied to each pair.

$$y_{ijkl} = u + a_i + c_k + d_{ij} + e_{ijkl}$$

where $y_{ijkl}$ is the Log2 normalized hybridization intensity of probe k in replicate $l$ strain $i$ [BY or RM] in experiment type $j$ (nucleosome positioning or ChIP-CHIP), $u$ is the Global mean of the microarray signal, $a_i$ is the yeast strain effect (BY or RM), $b_j$ is the Experiment type effect (nucleosome positioning or ChIP-Chip), $c_k$ is the Probe effect, $d_{ij}$ is the Interaction term between strain and experiment, $e_{ijkl}$ is the residual.

Interestingly,distinct levels of acetylation at H3K14 in nucleosomes of two strains were clearly visible. Out of the 58,694 nucleosomes, 5442 differed in their levels of acetylation between BY and RM strains (Fig. 3). The SNEPs accounted for nearly 10 % of the nucleosomes investigated and were random in distribution. The possibilities of epi-polymorphisms in nucleosomes adjacent to SNEPs were higher indicating positional effect of H3K14ac SNEP. Acetylation variability did not correlate with transcriptional variability of the two strains. However, abundance of acetylation was associated with genes with higher transcription variability. Genes

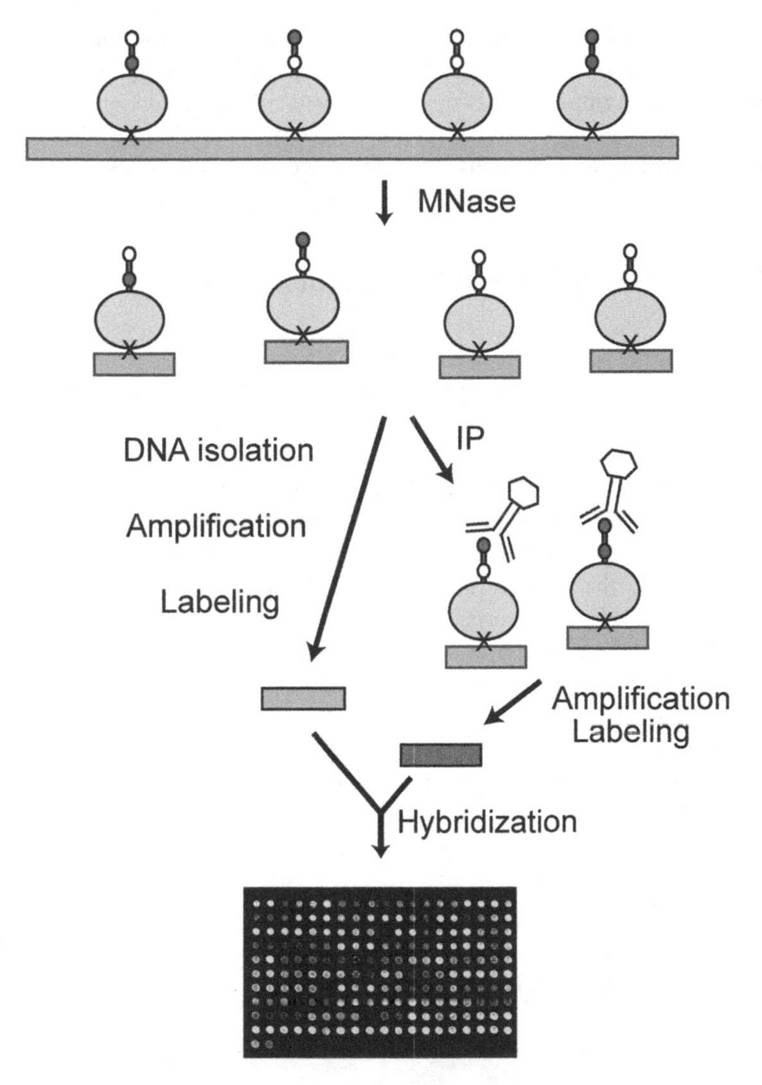


**Fig. 2** Overview of chromatin immunoprecipitation assay. Chromatin immunoprecipitation (ChIP) has turned out to be an efficient assay that detects genomic locations linked with proteins of interest and is extensively being applied in epigenomic research. The assay begins by treating cells with formaldehyde to bring about a tight in situ cross link between protein and chromatin followed by fragmentation by MNase treatment. MNase activity releases raw mononucleosomes and they are processed through DNA extraction protocol. Some of the antibodies are efficient enough to target various histone tail post translational modifications, which are one of the principal causes for epigenetic variation. They detect and bind with specific histone modification. The cross linking is reversed to digest the protein and to release purified nucleosomal DNA associated with specific modification. The resulting mononucleosomal DNA fragments are identified using either microarray (ChIP-CHIP) or high throughput sequencing (ChIP-Seq) methods. However, among the techniques ChIP-Seq is being widely used due to its high resolution, low background noise, and high genomic coverage and has firmly rooted as a pivotal assay in epigenomics. The ChIP-Seq data are analyzed using various bioinformatics tools to identify the positions of modifications on the reference nucleosome map (Adapted from [19])

with SNEPs were found to be more flexible in terms of transcription levels than those without H3K14 modification. The number of epi-polymorphisms was high at conserved DNA region. It opens up a new avenue "*population epigenomics*" and raises important questions on the origin of epigenomic variability and its role in the intra species variability in response to local environmental changes.

## 9 Types of Single Natural Epi-Polymorphisms

The identification of SNEPs within natural population has brought an abrupt inclination in researchers to find out the origin of such differences and their influence on the physiology. Notwithstanding, studying the inter-individual differences at epigenome level in a population are more strenuous in contrast to genetic variability studies as the epigenome marks are self-regulating. Beside this fact, environmental factors or the information coded in the DNA sequences can also regulate epigenome modifications or such modifications may inherit epigenetically. Therefore, it is essential to study the stability of such intra specific epi-polymorphisms and its correlation with DNA polymorphisms. Abraham et al. [1] studied the inheritance of induced histone acetylation in the yeast strains by exposing the cells to an epi-drug. Acetylation variations at some nucleosomes were found to be persistent but others were lost after a few generations, therefore they were known as "persistent" and "labile" epi-polymorphisms," respectively (Fig. 4). Labile epi-polymorphisms were associated with poor genetic control in contrast to persistent epi-polymorphism

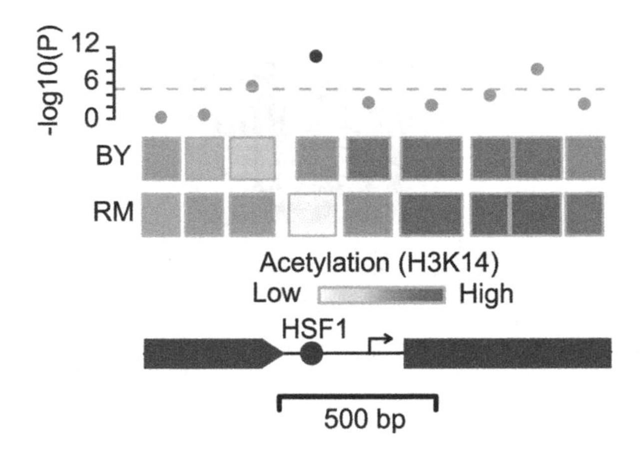


**Fig. 3** Single Nucleotide Epi-Polymorphisms (SNEPs) in yeast *S. cerevisiae*. Schematic representation of nucleosome organization in the regulatory region of *AHA1* gene. The *rectangles* represent nucleosomes. The nucleosomes are colored according to the mean *log(ac/nuc)* value across all probes of the nucleosome. The SNEP detection ($-\log_{10}$(P-value)) is indicated on each nucleosome. The *blue spot*, a highly significant SNEP covers the DNA binding site for HSF1 transcription factor. *Arrow* indicates the transcription start site. *Brown box* indicates beginning of coding sequence (Adapted from [21])

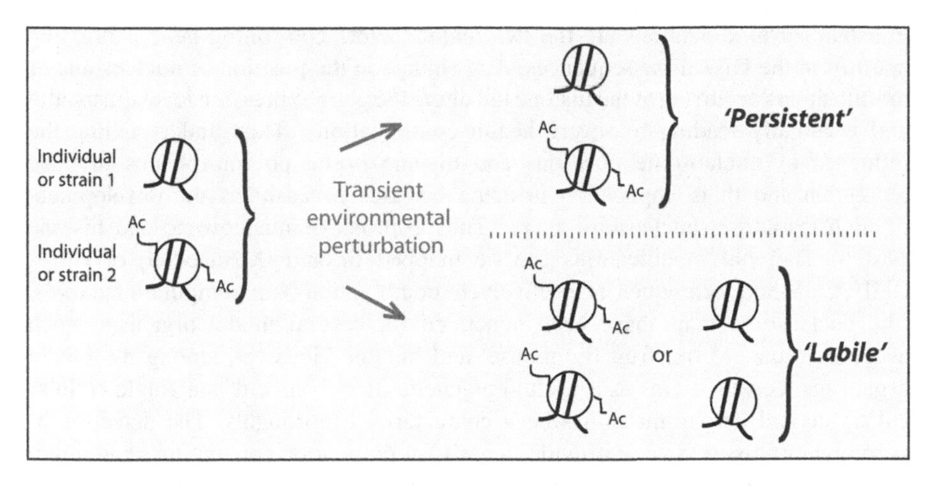


**Fig. 4** Types of Single Nucleotide Epi-Polymorphisms (SNEPs). SNEPs initially present but may be lost after undergoing perturbing environmental conditions called "labile." The SNEPs present after undergoing perturbing environmental conditions are called "persistent" (Adapted from [1])

that showed a tight link with genetic modifiers. The ephemeral nature of labile epi-polymorphism aroused due to exposure to different environmental conditions and these polymorphisms were lost after a few generations. But the very stable nature of persistent SNEPs endured in the descendants providing a strong intimation of DNA encoded determinism of epi-polymorphisms. The confounding and reversible nature of some of the epi-mutations limits the further research.

## 10 Complex Pattern of Single Natural Epi-Polymorphism

Filleton et al. [8] studied the epigenomic variations of H3K4me3, H3K9ac, H3K14ac, H4K12ac, H3K4me1 marks in three unrelated yeast strains (BY4716, RM11-1a, and YJM789) using ChIP-Seq technique. The epigenetic variation was abundant and complex for every mark due to the influence one modification over the others. The existence of complex pattern of epi-polymorphism in unicellular eukaryotic yeast suggests that a much higher level of complex pattern can be expected in case of multicellular organisms.

## 11 Conclusion

Epigenomics research has provided profound insights into disease development and diagnosis and has expanded novel clinical interventions to defeat the illness in a more successful way. Nucleosome positions and the flexibility of histone post translational modifications are the two major factors controlling gene expression apart from the DNA base sequences. Any change in the position of nucleosome or modifications occurring at the histone tail alters the gene expression level drastically and eventually leading to severe health complications. Thus understanding the influence of nucleosome positions and histone borne polymorphisms in gene regulation and their implication in acute diseases necessitates the development of high resolution nucleosome maps. The positions of nucleosome and histone post translational modifications can be mapped through MNase-Seq or ChIP-CHIP/ChIP-Seq techniques, respectively, in combination with computational tools. The nucleosome maps have been generated for several model organisms such as yeast, round worm, fruit fly, mouse, and human. However, among the model organisms, yeast remains as a useful epigenetic model since it is a single cellular eukaryote and its chromatin has been characterized thoroughly. The detection of epi-polymorphisms in yeast provides a basis for population epigenomics and might contribute to strategies in developing robust personalized medicines in future.

**Acknowledgments** The first author is grateful to Science and Engineering Research Board (SERB), Department of Science and technology, Government of India, New Delhi for financial Assistance (SR/S0/AS-84/2012).

## References

1. Abraham AL, Nagarajan M, Veyrieras JB et al (2012) Genetic modifiers of chromatin acetylation antagonize the reprogramming of epi-polymorphisms. PLoS Genet 8(9): e1002958.
2. Alharbi BA, Alshammari TH, Felton NL (2014) nuMap: a web platform for accurate prediction of nucleosome positioning. Genomics Proteomics Bioinformatics 12(5): 249–253
3. Becker J, Yau C, Hancock JM, Holmes CC (2013) NucleoFinder: a statistical approach for the detection of nucleosome positions. Bioinformatics 29(6): 711–6
4. Bell O, Tiwari VK, Thoma NH et al (2011) Determinants and dynamics of genome accessibility. Nat Rev Genet 12(8): 554–564.
5. Chen K, Xi Y, Pan X, Li Z, Kaestner K, Tyler J, Dent S, He X, Li W (2013) DANPOS: dynamic analysis of nucleosome position and occupancy by sequencing. Genome Res 23(2): 341–351
6. Chen W, Liu Y, Zhu S, Green CD, et al (2014) Improved nucleosome-positioning algorithm iNPS for accurate nucleosome positioning from sequencing data. Nat. Commun 5: 4909
7. Chen W, Lin H, Feng P-M, et al (2012) iNuc-PhysChem: A Sequence-Based Predictor for Identifying Nucleosomes via Physicochemical Properties. PLoS ONE 7(10): e47843.
8. Filleton F, Chuffart F, Nagarajan M et al (2015) The complex pattern of epigenomic variation between natural yeast strains at single-nucleosome resolution. Epigenetics & Chromatin 8: 26
9. Furey TS (2012) ChIP-seq and Beyond: new and improved methodologies to detect and characterize protein-DNA interactions. Nat Rev Genet 13(12): 840–852.
10. Gabdank I, Barash D, Trifonov EN (2010) FineStr: a web server for single-base-resolution nucleosome positioning. *Bioinformatics* 26 (6): 845–846.
11. Simon-Pierre Guay SP, Cécilia Légaré C, Houde A-A et al (2014) Acetylsalicylic acid, aging and coronary artery disease are associated with *ABCA1* DNA methylation in men. Clinical Epigenetics 6:14
12. Guertin MJ, Lis JT (2013) Mechanisms by which transcription factors gain access to target sequence elements in chromatin. Curr Opin Genet Dev 23(2): 116–123.

13. Guo SH, Deng EZ, Xu LQ, et al (2014) iNuc-PseKNC: a sequence-based predictor for predicting nucleosome positioning in genomes with pseudo k-tuple nucleotide composition. *Bioinformatics* 30 (11): 1522–1529.
14. Jiang C, Pugh BF (2009) Nucleosome positioning and gene regulation: advances through genomics. Nat Rev Genet 10(3): 161–172.
15. Kaplan N, Moore IK, Mittendorf-Fondufe Y et al (2009) The DNA-encoded nucleosome organization of a eukaryotic genome. Nature 458(7236): 362–366.
16. Kelly TK, Liu Y, Lay FD et al (2012) Genome-wide mapping of nucleosome positioning and DNA methylation within individual DNA molecules. Genome Res 22(12): 2497–2506.
17. Lee W, Tillo D, Bray N et al (2007) A high-resolution atlas of nucleosome occupancy in yeast. Nat Genet 39(10): 1235–1244.
18. Lieleg C, Ketterer P, Nuebler J et al (2015) Nucleosome spacing generated by ISWI and CHD1 remodelers is constant regardless of nucleosome density. Mol Cell Biol 35(9): 1588–1605.
19. Liu CL, Kaplan T, Kim M et al (2005) Single-nucleosome mapping of histone modifications in *S. cerevisiae*. PLoS Biol 3(10): e328.
20. Mavrich TN, Jiang C, Loshikhes IP et al (2008) Nucleosome organization in the Drosophila genome. Nature 453(7193): 358–362.
21. Nagarajan M, Veyrieras J-B, Dieuleveult Md et al (2010) Natural single-nucleosome epi-polymorphisms in yeast. PLoS Genet 6(4): e1000913.
22. Peckham HE, Thurman RE, Fu Y et al (2007) Nucleosome positioning signals in genomic DNA. Genome Res 17(8): 1170–1177.
23. Polishko A Ponts N, Le Roch KG (2012) NOrMAL: accurate nucleosome positioning using a modified Gaussian mixture model. *Bioinformatics* 28 (12): i242-i249.
24. Quintales L, Vázquez E, Antequera F (2015) Comparative analysis of methods for genome-wide nucleosome cartography. Brief Bioinform 16(4): 576–587.
25. Ramachandran S, Zentner GE, Henikoff S (2015) Asymmetric nucleosomes flank promoters in the budding yeast genome. Genome Res 25(3): 381–390.
26. Rando OJ, Chang HY (2009) Genome-wide views of chromatin structure. Annu Rev Biochem 78(1): 245–271
27. Rivera CM and Ren B (2013) Mapping Human Epigenomes. Cell 155(1): 39–55.
28. Rizzo JM, Bard JE, Buck MJ (2012) Standardized collection of MNase-seq experiments enables unbiased dataset comparisons. BMC Mole Bio 13(1): 15.
29. Schep AN, Buenrostro JD, Denny SK et al (2015) Structured nucleosome fingerprints enable high-resolution mapping of chromatin architecture within regulatory regions. Genome Res 25(11): 1757–1770.
30. Schones DE, Cui K, Cuddapah S et al (2008) Dynamic regulation of nucleosome positioning in the human Genome. Cell 132(5): 887–898.
31. Stolz RC, Bishop TC. *(2010)* ICM Web: the interactive chromatin modeling web server. *Nucl Acids Res* 38 (2): W254-W261.
32. Valouev A, Ichikawa J, Tonthat T et al (2008) A high-resolution nucleosome position map of *C. elegans* reveals a lack of universal sequence-dictated positioning. Genome Res 18(7): 1051–63.
33. Weiner A, Hughes A, Yassour M et al (2010) High- resolution nucleosome mapping reveals transcription-dependent promoter packaging. Genome Res 20(1): 90–100.
34. Yassour M, Kaplan T, Jaimovich A et al (2008) Nucleosome positioning from tiling microarray data. Bioinformatics 24(13): i139-i146.
35. Yuan G-C, Liu YJ, Dion MF et al (2005) Genome-scale identification of nucleosome positions in *S. cerevisiae*. Science 309(5734): 626–630.
36. Zhang X, Robertson G, Woo S, et al (2012) Probabilistic Inference for Nucleosome Positioning with MNase-Based or Sonicated Short-Read Data. PLoS ONE 7(2): e32095.
37. Zentner GE and Henikoff S (2012) Surveying the epigenomic landscape, one base at a time. Genome Biology 13(10): 250.

# Part III
# Cancer Analytics

# Perspectives of Machine Learning Techniques in Big Data Mining of Cancer

**Archana Prabahar and Subashini Swaminathan**

**Abstract** Advancements in cancer genomics and the emergence of personalized medicine hassle the need for decoding the genetic information obtained from various high-throughput techniques. Analysis and interpretation of the immense amount of data that gets produced from clinical samples is highly complicated and it remains as a great challenge. The future of cancer medical discoveries will mostly depend on our ability to process and analyze large genomic data sets by relating the profiles of the cancer genome to direct rational and personalized cancer therapeutics. Therefore, it necessitates the integrative approaches of big data mining to handle this large-scale genomic data, to deal with high complexity somatic genomic alterations in cancer genomes and to determine the etiology of a disease to determine drug targets. This demands the progression of robust methods in order to interrogate the functional process of various genes identified by different genomics efforts. This might be useful to understand the modern trends and strategies of the fast evolving cancer genomics research. In the recent years, parallel, incremental, and multi-view machine learning algorithms have been proposed. This chapter addresses the perspectives of machine learning algorithms in cancer genomics and gives an overview of state-of-the-art techniques in this field.

**Keywords** Cancer genomics • Big data mining • Machine learning • Genomic data

A. Prabahar
Department of Plant Molecular Biology and Bioinformatics, Centre for Plant Molecular Biology and Biotechnology, Tamil Nadu Agricultural University, Coimbatore 641 003, Tamilnadu, India

Data Mining and Text Mining Lab, Department of Bioinformatics, School of Life Sciences, Bharathiar University, Coimbatore 641 046, Tamilnadu, India
e-mail: archana.prabahar@gmail.com

S. Swaminathan (✉)
Department of Plant Molecular Biology and Bioinformatics, Centre for Plant Molecular Biology and Biotechnology, Tamil Nadu Agricultural University, Coimbatore 641 003, Tamilnadu, India
e-mail: subabioinfo@gmail.com



K.-C. Wong (ed.), *Big Data Analytics in Genomics*,
DOI 10.1007/978-3-319-41279-5_9

# 1 Introduction

Novel research in the field of biomedical science is mainly driven by high-throughput experiments, such as gene chips from microarray which measures the expression levels of all human genes, or proteomics data using mass spectrometry that detects several thousands of proteins in a sample. Most advanced techniques include high-throughput sequencing (HTS) techniques such as next-generation sequencing (NGS) data that are inevitable to obtain huge volume of data on cells, tissues, disease, and individual genome and that uncover the characteristics of thousands of entities in a single experiment [50]. With the increasing growth of experiments and its published data in the field of biomedicine, there is huge demand to develop novel techniques for better accessibility of information from biomedical data. A vast amount of this information from several experimental findings has been stored in public repositories such as NCBI, EBI, and Sanger database. Relevant scientific articles supporting these experimental evidences are stored in literature databases such as MEDLINE, PubMed Central, BioMed Central, etc. On the downside, experimental data remain along with high levels of noise which require complex statistical measures and various normalization techniques. Typically handling large amounts of data from the HTS experiments can be a real challenge [47]. Research articles published as an outcome of HTS experiments such as NGS and gene expression data are shown in Fig. 1a, b.

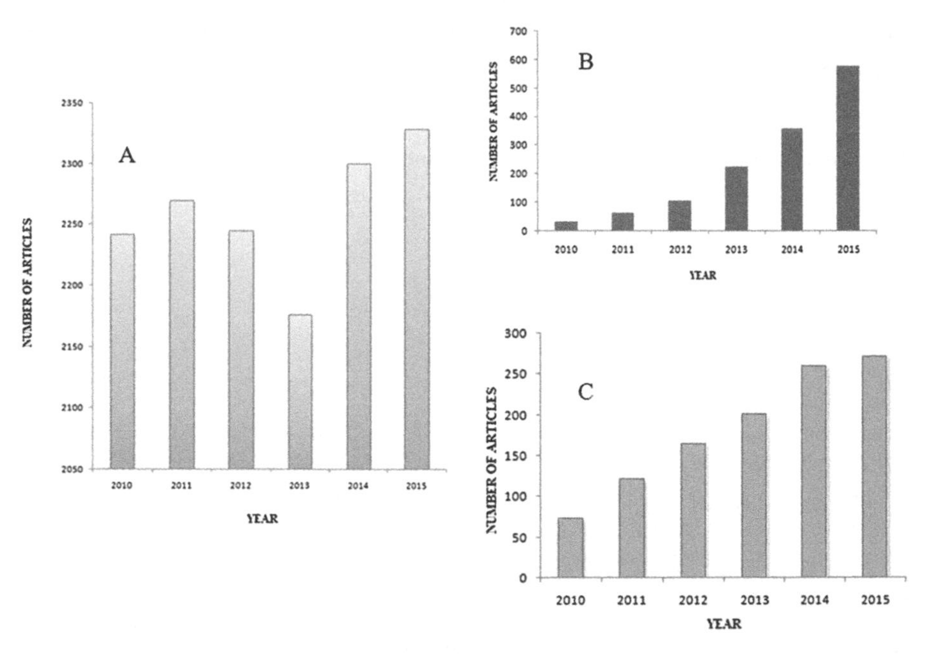


**Fig. 1** Literature growth of PubMed articles. (**a**) PubMed articles on microarray data. (**b**) PubMed articles on NGS data. (**c**) PubMed articles on machine learning

Challenges that emerge from this flood of data include parallelization of algorithms, compression of genomic sequences, and cloud-based execution of complex scientific workflows. The field of biomedical informatics is on the cusp of entering a new era where big data handling emerged, thus providing room for unlimited information growth. Data mining and big data analytics remain as one of the best resources for diagnosis, prognosis, treatment, and cure of all patients in need of medical care. There are numerous current areas of research within the realm of healthcare including Bioinformatics, Image Informatics (e.g., Neuroinformatics), and Clinical Informatics and Translational BioInformatics (TBI). There is an increasing demand for translational medicine, i.e., the accelerated transition of biomedical research results into medical practice.

The volume of data being produced is increasing at an exponential rate and hence storage and sharing of these vast amounts of data is computationally expensive. In this regard, machine learning (ML) algorithms would be the right choice to extract information from these large volumes of data [63]. Literature growth in the number of machine learning algorithms in cancer is shown in Fig. 1c. Despite the fact that these algorithms are computationally expensive, the computational requirements give the impression that it is equal to the amount of data being processed. Therefore, the scientific breakthroughs of the future will undoubtedly be powered by advanced computing capabilities that will allow researchers to manipulate and explore massive datasets. Somehow, the pressure is to shift development toward high-throughput parallel architectures, crucial for real-world applications.

In this post genomic revolution, several studies have been reported in the literature that is based on different strategies to enable the early cancer diagnosis and prognosis. These studies mainly explain the biomarkers and differentially expressed coding and non-coding RNAs. Dataset generated from these studies are the most promising entities for cancer detection and identification. Patient samples from various cancer tissues and control samples are subjected to differential expression studies and biomarkers of cancer are identified to enable screening at early stages. These studies also aid in discriminating benign from malignant tumors. Various aspects regarding the prediction of cancer outcome based on gene expression signatures are the outcome of high-throughput techniques such as microarrays, NGS, etc. Even though gene signatures could significantly improve our ability for prognosis in cancer patients, there is possibility of false predictions and hence there is huge need for machine learning analytics in cancer detection.

## 2 Need for Big Data in Oncology

Big data has set its path many fields, including oncology where data from next-generation DNA sequencing to electronic medical records to linked registry and insurance claims data. For research outcome, the development of big data infrastructure holds immense promise because it enables research involving patients who are not in clinical trials. Despite the fact that only 3 % of the cancer patients are willing to clinical trials, big data resources now allow researchers to observe large, retrospective, and heterogeneous cohorts of cancer patients from screening to end-of-life care [8]. A clinical trial does not have a sound proof of real-world data whereas the big data infrastructure, with data on real-world patients and practices might be a solution for answering questions regarding treatment effectiveness and long-term outcomes for cancer cure.

## *2.1 Big Data Era: A Challenge in Oncology Practice and Research*

Big data contributes to make decisions regarding treatment of cancer patients for personalized medicine at the bedside and the rapid pace of scientific discovery. This remains as one of the challenges for oncologists without research that disseminate evidence from existing data. New, interdisciplinary science collaborators and tools are being developed to transform terabytes, petabytes, and exabytes of data into useful research data sets that represent discrete patient populations. Handling of these intensive datasets in large scale can be achieved by means of machine learning. Therefore, clinical practitioners who generate and collect data medical records will play an increasingly vital role in the future of big data [61]. Models and diagnostic tools developed using these stored data will remain as the hub between medical treatment and cure of the diagnosed cancer patient.

## *2.2 New Opportunities for Big Data*

Several national policies have recently advanced the big data research agenda, in America and other countries in the world including the American Recovery and Reinvestment Act of 2009 and the Patient Protection and Affordable Care Act. Several ambitious and exciting investments in big data include the US Food and Drug Administration's Mini-Sentinel, PCORNet (the National Patient-Centered Clinical Research Network), Oncology Cloud (the cloud-based data repository), and the American Society of Clinical Oncology's CancerLinq. Next-Generation Sequencing (NGS) shifts the bottleneck in sequencing processes from experimental data production to computationally intensive informatics-based data analysis at comparatively higher and reduced cost. Advent of this technique aids the identification and storage of various diseases and normal genes of several patients. Research in these areas can provide great insight into the intricate regulations underlying different diseases and uncover connections to those diseases generate new leads for effective therapy. High-throughput data analysis and their implementation using recent machine learning approaches would aid novel discovery of novel targets for diagnosis and treatment [14].

### 2.3 Emergence of Big Data in Cancer Informatics

The field of oncology is on the cusp of its most essential period where it enters a new era to start handling Big Data for bringing about unlimited potential in the field of informatics. Data mining and Big Data analytics are helping to realize the goals of diagnosing, treating, and healing all patients in need of healthcare [11]. Cancer Informatics is a combination of information science and computer science within the realm of cancer. Several research fields such as Health Informatics including Bioinformatics, Image Informatics (e.g. Neuroinformatics), Clinical Informatics, Public Health Informatics, and also Translational BioInformatics (TBI) are playing a crucial role in the cancer research. Research in this field can range from data acquisition, retrieval, storage, analytics employing data mining techniques, and prediction. Bioinformatics uses molecular level data, Neuroinformatics employs tissue level data, Clinical Informatics applies patient level data, Image informatics utilizes images from various diagnostic scans such as mammography, CT (computerized tomography) scan, MRI (magnetic resonance index), PET (positron emission tomography), etc., and Public Health Informatics utilizes population data (either from the population or on the population) [25].

## 3 Big Data for Cancer Informatics

The digitization of hospital medical records of cancer patients and huge amount of data revolved to the top companies such as Microsoft, SAS, IBM, Dell (DELL), and Oracle (ORCL) since expertise in data-mining, helps the providers of medical oncology and additionally improving the cancer cure for the people. This explosion of data has been proven by International Data Corporation (IDC) that the data in worldwide reached about 0.8 ZB in 2009 and predicted to reach up to 40 ZB by the year of 2020. For accessing these data sets for analysis, the users can hire the infrastructure on the basis of “Pay as you Go” which avoids large capital infrastructure and maintenance cost.

Cancer related data generated from high-throughput techniques include (1) gene expression data (Microarray) (2) NGS data (3) protein–protein interaction (PPI) data, (4) pathway annotation data (5) gene ontology (GO) information (6) gene–disease association data, etc. These data are highly important for many research directions including cancer diagnosis and treatment.

### 3.1 Gene Expression Data

In microarray analysis, the expression levels of thousands of normal and cancer genes are analyzed over different conditions, such as developmental stages of cancer diagnosis, treatment response, metastasis, etc. Microarray-based gene expression profiling is used to record the expression levels for analysis. The analysis results may be used to suggest biomarkers for cancer diagnosis, prevention, and treatment. There are many public sources for microarray databases, such as Array Express (www.ebi.ac.uk/arrayexpress), Gene Expression Omnibus (www.ncbi.nlm.nih.gov/geo), and Stanford Microarray Database (smd.princeton.edu). In addition to the common microarray repositories, cancer specific gene expression data are also stored by The Cancer Genome Atlas Research Network (TCGA), ONCOMINE, etc.

## 3.2 *Next-Generation Sequencing (NGS) Data*

In NGS data several next-generation sequence data such as DNA, RNA, small non-coding RNAs such as microRNA, siRNA, etc. are produced. Various analytical methods are used to understand their features, functions, structures, and evolution. Although RNA sequencing is mainly used as an alternative for microarrays, it can be used for additional purposes also, such as cancer gene mutation identification, identification of post-transcriptional mechanisms, detection of viruses and exogenous RNAs, and identification of polyadenylation. Important sequence databases include NCBI (www.ncbi.org), DNA Data Bank of Japan (www.ddbj.nig.ac.jp) and non-coding RNA databases such as miRBase (www.mirbase.org), SILVIA (www.arb-silva.de/), etc. Sequence Read Archive (www.ncbi.nlm.nih.gov/sra) and European Nucleotide Archive (www.ebi.ac.uk/ena) are the repositories of NGS sequences from various HTS experiments. National Cancer Institute (NCI) has developed Cancer Genomics Hub (CGHub) for storing, cataloging, and accessing cancer specific NGS sequencing data.

## 3.3 *Protein–Protein Interaction Data*

PPIs provide crucial information regarding all biological processes. PPI networks will not only determine the functional role of genes, but will also determine the biomarkers involved in the dysregulatory function of a disease. Databases such as Human Protein Reference Database (HPRD) data [33], STRING (string.embl.de), and BioGRID (thebiogrid.org) are important repositories of PPI data. PPIs in the interactome also enables the discovery of various disease mechanisms and functional associations in cancer.

## 3.4 *Pathway Annotation Data*

Pathway analysis is useful for understanding molecular basis of a disease and various enzymes and reactions involved in their regulatory pathway mechanism. Disease etiology and various biological components (genes and proteins) associated with cancer are thoroughly understood from the pathways. Signaling pathways such as transforming growth factor beta (TGFβ) pathways, tumor necrosis factor (TNF) pathways involved in various cancer types function either in tumor growth suppression or promote late stage progression. Pathway analysis can also be used to predict cancer drug targets, and helps to integrate diverse biological information and assign functions to genes. Pathway database includes KEGG (www.kegg.org), Reactome, and Pathway Commons.

## 3.5 *Gene Ontology (GO) Information*

The GO (www.geneontology.org) database [23] provides gene ontologies of various biological processes, cellular components, and molecular functions of the gene. Several tools utilize the GO database for bioinformatics research. Tools such as AmiGO, DAVID [26], etc. are some of the GO annotation tools. GO also builds ontologies for anatomies, to validate semisupervised and unsupervised analytics results from data. Generally GO covers three domains, viz., cellular component, molecular function, and biological process. GO enrichment analysis involves identification of GO terms that are significantly overrepresented in a given set of genes using statistical hypergeometric test. Tools such as The Database for Annotation, Visualization and Integrated Discovery (DAVID) (david.ncifcrf.gov), Gene Set Enrichment Analysis (GSEA) (http://www.broadinstitute.org/gsea) are developed for enrichment analysis. Methods of network analysis and enrichment analysis have been used to identify targets of various cancers.

## 3.6 *Gene–Disease Association Data*

Gene disease association data include genes that are responsible for biomarkers of diseases such as cancer based on their function, mutation, and causative agents. Databases such as Online Mendelian Inheritance in Man (OMIM) [22] and Comparative Toxicogenomics Database (CTD) [15] are the major gene disease association database. OMIM database provides phenotype description (molecular basis known, molecular basis unknown, gene and phenotype combined) of the disease genes. These associations play a major role in the cancer diagnosis.

Storage of this vast amount of data is computationally expensive and hence several tools emerged with the emergence of big data analytics. Tools and web services available for big data storage and utility are shown in Table 1.

## 4 Techniques for Big Data Analytics

Supervised, unsupervised, and hybrid machine learning approaches are widely used tools for big data analytics. Machine learning techniques are shown as an illustration in Fig. 2. The problem of big data volume can be somewhat minimized by dimensionality reduction which can be performed by feature selection and reduction methods such as principal component analysis (PCA) and singular value decomposition (SVD). Another important tool used in big data analytics is mathematical optimization methods such as multi-objective and multi-modal optimization methods, namely pareto optimization and evolutionary algorithms, respectively.

Big data analytics has a close proximity to data mining approaches. Mining big data is more challenging than traditional data mining due to massive volume of data. Some of the clustering algorithms include CLARA (Clustering LARge Applications) and BIRCH (Balanced Iterative Reducing using Cluster Hierarchies). Reduction of computational complexity of data mining algorithms can be performed by spectral regression discriminant analysis that significantly reduces the time and space complexity. In the recent years, distributed and parallel computing technologies have provided the best solution to large-scale computing problems, due to their scalability, performance, and reliability. Hence distributed computing is performed for big data analytics. Mining of distributed data remains as the new paradigm of data analytics. Further, cloud computing infrastructure-based systems are utilized for performing distributed machine learning, such as the Distributed GraphLab framework.

Cloud computing infrastructure and distributed processing platforms such as MapReduce are widely used for big data analytics. In the recent years, distributed file system technologies, such as HDFS [64] and QFS [54], as well as NoSQL databases for unstructured data, such as MongoDB15 and CouchDB16 have been widely used for big data analytics. Machine learning libraries such as Apache Mahout [55], Apache Spark platform, a MapReduce variant for iterative and fast computations on big data are also used. MapReduce is a data-parallel architecture, originally developed by Google [16]. Apache Hadoop (hadoop.apache.org) is a highly used open-source implementation of MapReduce. MapReduce is a functional paradigm where the input data is fed to the map function and the resultant temporary data to a reduce function. Hadoop, an actual implementation of MapReduce handles the process of large datasets in a distributed computing environment. For example, chromatin immune precipitation (ChIP) exploits this reduce function for integrating the heights of pileups read across Loci for the detection of the transcriptional regulation.

**Table 1** Tools and web services for big data analytics

| Functional role | Algorithm: description |
|---|---|
| Genomic sequence mapping | CloudAligner: A MapReduce based application for mapping short reads generated by next-generation sequencing [51] CloudBurst: A parallel read-mapping algorithm used for mapping next-generation sequence data to the human genome and other genomes [60] BlastReduce: A parallel short DNA sequence read mapping algorithm for aligning sequence data for use in SNP discovery, personal genomics (www.cbcb.umd.edu/software/blastreduce) Rainbow: A cloud-based software package for the automation of large-scale whole-genome sequencing (WGS) data analyses [12] SEAL: A suite of distributed applications for aligning, manipulating, and analyzing short DNA sequence reads [44] DistMap is a toolkit for distributed short-read mapping on a Hadoop cluster. (The nine supported mappers are BWA, Bowtie, Bowtie2, GSNAP, SOAP, STAR, Bismark, BSMAP, and TopHat.) [56] |
| Genomic sequencing analysis | Crossbow: A scalable software pipeline that combines Bowtie and SoapSNP for whole genome resequencing analysis [38] Contrail: An algorithm for de novo assembly of large genomes from short sequencing reads [59] CloudBrush: A distributed genome assembler based on string graphs [10] SOAP3: Short sequence read alignment algorithm that uses the multi-processors in a graphic processing unit to achieve ultra-fast alignments [43] |
| RNA sequence analysis | Myrna: A cloud computing pipeline for calculating differential gene expression in large RNA sequence datasets [37] FX RNA: Sequence analysis tool for the estimation of gene expression levels and genomic variant calling [24] Eoulsan: An integrated and flexible solution for RNA sequence data analysis of differential expression [28] GATK: A gene analysis toolkit for next-generation resequencing data [49] Nephele: A set of tools, which use the complete composition vector algorithm in order to group sequence clustering into genotypes based on a distance measure [13] |
| Sequence filemanagement | Hadoop-BAM: A novel library for scalable manipulation of aligned next-generation sequencing data [52] SeqWare: A tool set used for next-generation genome sequencing technologies which includes an LIMS, Pipeline and Query Engine [53] |
| Sequence and Alignment tool | GPU-BLAST: An accelerated version of NCBI-BLAST which uses general purpose graphics processing unit (GPU), designed to rapidly manipulate and alter memory to accelerate overall algorithm processing [67] BioDoop: A set of tools which modules for handling Fasta streams, wrappers for Blast, converting sequences to the different formats and so on [41] |
| Search engine | Hydra: A protein sequence database search engine specifically designed to run efficiently on the Hadoop MapReduce Framework [42] CloudBlast: Scalable BLAST in the cloud [48] |
| Cancer Genome Repositories | Cancer Genome Anatomy (CGA) Project, National Cancer Institute, USA (http://cancer-genetics.org) The Cancer Genome Atlas Research Network (TCGA) (http://cancergenome.nih.gov/) SNP500Cancer (http://snp500cancer.nci.nih.gov): Sequence and genotype verification of SNPs |

(continued)

**Table 1** (continued)

| Functional role | Algorithm: description |
|---|---|
| Mutation and SNP analysis | BlueSNP: An algorithm for computationally intensive SNP analyses [27] COSMIC: Catalogue of Somatic Mutations in Cancer |
| Clustering | CLARA (Clustering LARge Applications) [31] BIRCH (Balanced Iterative Reducing using Cluster Hierarchies) [71] CLIQUE [7] a graph theoretical clustering method |
| Distributed file system technologies | HDFS [20] and QFS [62] NoSQL databases for unstructured data. MongoDB15 (www.mongodb.org) and CouchDB16 (couchdb.apache.org) |
| Machine learning library for big data analytics | Apache Mahout [35], which contains implementations of various machine learning techniques, such as classifiers and clusteringMLlib17 (spark.apache.org/mllib) is a similar library to perform machine learning on big data on the Apache Spark platform, a MapReduce variant for iterative and fast computations on big data |

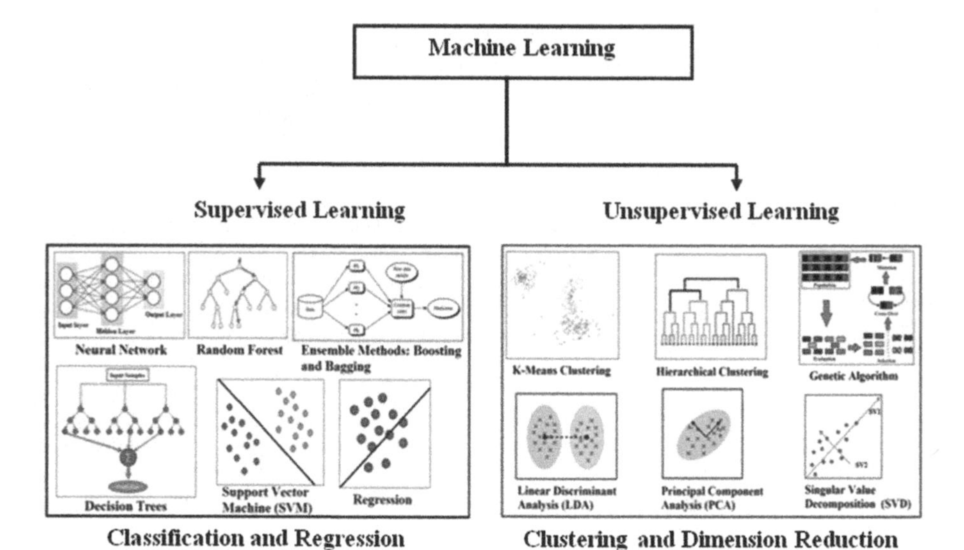


**Fig. 2** Machine learning methods

## *4.1 Big Data Storage*

Novel storage solutions have increased the availability of cancer genomics data sets to the research community. For example, the International Cancer Genome Consortium (http://cancergenome.nih.gov) and The Cancer Genome Atlas (http://cancergenome.nih.gov/) each has the capability to store over two petabytes of genomic data across 34 cancer types. They provide wide applications related to mutational signature analysis [1] and pan-cancer analysis [66]. These studies provide essential link between large-scale genomics and translational research.

Other services like Cancer Genomics Hub [69], the Database of Genotypes and Phenotypes [46], the European Genome Archive [39], and the European Nucleotide Archive [40] provide access to users to download the data of interest and databases like Catalogue of Somatic Mutations in Cancer [5] and cBioPortal for Cancer Genomics [9] provide all the curated published data. Cloud computing might enable the access of these data freely available to the public.

### *4.2 Cloud Computing Services*

Cloud computing also could enable the options for the users to (1) Run applications on the cloud, (2) Analyze and download the data, (3) Perform downstream analysis with the results obtained. Some of the service providers include the Globus Genomics System [45], an Amazon cloud-based analysis and data management client built on the open-source, web-based Galaxy platform [21]. Some of the data management systems allow users to integrate large-scale genomics data and other metadata which includes TranSMART [2], BioMart [30], and the Integrated Rule-Oriented Data System (iRODS) (http://irods.org/). These platforms enable the use of cloud computing and perform the analysis at reduced cost more efficiently at a faster rate.

## 5 Machine Learning for Cancer Using Big Data

Machine learning is a field of computer science that studies the computational methods that learn from data. There are mainly two types of learning methods in machine learning, such as supervised and unsupervised learning methods. In supervised learning a set of objects with class label (training set) is used. The knowledge obtained from training data is further used to assign label to unknown objects called the test. On the other hand, unsupervised learning methods do not depend on training instances with class labels. One of the most powerful and efficient preprocessing tasks is the feature selection. Hybrid learning methods such as Deep learning have become popular in the recent years and provide significantly high accuracy. For machine learning using big data properties such as Scalability, Robustness, Transparency, and distributed data processing is highly needed. Figure 3 shows the machine learning for big data from biological samples.

Recent advancements in high-throughput technologies (HTTs) have produced huge amounts of cancer data that are collected and stored in various public and private repositories and are available to the medical research community. The development of a community resource projects such as TCGA has the potential support for personal medicine as it provides large-scale genomic data about specific cancer. TCGA provides various clinical cancer data in order to understand the molecular basis of cancer through the application of high-throughput genome

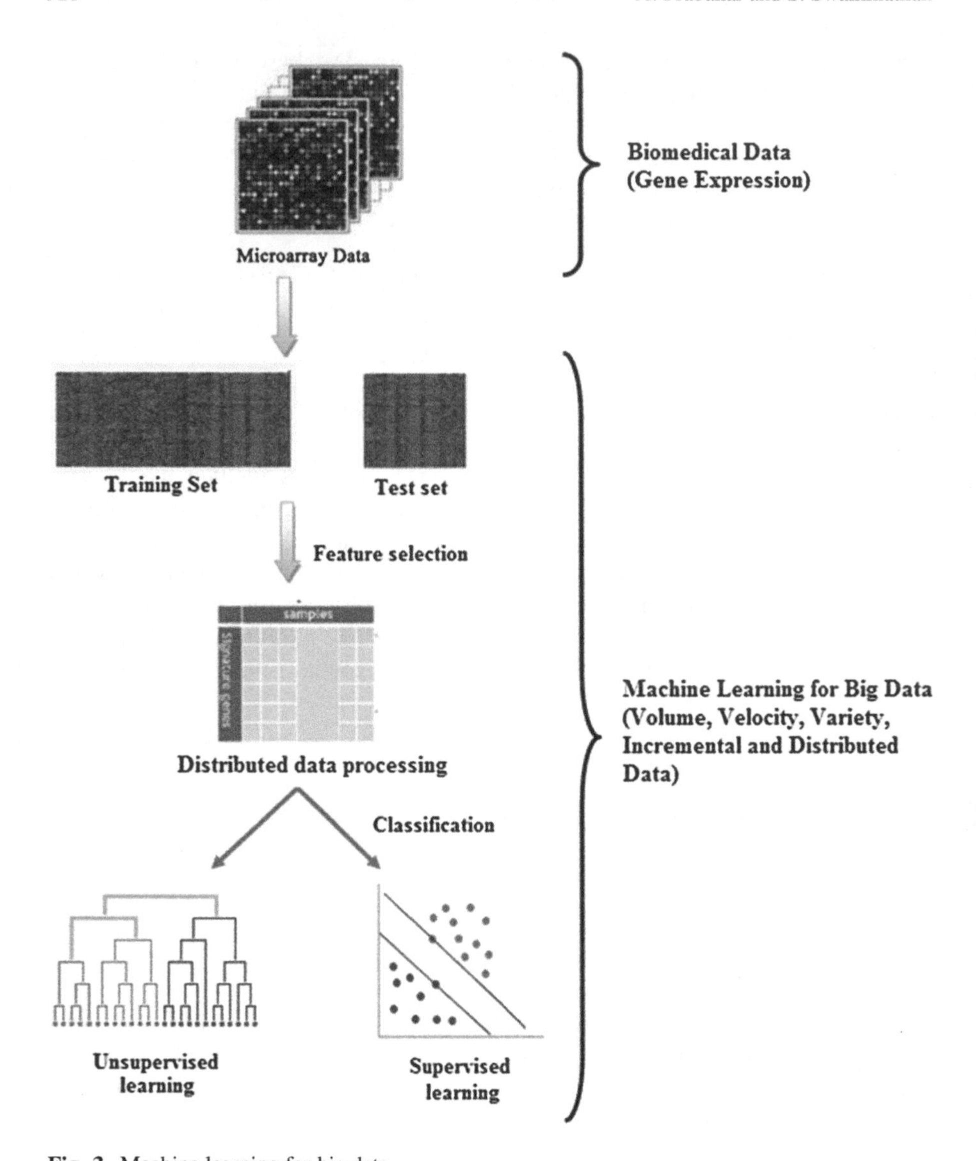


**Fig. 3** Machine learning for big data

technologies. NCBI databases such as SRA and GEO also store the HTS data for cancer along with EBI Array Express. However, the accurate prediction of a disease outcome and diagnosis is the most interesting and challenging task for physicians. Hence, ML methods have become a popular technique to advocate this challenge. Additionally, feature selection methods have been published in the literature with their application in cancer [57, 68].

## 5.1 Feature Selection

Feature selection is very important for big data analytics due to semi-infinite programming (SIP) problem [36]. The SIP is an optimization problem that can either be associated with a finite number of variables and an infinite number of constraints, or an infinite number of variables and a finite number of constraints. To address the SIP problem, various researchers have proposed novel feature selection techniques and provided various solutions. Tan et al. [65] proposed a method for multiple kernel learning (MKL) subproblems with low bias on feature selection of the dataset. In bioinformatics, protein sequence analysis and PPI analysis are complex problems in functional genomics to be reduced dimensionally and handled with feature vectors. Despite the reduction in complexity, prediction accuracy may be reduced by selecting lower number of feature vectors. To overcome this problem, Bagyamathi and Inbarani [4] proposed a new feature selection method by combining improved harmony search algorithm with rough set theory to tackle the feature selection problem in big data. Barbu et al. [6] proposed a novel feature selection method with annealing technique for big data learning. Zeng et al. [70] proposed an incremental feature selection method called FRSA-IFS-HIS (AD) using fuzzy-rough set theory on hybrid information systems.

## 5.2 Machine Learning Based Cancer Prediction

Machine learning based cancer prediction enables the diagnosis of cancer patients more accurately and hence can be used for the estimation of various cancers. One of the recent studies involves the breast cancer risk estimation by means of artificial neural networks (ANNs) and it achieves best results with area under the curve (AUC) value of 0.965 using tenfold cross validation [3]. Breast cancer recurrence prediction was performed by Eshlaghy et al. [18] using support vector machine (SVM) classification achieving an accuracy of 95 %. Breast cancer survival prediction by Delen et al. [17] achieved an accuracy of 93 % using decision tree machine learning method. Several features such as age at diagnosis, tumor size, number of nodes, histology, mammography findings were generated as features for classification to enable higher accuracy and increase the prediction. With the advancement of machine learning techniques, big data applications enable faster prediction of samples. The Spark system contains one of the most reliable advanced analytics tools, named MLlib (machine learning library) and SparkR, which enables modelling the new framework to handle big data. Workflow of cancer prediction using machine learning approach is illustrated in Fig. 4.

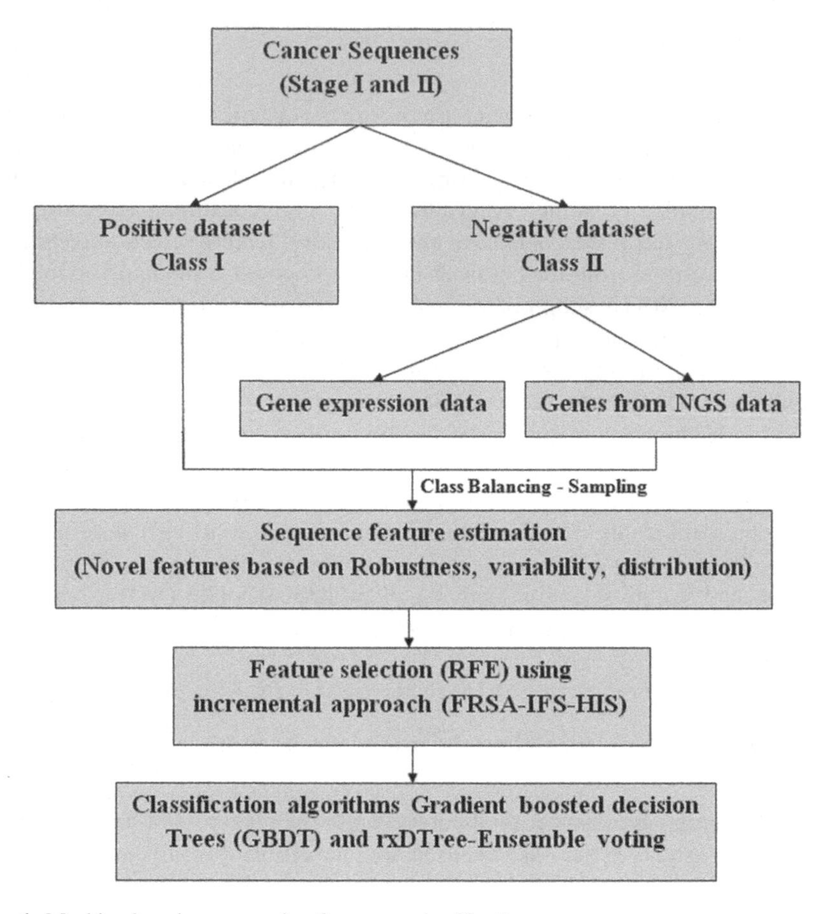


**Fig. 4** Machine learning approaches for cancer classification

## 5.3 *Deep Learning*

Deep learning attempts to model high-level abstractions in data using supervised and/or unsupervised learning algorithms, in order to learn from multiple levels of abstractions. It uses hierarchical representations for data classification. Deep learning methods have been used in many applications, viz., pattern recognition, computer vision, natural language processing, and sequence analysis of HTS data. Due to exponential increase of data in these applications, deep learning is useful for accurate prediction from voluminous data. Recent research development aids in the effective and scalable parallel algorithms for training deep models. In deep learning, Input data is partitioned into multiple samples for data abstractions. The intermediate layers are used to process the features at multiple levels for prediction from data. The final prediction is performed at the output layer using the outputs of its immediate upper layer. Figure 5 illustrates the architecture of deep learning.

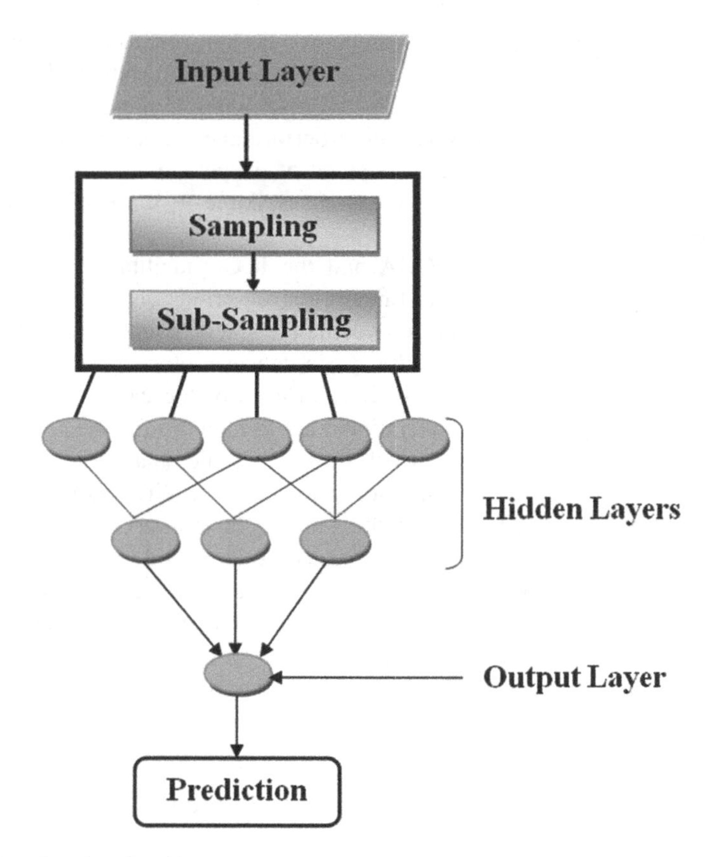


**Fig. 5** Deep learning algorithm

## 6 Challenges in Big Data Analytics of Cancer

Scientific community faces several challenges of Big Data analytics such as scalable infrastructure for parallel computation, management and storage of large-scale datasets, and data-mining analytics. The Apache Hadoop ecosystem, Spark and other data management systems provide libraries and tools for data storage, access to the data and enable parallel processing of Big Data. These efforts meet the first two infrastructural challenges. Machine learning and artificial intelligence enable the exploitation of information hidden in Big Data which solves the problem regarding big data analytics.

Big Data is high dimensional, heterogeneous, complex or unstructured, massive and has various challenges when traditional statistical approaches are applied as they are mainly designed for analyzing relatively smaller samples. Biological systems are so complex that they cannot be adequately described by traditional statistical methods (e.g., classical linear regression analysis, and correlation-based statistical analysis) whereas many modern data-driven learning techniques such as SVM, decision trees, and boosting could provide more feasible solutions. In biology, most of the methods used in genome-wide research are based on statistical testing and computation of probabilistic scoring. In this post genomic era, machine learning has been envisaged by biologists as a high-performance scalable learning system for data-driven biology. With the emergence of encyclopedia of DNA elements (ENCODE) project [32], numerous data are available to the data scientists for machine learning analysis.

The recent establishment of TCGA and the ICGC facilitates the integration and sharing of cancer genome data. Integration of various omics data referred to as integromics could define tumor heterogeneity, genetic events such as mutation and explain the role of signaling pathways in cancer biology [58]. Drug activity studies and further data on drug sensitivity profiles of the cancer genome could provide a powerful platform to guide rational and personalized cancer therapeutics. In the recent years, there is huge growth in the genomic data which enables new integromics study, particularly related to tumor heterogeneity and biomarker discovery. In addition to the identification of candidate biomarkers involved in cancer, functional genetic alterations, i.e. driver events are identified by the evaluation of both clinical and functional data. Identification of these integromics data and further detection of driver mutations in cancer is of high challenge in this omics year. Big data analytics approaches enable the possibility of prediction of these events and enter the path of translational medicine to achieve complete cure for cancer survival.

## 7 Limitations of Big Data Analytics in Cancer

Big data analytics in cancer genomics has quite a large number of limitations in terms of number of patient samples (sample size), missing values, selection bias, and other data handling methods. Some of the major issues involved in these big data limitations are as follows:

(1) The sample size of the cancer patients may not be big when compared with the number of attributes (features) to perform statistical analysis. This may be due to the lack of availability of samples, cost of high-throughput experiments, etc.
(2) Selection of dataset for the study from the raw data is more difficult. Hence, selection bias may occur while selecting a subset of data from real data for analysis besides the size of the data.
(3) Missing values (MVs) is a common problem in big data mining. MVs may occur as a form of outliers or even wrong data. In many cases the problem can be significant and hence missing values are generated using imputation methods either through generation [19] or elimination of data [29].
(4) Data handling is a major limitation in big data due to the underdevelopment of more algorithms to some extent and the lack of facilities for big data handling. Data handling requires more attention in this aspect.

(5) Next major limitation involves the possibility of false positives in statistical analyses which occur due to data handling errors and data bias in machine learning techniques for cancer.
(6) Lack of knowledge in handling of data and novel algorithms in big data result in wrong predictions during the interpretation of results.

# 8 Conclusion

Perspectives of machine learning techniques in big data analytics of cancer are well depicted in this chapter. The complexity of omics data in cancer biology needs to be integrated to provide complete solution to disease diagnosis and treatment. Functional evaluation should be considered even for clinical as well as mechanistic applications of cancer genome data. A deep understanding of both cancer biology and machine learning approaches will facilitate the designing of novel diagnostic tools and enable the possibility of personal genomics. Moving forward, it is clear that progress will accelerate the translational research and lead to the demonstrated success in personalized medicine during the near future. The big data processing capabilities of cloud computing will facilitate large-scale, multi-disciplinary data sharing and utility in this new area of big data omics. Data from cancer genomics can facilitate the paradigm shift from hypothesis-driven to hypothesis-generating research. Next-generation sequencing analysis can be performed from the cloud-based computational biology and has the potential to revolutionize the medical field. From the current approaches in big data analytics, it is evident that the integration of multidimensional heterogeneous data from various high-throughput techniques demands statistical exploration and model generation using feature selection and classification algorithms in machine learning to provide promising tools for inference in the cancer domain.

# References


1. Alexandrov LB, Nik-Zainal S, Wedge DC, Aparicio SA, Behjati S, et al. (2013) Signatures of mutational processes in human cancer. Nature 500(7463): 415–421.
2. Athey BD, Braxenthaler M, Haas M, Guo Y. (2013) tranSMART: an open source and community-driven informatics and data sharing platform for clinical and translational research. AMIA Jt Summits Transl Sci Proc. 6–8.
3. Ayer T, Alagoz O, Chhatwal J, Shavlik JW, Kahn CE, Burnside ES. (2010) Breast cancer risk estimation with artificial neural networks revisited. Cancer. 116:3310–21.
4. Bagyamathi M and Inbarani HH. (2015) A novel hybridized rough set and improved harmony search based feature selection for protein sequence classification. Big Data in Complex Systems. Springer. 173–204.
5. Bamford S, Dawson E, Forbes S, Clements J, Pettett R, Dogan A, Flanagan A, Teague J, Futreal PA, Stratton MR, Wooster R. (2004) The COSMIC (Catalogue of Somatic Mutations in Cancer) database and website. Br J Cancer. 91(2): 355–358.

6. Barbu A, She Y, Ding L, and Gramajo G. (2013) Feature selection with annealing for big data learning. arXiv preprint:1310. 2880.
7. Berkhin P. (2006) A survey of clustering data mining techniques. Grouping multidimensional data. Springer. 25–71.
8. Berman JJ. (2013) Principles of Big Data: Preparing, Sharing, and Analyzing Complex Information. Elsevier.
9. Cerami E, Gao J, Dogrusoz U, Gross BE, Sumer SO, Aksoy BA, Jacobsen A, Byrne CJ, Heuer ML, Larsson E, Antipin Y, Reva B, Goldberg AP, Sander C, Schultz N. (2012) The cBio cancer genomics portal: an open platform for exploring multidimensional cancer genomics data. Cancer Discov. 2(5): 401–404
10. Chang YJ, Chen CC, Chen CL, Ho JM. (2012) A de novo next generation genomic sequence assembler based on string graph and MapReduce cloud computing framework. BMC Genomics. 13 (Suppl 7):S28
11. Chen J, Qian F, Yan W, Shen B (2013) Translational biomedical informatics in the cloud: present and future. BioMed Res Int. 2013:1–8
12. Chong Z, Ruan J, Wu CI. (2012) Rainbow: an integrated tool for efficient clustering and assembling RAD-seq reads. Bioinformatics. 28(21):2732–7.
13. Colosimo ME, Peterson MW, Mardis S, Hirschman L. (2011) Nephele: genotyping via complete composition vectors and MapReduce. Source Code Biol Med. 6:13
14. Cruz JA, Wishart DS. (2006) Applications of machine learning in cancer prediction and prognosis. Cancer Informat. 2:59.
15. Davis AP, Murphy CG, Johnson R, Lay JM, Lennon-Hopkins K, Saraceni-Richards C, Sciaky D, King BL, Rosenstein MC, Wiegers TC, Mattingly CJ. (2013) The Comparative Toxicogenomics Database: update 2013. Nucleic Acids Res. 41(Database issue):D1104–14.
16. Dean J and Ghemawat S. (2008) Mapreduce: simplified data processing on large clusters. Communications of the ACM, 51(1): 107–113.
17. Delen D, Walker G, Kadam A. (2005) Predicting breast cancer survivability: a comparison of three data mining methods. Artif Intell Med. 34:113–27.
18. Eshlaghy AT, Poorebrahimi A, Ebrahimi M, Razavi AR, Ahmad LG. (2013) Using three machine learning techniques for predicting breast cancer recurrence. J Health Med Inform 4:124.
19. Farhangfar A, Kurgan L, Dy J (2008) Impact of imputation of missing values on classification error for discrete data. Pattern Recognition 41: 3692–3705
20. Fusaro V, Patil P, Gafni E, Wall D, Tonellato P. (2011) Biomedical cloud computing with amazon web services. PLOS Comput Biol. 7(8):e1002147
21. Goecks J, Nekrutenko A, Taylor J (2010) Galaxy: a comprehensive approach for supporting accessible, reproducible, and transparent computational research in the life sciences. Genome Biol. 11(8): R86.
22. Hamosh A, Scott AF, Amberger JS, Bocchini CA, McKusick VA (2005) Online Mendelian Inheritance in Man (OMIM), a knowledgebase of human genes and genetic disorders, Nucleic Acids Res. 33: D514–517.
23. Harris MA, Clark J, Ireland A, Lomax J, Ashburner M, Foulger R, et al. (2004) The Gene Ontology (GO) database and informatics resource. Nucleic Acids Res. 32: D258–61.
24. Hong D, Rhie A, Park SS, Lee J, Ju YS, Kim S, et al. (2012) FX: an RNA-Seq analysis tool on the cloud. Bioinformatics. 28: 721–723
25. Hsu W, Markey MK and Wang MD. (2013) Biomedical imaging informatics in the era of precision medicine: progress, challenges, and opportunities. J Am Med Inform Assoc. 20(6): 1010–1013.
26. Huang da W, Sherman BT, Lempicki RA (2009) Systematic and integrative analysis of large gene lists using DAVID bioinformatics resources, Nat. Protoc. 4:44–57.
27. Huang HL, Tata S, Prill RJ. (2013) BlueSNP: R package for highly scalable genome-wide association studies using Hadoop clusters. Bioinformatics, 29:135–136
28. Jourdren L, Bernard M, Dillies MA, Le Crom S (2012) Eoulsan: a cloud computing-based framework facilitating high throughput sequencing analyses. Bioinformatics. 28(11):1542–3

29. Kantardzic M (2003) Data Mining - Concepts, Models, Methods, and Algorithms, IEEE. 165–176.
30. Kasprzyk A (2011) BioMart: driving a paradigm change in biological data management. Database (Oxford) 2011: bar049.
31. Kaufman L and Rousseeuw PJ (2005) Finding groups in data. An introduction to cluster analysis, Wiley Series in Probability and Statistics, New York. 1–368
32. Kellisa M, Wold B, Snyderd MP, Bernsteinb BE et al. (2014) Defining functional DNA elements in the human genome. 111(17): 6131–6138.
33. Keshava Prasad TS, Goel R, Kandasamy K, Keerthikumar S, Kumar S, et al. (2009) Human Protein Reference Database–2009 update, Nucleic Acids Res. 37: D767–772.
34. Kim J, Shin H. (2013) Breast cancer survivability prediction using labeled, unlabeled, and pseudo-labeled patient data. J Am Med Inform Assoc. 20:613–8.
35. Krampis K, Booth T, Chapman B, Tiwari B, Bicak M, Field D, et al. (2012) Cloud BioLinux: pre-configured and on-demand bioinformatics computing for the genomics community. BMC Bioinformatics. 13:42.
36. López M and Still G. (2007) Semi-infinite programming. European Journal of Operational Research. 180(2): 491–518.
37. Langmead B, Hansen KD, Leek JT. (2010) Cloud-scale RNA-sequencing differential expression analysis with Myrna. Genome Biol, 11:R83.
38. Langmead B, Schatz MC, Lin J, Pop M, Salzberg SL. (2009) Searching for SNPs with cloud computing. Genome Biol. 10: R134.
39. Lappalainen I, Almeida-King J, Kumanduri V, Senf A, Spalding JD, ur-Rehman S, Saunders G, Kandasamy J. (2015) The European Genome-phenome Archive of human data consented for biomedical research. Nat Genet 47(7): 692–695.
40. Leinonen R, Akhtar R, Birney E, Bower L, Cerdeno-Tarraga A, et al. (2011) The European Nucleotide Archive. Nucleic Acids Res (Database issue) 39: D28–D31.
41. Leo S, Santoni F, Zanetti G. (2009) Biodoop: bioinformatics on hadoop. Parallel processing workshops. International Conference on ICPPW 09. 415–22.
42. Lewis S, Csordas A, Killcoyne S, Hermjakob H, Hoopmann MR, Moritz RL, et al. (2012) Hydra: a scalable proteomic search engine which utilizes the Hadoop distributed computing. BMC Bioinformatics. 13:324
43. Liu CM, Wong T, Wu E, Luo RB, Yiu SM, Li YR, et al. (2012) SOAP3: ultra-fast GPU-based parallel alignment tool for short reads. Bioinformatics. 28: 878–879
44. Luca Pireddu, Simone Leo, and Gianluigi Zanetti. (2011) SEAL: a distributed short read mapping and duplicate removal tool. Bioinformatics. 27(15): 2159–2160.
45. Madduri RK, Sulakhe D, Lacinski L, Liu B, Rodriguez A, Chard K, Dave UJ, Foster IT (2014) Experiences building Globus Genomics: a next-generation sequencing analysis service using Galaxy, Globus, and Amazon Web Services. Concurr Comput. 26(13): 2266–2279.
46. Mailman MD, Feolo M, Jin Y, Kimura M, Tryka K, Bagoutdinov R, Hao L, Kiang A, Paschall J, Phan L, Popova N. (2007) The NCBI dbGaP database of genotypes and phenotypes. Nature Genetics. 39(10):1181–6.
47. Marx V (2013) The big challenges of big data. Nature. 498(7453): 255–260.
48. Matsunaga A, Tsugawa M, and Fortes J. (2008) CloudBLAST: Combining MapReduce and Virtualization on Distributed Resources for Bioinformatics Applications. IEEE Fourth International Conference on eScience. 222–229.
49. McKenna A, Hanna M, Banks E, Sivachenko A, Cibulskis K, Kernytsky A, Garimella K, Altshuler D, Gabriel S, Daly M, DePristo MA. (2010) The genome analysis toolkit: a MapReduce framework for analysing next-generation DNA sequencing data. Genome Res. 20(9):1297–303
50. Nekrutenko A and Taylor J. (2012) Next-generation sequencing data interpretation: enhancing reproducibility and accessibility. Nature Reviews Genetics. 13(9): 667–672.
51. Nguyen T, Shi W, Ruden D (2011) CloudAligner: a fast and full-featured MapReduce based tool for sequence mapping, BMC Res Notes. 4: 171

52. Niemenmaa M, Kallio A, Schumacher A, Klemela P, Korpelainen E, Heljanko K. (2012) Hadoop-BAM: directly manipulating next generation sequencing data in the cloud. Bioinformatics. 28(6):876–7
53. O'Connor BD, Merriman B, Nelson BF. (2010) SeqWare Query Engine: storing and searching sequence data in the cloud. BMC Bioinform, 11(12):1
54. Ovsiannikov M, Rus S, Reeves D, Sutter P, Rao S, and Kelly J. (2013) The quantcast file system. Proceedings of the VLDB Endowment. 6(11): 1092–1101.
55. Owen S, Anil R, Dunning T, and Friedman E. (2011) Mahout in action. Manning. 145–182
56. Ram Vinay Pandey and Christian Schlötterer. (2013) DistMap: A Toolkit for Distributed Short Read Mapping on a Hadoop Cluster. PLoS One. 8(8): e72614.
57. Ren X, Wang Y, Zhang X-S, Jin Q. (2013) iPcc: a novel feature extraction method for accurate disease class discovery and prediction. Nucleic Acids Res: gkt343.
58. Rhodes DR, Chinnaiyan AM. (2005) Integrative analysis of the cancer transcriptome. Nat Genet 37: S31–S37
59. Schatz M, Sommer D, Kelley D, Pop M. (2010) De Novo assembly of large genomes with cloud computing. In Proceedings of the Cold Spring Harbor Biology of Genomes.
60. Schatz MC. (2009) CloudBurst: highly sensitive read mapping with MapReduce. Bioinformatics. 25: 1363–1369
61. Schatz, M.C. (2012) Computational thinking in the era of big data biology. Genome Bio. 13: 177
62. Shachak A, Shuval K, Fine S. (2007) Barriers and enablers to the acceptance of bioinformatics tools: a qualitative study. J Med Libr Assoc. 95: 454–458
63. Shi W, Guo YF, Jin C, and Xue X (2008) An improved generalized discriminant analysis for large-scale data set. Machine Learning and Applications. ICMLA'08. Seventh International Conference on. 769–772.
64. Shvachko K, Kuang H, Radia S, and Chansler R. (2010) The hadoop distributed file system. Mass Storage Systems and Technologies (MSST) on IEEE 26th Symposium. IEEE. 1–10.
65. Tan M, Tsang IW, and Wang L. Towards ultrahigh dimensional feature selection for big data. (2014) The Journal of Machine Learning Research. 15(1): 1371–1429.
66. The Cancer Genome Atlas Research Network, Weinstein JN, Collisson EA, Mills GB, Shaw KRM, Ozenberger BA, Ellrott K, Shmulevich I, Sander C, Stuart JM. (2013) The Cancer Genome Atlas Pan-Cancer analysis project. Nat Genet 45(10): 1113–1120.
67. Vouzis PD, Sahinidis NV. (2011) GPU-BLAST: using graphics processors to accelerate protein sequence alignment. Bioinformatics, 27: 182–188
68. Wang Y, Wu Q-F, Chen C, Wu L-Y, Yan X-Z, Yu S-G, et al. (2012) Revealing metabolite biomarkers for acupuncture treatment by linear programming based feature selection. BMC Syst Biol. 6:S15.
69. Wilks C, Cline MS, Weiler E, Diehkans M, Craft B, Martin C, et al. (2014) The Cancer Genomics Hub (CGHub): overcoming cancer through the power of torrential data. Database (Oxford) bau093: 1–10
70. Zeng A, Li T, Liu D, Zhang J, and Chen H (2015) A fuzzy rough set approach for incremental feature selection on hybrid information systems. Fuzzy Sets and Systems. 258: 39–60.
71. Zhang T, Ramakrishnan R and Livny M (1996) Birch: an efficient data clustering method for very large databases. In ACM SIGMOD Record. 25(2): 103–114.

# Mining Massive Genomic Data for Therapeutic Biomarker Discovery in Cancer: Resources, Tools, and Algorithms

**Pan Tong and Hua Li**

**Abstract** Cancer research is experiencing an evolution empowered by high-throughput technologies that makes it possible to collect molecular information for the entire genome at the DNA, RNA, protein, and epigenetic levels. Due to the complex nature of cancer, several organizations have launched comprehensive molecular profiling for thousands of cancer patients using multiple high-throughput technologies to investigate cancer genomics, transcriptomics, proteomics, and epigenomics. To speed up the bench-to-bedside translation, additional efforts have been made to profile hundreds of preclinical cell line models coupled with systematic screening of anticancer agents. This leads to an explosion of massive genomic data that shifts the bottleneck from data generation to data analytics. In this chapter, we will first introduce different types of genomic data as well as resources from publicly accessible data repositories that can be utilized to search for therapeutic targets for cancer treatment. We then introduce software tools frequently used for genomic data mining. Finally, we summarize working algorithms for the discovery of therapeutic biomarkers.

**Keywords** Genomics • Transcriptomics • Proteomics • Epigenomics • Biomarker discovery • Cancer

## 1 Introduction

Cancer is a disease of genetics involving dynamic changes of the genome [1]. Multiple genetic alterations have been identified in cancer including somatic mutations, DNA copy number change, epigenetic modifications, and dysregulated gene expression. Systematic discovery of cancer-driven alterations not only helps

P. Tong
Department of Bioinformatics and Computational Biology, The University of Texas MD Anderson Cancer Center, Houston, TX, USA

H. Li (✉)
School of Biomedical Engineering, Bio-ID Center, Shanghai Jiao Tong University, Shanghai, China
e-mail: kaixinsjtu@hotmail.com



K.-C. Wong (ed.), *Big Data Analytics in Genomics*,
DOI 10.1007/978-3-319-41279-5_10

us better understand tumorigenesis but also plays crucial roles in developing biomarkers for cancer detection, diagnosis, and prognosis. Over the last decade, there has been a dramatic advance in technologies allowing holistic interrogations of various aspects of cellular process including mRNAs (transcriptome), proteins (proteome), sequence and structural variations (genomics), metabolites (metabolomics), and interactions (interactome). While data from individual assay type is informative for certain aspects of biology, integrative analysis using multi-assay data sets is more powerful and provides deeper insights into understanding complex biological systems and diseases. As a result, there is an increasing trend for both individual laboratories and large consortia to generate multi-assay genomic profiling of cancer patients. For example, pioneering studies from the Sanger Institute and the Johns Hopkins Hospital identified frequent mutations in melanoma and colon cancer [2, 3]. Later studies in Boston and New York uncovered activating mutations in lung cancer which predicted response to kinase inhibitors [4–6]. Soon thereafter, the Human Cancer Genome Project was proposed by US National Cancer Institute which was later called The Cancer Genome Atlas (TCGA) [7]. In parallel, the International Cancer Genome Consortium (ICGC) was launched to foster international collaborations for large-scale cancer genomics studies [8]. To speed up the transition from bench to bedside, cell line models derived from cancer patients have been under extensive investigation. Several studies have generated comprehensive genomic characterizations of hundreds of cell line models coupled with drug screening enabling us to generate predictors of drug sensitivity based on genomic information [9–12].

While genomic data is now generated faster and cheaper than ever before, our ability to manage, analyze, and interpret it has not paced with the data deluge. Consequently, for the first time in history, the bottleneck in cancer research is shifting from data collection to data mining [13]. The objective of this chapter is to bridge the gap between advances in high-throughput genomics and our ability to manage, integrate, and analyze genomic data with special focus on therapeutic biomarker discovery. We begin with the definition of biomarker and an overview of different types of genomic data. We then summarize resources from publicly data repositories that can be utilized to search for therapeutic targets. We further introduce software tools frequently used for genomic data mining. Finally, we discuss working algorithms for the discovery of biomarkers.

## 2 Biomarkers in Cancer

According to National Cancer Institute, a biomarker is defined as a molecule found in blood or other body fluids that is objectively measured and evaluated as an indicator of disease status, pathogenic processes, or pharmacologic responses to therapeutic agents [14]. Biomarkers have been utilized for various applications including (1) measuring the natural history of disease, (2) correlating with clinical outcomes, (3) determining the biological effect of a therapeutic intervention, and

(4) serving as surrogate endpoints in clinical trials [15, 16]. Based on their utility, several types of biomarkers exist. Diagnostic biomarkers are used for early disease detection. Predictive biomarkers can infer the efficacy or toxicity of a drug. Prognostic biomarkers are used to assess if a patient receiving treatment will perform well or whether more aggressive treatment is needed to prevent recurrence. Staging biomarkers are used to determine the stage of progression of a disease.

Several methods can be used to identify candidate biomarkers. The classic approach is to identify biomarkers based on tumor biology where pivotal molecules in regulatory pathways are selected as candidates. However, such approach is time-consuming giving the large number of molecules and metabolites to be searched for. Recent development of high-throughput technologies has brought biomarker discovery into the "omics" era enabling simultaneous measurement of thousands of molecules. Genomics studies involving genotyping and next-generation sequencing have identified a considerable amount of biomarkers (such as single-nucleotide polymorphisms and structural variations) associated with drug efficacy and disease progression [17, 18]. Similarly, transcriptome and proteome profiling have also revealed biomarkers (such as dysregulated expression of RNA and proteins) that are highly correlated with clinical outcomes [19, 20].

## 3 High-Throughput Genomics

Over the past decade, there has been a dramatic advance in technologies enabling genome-scale data collection regarding various aspects of cellular process including sequence and structural variation, transcriptome, epigenome, and proteome. These technologies generate massive amounts of genomic data faster than ever before. Below we summarize major types of genomics data in cancer research and related technologies used to collect such data.

### *3.1 Transcriptomics*

Transcriptomics is the study of the complete set of RNA transcripts produced by the genome. Data collected for transcriptome starts with DNA microarrays using either spotted oligonucleotides or in situ synthesized probes to quantitatively measure mRNA levels of a large number of genes. The emergence of low-cost short read sequencing, also known as next-generation sequencing (NGS) technology, escalates transcriptomics studies to a new level by overcoming many drawbacks inherent in microarray such as requirement of carefully designed probes, cross-hybridization, high background noise, and low resolution [21]. In addition to provide fast and accurate measurement of transcripts, NGS RNA sequencing also facilitates deeper understanding of the transcriptome including alternative splicing, gene fusion, and isoform expression. It is worth noting that transcriptomics studies are not limited to the investigation of messenger RNA. For example, whole genome profiling of microRNAs and other noncoding RNAs is usually employed to decipher post-transcriptional regulation of gene expression [22].

### 3.2 Proteomics

Proteomics is the large-scale study of proteins including protein abundance, modifications, localizations and interactions. The growth of proteomics studies owes to the advances in protein technologies such as capillary electrophoresis, high performance liquid chromatography (HPLC), matrix-assisted laser desorption/ionization (MALDI), and mass spectrometry [23]. The reverse-phase protein array (RPPA) first introduced in the early 2000s is widely used in biomarker discovery, therapeutic target evaluation, and cancer research. It now becomes a promising tool for clinical trials [24].

### 3.3 Epigenomics

Epigenomics refers to the study of epigenetic modifications in the DNA sequence as well chromatin including DNA methylation, covalent modifications of cytosine, and post-translational modifications of histones such as methylation, acetylation, and phosphorylation [25]. Functionally, epigenetic modifications are involved in regulation of gene expression, gene dosage, chromosome inactivation, and genome imprinting. It has been found that changes in epigenomics have been implicated in multiple diseases including cancer [26]. Epigenomics can be studied using DNA methylation array or next-generation sequencing with chemically treated DNA [27].

### 3.4 Sequence Variation

Genomic sequence variation includes single-nucleotide polymorphisms (SNP), mutations, copy number variations, and structural variations. The ultimate goal of human genetics is to identify all genomic sequence variation and deciphers how they contribute to phenotype and diseases. Currently, SNP arrays are cost-effective instruments to identify SNPs and copy number variations. In contrast, NGS technologies can be applied to interrogate all the genomic variations and provide higher resolution data for downstream functional studies [28].

# 4 Resources for Genomic Data

There is a rich source of public genomic data which provides unprecedented possibilities for hypothesis generation and data mining.

## *4.1 Gene Expression Omnibus (GEO)*

GEO (http://www.ncbi.nlm.nih.gov/geo) is the largest public repository for high-throughput gene expression data [29]. GEO archives and freely distributes microarray, next-generation sequencing, and other forms of high-throughput functional genomic data generated by the scientific community. There are three main entities in GEO database: Platform, Sample, and Series [30]. A Platform record includes a summary description of the array or sequencer and an additional table providing probe annotation or sequence information. Each Platform record is assigned a unique GEO accession number with prefix GPL. A Sample record provides all information related to a sample including phenotype information, experimental protocol, and abundance measurements for each feature recorded in the Platform. Sample accession numbers have a GSM prefix. A Series record defines a set of Samples related to a particular study and provides a description of overall study design. Series records have a prefix GSE. As of 2013, the GEO database hosts >32,000 records of Series submitted by around 13,000 laboratories, corresponding to 800,000 samples derived from over 1600 organisms [31]. Genomic data is worthless without contextual biological details and analysis methodologies for preprocessing. To ensure important information is preserved, scientific reporting standards have been proposed such as MIAME (Minimum Information About a Microarray Experiment) for data annotation and MINiML (MIAME Notation in Markup Language) for XML based data exchange [31]. The GEO database is in compliance with both MIAME and MINiML standards which greatly facilitates data retrieval. In addition to provide a searchable database for data retrieval, GEO now includes basic data mining and visualization functionalities. Users can compare two sets of samples with specified statistical parameters, construct clustered heatmaps, retrieve profiles with similar patterns of expression, and identify profiles belonging to the same homologs [32]. A major update recently was the release of GEO2R web application which allows users to perform sophisticated analysis using R [31]. Once a user specifies a Series number to be analyzed, GEO will retrieve the data using GEOquery [33] in the backend. The retrieved data is then subjected to analysis specified by user or from default settings. Results computed from the server are transferred to user using JSON and rendered as HTML page. Since R script is provided, users can always reproduce the analysis and fine-tune it offline.

## 4.2 ArrayExpress

ArrayExpress (https://www.ebi.ac.uk/arrayexpress/) is the European counterpart of GEO. ArrayExpress complies with the MIAME and MAGE-ML (Microarray Gene Expression Markup Language) standards to ensure data consistency. Currently, ArrayExpress contains more than 1.8 million samples with high-throughput assays across over 62,000 studies with a total file size of 40 TB. Programmatic access of the ArrayExpress data is available through the ArrayExpress Bioconductor package [34].

## 4.3 The Cancer Genome Atlas (TCGA)

The first public repository dedicated to cancer is TCGA (https://tcga-data.nci.nih.gov/tcga/). The overall goal of TCGA is to generate comprehensive, multi-dimensional profiling of genomic alterations in major cancer types. TCGA is organized by different centers responsible for sample collection, processing, and analysis. First, Tissue Source Sites (TSSs) collect biospecimens from eligible cancer patients and deliver them to Biospecimen Core Resources (BCRs). BCRs then catalogue, process, and verify the received samples before submitting to Data Coordinating Centers (DCC). DCCs provide molecular analytes for the Genome Characterization Centers (GCCs) and Genome Sequencing Centers (GSCs) for genomic characterization. The generated genomic data is passed to Genome Data Analysis Centers (GDACs) for information processing, analysis, and tool development. All data generated is made publicly available through TCGA Data Portal (https://tcga-data.nci.nih.gov/tcga/tcgaDownload.jsp) and CGHub (https://cghub.ucsc.edu/). TCGA employs several high-throughput technologies including microarrays, next-generation sequencing, and reverse-phase protein array (RPPA) to interrogate global alterations at DNA, RNA, and protein levels. In particular, RNA sequencing provides transcriptomic monitoring of gene expression, isoforms, gene fusions, and noncoding RNAs. DNA sequencing determines genetic alterations such as insertions, deletions, polymorphisms, and copy number variations. SNP-based platforms assess single-nucleotide polymorphisms (SNPs), copy number variations, and loss of heterozygosity (LOH). Array-based methylation provides epigenetic information at CpG sites. Bisulfite sequencing characterizes DNA methylation at single nucleotide resolution. RPPA provides quantitative measurements of protein expression with high sensitivity. Since its inaugural in 2006, TCGA has comprehensively profiled more than 10,000 samples across 33 cancer types.

### 4.4 International Cancer Genome Consortium (ICGC)

While TCGA provides comprehensive genomic characterization for cancer patients in the USA, the ICGC project (https://icgc.org/) aims to generate an extensive catalogue of genomic abnormalities for cancer patients throughout the world contributed by different participating countries. Currently, the ICGC data portal records 78 projects covering 50 different cancer types. The ICGC data portal periodically updates with newly generated data and provides tools for data downloading, visualizing, and querying. Due to the large size of data which may take months to download, ICGC partners with Amazon Web Services to facilitate data access through the cloud. ICGC also releases analytic workflows so that users can analyze their data using the same workflows after initiating an Amazon machine.

### 4.5 The NCI-60 Cell Line Panel

Immortalized cell lines derived from human cancer have made significant contributions to cancer biology and formed the basis of current understanding of drug sensitivity and resistance. Therefore, systematic genomic characterization of cell line models coupled with pharmacological interrogation would greatly facilitate biomarker identification and drug development. One of the early endeavors is the NCI-60 project which has released a panel of 60 cell lines with high-throughput gene genomic profiling including DNA copy number, gene expression, protein expression, and mutation and additional anticancer drug screening [11]. The NCI-60 panel quickly becomes a rich source of information to investigate mechanisms of drug resistance. A major discovery using the NCI-60 data set has been the linkage between P-glycoprotein expression and multi-drug resistance [35].

### 4.6 Cancer Cell Line Encyclopedia (CCLE)

Following the success of NCI60, the CCLE project (http://www.broadinstitute.org/ccle/home) [9] has extended genomic characterization to around 1000 cell lines using gene expression, chromosomal copy number, and massively parallel sequencing technology. CCLE has also screened 24 anticancer drugs from 479 cell lines using an automated compound-screening platform [9]. An integrative analysis of the CCLE data identified genetic, lineage and gene expression based predictors of pharmacological vulnerabilities which had important clinical implications for personalizing cancer therapy [9].

### 4.7 Genomics of Drug Sensitivity in Cancer (GDSC)

Similar to CCLE, the GDSC (http://www.cancerrxgene.org/) database has profiled 138 anticancer drugs encompassing both targeted agents and cytotoxic therapeutics across 700 cancer cell lines [12]. Initial analysis using GDSC data found that mutated cancer genes were markers of sensitivity or resistance to a broad range of anticancer drugs. Further, the mutated cancer genes mostly associated with sensitivity were found to be oncogenes that were direct targets of the drug [12]. On the other hand, inactivating mutations in tumor suppressors were associated with drug resistance [12]. For example, mutations in BRAF, an oncogene responsible for protein kinase signaling, were associated with sensitivity of BRAF inhibitors MEK1 and MEK2. In contrast, mutations in TP53, an important tumor suppressor responsible for apoptosis, conferred resistance to nutlin-3a, which was an inhibitor of MDM2 that negatively regulated p53 protein [36].

### 4.8 Cancer Therapeutics Response Portal (CTRP)

In addition to identify biomarkers of drug sensitivity, genomic characterization coupled with drug screening can also shed light on mechanisms of action (MoA). Recently, the CTRP (http://www.broadinstitute.org/ctrp/) database published high quality screening data of 481 compounds across 860 cancer cell lines spanning 23 lineages [10]. By comparing the sensitivity pattern of compounds targeting the same gene, targeting genes in the same pathway and targeting genes that metabolically process the compounds, the authors observed that sensitivity may depend on metabolic activation, import of the compound, the presence of target-drug complex, and the presence of target expression. On the other hand, drug resistance was linked to drug inactivation or an efflux mechanism that depleted drug from the cell [10].

### 4.9 Project Achilles

In an effort to identify genes essential for cell proliferation and viability in cancer cell lines, Project Achilles (https://www.broadinstitute.org/achilles) employs genome-wide genetic perturbation experiments using pooled shRNA technology. The screening pipeline uses around 54,000 shRNA plasmids targeting 11,000 genes with a minimum representation of 200 cells per shRNA [37]. The pooled shRNA screens are able to silence or knock-out genes and thus identify genes essential for growth and survival. After incubation for a certain period of time, the cell lines are harvested to determine relative levels of shRNA plasmids using Illumina sequencing technology. When linked with genetic characteristics of the cell lines, Project Achilles provides valuable information for prioritizing targets for therapeutic drug development.

While individual data resource introduced here can be helpful in addressing different questions, it is usually more valuable to integrate across different resources since they largely provide complementary information. For example, candidate biomarkers overexpressed in cancers can be identified using the TCGA data. The therapeutic relevance of such biomarkers in terms of in vitro drug sensitivity can then be evaluated using the NCI-60 panel, the CCLE, and the GDSC database. Finally, essentiality of these biomarkers from knock-out experiments can be extracted from the Project Achilles data. Such an integrated analysis not only provides a full picture of the utility associated with identified biomarkers, but greatly narrows down the number of candidates and thus can greatly reduce costs in validating the biomarkers.

# 5 Tools for Mining Genomic Data

Choosing the right set of tools is vital for genomic data mining. One of the most popular tools is the R programming language, an open source environment for statistical computing. R has strong support for statistical analysis including linear and nonlinear modeling, hypothesis testing, time series analysis, spatial analysis, clustering, and classification. R also provides various facilities for data manipulation, calculation, and visualization [38]. Further, R is highly extensible with lots of packages contributed by users in the R community. Among the various packages dedicated to high-throughput genomics, Bioconductor is one of the most comprehensive and versatile tools [39]. It greatly facilitates rapid creation of pipelines by combining multiple procedures. Bioconductor includes tools for all stages of analysis ranging from data generation to final presentation. Bioconductor also has high quality documentation through three levels: vignettes that provide example usages of a particular package; manual pages that precisely describe inputs, outputs and examples of a function; and workflows that showcase complete analysis spanning multiple tools and packages. Recently, Bioconductor has enhanced its functionality by enabling analysis of next-generation sequencing data. Core infrastructure includes *Biostrings* for DNA and amino acid sequence manipulation, *ShortRead* for FASTQ files, *IRanges* and *GenomicRanges* for genome coordinate calculation, *GenomicAlignments* and *Rsamtools* for aligned sequencing data, *BSgenome* for curated whole-genome sequence, and *rtracklayer* for integration of genome browsers with experimental data. Currently, Bioconductor has 1104 packages, 895 annotation databases, and 257 packaged experimental data and the functionality is still expanding.

Genomic data will be useless if no metadata is given regarding the entities measured such as gene symbols, probe ID, genomic coordinates, and genome versions. Public service providers and instrument vendors have websites from which users can download relevant information for offline data wrangling. However, this process is time-consuming, error-prone, and irreproducible. The *biomaRt* package hosted on the Bioconductor repository provides a unified interface for accessing a large collection of databases including NCBI (National Center for Biotechnology Information), Ensembl, UCSC (University of California, Santa Cruz), COSMIC (Catalogue of Somatic Mutations in Cancer), Uniprot (Universal Protein Resource), HGNC (HUGO Gene Nomenclature Committee providing official gene names), and Reactome (curated biological pathways) [40]. *BiomaRt* allows seamless integration of identifier mapping and annotation into data analysis, creating a powerful platform for biological data mining [41].

Although R combined with Bioconductor proves to be a powerful computing engine for genomic analysis, users are required to have reasonable programming skills to fully unleash its power. Alternately, there are web-based tools suited for both experimentalists and computational colleagues where analysis can be performed with mouse click. Galaxy is one of such tools with a web-based graphical user interface for accessible, reproducible, and transparent genomic data mining [42]. By encapsulating high-end computation tools while hiding the technical details of computation and storage, the Galaxy software becomes highly accessible to users without programming skills. By automatically tracking metadata regarding input data sets, analysis parameters, analytic components and output data, and by supporting user specified annotations and tags, Galaxy makes it easy to assemble and reproduce any given analysis [43]. Galaxy also makes analysis transparent by allowing users to share their analysis using Galaxy's sharing model. This includes a web-based framework for displaying results, customizable web pages that users

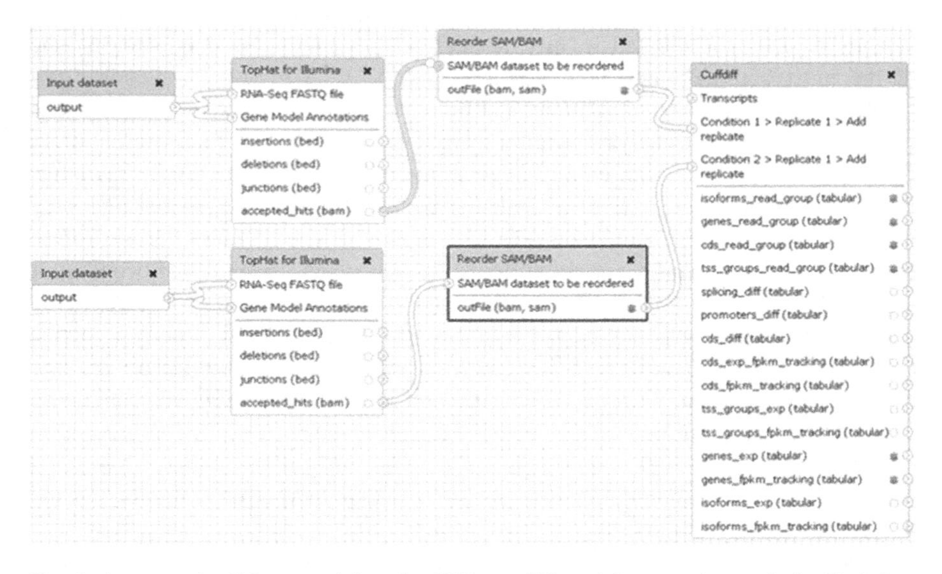


**Fig. 1** An example Galaxy workflow for RNAseq differential expression analysis. Each box represents a tool with input and output files. Users can connect the output files of one component to the input of another component to form a complete analysis. A compatible link between components will become green to aid users. Here input FASTQ files are fed into Tophat for alignment. The resulting bam file is then sorted and indexed for Cuffdiff differential gene expression analysis

can freely modify, and a public repository hosting published items such as datasets, histories, and workflows [42].

The Galaxy workflow greatly enhances usability by providing a drag-and-drop interface for building analytic pipelines. Figure 1 demonstrates a workflow for RNAseq differential expression analysis. Users insert different analytic components into the workflow canvas and connect them to form a complete analysis. The workflow editor verifies each link between the tools for compatibility. Compatible links will turn green transiently to visually aid workflow construction. To further simplify the creation of workflows, Galaxy allows users to create workflows from analysis history. This feature greatly simplifies workflow usage since users do not need to plan analysis upfront. In addition, Galaxy is highly extensible. Any piece of software written in any language can be integrated into the Galaxy workflow. To add a new tool to Galaxy, users only need to specify a configuration file dictating how to run the tool. Additionally, users need to describe input and output parameters so that the Galaxy framework knows how to work with the tool abstractly and automatically generates a web interface for it.

An alternative tool to Galaxy is the GenePattern software (www.genepattern.org/) which also does not require programming skills. GenePattern is claimed as a pipeline builder providing form-based methods for data preprocessing, analysis, and visualization [44]. GenePattern hosts different modules through a centralized repository so that users can download or upgrade when needed. In addition to the graphical user interface, GenePattern also allows command line access which makes automatic batch processing possible. Currently, users can access GenePattern through R, Matlab, and Java by invoking a local GenePattern instance. The combination of a graphical user interface with a programmatic console becomes a unique feature of GenePattern. Since its first release in 2004, GenePattern has over 23,000 registered users from over 2900 commercial and non-profit organizations worldwide.

There are also tools built on well-curated cancer genomic data. Here we illustrate two examples: cBioPortal (http://www.cbioportal.org/) and Oncomine (http://www.oncomine.org). cBioPortal provides a web interface for exploring, visualizing, analyzing, and downloading multi-platform cancer genomic data [45]. By hosting a large set of well-curated cancer genomic data including somatic mutation, mRNA and microRNA expression, protein expression, DNA copy number, and DNA methylation, cBioPortal greatly facilitates integrative genomic analysis by allowing users to query multiple data types and their associations at individual gene level. Further, cBioPortal also supports mutual exclusivity analysis for genomic alterations, survival analysis, co-expression analysis, enrichment analysis, and network analysis. While cBioPortal focuses on multi-platform cancer genomic data, Oncomine specializes in microarray gene expression data. Currently, Oncomine hosts 715 datasets consisting of 86,733 samples. Oncomine also provides a web interface for users to perform differential gene expression, co-expression, interaction network, cancer outlier profile analysis (COPA), and molecular concept analysis [46].

# 6 Algorithms for Cancer Biomarker Discovery

To identify biomarkers of various utilities, both supervised and unsupervised methods can be used. For supervised biomarker discovery, an outcome variable associated with each sample is required so that candidate markers predictive to the outcome variable can be identified. In comparison, unsupervised methods rely on the genomic assays only and search candidate biomarkers by modeling the signals from genomic measurements. Below we introduce several popular algorithms for both supervised and unsupervised biomarker discovery.

## 6.1 *Supervised Methods*

A common task of genomic biomarker discovery is to compare the gene expression levels (e.g., transcriptomic expression, proteomic expression, or microRNA expression) of samples under different treatment conditions or at different time points. This task is usually called a differential gene expression (DEG) analysis. Numerous methods have been published for DEG analysis using high-throughput genomics data. A straightforward approach is to use a two-sample *t*-test (in the case of binary outcome) or a linear regression framework (in the case of categorical or continuous outcome). However, genomic data may contain outlier measurements that violate the underlying statistical assumptions. Therefore, various variants of these methods have been developed by considering statistical robustness. One of the most popular methods used in gene expression analysis is the significance analysis of microarray (SAM) software developed by Tusher et al. [47]. For a two-sample comparison, SAM computes a "relative difference" metric $d(i)$ for each gene $i$:

$$d(i) = \frac{\overline{x}_1(i) - \overline{x}_2(i)}{s(i) + s_0}$$

Here $\overline{x}_1(i)$ and $\overline{x}_2(i)$ are the average expression in the two groups, $s(i)$ is the gene specific standard deviation of the repeated measurements, and $s_0$ is a stabilizing constant chosen to penalize uninteresting genes with poor signal to noise ratio. Since a theoretical null distribution of $d(i)$ is difficult to obtain, SAM instead resorts to a permutation based approach to assess statistical significance.

The original SAM software only dealt with binary outcome. Later versions of SAM allowed the analysis of data with multiclass outcome, continuous outcome, and censored survival time. To do this, the authors extended the definition of $d(i)$ as following:

$$d(i) = \frac{r_i}{s(i) + s_0}$$

Here $r_i$ is a score defined differently for different types of outcome. More details about this extension can be found on the SAM manual (http://statweb.stanford.edu/~tibs/SAM/).

An important concept for biomarker discovery is that statistical significance does not ensure biological significance. Clinically useful biomarkers need to have strong dynamic range and a manageable gene size so that they can be easily assayed in a single panel. The SAM method computes a score for each gene and additional filtering step is needed to narrow down the gene list. The Top Scoring Pair (TSP) method uses a different strategy by comparing the relative expression of every possible gene pairs [48]. It effectively reduces the number of biomarkers to two and limits the selected genes to have a strong contrast easy to quantify [48]. Let the expression of a particular gene in sample $i$ be $x_i$ and let the class label associated with each sample be $c$ which can be any value in $\{1, 2, \ldots, C\}$. TSP computes the frequency of observing $x_i < x_j$ for each class c as $p_{ij}(\mathrm{c}) = \mathrm{P}\left(x_i < x_j \middle| \mathrm{c}\right)$. In the case of $\mathrm{C} = 2$, the TSP score $\Delta_{ij}$ is defined as following although this metric can be extended to higher number of classes:

$$\Delta_{ij} = \left|p_{ij}(1) - p_{ij}(2)\right|$$

The TSP method selects genes based on their relative expression, which is different from other approaches used in DEG analysis. Further, a TSP pair provides a simple rule to classify samples into different classes. For example, if gene $i$ has higher expression than gene $j$ in a TSP pair, the sample will be classified as class 1 or class 2 otherwise depending on the relative conditional probabilities. Notice that this classification rule only requires relative expression between the two genes, which will make such biomarkers more robust and easy to interpret. Various studies have reported the success of TSP as a two-gene classifier [49–51]. However, for data sets with a complex phenotype, a single TSP pair may not be sufficient. The so-called k-TSP method has been proposed to make use of top k scoring pairs [52, 53]. Although a majority vote can be used to obtain a final classification, other supervised machine learning methods have been used and benchmarked including support vector machine, decision trees, naive Bayes classifier, k-nearest neighbor (k-NN), and prediction analysis of microarray (PAM) [52, 53].

A major approach to narrow down selected biomarkers is through variable selection in the framework of linear regression. The traditional stepwise variable selection approach only works for data with a small set of features and becomes computationally infeasible for big data such as microarray or RNAseq. Shrinkage estimators such as lasso (least absolute shrinkage and selection operator) have been developed to efficiently deal with such high-dimensional genomic data. Later efforts have extended the original lasso method including the grouped lasso by Yuan et al. where variables are selected or excluded in groups [54], the elastic net by Zou et al. which deals with correlated variables through a hybrid penalty [55], the graphical lasso by Friedman et al. for space covariance estimation [56], and the regularization paths for support vector machine [57].

Here we summarize the algorithms for lasso and closely related methods including elastic net and ridge regression. Given the response variable $Y \in R$ and a predictor vector $X \in R^p$ in a p dimensional space, we can approximate a linear function through $E\left(Y \middle| X = x\right) = \beta_0 + x^T\beta$ after observing N observation pairs $(x_i, y_i)$ for $i = 1, 2, \ldots, N$ by solving the following optimization problem [58]:

$$\min_{(\beta_0,\beta)\in R^{p+1}} R_\lambda(\beta_0,\beta) = \min_{(\beta_0,\beta)\in R^{p+1}} \left[\frac{1}{2N}\sum_{i=1}^{N}\left(y_i - \beta_0 + x_i{}^T\beta\right)^2 + \lambda P_\alpha(\beta)\right]$$

where $P_\alpha(\beta)$ is the elastic net penalty term [55] defined as

$$P_\alpha(\beta) = (1-\alpha)\,\tfrac{1}{2}||\beta||_{\ell_2}^2 + \alpha||\beta||_{\ell_1}^1.$$

Here $\lambda$ is a tuning parameter that users can specify or can be automatically calculated using cross validation based on prediction error.

Both the lasso algorithm (when $\alpha = 1$) and ridge regression (when $\alpha = 0$) are special cases of the elastic net method. Lasso provides coefficient estimates as either zero (for excluded variables) or nonzero (for selected variables) which is quite appealing for big genomic data. Ridge regression, on the other hand, only shrinks the coefficients and provides nonzero estimates only. For correlated variables, lasso tends to just pick one while ignore others. On the other hand, ridge regression allows borrowing information across all variables but retains all variables in the model. The elastic net with $0 < \alpha < 1$ enjoys the nice properties of both and usually performs better in genomic data. Elastic net has been efficiently implemented using cyclical coordinate descent and is publicly available in the R package glmnet.

Traditional strategies for biomarker discovery have focused on individual genes. However, tumorigenesis is a multi-step process involving sequential acquisitions of multiple genomic alterations regulated by different pathways and regulatory networks [62]. It is therefore appealing to identify biomarkers as sets of genes. According to Huang et al., three types of gene set analysis tools are available [63]. The first type is called singular enrichment analysis (SEA) which takes a preselected gene list as input and iteratively computes statistical enrichment of annotated gene sets by comparing them to random gene sets. The second type is called modular enrichment analysis (MEA) which considers inter-relationships as well as redundancies among annotated gene sets. MEA extends enrichment analysis from gene-centric or term-centric analysis to module-centric analysis which is more biologically plausible. The third type is called gene set enrichment analysis (GSEA). Different from SEA or MEA which requires a filtered gene list as input, GSEA takes into account all genes available and thus avoids the need of arbitrary cutoff for gene filtering. Different tools have different advantages and limitations. Users need to choose a tool that best fits their needs by considering the underlying statistical model, gene set annotation source, programming requirement, and output format.

Algorithms to identify cancer biomarkers are not limited to deal with a single data type. Several methods have been developed to integrate information across data types. For example, integIRTy (integration using item response theory) is able to identify altered genes from multiple assay types accounting for multiple mechanisms of alteration [64]. integIRTy applies a latent variable approach to adjust for heterogeneity among different assay types for accurate inference. RABIT (regression analysis with background integration) is able to integrate public transcription factor (TF) binding profiles with tumor-profiling datasets [65]. RABIT controls confounding effects from copy number alteration, DNA methylation, and TF somatic mutation to identify cancer-associated TFs using a regression framework. Another interesting method is PARADIGM (PAthway Recognition Algorithm using Data Integration on Genomic Models) that integrates different genomic information based on pathway activity [66]. PARADIGM uses a factor graph to represent NCI pathway information which makes it effective to model different types of genomic data and various regulatory relationships.

## 6.2 Unsupervised Methods

The aforementioned methods are supervised since they require an outcome variable. There are also unsupervised methods for cancer biomarker discovery. Motivations for these methods originate from the fact that certain perturbations in the genome such as focal copy number change, gene fusions and mutations may lead to marked over-expression of oncogenes in a subset of samples. Since these oncogene activation events do not necessarily occur across all samples, traditional analytical approaches based on mean expression will fail [59]. Therefore, several methods have been proposed for this situation. For example, cancer outlier profile analysis (COPA) was developed to discover oncogenic chromosomal aberrations from outlier profiles based on median and median absolute deviation of gene expression. COPA identified the fusion of ERG and ETV1 which led to marked over-expression in 57 % of prostate cancer patients [59]. Later, a method called PACK (profile analysis using clustering and kurtosis) showed improved result by using Bayesian information criterion (BIC) and kurtosis [60]. Tong et al. developed SIBER (systematic identification of bimodally expressed genes using RNAseq data) using mixture model [61]. SIBER compares favorably to other methods and enjoys nice properties such as robustness, increased statistical power, and invariance to transformation [61]. We briefly summarize the SIBER algorithm here. Suppose the expression of a gene in sample $s$ is $e_s$, SIBER models the distribution of gene expression $Pr(\mathrm{e_s})$ using a two-component mixture model each with mean expression $\mu_1$, $\mu_2$ and a shared dispersion parameter $\phi$ as following:

$$Pr\left(\mathrm{e_s}\right) = \pi f\left(\mathrm{e_s}; \mu_1, \phi\right) + (1 - \pi) f\left(\mathrm{e_s}; \mu_2, \phi\right)$$

where $\pi$ is the proportion of samples coming from the first component with density function $f(\mathrm{e_s}; \mu_1, \phi)$. The density function frequently used to model RNAseq data can be negative Binomial, generalized Poisson or log-normal distribution. After estimating the parameters $(\pi, \mu_1, \mu_2, \phi)$, SIBER computes a generalized bimodality index BI as following:

$$BI = \sqrt{\pi(1-\pi)}\frac{\left|\mu_1 - \mu_2\right|}{\sqrt{(1-\pi)\sigma_1^2 + \pi\sigma_2^2}}$$

where $\sigma_1^2, \sigma_2^2$ were the variance of the two components. Through extensive simulation and real data analysis, Tong et al. showed that SIBER was a robust and powerful method to identify biomarkers with switch-like expression pattern [61].

## 7 Concluding Remarks

With recent advances in genomic technologies, the accumulation of genomic data is far exceeding Moore's law leading to the genomic data deluge. This represents a clear opportunity as well as pressing challenge for computational scientists to wade through the huge amount of data for biological insights. To identify biomarkers for cancer therapeutics, we should be familiar with relevant data resources and equip ourselves with effective computational tools. Given the extreme challenges for genomic data, the future success of cancer genomic research requires a continuous refinement and expansion of software tools and algorithms for the management, analysis, integration, and interpretation of high-throughput data.

## References

1. Hanahan, D. and R.A. Weinberg, *The hallmarks of cancer.* cell, 2000. **100**(1): p. 57–70.
2. Davies, H., et al., *Mutations of the BRAF gene in human cancer.* Nature, 2002. **417**(6892): p. 949–954.
3. Samuels, Y., et al., *High frequency of mutations of the PIK3CA gene in human cancers.* Science, 2004. **304**(5670): p. 554–554.
4. Lynch, T.J., et al., *Activating mutations in the epidermal growth factor receptor underlying responsiveness of non–small-cell lung cancer to gefitinib.* New England Journal of Medicine, 2004. **350**(21): p. 2129–2139.
5. Paez, J.G., et al., *EGFR mutations in lung cancer: correlation with clinical response to gefitinib therapy.* Science, 2004. **304**(5676): p. 1497–1500.
6. Pao, W., et al., *EGF receptor gene mutations are common in lung cancers from "never smokers" and are associated with sensitivity of tumors to gefitinib and erlotinib.* Proceedings of the National Academy of Sciences of the United States of America, 2004. **101**(36): p. 13306–13311.
7. Weiss, R. *NIH Launches Cancer Genome Project*. 2005; Available from: http://www.washingtonpost.com/wp-dyn/content/article/2005/12/13/AR2005121301667.html.
8. Hudson, T.J., et al., *International network of cancer genome projects.* Nature, 2010. **464**(7291): p. 993–998.

9. Barretina, J., et al., *The Cancer Cell Line Encyclopedia enables predictive modelling of anticancer drug sensitivity.* Nature, 2012. **483**(7391): p. 603–607.
10. Rees, M.G., et al., *Correlating chemical sensitivity and basal gene expression reveals mechanism of action.* Nature chemical biology, 2015.
11. Shoemaker, R.H., *The NCI60 human tumour cell line anticancer drug screen.* Nature Reviews Cancer, 2006. **6**(10): p. 813–823.
12. Yang, W., et al., *Genomics of Drug Sensitivity in Cancer (GDSC): a resource for therapeutic biomarker discovery in cancer cells.* Nucleic acids research, 2013. **41**(D1): p. D955–D961.
13. Ding, L., et al., *Expanding the computational toolbox for mining cancer genomes.* Nature Reviews Genetics, 2014. **15**(8): p. 556–570.
14. Colburn, W., et al., *Biomarkers and surrogate endpoints: Preferred definitions and conceptual framework. Biomarkers Definitions Working Group.* Clinical Pharmacol & Therapeutics, 2001. **69**: p. 89–95.
15. Frank, R. and R. Hargreaves, *Clinical biomarkers in drug discovery and development.* Nature Reviews Drug Discovery, 2003. **2**(7): p. 566–580.
16. Liang, M.H., et al., *Methodologic issues in the validation of putative biomarkers and surrogate endpoints in treatment evaluation for systemic lupus erythematosus.* Endocrine, metabolic & immune disorders drug targets, 2009. **9**(1): p. 108.
17. Leary, R.J., et al., *Development of personalized tumor biomarkers using massively parallel sequencing.* Science translational medicine, 2010. **2**(20): p. 20ra14–20ra14.
18. Ji, Y., et al., *Glycine and a Glycine Dehydrogenase (GLDC) SNP as Citalopram/Escitalopram Response Biomarkers in Depression: Pharmacometabolomics-Informed Pharmacogenomics.* Clinical Pharmacology & Therapeutics, 2011. **89**(1): p. 97–104.
19. CHEN, H.Y., et al., *Biomarkers and transcriptome profiling of lung cancer.* Respirology, 2012. **17**(4): p. 620–626.
20. Zhao, L., et al., *Identification of candidate biomarkers of therapeutic response to docetaxel by proteomic profiling.* Cancer research, 2009. **69**(19): p. 7696–7703.
21. Wang, Z., M. Gerstein, and M. Snyder, *RNA-Seq: a revolutionary tool for transcriptomics.* Nature Reviews Genetics, 2009. **10**(1): p. 57–63.
22. Pritchard, C.C., H.H. Cheng, and M. Tewari, *MicroRNA profiling: approaches and considerations.* Nature Reviews Genetics, 2012. **13**(5): p. 358–369.
23. Wright, P., et al., *A review of current proteomics technologies with a survey on their widespread use in reproductive biology investigations.* Theriogenology, 2012. **77**(4): p. 738–765. e52.
24. Mueller, C., L.A. Liotta, and V. Espina, *Reverse phase protein microarrays advance to use in clinical trials.* Molecular oncology, 2010. **4**(6): p. 461–481.
25. Strahl, B.D. and C.D. Allis, *The language of covalent histone modifications.* Nature, 2000. **403**(6765): p. 41–45.
26. Lund, A.H. and M. van Lohuizen, *Epigenetics and cancer.* Genes & development, 2004. **18**(19): p. 2315–2335.
27. Zuo, T., et al., *Methods in DNA methylation profiling.* Epigenomics, 2009. **1**(2): p. 331–345.
28. Soon, W.W., M. Hariharan, and M.P. Snyder, *High-throughput sequencing for biology and medicine.* Molecular systems biology, 2013. **9**(1): p. 640.
29. Barrett, T., et al., *NCBI GEO: mining tens of millions of expression profiles—database and tools update.* Nucleic acids research, 2007. **35**(suppl 1): p. D760–D765.
30. Barrett, T. and R. Edgar, *Gene Expression Omnibus: Microarray Data Storage, Submission, Retrieval, and Analysis.* Methods in enzymology, 2006. **411**: p. 352–369.
31. Barrett, T., et al., *NCBI GEO: archive for functional genomics data sets—update.* Nucleic acids research, 2013. **41**(D1): p. D991–D995.
32. Wilhite, S.E. and T. Barrett, *Strategies to explore functional genomics data sets in NCBI's GEO database*, in *Next Generation Microarray Bioinformatics*. 2012, Springer. p. 41–53.
33. Davis, S. and P.S. Meltzer, *GEOquery: a bridge between the Gene Expression Omnibus (GEO) and BioConductor.* Bioinformatics, 2007. **23**(14): p. 1846–1847.
34. Kauffmann, A., et al., *Importing arrayexpress datasets into r/bioconductor.* Bioinformatics, 2009. **25**(16): p. 2092–2094.

35. Wu, L., et al., *Multidrug-resistant phenotype of disease-oriented panels of human tumor cell lines used for anticancer drug screening.* Cancer research, 1992. **52**(11): p. 3029–3034.
36. Garnett, M.J., et al., *Systematic identification of genomic markers of drug sensitivity in cancer cells.* Nature, 2012. **483**(7391): p. 570–575.
37. Cowley, G.S., et al., *Parallel genome-scale loss of function screens in 216 cancer cell lines for the identification of context-specific genetic dependencies.* Scientific data, 2014. **1**.
38. Team, R.C., *R: A language and environment for statistical computing. R Foundation for Statistical Computing, Vienna, Austria, 2012.* 2014, ISBN 3-900051-07-0.
39. Huber, W., et al., *Orchestrating high-throughput genomic analysis with Bioconductor.* Nature methods, 2015. **12**(2): p. 115–121.
40. Durinck, S., et al., *Mapping identifiers for the integration of genomic datasets with the R/Bioconductor package biomaRt.* Nature protocols, 2009. **4**(8): p. 1184–1191.
41. Durinck, S., et al., *BioMart and Bioconductor: a powerful link between biological databases and microarray data analysis.* Bioinformatics, 2005. **21**(16): p. 3439–3440.
42. Goecks, J., A. Nekrutenko, and J. Taylor, *Galaxy: a comprehensive approach for supporting accessible, reproducible, and transparent computational research in the life sciences.* Genome Biol, 2010. **11**(8): p. R86.
43. Blankenberg, D., et al., *Galaxy: a web-based genome analysis tool for experimentalists.* Current protocols in molecular biology, 2010: p. 19.10. 1–19.10. 21.
44. Reich, M., et al., *GenePattern 2.0.* Nature genetics, 2006. **38**(5): p. 500–501.
45. Gao, J., et al., *Integrative analysis of complex cancer genomics and clinical profiles using the cBioPortal.* Science signaling, 2013. **6**(269): p. pl1.
46. Rhodes, D.R., et al., *Oncomine 3.0: genes, pathways, and networks in a collection of 18,000 cancer gene expression profiles.* Neoplasia, 2007. **9**(2): p. 166-180.
47. Tusher, V.G., R. Tibshirani, and G. Chu, *Significance analysis of microarrays applied to the ionizing radiation response.* Proceedings of the National Academy of Sciences, 2001. **98**(9): p. 5116–5121.
48. Geman, D., et al., *Classifying gene expression profiles from pairwise mRNA comparisons.* Statistical applications in genetics and molecular biology, 2004. **3**(1): p. 1–19.
49. Youssef, Y.M., et al., *Accurate molecular classification of kidney cancer subtypes using microRNA signature.* European urology, 2011. **59**(5): p. 721–730.
50. Price, N.D., et al., *Highly accurate two-gene classifier for differentiating gastrointestinal stromal tumors and leiomyosarcomas.* Proceedings of the National Academy of Sciences, 2007. **104**(9): p. 3414–3419.
51. Xu, L., et al., *Robust prostate cancer marker genes emerge from direct integration of inter-study microarray data.* Bioinformatics, 2005. **21**(20): p. 3905–3911.
52. Shi, P., et al., *Top scoring pairs for feature selection in machine learning and applications to cancer outcome prediction.* Bmc Bioinformatics, 2011. **12**(1): p. 375.
53. Tan, A.C., et al., *Simple decision rules for classifying human cancers from gene expression profiles.* Bioinformatics, 2005. **21**(20): p. 3896–3904.
54. Yuan, M. and Y. Lin, *Model selection and estimation in regression with grouped variables.* Journal of the Royal Statistical Society: Series B (Statistical Methodology), 2006. **68**(1): p. 49–67.
55. Zou, H. and T. Hastie, *Regularization and variable selection via the elastic net.* Journal of the Royal Statistical Society: Series B (Statistical Methodology), 2005. **67**(2): p. 301–320.
56. Friedman, J., T. Hastie, and R. Tibshirani, *Sparse inverse covariance estimation with the graphical lasso.* Biostatistics, 2008. **9**(3): p. 432–441.
57. Hastie, T., et al., *The entire regularization path for the support vector machine.* The Journal of Machine Learning Research, 2004. **5**: p. 1391–1415.
58. Friedman, J., T. Hastie, and R. Tibshirani, *Regularization paths for generalized linear models via coordinate descent.* Journal of statistical software, 2010. **33**(1): p. 1.
59. Tomlins, S.A., et al., *Recurrent fusion of TMPRSS2 and ETS transcription factor genes in prostate cancer.* Science, 2005. **310**(5748): p. 644–648.

60. Teschendorff, A.E., et al., *PACK: Profile Analysis using Clustering and Kurtosis to find molecular classifiers in cancer.* Bioinformatics, 2006. **22**(18): p. 2269–2275.
61. Tong, P., et al., *SIBER: systematic identification of bimodally expressed genes using RNAseq data.* Bioinformatics, 2013. **29**(5): p. 605–613.
62. Hanahan, D. and R.A. Weinberg, *Hallmarks of cancer: the next generation.* cell, 2011. **144**(5): p. 646–674.
63. Huang, D.W., B.T. Sherman, and R.A. Lempicki, *Bioinformatics enrichment tools: paths toward the comprehensive functional analysis of large gene lists.* Nucleic acids research, 2009. **37**(1): p. 1–13.
64. Tong, P. and K.R. Coombes, *integIRTy: a method to identify genes altered in cancer by accounting for multiple mechanisms of regulation using item response theory.* Bioinformatics, 2012. **28**(22): p. 2861–2869.
65. Jiang, P., et al., *Inference of transcriptional regulation in cancers.* Proceedings of the National Academy of Sciences, 2015. **112**(25): p. 7731–7736.
66. Vaske, C.J., et al., *Inference of patient-specific pathway activities from multi-dimensional cancer genomics data using PARADIGM.* Bioinformatics, 2010. **26**(12): p. i237–i245.

# NGS Analysis of Somatic Mutations in Cancer Genomes

T. Prieto, J. M. Alves, and D. Posada

**Abstract** The emergence of next-generation sequencing (NGS) technologies has facilitated the accumulation of large genomic datasets for most types of cancer. The analysis of these data has confirmed the early predictions of extensive sequence and structural diversity of cancer genomes, fueling the development of new computational approaches to decipher inter- and intratumoral somatic variation within and among cancer patients. Overall, these techniques have led to a better understanding of the disease as well as to relevant improvements in the diagnosis and therapy of cancer. In this chapter, we review current approaches for the analysis of somatic mutations in cancer genomes using NGS.

**Keywords** Cancer genomics • Somatic mutations • Driver mutations • Somatic variant calling • Tumor clones • Intra-tumor heterogeneity • Inter-tumor heterogeneity • Clonal inference • Tumor phylogenetic reconstruction

## 1 Introduction

It has long been recognized that the large majority of cancer cells in a tumor are genomically heterogeneous despite its monoclonal origin. After malignant transformation of a healthy cell (tumor initiation), the continuous growth of the tumor mass (tumor progression) contributes to the accumulation of *somatic mutations* within the malignant progeny, promoting the proliferation through time of distinct genetic lineages or *clones*. From an evolutionary perspective, a tumor may be interpreted as an uncontrollable expansion of a cell population where mainly Darwinian selection, but also random genetic drift and migration, coupled with the genomic instability of cancer cells, play major roles in shaping the distribution and frequency of its clones [26].

T. Prieto • J.M. Alves • D. Posada (✉)
Department of Biochemistry, Genetics and Immunology, University of Vigo, Spain

Galicia Sur Health Research Institute (IISGS), Spain
e-mail: dposada@uvigo.es



K.-C. Wong (ed.), *Big Data Analytics in Genomics*,
DOI 10.1007/978-3-319-41279-5_11

Exploring the extent of somatic alterations occurring in the genomes of cancer cells is critical, for both basic and translational research, allowing for a better understanding of the disease and providing better diagnoses and therapies in a clinical setting [66, 70]. Fortunately, with the recent improvements in next-generation sequencing (NGS) technologies, it is becoming increasingly affordable to obtain high-throughput data from cancer genomes, and, with the emergence of new molecular and computational methods, several solutions are now available to explore both the mutational and phenotypic landscape of tumors [44].

Throughout this chapter, we review the current strategies and challenges for the analysis of NGS data from cancer genomes. In particular, we focus on the detection of somatic variants, on the recognition of clones and their evolutionary relationships, and on the identification of mutations associated with tumorigenesis.

## 2 Sequencing Approaches in Cancer Genomics

A critical step in cancer genomics is the identification of the genetic mutations that have accumulated during tumor development. On this basis, somatic mutation profiling is generally carried out by comparing sequence information from both healthy and tumor samples of a given patient (*paired* or *matched samples*) (Fig. 1a). This fairly simple workflow is, however, made complicated by specificities of the different high-throughput sequencing applications available. At present, most sequencing efforts are being directed towards the identification of somatic variants in protein-coding sites, either by targeting specific genes or extending the analysis over the entire exome (about 2 % of the human genome).

In theory, *targeted sequencing* strategies (e.g., Whole-Exome Sequencing or WES) represent a cost-effective approach to study cancer genomes, as it restricts the analysis to the genomic regions that are expected to be functionally relevant. Moreover, given its relatively low cost compared to whole-genome surveys, targeted sequencing data could in principle be obtained from multiple tumor samples, providing more accurate estimates of the overall genetic diversity. However, targeted approaches have been found to (1) preferentially capture *reference alleles* at heterozygous positions, and (2) provide uneven *coverage depth* across the targeted regions [45]. The latter can, for example, compromise the inference of copy number states, which in turn can affect variant discovery and posterior inferences based on these variants.

*Whole-Genome sequencing* (*WGS*), on the other hand, consists of a broader methodology designed to sequence the entire genome of an individual sample, thus allowing a more comprehensive characterization of the sequence and structural plasticity of cancer genomes. In addition, WGS typically yields a homogeneous genome coverage, making it more suitable for copy number variant detection than WES [87]. Whole-genome sequencing is becoming increasingly popular in cancer genomics, due to its ability to survey multiple classes of genomic variants in a single assay, coupled with higher detection accuracy [4]. Indeed, as sequencing costs continue to decrease, and more robust methods to analyze complex genome structures emerge, the cancer genomics community will surely begin deciphering the functional implications of non-coding and structural variants in cancer.

While the scope of the current chapter falls exclusively on the sequence variability of cancer genomes, it is important to note that other layers of heterogeneity can be explored in cancer using NGS-based strategies, like variation in DNA methylation, gene expression, or metabolic pathways [11].

# 3 Analyzing Cancer NGS Data

As in most NGS studies, cancer NGS data from both healthy and tumor samples are typically provided in a raw state to the user. As a consequence, a series of processing steps are essential before further analyses, including (1) aligning the reads to a genome reference, (2) marking reads that have been sequenced multiple times (i.e., de-duplication), (3) realigning reads around known indels to exclude potential mapping artifacts, and (4) recalculating per-base quality scores, as estimates from sequencing platforms are occasionally inaccurate and can be biased [75]. In recent years, a variety of statistical algorithms have been developed to perform such steps in a fairly robust manner [81].

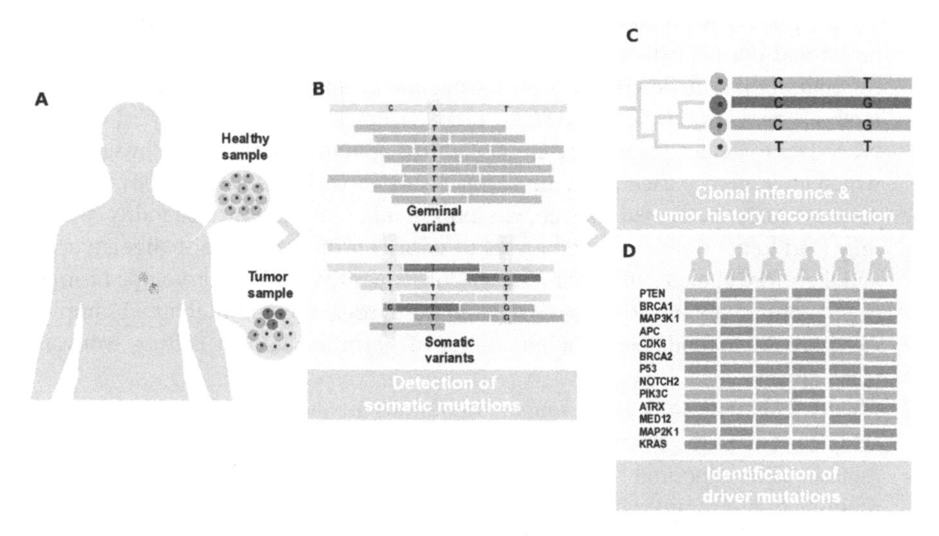


**Fig. 1** Workflow of a hypothetical cancer NGS study. (**a**) Independent sequencing of paired healthy and tumor samples from a single cancer patient. (**b**) Detecting germline and somatic mutations (i.e., specific to tumor samples). (**c**) Estimating the number of clones and their phylogenetic history. (**d**) Distinguishing between driver and passenger mutations using the frequency of somatic mutations over many cancer patients

It should be emphasized that the standard strategy in cancer research generally relies on population-level sequencing from bulk tissue samples. That is, the NGS reads obtained correspond to DNA fragments derived from multiple cells (typically hundreds of thousands or more) mixed together (i.e., we do not know which reads are from which cell). Given the extensive diversity within individual tumors, this concept resembles "pooling" experiments (i.e., *pool sequencing* or *pool-seq*) [63] as each tumor sample will contain a heterogeneous combination of normal cells and one or more cancer clones (Fig. 1b). Analyzing cancer NGS data is not straightforward, ultimately requiring ad hoc bioinformatic tools for proper identification of somatic variants segregating in the tumoral population.

## 3.1 Profiling Somatic Variation

As stated above, genomic information of tumors often stems directly from a heterogeneous population of distinct cell types (healthy and malignant) which may affect, to a large extent, the downstream interpretation of the data. In consequence, specific features of cancer genomes should be considered while calling genetic variants in order to reduce potential sequencing artifacts that could be confounded with real biological variation. These include *tumor purity*, *tumor ploidy*, and *intratumoral heterogeneity* (*ITH*).

1. Tumor purity, also called *tumor cellularity*, reflects the relative proportion of tumor and normal cells in a sample. Traditionally, purity scores were estimated by pathological review of sectioned specimens. Interestingly, recent in silico methods as *ABSOLUTE* [8], *ASCAT* [76], *THetA* [53], and *TITAN* [27] rely on NGS data to determine the degree of healthy "contamination" in tumor samples, which may then be used as an initial parameter in variant detection [68].
2. As previously mentioned, cancer cells exhibit inherent genomic instability, which may lead to aberrant gains and losses of entire chromosomes in malignant cells and therefore changes in tumor ploidy. Consequently, distinct karyotype profiles (i.e., genomic organization) are expected to segregate within individual samples [52], breaking fundamental assumptions of germline variant calling software (see below).
3. Finally, ITH will often cause somatic mutations at variable frequencies in tumor samples. Given the potential functional significance of low-frequency somatic mutations, it is important to distinguish them from sequencing errors. Sequencing multiple tumor samples from the same patient (*multiregion sequencing*) may help circumvent these limitations [24, 69].

In addition, it is important to note that most variant calling software tools are limited to a specific type of genomic variation (see Table 1). On this basis, different calling algorithms should be applied depending on the type of variants under interrogation. Continuous advances of NGS-based methods towards variant detection have considerably improved our ability to interpret genomic information

**Table 1** Computational methods commonly used in cancer studies for detecting somatic variants

| Variant caller | Joint analysis | SNV | Indel | Unbalanced SVs | Balanced SVs | References |
|---|---|---|---|---|---|---|
| *Platypus* | | √ | √ | | | Rimmer et al. [59] |
| *HaplotypeCaller* | | √ | √ | | | McKenna et al. [43] |
| *FreeBayes* | | √ | √ | | | Garrison and Marth [22] |
| *Samtools mpileup* | | √ | √ | | | Li [36] |
| *JointSNVMix* | √ | √ | | | | Roth et al. [60] |
| *DeepSNV* | √ | √ | | | | Gerstung et al. [25] |
| *FaSD-somatic* | √ | √y | | | | Wang et al. [79] |
| *SomaticSniper* | √ | √ | | | | Larson et al. [32] |
| *Shimmer* | √ | √ | | | | Hansen et al. [28] |
| *Dindel* | | | √ | | | Albers et al. [1] |
| *Strelka* | √ | √ | √ | | | Saunders et al. [62] |
| *Mutect* | √ | √ | √ | | | Cibulskis et al. [13] |
| *EBCall* | √ | √ | √ | | | Shiraishi et al. [65] |
| *HapMuc* | √ | √ | √ | | | Usuyama et al. [74] |
| *VarScan* | √ | √ | √ | √ | | Koboldt et al. [30] |
| *Seraut* | √ | √ | √ | √ | √ | Christoforides et al. [12] |
| *SMUFIN* | √ | √ | √ | √ | √ | Moncunill et al. [47] |
| *BreakDancer* | | | √ | √ | √ | Fan et al. [20] |
| *HMMcopy* | √ | | | √ | | Lai et al. [31] |
| *Patchwork* | √ | | | √ | | Mayrhofer et al. [41] |
| *OncoSNP-SEQ* | √ | | | √ | | Yau [83] |
| *Control-FREEC* | √ | | | √ | | Boeva et al. [5] |
| *SegSeq* | √ | | | √ | | Chiang et al. [10] |
| *Pindel* | | | | √ | √ | Ye et al. [84] |
| *CREST* | √ | | | √ | √ | Wang et al. [78] |
| *DELLY* | √ | | | √ | √ | Rausch et al. [58] |
| *GASV-Pro* | | | | √ | √ | Sindi et al. [67] |
| *TIGRA* | | | | √ | √ | Chen et al. [9] |
| *Hydra* | | | | √ | √ | Quinlan et al. [57] |
| *Meerkat* | √ | | | √ | √ | Yang et al. [82] |
| *LUMPY* | √ | | | √ | √ | Layer et al. [34] |

and are now being widely applied to study cancer genomic diversity and evolution. However, due to the systematic errors of sequencing technologies, it should be highlighted that *orthogonal validation* is ultimately required to confirm that the variants identified are real.

### 3.1.1 Single Nucleotide Variants

*Single-nucleotide variants* (SNVs) represent the most frequent class of somatic variation in cancer [49]. This has motivated the development of a large variety of statistical tools to detect somatic point mutations from cancer genomes. At present, two main strategies are routinely used in cancer NGS variant calling, namely *independent* or *joint* analysis [79]. Arguably, independent analysis of paired normal and tumor samples is still the most frequent approach, where genotype information from healthy and tumoral tissues from the same patient is initially inferred independently, being subsequently compared in order to distinguish germline from de novo somatic variation. Independent SNV-calling software include the *Genome Analysis Toolkit* (*GATK*) [43], *FreeBayes* [22], *Samtools* [36], and *Platypus* [59]. However, these methods have been developed for germline variation and therefore assume a homogenous population of diploid cells, which is not usually the case for tumors.

Conversely, joint variant calling analysis represent an improved statistical approach towards somatic variant discovery, where tumor and matched normal variants are simultaneously called, using a frequentist or a Bayesian approach. The former makes use of allele frequency estimates between paired samples to classify germline and somatic mutations, whereas Bayesian methods usually incorporate a prior somatic mutation rate. Table 1 summarizes some of the most common computational tools used for somatic SNV discovery.

### 3.1.2 Insertion and Deletions

Perhaps owing to the difficulty of mapping NGS reads that overlap small insertion or deletion events (*indels*), the characterization of these variants in cancer genomes has so far received little attention. Nevertheless, new approaches are becoming available that allow for gapped alignment and local de novo assemblies around potential indel sites, significantly improving the reliability of indel calls [37, 43, 48]. Interestingly, these methods are being implemented in most SNV-calling tools, allowing the parallel analysis of both types of variant (Table 1).

### 3.1.3 Structural Variants

*Structural variants* (*SVs*) can be defined as a wide collection of genomic rearrangements, which may involve the loss or gain of genetic material. These rearrangements range from *balanced structural changes*, such as inversions and translocations, to *unbalanced alterations*, such as *copy number variants/aberrations* (*CNVs/CNAs*). While the process by which SVs accumulate in cancer genomes remains unclear—i.e., gradual acquisition vs. *chromothripsis/chromoplexy*; see [86] for a detailed review, new and robust NGS-based inferential methods are boosting our ability to understand the architecture of cancer genomes. In Table 1 we list some of these methods, which make use of mapping distance and orientation information from paired-end reads and split reads, as well as differences in depth of sequencing coverage, to predict SVs.

### *3.2 Understanding Clonal Composition and Tumor Evolution*

Patterns of genetic variation, within and between populations, have been traditionally used by evolutionary geneticists to infer the demographic, mutational, and/or selective processes shaping the evolution of species. A similar *rationale* may be applied in cancer research, where the analysis of intra- and inter-tumoral genetic heterogeneity is expected to provide relevant insights into the biological processes driving cancer development [42]. Indeed, obtaining a snapshot of the genomic intratumoral diversity (i.e., number and distribution of clones) at a certain point in cancer progression, identifying clones with metastatic potential or drug resistance, detecting the presence or absence of a given clone in longitudinal samples, different tumor regions or metastases and, in general, reconstructing clonal evolution is crucial for a better understanding of cancer, but also to improve clinical diagnosis and treatment strategies [56].

Interestingly, different algorithms have been recently designed to reconstruct the clonal composition of pooled tumor samples from allele frequency estimates and/or copy number profiles (Table 2). By assuming an *infinite-sites model* (i.e., mutations never happen twice at the same position), these methods apply the *pigeonhole principle* (Fig. 2) for clustering mutations segregating at similar frequencies, allowing the construction of mutation profiles of distinct clones and

**Table 2** Computational methods for studying clonal evolution

| Software | Variant type | Clonal tree reconstruction | References |
|---|---|---|---|
| *THetA* | CNVs | | Oesper et al. [53] |
| *Pyclone* | SNVs | | Roth et al. [61] |
| *Clomial* | SNVs | | Zare et al. [85] |
| *SciClone* | SNVs | | Miller et al. [46] |
| *CloneHD* | SNVs, CNVs | | Fischer et al. [21] |
| *SubcloneSeeker* | SNVs, CNVs | √ | Qiao et al. [56] |
| *PhyloWGS* | SNVs, CNVs | √ | Deshwar et al. [16] |
| *AncesTree* | SNVs | √ | El-Kebir et al. [19] |
| *CITUP* | SNVs | √ | Malikic et al. [39] |
| *LICHeE* | SNVs | √ | Popic et al. [54] |
| *PhyloSub* | SNVs | √ | Jiao et al. [29] |
| *Trap* | SNVs | √ | Strino et al. [71] |
| *TuMult* | CNVs | √ | Letouzé et al. [35] |
| *MEDICC* | CNVs | √ | Schwarz et al. [64] |

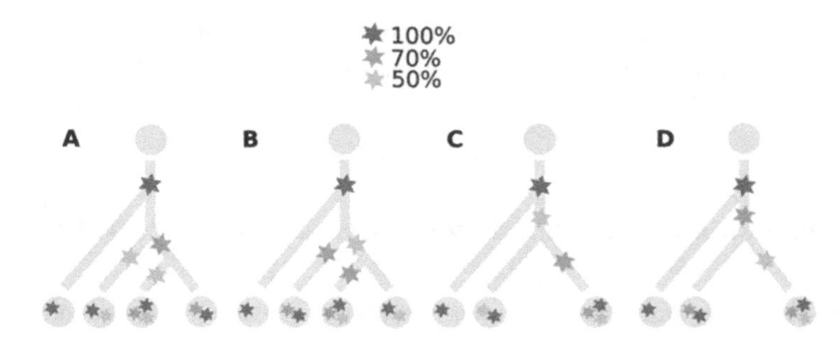


**Fig. 2** Applying the "pigeonhole principle" for reconstructing clonal evolution from somatic mutations. Consider the example above, where three mutations (*star shapes*) are present at variable frequencies in a tumor sample: one mutation (*red*) is fixed in the tumor cell population (100 %), another mutation (*orange*) is present in the large majority of tumor cells (70 %), and a third mutation (*light blue*) is present in only half of the tumor cells (50 %). The "pigeonhole principle" reasoning implies that when the sum of allele frequencies of any two mutations is above 100 % (e.g., 70 % + 50 % = 120 %), there must be some cells harboring both mutations. This principle thus provides the potential genotypes that might exist in the population. Moreover, since each mutation is not allowed to occur multiple times during evolution (infinite-sites model), tree *A* and *B* represent unfeasible scenarios, as both require the same mutation to occur in different branches of the tree. Finally, as mutations present in a greater fraction of tumor cells are expected to occur earlier in the tumor phylogeny, tree *D* represents the most likely scenario of clonal evolution

inferring their *cell prevalence*, genealogy, and geographical distribution (Fig. 1c). Importantly, *multiregion sequencing* from (1) multiple sites from a primary tumor [38], or (2) primary and metastatic sites [24], is becoming common practice in cancer research, providing detailed information about clonal stratification and dissemination potential.

## *3.3 Identifying Driver Mutations*

Tumor evolution has been traditionally seen as the consequence of clonal competition and adaptation by natural selection, although this paradigm has been recently challenged for some solid tumors [38, 69]. In any case, a fundamental line of research in cancer genomics has been the identification of *driver mutations* (Fig. 1d), the somatic variants that confers selective growth advantage to the cancer cell bearing it, and of the *driver genes* where these mutations take place. On the other hand, a mutation that does not provide a cancer cell with any selective advantage is called a *passenger mutation*. Determining whether a somatic mutation is a driver or a passenger remains challenging, as it implies detecting the footprint of natural selection, a recurrent issue in evolutionary biology. In recent years, several computational approaches have been developed in order to detect driver genes in cancer. These methods could be grouped in five main categories:

1. Strategies based on functional impact. A somatic mutation can be synonymous, nonsynonymous, disrupt the expression product, change the reading frame, upregulate or downregulate the expression of genes, etc. (i.e., it can have distinct functional consequences). Those mutations responsible for a significant phenotypic impact are thought to be drivers. Different tools are currently available to predict driver events by estimating the functional relevance of somatic mutations. For example, *CHASM software* [7] uses a machine learning method together with well-curated databases (e.g., COSMIC) to quantify the potential functional significance of somatic mutations.
2. Strategies based on dN/dS ratio. An excess of nonsynonymous vs. synonymous (dN/dS) somatic mutations is considered an evidence of positive selection [3].
3. Strategies based on mutation recurrence. Somatic mutations detected more frequently than expected by chance in independent tumors of several patients are thought to be drivers. Software as *MuSiC* [15] or *MutSigCV* [33] apply this concept in which mutation profiles of multiple cancers are integrated to predict driver mutation events.
4. Strategies based on mutation clustering. An excess of somatic mutations in a given region of the genome can be a signal of natural selection acting over that genomic region. *OncodriveCLUST* [73] explores this idea by searching for mutation clusters in protein-coding sequences.
5. Strategies based on pathways and networks. Somatic mutations can be annotated as drivers when associated with a pathway known to be targeted in cancer [18]. A common approach is to analyze whether somatic mutations are included in cancer-related gene sets or are part of biological interaction networks.

As with the somatic mutation calling algorithms, an ideal solution is yet to be developed. Meanwhile, performing an integrated meta-analysis using multiple methods seems to be a good choice [18].

Apart from determining whether a given somatic mutation is a driver event in tumorigenesis, some researchers have also tried to represent the order of accumulation of driver mutations using *oncogenetic trees*. Unlike phylogenetic trees, which represent evolutionary relationships among samples, oncogenetic trees make use of tumor mutation profiles from different patients in order to reconstruct the sequential accumulation of driver events in cancer, therefore identifying common evolutionary pathways and mutational dependencies [3, 17, 72].

## 4 Single-Cell Genomics

A recent strategy that has been gaining *momentum* in the cancer community is based on sequencing genomes from single cells. Originally introduced by Navin in 2011 [51], single-cell sequencing approaches allow researchers to assess the genomes of individual malignant cells, either from the tumor or from *circulating tumor cells* (*CTCs*), thus significantly improving our ability to assess genomic heterogeneity within tumors [50, 80]. Nevertheless, given the limited amount of starting material, multiple rounds of whole-genome amplification are often needed before extensive sequencing, which introduces a high number of sequence artifacts that can be confounded with genuine biological variation [77]. Other technical errors, such as uneven genome amplification, *allelic dropout* (*ADO*) and low *coverage breadth*, may also generate substantial artificial variability in cancer genomes, compromising the power to detect real somatic events from single-cell data [23].

Nevertheless, it is important to highlight that NGS data analyses based on single-cell sequencing information circumvent the cell heterogeneity issue of bulk sequencing approaches. Thus, if one is willing to ignore variations due to tumor ploidy or ADO, traditional "germline" algorithms can be applied to detect somatic variants, as each sample represents individual cells for which genotypes can be directly inferred [51, 80]. Similarly, single-cell genomics opens the door to classical phylogenetic and population genetics methods for the study of tumor evolution [55].

Therefore, with the continuous improvements in sequencing technologies, it seems obvious that single-cell approaches will become a key—perhaps the default—sequencing strategy for tumors in the future. However, given the high amount of technical errors currently associated with single-cell methods, independent validation of the mutations events is often needed, largely impacting the economical cost.

## 5 Future Prospects in Cancer Genomics

Notwithstanding its recent emergence, the field of cancer genomics has evolved rapidly. At the beginning, most of the progress was a consequence of international projects and associations as The Cancer Genome Atlas (http://cancergenome.nih.gov/) and the International Cancer Genome Consortium (https://icgc.org/), which by releasing data and dedicated software for the scientific community, have already contributed to the design of new therapeutic targets and improved treatment plans [6].

In the future, it seems clear that cancer genomics will benefit the most from the implementation of single-cell genomics and further progress in NGS technologies. As just mentioned, single-cell sequencing studies will allow for a more detailed description of ITH [51, 80]. On the other hand, a shift towards third-generation sequencing technologies should significantly reduce the computational demands of most genomic analyses, allowing the development of alternative and more flexible strategies to study cancer genomes. For instance, longer reads provided by third-generation technologies are expected to considerably improve haplotype phasing, a key point to understand epistatic interactions in cancer cells.

From a clinical perspective, the adoption of NGS technologies in cancer research is also expected to significantly improve precision medicine and personalized care [66]. Tumor dynamics and clones identified from NGS can help patient prognosis [2]. Indeed, once sequencing of matched samples becomes routinely implemented in clinical settings, patients' specific treatments could be considered in order to reduce many present-day side effects, while allowing the discovery of new therapeutic targets and/or the identification of clonal lineages capable of driving recurrent events.

In summary, as cancer NGS data continues to accumulate in publicly available databases, big data approaches are crucially needed for proper characterization of cancer genomes. At present, even though most methodologies still lack a comprehensive benchmarking, big efforts are being made in order to circumvent the computational burden of NGS data handling and processing (e.g., CRAMTools) [14]. In addition, cloud computing is becoming a valuable resource for many cancer research centers that lack the physical infrastructure to store the massive amounts of data being generated [40]. Driven by these technological improvements, together with the reduction of sequencing costs, addressing important questions requiring larger sample sizes in cancer genomics will certainly become possible.

## Glossary

*Allelic dropout* (*ADO*): lack of amplification of one of the two alleles at a site.

*Balanced structural variants*: a class of genomic rearrangements characterized by changes in the linear sequence of a genome, without altering the overall content of cellular DNA. It includes chromosomal inversions and translocations.

*Cell prevalence*: relative fraction of tumor cells carrying a given somatic mutation.

*Chromothripsis*: a single catastrophic event that comprises multiple structural changes in a limited genomic region (e.g., arm of a given chromosome).

*Chromoplexy*: large-scale changes in the structural architecture of the whole cancer genome where broken DNA segments from one chromosome may get incorporated by different chromosomes in a completely distinct configuration.

*Circulating tumor cells* (*CTCs*): cancer cells which have been released from the primary tumor and can be isolated from peripheral blood.

*Clone*: set of genetically identical cells that descend from a common ancestral cell. In practice, a clone represents a collection of cells that harbor the same somatic genomic variants, including SNVs, CNAs, SVs, studied.

*Copy number variants/aberrations* (*CNVs/CNAs*): a DNA segment of a cancer genome typically larger than 1 kb, that differs in terms of its copy number with respect to the healthy genome.

*Coverage depth*: number of reads which align to a given position of the reference genome. Sometimes coverage depth is simply referred to as "coverage" or "sequencing depth." It provides the redundancy needed to ensure data accuracy.

*Coverage breadth*: percentage of the genome sequence which is covered by at least one sequencing read.

*Driver mutations*: selectively advantageous somatic mutations. They are thought to be responsible for clonal expansions.

*Driver genes*: mutated genes involved in pathways implicated in cancer progression.

*Infinite-sites model*: a description of the process of mutation accumulation in populations in which mutations can only happen once at a given position of the DNA sequence.

*Insertion and deletion events* (*indels*): relatively small gain or loss of nucleotides.

*Intratumoral heterogeneity* (*ITH*): genomic variation harbored by cancer cells within a given tumor.

*Multiregion sequencing*: a study design in which different regions of the same tumor are sampled, sequenced, and compared.

*Oncogenetic tree*: a graphical representation of the order in which driver events can take place.

*Orthogonal validation*: validation of variants usually performed with a different technology. In the context of NGS, orthogonal validation could be carried out, for instance, using a hybridization-based method (e.g., DNA microarrays).

*Paired* or *matched samples*: healthy and tumor samples obtained from the same patient.

*Passenger mutations*: neutral mutations acquired during tumor development.

*Pigeonhole principle*: also called Dirichlet's box, states that if frequencies of two different somatic mutations which are present in a tumor cell population sum up to more than 100 %, then some tumor cells must harbor both mutations.

*Pool sequencing* (*Pool-seq*): sequencing of the DNA molecules from a pool of cells. In cancer research, it is usually referred to as *bulk-seq*.

*Reference alleles*: alleles matching the genome reference.

*Single nucleotide variant* (*SNV*): point mutation at a specific base position of the genome.

*Somatic mutations*: mutations which are not inherited but appear de novo during the development of a multicellular organism.

*Structural variants* (*SVs*): genomic structural rearrangements which span from 1 kb to many megabases.

*Targeted sequencing*: sequencing of specific genomic regions (i.e., targets) from a given sample.

*Tumor ploidy*: refers to the number of chromosome sets present in malignant cells.

*Tumor purity/cellularity*: proportion of malignant cells in a given sample.

*Unbalanced structural variants*: usually defined as relatively large genomic rearrangements (>1 kb), involving loss or gain of genetic material. Generally described as copy number variants (CNVs), it includes insertions, deletions, and/or duplications of different sizes.

*Whole-genome sequencing* (*WGS*): sequencing of the complete genomic sequence contained in a sample.

**Acknowledgments** This work was supported by the European Research Council (ERC-617457-PHYLOCANCER to D.P.). T.P. is supported by a PhD fellowship from the Galician Government (ED481A-2015/083).

## References

1. Albers CA, Lunter G, MacArthur DG, et al (2011) Dindel: accurate indel calls from short-read data. Genome Res 21:961–973.
2. Andor N, Graham TA, Jansen M, et al (2016) Pan-cancer analysis of the extent and consequences of intratumor heterogeneity. Nat Med 22:105–113.
3. Beerenwinkel N, Schwarz RF, Gerstung M, Markowetz F (2015) Cancer evolution: mathematical models and computational inference. Syst Biol 64:e1–25.
4. Belkadi A, Bolze A, Itan Y, et al (2015) Whole-genome sequencing is more powerful than whole-exome sequencing for detecting exome variants. Proc Natl Acad Sci U S A 112: 5473–5478.
5. Boeva V, Zinovyev A, Bleakley K, et al (2011) Control-free calling of copy number alterations in deep-sequencing data using GC-content normalization. Bioinformatics 27:268–269.
6. Cancer Genome Atlas Network (2015) Comprehensive genomic characterization of head and neck squamous cell carcinomas. Nature 517:576–582.
7. Carter H, Chen S, Isik L, et al (2009) Cancer-specific high-throughput annotation of somatic mutations: computational prediction of driver missense mutations. Cancer Res 69:6660–6667.
8. Carter SL, Cibulskis K, Helman E, et al (2012) Absolute quantification of somatic DNA alterations in human cancer. Nat Biotechnol 30:413–421.
9. Chen K, Chen L, Fan X, et al (2014) TIGRA: a targeted iterative graph routing assembler for breakpoint assembly. Genome Res 24:310–317.
10. Chiang DY, Getz G, Jaffe DB, et al (2009) High-resolution mapping of copy-number alterations with massively parallel sequencing. Nat Methods 6:99–103.
11. Chmielecki J, Juliann C, Matthew M (2014) DNA Sequencing of Cancer: What Have We Learned? Annu Rev Med 65:63–79.
12. Christoforides A, Carpten JD, Weiss GJ, et al (2013) Identification of somatic mutations in cancer through Bayesian-based analysis of sequenced genome pairs. BMC Genomics 14:302.
13. Cibulskis K, Lawrence MS, Carter SL, et al (2013) Sensitive detection of somatic point mutations in impure and heterogeneous cancer samples. Nat Biotechnol 31:213–219.
14. Cochrane G, Cook CE, Birney E (2012) The future of DNA sequence archiving. Gigascience 1:2.
15. Dees ND, Zhang Q, Kandoth C, et al (2012) MuSiC: identifying mutational significance in cancer genomes. Genome Res 22:1589–1598.
16. Deshwar AG, Vembu S, Yung CK, et al (2015) PhyloWGS: reconstructing subclonal composition and evolution from whole-genome sequencing of tumors. Genome Biol 16:35.
17. Desper R, Jiang F, Kallioniemi OP, et al (1999) Inferring tree models for oncogenesis from comparative genome hybridization data. J Comput Biol 6:37–51.
18. Ding L, Wendl MC, McMichael JF, Raphael BJ (2014) Expanding the computational toolbox for mining cancer genomes. Nat Rev Genet 15:556–570.
19. El-Kebir M, Oesper L, Acheson-Field H, Raphael BJ (2015) Reconstruction of clonal trees and tumor composition from multi-sample sequencing data. Bioinformatics 31:i62–70.
20. Fan X, Xian F, Abbott TE, et al (2014) BreakDancer: Identification of Genomic Structural Variation from Paired-End Read Mapping. In: Current Protocols in Bioinformatics. pp 15.6.1–15.6.11
21. Fischer A, Vázquez-García I, Illingworth CJR, Mustonen V (2014) High-definition reconstruction of clonal composition in cancer. Cell Rep 7:1740–1752.
22. Garrison E, Marth G (2012) Haplotype-based variant detection from short-read sequencing.
23. Gawad C, Charles G, Winston K, Quake SR (2016) Single-cell genome sequencing: current state of the science. Nat Rev Genet 17:175–188.
24. Gerlinger M, Rowan AJ, Horswell S, et al (2012) Intratumor heterogeneity and branched evolution revealed by multiregion sequencing. N Engl J Med 366:883–892.
25. Gerstung M, Beisel C, Rechsteiner M, et al (2012) Reliable detection of subclonal single-nucleotide variants in tumour cell populations. Nat Commun 3:811.
26. Greaves M, Maley CC (2012) Clonal evolution in cancer. Nature 481:306–313.

27. Ha G, Roth A, Khattra J, et al (2014) TITAN: inference of copy number architectures in clonal cell populations from tumor whole-genome sequence data. Genome Res 24:1881–1893.
28. Hansen NF, Gartner JJ, Mei L, et al (2013) Shimmer: detection of genetic alterations in tumors using next-generation sequence data. Bioinformatics 29:1498–1503.
29. Jiao W, Vembu S, Deshwar AG, et al (2014) Inferring clonal evolution of tumors from single nucleotide somatic mutations. BMC Bioinformatics 15:35.
30. Koboldt DC, Zhang Q, Larson DE, et al (2012) VarScan 2: somatic mutation and copy number alteration discovery in cancer by exome sequencing. Genome Res 22:568–576.
31. Lai D, Ha G, Shah S (2012) HMMcopy: Copy number prediction with correction for GC and mappability bias for HTS data.
32. Larson DE, Harris CC, Chen K, et al (2012) SomaticSniper: identification of somatic point mutations in whole genome sequencing data. Bioinformatics 28:311–317.
33. Lawrence MS, Stojanov P, Polak P, et al (2013) Mutational heterogeneity in cancer and the search for new cancer-associated genes. Nature 499:214–218.
34. Layer RM, Chiang C, Quinlan AR, Hall IM (2014) LUMPY: a probabilistic framework for structural variant discovery. Genome Biol 15:R84.
35. Letouzé E, Allory Y, Bollet MA, et al (2010) Analysis of the copy number profiles of several tumor samples from the same patient reveals the successive steps in tumorigenesis. Genome Biol 11:R76.
36. Li H (2011) A statistical framework for SNP calling, mutation discovery, association mapping and population genetical parameter estimation from sequencing data. Bioinformatics 27: 2987–2993.
37. Li H (2012) Exploring single-sample SNP and INDEL calling with whole-genome de novo assembly. Bioinformatics 28:1838–1844.
38. Ling S, Shaoping L, Zheng H, et al (2015) Extremely high genetic diversity in a single tumor points to prevalence of non-Darwinian cell evolution. Proceedings of the National Academy of Sciences 201519556.
39. Malikic S, McPherson AW, Donmez N, Sahinalp CS (2015) Clonality inference in multiple tumor samples using phylogeny. Bioinformatics 31:1349–1356.
40. Marx V, Vivien M (2013) Biology: The big challenges of big data. Nature 498:255–260.
41. Mayrhofer M, DiLorenzo S, Isaksson A (2013) Patchwork: allele-specific copy number analysis of whole-genome sequenced tumor tissue. Genome Biol 14:R24.
42. McGranahan N, Swanton C (2015) Biological and therapeutic impact of intratumor heterogeneity in cancer evolution. Cancer Cell 27:15–26.
43. McKenna A, Hanna M, Banks E, et al (2010) The Genome Analysis Toolkit: a MapReduce framework for analyzing next-generation DNA sequencing data. Genome Res 20:1297–1303.
44. Meyerson M, Gabriel S, Getz G (2010) Advances in understanding cancer genomes through second-generation sequencing. Nat Rev Genet 11:685–696.
45. Meynert AM, Morad A, FitzPatrick DR, Taylor MS (2014) Variant detection sensitivity and biases in whole genome and exome sequencing. BMC Bioinformatics 15:247.
46. Miller CA, White BS, Dees ND, et al (2014) SciClone: inferring clonal architecture and tracking the spatial and temporal patterns of tumor evolution. PLoS Comput Biol 10:e1003665.
47. Moncunill V, Gonzalez S, Beà S, et al (2014) Comprehensive characterization of complex structural variations in cancer by directly comparing genome sequence reads. Nat Biotechnol 32:1106–1112.
48. Mose LE, Wilkerson MD, Hayes DN, et al (2014) ABRA: improved coding indel detection via assembly-based realignment. Bioinformatics 30:2813–2815.
49. Mwenifumbo JC, Marra MA (2013) Cancer genome-sequencing study design. Nat Rev Genet 14:321–332.
50. Navin NE (2014) Cancer genomics: one cell at a time. Genome Biol. doi: 10.1186/s13059-014-0452-9
51. Navin N, Kendall J, Troge J, et al (2011) Tumour evolution inferred by single-cell sequencing. Nature 472:90–94.

52. Nicholson JM (2013) Will we cure cancer by sequencing thousands of genomes? Mol Cytogenet 6:57.
53. Oesper L, Mahmoody A, Raphael BJ (2013) THetA: inferring intra-tumor heterogeneity from high-throughput DNA sequencing data. Genome Biol 14:R80.
54. Popic V, Salari R, Hajirasouliha I, et al (2015) Fast and scalable inference of multi-sample cancer lineages. Genome Biol 16:91.
55. Posada D (2015) Cancer Molecular Evolution. J Mol Evol 81:81–83.
56. Qiao Y, Quinlan AR, Jazaeri AA, et al (2014) SubcloneSeeker: a computational framework for reconstructing tumor clone structure for cancer variant interpretation and prioritization. Genome Biol 15:443.
57. Quinlan AR, Clark RA, Sokolova S, et al (2010) Genome-wide mapping and assembly of structural variant breakpoints in the mouse genome. Genome Res 20:623–635.
58. Rausch T, Zichner T, Schlattl A, et al (2012) DELLY: structural variant discovery by integrated paired-end and split-read analysis. Bioinformatics 28:i333–i339.
59. Rimmer A, Phan H, Mathieson I, et al (2014) Integrating mapping-, assembly- and haplotype-based approaches for calling variants in clinical sequencing applications. Nat Genet 46: 912–918.
60. Roth A, Ding J, Morin R, et al (2012) JointSNVMix: a probabilistic model for accurate detection of somatic mutations in normal/tumour paired next-generation sequencing data. Bioinformatics 28:907–913.
61. Roth A, Khattra J, Yap D, et al (2014) PyClone: statistical inference of clonal population structure in cancer. Nat Methods 11:396–398.
62. Saunders CT, Wong WSW, Swamy S, et al (2012) Strelka: accurate somatic small-variant calling from sequenced tumor-normal sample pairs. Bioinformatics 28:1811–1817.
63. Schlötterer C, Christian S, Raymond T, et al (2014) Sequencing pools of individuals — mining genome-wide polymorphism data without big funding. Nat Rev Genet 15:749–763.
64. Schwarz RF, Trinh A, Sipos B, et al (2014) Phylogenetic quantification of intra-tumour heterogeneity. PLoS Comput Biol 10:e1003535.
65. Shiraishi Y, Sato Y, Chiba K, et al (2013) An empirical Bayesian framework for somatic mutation detection from cancer genome sequencing data. Nucleic Acids Res 41:e89.
66. Simon R, Roychowdhury S (2013) Implementing personalized cancer genomics in clinical trials. Nat Rev Drug Discov 12:358–369.
67. Sindi SS, Onal S, Peng LC, et al (2012) An integrative probabilistic model for identification of structural variation in sequencing data. Genome Biol 13:R22.
68. Song S, Nones K, Miller D, et al (2012) qpure: A tool to estimate tumor cellularity from genome-wide single-nucleotide polymorphism profiles. PLoS One 7:e45835.
69. Sottoriva A, Kang H, Ma Z, et al (2015) A Big Bang model of human colorectal tumor growth. Nat Genet 47:209–216.
70. Stratton MR (2011) Exploring the genomes of cancer cells: progress and promise. Science 331:1553–1558.
71. Strino F, Parisi F, Micsinai M, Kluger Y (2013) TrAp: a tree approach for fingerprinting subclonal tumor composition. Nucleic Acids Res 41:e165.
72. Szabo A, Boucher K (2002) Estimating an oncogenetic tree when false negatives and positives are present. Math Biosci 176:219–236.
73. Tamborero D, Gonzalez-Perez A, Lopez-Bigas N (2013) OncodriveCLUST: exploiting the positional clustering of somatic mutations to identify cancer genes. Bioinformatics 29: 2238–2244.
74. Usuyama N, Shiraishi Y, Sato Y, et al (2014) HapMuC: somatic mutation calling using heterozygous germ line variants near candidate mutations. Bioinformatics 30:3302–3309.
75. Van der Auwera GA, Carneiro MO, Hartl C, et al (2013) From FastQ data to high confidence variant calls: the Genome Analysis Toolkit best practices pipeline. Curr Protoc Bioinformatics 11:11.10.1–11.10.33.
76. Van Loo P, Nordgard SH, Lingjaerde OC, et al (2010) Allele-specific copy number analysis of tumors. Proceedings of the National Academy of Sciences 107:16910–16915.

77. Van Loo P, Voet T (2014) Single cell analysis of cancer genomes. Curr Opin Genet Dev 24: 82–91.
78. Wang J, Mullighan CG, Easton J, et al (2011) CREST maps somatic structural variation in cancer genomes with base-pair resolution. Nat Methods 8:652–654.
79. Wang W, Wang P, Xu F, et al (2014a) FaSD-somatic: a fast and accurate somatic SNV detection algorithm for cancer genome sequencing data. Bioinformatics 30:2498–2500.
80. Wang Y, Jill W, Leung ML, et al (2014b) Clonal evolution in breast cancer revealed by single nucleus genome sequencing. Nature 512:155–160.
81. Watson M, Mick W (2014) Quality assessment and control of high-throughput sequencing data. Front Genet. doi: 10.3389/fgene.2014.00235
82. Yang L, Luquette LJ, Gehlenborg N, et al (2013) Diverse mechanisms of somatic structural variations in human cancer genomes. Cell 153:919–929.
83. Yau C (2014) Accounting for sources of bias and uncertainty in copy number-based statistical deconvolution of heterogeneous tumour samples.
84. Ye K, Schulz MH, Long Q, et al (2009) Pindel: a pattern growth approach to detect break points of large deletions and medium sized insertions from paired-end short reads. Bioinformatics 25:2865–2871.
85. Zare H, Wang J, Hu A, et al (2014) Inferring clonal composition from multiple sections of a breast cancer. PLoS Comput Biol 10:e1003703.
86. Zhang C-Z, Leibowitz ML, Pellman D (2013) Chromothripsis and beyond: rapid genome evolution from complex chromosomal rearrangements. Genes Dev 27:2513–2530.
87. Zhao M, Wang Q, Wang Q, et al (2013) Computational tools for copy number variation (CNV) detection using next-generation sequencing data: features and perspectives. BMC Bioinformatics 14 Suppl 11:S1.

# OncoMiner: A Pipeline for Bioinformatics Analysis of Exonic Sequence Variants in Cancer

**Ming-Ying Leung, Joseph A. Knapka, Amy E. Wagler, Georgialina Rodriguez, and Robert A. Kirken**

**Abstract** With recent developments in high-throughput sequencing technologies, whole exome sequencing (WES) data have become a rich source of information from which scientists can explore the overall mutational landscape in patients with various types of cancers. We have developed the OncoMiner pipeline for mining WES data to identify exonic sequence variants, link them with associated research literature, visualize their genomic locations, and compare their occurrence frequencies among different groups of subjects. This pipeline, written in Python on an IBM High-Performance Cluster, HPC Version 3.2, is accessible at oncominer.utep.edu. It begins with taking all the identified missense mutations of an individual and translating the affected genes based on Genome Reference Consortium's human genome build 37. After constructing a list of exonic sequence variants from the individual, OncoMiner uses PROVEAN scoring scheme to assess each variant's functional consequences, followed by PubMed searches to link the variant to previous reports. Users can then select subjects to visualize their PROVEAN score profiles with Circos diagrams and to compare the proportions of variant occurrences between different groups using Fisher's exact tests. As such statistical comparisons typically involve many hypothesis tests, options for multiple-test corrections are included to control familywise error or false discovery rates. We have used OncoMiner to analyze variants of cancer-related genes in 14 samples

M.-Y. Leung (✉)
Department of Mathematical Sciences, Bioinformatics and Computational Science Programs, and Border Biomedical Research Center, The University of Texas at El Paso, El Paso, TX, USA
e-mail: mleung@utep.edu

J.A. Knapka
Bioinformatics Program, The University of Texas at El Paso, El Paso, TX, USA
e-mail: jknapka@kneuro.net

A.E. Wagler
Department of Mathematical Sciences, Computational Science Program, and Border Biomedical Research Center, The University of Texas at El Paso, El Paso, TX, USA
e-mail: awagler2@utep.edu

G. Rodriguez • R.A. Kirken
Department of Biological Sciences and Border Biomedical Research Center, The University of Texas at El Paso, El Paso, TX, USA
e-mail: grodriguez@utep.edu; rkirken@utep.edu



K.-C. Wong (ed.), *Big Data Analytics in Genomics*,
DOI 10.1007/978-3-319-41279-5_12

taken from patients with cancer, six from cancer cell lines, and ten from normal individuals. Variants showing significant differences between the cancer and control groups are identified and experiments are being designed to elucidate their roles in cancer.

**Keywords** Computational pipeline • Cancer research • Exome • Exonic sequence variants • Bioinformatics

# 1 Introduction

Advances of next-generation sequencing technologies in the past few years have greatly facilitated research studies on many human diseases at the genomic level. In a genome, the collection of all protein-coding regions, known as exons, is called the exome. Although the whole exome represents only less than 3 % of the entire human genome [1], somatic mutations within the exome can lead to serious genetic disorders and diseases like cancers. In many recent studies, analyses of whole exome sequencing (WES) data have proven to be an efficient way of identifying novel genetic alterations associated with various types of cancer. For example, WES of a rare case of familial childhood acute lymphoblastic leukemia has revealed several putative predisposing mutations in Fanconi anemia genes [2]. Using WES, Li et al. [3] have identified mutations in cell–cell adhesion genes in Chinese patients with lung squamous cell carcinoma, while Robles et al. [4] have found that colorectal tumors associated with inflammatory bowel disease have distinct genetic features from sporadic colorectal tumors.

At the cancer bio-repository housed at The University of Texas at El Paso (UTEP), an increasing number of tissue samples from patients with cancer in local hospitals have been collected. Research projects using WES have been initiated in order to better understand the molecular mechanisms and cellular pathways involved in the disease development. However, one critical obstacle that arises from the use of WES is the massive dataset produced from a single tumor source. Bioinformatics methods that can filter through multiple mutations and identify a short list of exonic variants for focused experimental investigation in the wet lab would be required. Depending on the specific dataset and the goals of their investigations, researchers may have certain preferences for scoring schemes, visualization tools, and statistical comparison methods in the process of selecting this shortlist.

While such functionalities may be available separately in various existing WES data analysis tools, they are found in different software packages or web-servers, making them difficult to be applied directly by cancer researchers who generally would not have the necessary time or experience to deal with technical computing issues. The initial motivation for our group to develop OncoMiner was the need of an easy-to-use pipeline for local researchers to explore high-dimensional WES datasets with their selected approach of combining the evolution-based PROVEAN scoring scheme [5], Circos visualization tool [6], and statistical analysis with corrections of multiple testing by controlling familywise error or false discovery rate. It also provides a user interface for researchers to directly submit their WES data for analysis through the OncoMiner website and receive the results via email. Recognizing that the selection of these choices of variant prioritization, visualization, and statistical methodologies may need to be changed from time to time for different research questions and study design, OncoMiner is implemented with a flexible modular structure, allowing pipeline components to be modified and added in the future.

A brief review of recent WES data analysis programs is given in Sect. 2, followed by a description of the implementation of OncoMiner in Sect. 3, and a detailed explanation of the statistical analysis with multiple-testing corrections in Sect. 4. An application of OncoMiner to compare cancer and normal samples is presented in Sect. 5. Some future developments planned for OncoMiner are given in the concluding Sect. 6.

It should be noted that OncoMiner is different from the package ONCOMINE with a similar name. The latter, originally developed by Rhodes et al. [7] in 2004, is a database and data-mining platform for cancer microarray data and does not provide tools for WES data analysis.

## 2 Existing Exome Sequence Analysis Tools

Many different bioinformatics tools have been developed in the past few years to analyze WES data, with the aim of identifying disease-driving mutations and discovering biomarkers to help diagnosis and treatment selections. As WES datasets are typically large, complex, and noisy, they pose considerable computational challenges in all the steps involved in WES data management and analysis, which include data preprocessing, sequence alignment, post-alignment processing, variant calling, annotation, visualization, and prioritization.

The earlier bioinformatics tools for handling WES data include Variant Effect Predictor [8] distributed in Ensembl 2011 [9], SNPeffect [10], AnnTools [11], MuSiC [12], snpEff [13], VARIANT [14], as well as the annotation tool ANNOVAR [15], the probability-based disease gene finder tool VAAST [16], and the specialized phosphorylation-related polymorphism analysis tool PhosSNP [17].

After their first introduction, additional tools developed later have been incorporated to the widely used ANNOVAR and VAAST software. For example, a web-based version wANNOVAR has been developed by Chang and Wang [18], and step-by-step protocols for using ANNOVAR and wANNOVAR are provided by Yang and Wang [19]. Dong et al. [20] pre-computed a set of variant prioritization ensemble scores for over 87 million possible exome variants and made them publicly available through ANNOVAR. Vuong et al. [21] implemented AVIA by adopting the ANNOVAR framework and adding visualization functionality. To improve the performance of the original VAAST, a new VAAST 2.0 has been developed by Hu et al. [22]. Kennedy et al. [23] described the protocols of best practices in prioritizing variants using VAAST 2.0.

Aside from additions and updates to the more established software, new tools continue to be created. For example, Douville et al. [24] have developed the cancer-related analysis of variants toolkit CRAVAT and Nadaf et al. [25] have implemented ExomeAI for detecting recurrent allelic imbalance in tumors, while Hansen et al. [26] have written a set of novel scripts for variant annotation and selection in the environment of Mathematica [27]. Several comprehensive reviews and surveys of bioinformatics analysis tools for WES, or more generally, whole genome sequencing data have been given by Gnad et al. [28], Pabinger et al. [29], Bao et al. [30], and Raphael et al. [31]. These reviews, as well as other more specific studies, such as McCarthy et al. [32] and Granzow et al. [33], generally point to the lack of concordant results from these different tools. The development of new computational methodologies and exploration of new combinations of established approaches to analyze WES for identifying biologically or clinically important sequence variants are still ongoing.

OncoMiner is a tool designed for identifying possible cancer-associated variants from a list of single-nucleotide substitutions and small insertion/deletions reported as the results of WES. Analyses of copy-number aberrations or large-genome rearrangements are not included. As the major functions currently supported by OncoMiner involve variant scoring, visualization, and statistical comparison, we first give a brief review of other software that can, although separately, provide similar functions.

## 2.1 Variant Scoring

To prioritize the long list of exonic variants and select those most likely to have the greatest impacts in relation to a disease of interest, biomedical researchers would have to evaluate each listed variant by a score that indicates the deleteriousness of the resulting change in the amino acid. Dong et al. [20] have recently compared 18 current deleteriousness-scoring methods, including the popular PolyPhen-2 [34] and SIFT [35] function prediction scores and the support vector machine based CADD ensemble score [36]. The majority of these scoring methods are available in the ANNOVAR software mentioned above and the dbNSFP v2.0 database [37], where a few other scoring methods, including the evolution-based PROVEAN [5] score have been added in late 2015 [38]. As we have chosen to use PROVEAN in the OncoMiner pipeline, we will explain its scoring scheme in greater detail below.

PROVEAN stands for Protein Variation Effect Analyzer. It provides predictions of the functional effects of protein sequence variations, including single or multiple substitutions and insertions and deletions (indels) in amino acid sequences. It is the first established variant scoring scheme to consider indels in addition to amino acid substitutions. Like many other computational approaches, PROVEAN assesses the potential impact of amino acid changes on gene functions based on the principle that evolutionarily conserved amino acid positions across multiple species are likely to be functionally important. If an amino acid change occurs at conserved positions, it has higher chances of leading to deleterious effects on gene functions.

PROVEAN score calculations require a database containing a diverse set of naturally occurring homologous and distantly related sequences, e.g., the National Center for Biotechnology Information (NCBI) Non-Redundant (NR) Protein Database [39] and the UniProtKB/Swiss-Prot Database [40], to compare with the query sequence. Suppose a query sequence $Q$ is mutated to $Q'$ by the variant $v$. For each database sequence $S$, an alignment score $A(Q, S)$ between $Q$ and $S$ is obtained. Another similar alignment score $A(Q', S)$ is obtained between $S$ and the mutated query sequence $Q'$. The difference

$$\Delta\left(Q, v, S\right) = A\left(Q', S\right) - A\left(Q, S\right)$$

is called the "delta score." A negative delta score would imply that the variant causes the query sequence to become less similar to $S$. If we observe a significantly negative average delta scores for $Q$ when aligned with the sequences in the database, it suggests that the variant $v$ has a deleterious effect.

To avoid bias due to the presence of very large numbers of highly similar sequences in certain family of proteins in the database, PROVEAN calculates an unbiased average delta score for $Q$ by first forming clusters of highly similar sequences, obtains the average scores of the clusters that are sufficiently similar to $Q$, and then takes an overall average of those cluster averages. As this process of aligning and clustering sequences is computational intensive, PROVEAN score calculation for a large number of queries is generally very slow. However, the process can speed up substantially using parallel computing technologies or by pre-computing PROVEAN scores for the known variants to date and storing them systematically in a look-up table.

## 2.2 *Visualization*

The importance of visualization in genomics data analysis has long been recognized. Pabinger et al. [29] give a very informative survey of a total of 40 such software, most of which are also included in the review by Pavlopoulos et al. [41]. Some genomic sequence analysis software such as VAAST 2.0 [22], VARIANT [14], and VarSifter [42] also provide visualization functionalities. In addition, the output of CRAVAT [24] includes a formatted submission file for the interactive visualization tool Mutation Position Imaging Toolbox [43] that allows users to map single-nucleotide variants onto the coordinates of available 3D protein structures in the Protein Data Bank at rcsb.org [44].

Among a large variety of visualization tools available at many different internet sites, it is not an easy task for biomedical researchers to gain sufficient familiarity with them to decide which ones will best fit their specific diseases of interest and learn how to use them effectively. Fortunately, as pointed out by Bao et al. [30], there are already ongoing efforts to integrate these visualization tools. The Oncoprint tool in the cBio Cancer Genomics Portal [45] and its integration with the Integrative Genomics Viewer [46] are such examples.

In the article by Nielsen et al. [47] that provides a guide to genomic data visualization methods, the authors divide such software into three categories. First, there are the finishing tools for de novo or re-sequencing experiments. Second, there are tools for browsing annotations and experimental data in relation to reference genomes. Finally, the third category contains visualization tools to compare sequence data from different individuals. The visualization function of OncoMiner belongs to the third category. It focuses on the task of visualizing genomic locations of potential deleterious variants to help biomedical researchers to decide on a shortlist of variants for further investigation based on their specific research questions and previous knowledge. An existing program that provides variant location visualization function is AVIA [21], which includes the Circos visualization tool [6] to display the genomic locations of the variants and their deleterious scores. However, the use of Circos diagrams for the computationally intensive PROVEAN scores by OncoMiner is not part of the visualization options provided by AVIA.

### *2.3 Statistical Analysis*

One important goal of WES data analysis is to detect genetic variants that have influence on the risk of diseases such as cancer. This will require assessing the prevalence of a particular variant in a disease group versus the control group. If we are considering only a single variant, testing for statistical significant difference between the disease and control groups can be done by a straightforward two-sample $z$-test for comparing proportions when the sample sizes are sufficiently large. For small sample sizes, one can use the Fisher's exact test instead [48]. WES datasets, however, provide a large number of variants for each individual, leading to the need of testing thousands of hypotheses at the same time. In such settings, suitable control of the probability of false positive findings is an important issue. WES analysis tools require suitable p-value adjustments in their statistical analysis to make appropriate corrections for multiple testing, while keeping a reasonable level of statistical power. These issues have been mentioned in Raphael et al. [31], Sham and Purcell [49], and Wang et al. [50], among others. In Sects. 4 and 5 of this chapter, we will provide examples, using both simulated and real datasets to illustrate the impacts of the multiple-test corrections in the selection of potential disease associated variants.

## 3 The OncoMiner Pipeline

OncoMiner is developed for mining WES data with the aim of identifying exonic variants, linking them to associated research literature, visualizing their genomic locations, and comparing their frequencies of occurrence among different groups of subjects. The pipeline, implemented on UTEP's IBM High-Performance Cluster, HPC Version 3.2, is accessible at oncominer.utep.edu.

## 3.1 OncoMiner Overview

OncoMiner runs on an IBM Platform LSF (Load Sharing Facility), a cluster management product provided by IBM Platform Computing [51], for job scheduling services atop the Red Hat Enterprise Server Linux Operating System running on the HPC at UTEP.

The base hardware platform is a blade server equipped with a 12-core Intel Xeon E5-2630 processor and 64GB of RAM sharing a 15 TB storage array. At the time of writing, OncoMiner has exclusive access to nine blades with a total of 108 CPU cores and provisional access to additional blades depending on the cluster's computational load. New blades can be added to the cluster as funding permits, and OncoMiner can utilize all available CPUs, since all of its computational activity is managed via the Platform LSF scheduler. It is also possible to use another open source cluster management system, such as HTCondor [52], to implement OncoMiner on other clusters provided that all the compute nodes are configured to share a common file system.

The pipeline begins with taking individual datasets resulting from Otogenetics exome sequencing [53], identifies all the missense mutations, and translates the affected genes in all predicted splice variants based on the Genome Reference Consortium's human genome build 37 (GRCh37), retrievable from NCBI's Annotation Release 104 for CRCh37 [54]. OncoMiner currently supports three major functions on exonic variants:

1. Parallel computation of PROVEAN scores and PubMed search for a collection of individual datasets submitted by the user.
2. Simultaneous visualization of multiple PROVEAN score and PubMed publication profiles generated by Step 1 above.
3. Comparative analysis of two groups of multiple individual datasets to identify variants with statistically significant differences between the groups.

The OncoMiner user interface is a Python web service implemented using the Web Service Gateway Interface at wsgi.org [55] and the web.py web framework at webpy.org [56].

## 3.2 Parallel PROVEAN Scoring

Since the PROVEAN scoring scheme is computationally intensive, using it on typical WES datasets from a single individual generally would take over a week to complete the calculations on a regular desktop computer. For supporting high-volume PROVEAN scoring [5], OncoMiner allows the submission of full-exome sequencing results as CSV files containing a specified collection of named columns indicating the chromosome number, location, NCBI gene name, reference nucleotide, and the modified nucleotide(s) of each variant. One input file typically contains WES

**Table 1** Columns of an input file to OncoMiner

| Column name | Information to be entered | Example entry |
|---|---|---|
| Var_index | Unique numeric identifier of the variant | 123 |
| Chrom | Chromosome number on which the variant occurs | chr3 |
| Left | Locus of the variant from the 5′ end of the chromosome | 67936561 |
| Right | Locus immediately after the end of the variant in the reference (unmutated) chromosome sequence; will be left + 1 for single-nucleotide polymorphisms | 67936562 |
| Ref_seq | Nucleotide at the left locus in the reference chromosome | g |
| Var_seq1 | Nucleotide at the left locus in the first copy of the mutated chromosome | c |
| Var_seq2 | Nucleotide at the left locus in the second copy of the mutated chromosome | a |
| Var_score | Read score of the variant on a scale of 0–35 with 35 indicating the highest reliability | 34 |
| Gene_name | GenBank name of the gene where the variant occurs | AAK1 |
| Where_in_transcript | "CDS" indicates that the variant occurs in a coding sequence region. Rows with any other values are ignored | CDS |
| Change_type1 | "Synonymous" or "non-synonymous", indicating the impact of the first chromosome copy's nucleotide variant on the coded protein. Synonymous variants are not scored | Non-synonymous |
| Change_type2 | Same as in change_type1 but on the second chromosome copy | Non-synonymous |
| dbSNP (optional) | The dbSNP[a] ID of the nucleotide variant, if it is available; otherwise blank | rs1801058 |

The required column names are listed along with explanations of the information expected to be entered. The last column gives an example showing the entries corresponding to a heterozygous single-nucleotide substitution occurring on a coding region of the AAK1 gene, leading to non-synonymous variants on both copies of chromosome 3

[a]dbSNP is a database for short genetic variations maintained by NCBI [57]

data for a single individual with approximately 10,000 nucleotide variants in a text file of around 20 MB. Table 1 shows a list of columns needed to be entered in the input file, and Fig. 1 gives a very small example to show what the file looks like when opened in an Excel spreadsheet.

OncoMiner produces output files with the same structure as the input, except that new columns indicating the protein variation implied by a nucleotide variation and the PROVEAN score of each protein variation, as well as some annotation columns are added. All protein sequences are computed based on the GenBank annotation and nucleotide sequence data for human genome build 37 [54].

| var_index | chrom | left | right | ref_seq | var_seq1 | var_seq2 | var_score | gene_name | where_in_ | change_type1 | change_type2 | dbsnp |
|---|---|---|---|---|---|---|---|---|---|---|---|---|
| 1 | chr1 | 881627 | 881628 | g | a | a | 35 | NOC2L | CDS | Synonymous | Synonymous | rs2272757 |
| 2 | chr1 | 887801 | 887802 | a | g | g | 35 | NOC2L | CDS | Synonymous | Synonymous | rs3828047 |
| 3 | chr1 | 888639 | 888640 | t | c | c | 35 | NOC2L | CDS | Synonymous | Synonymous | rs3748596 |
| 4 | chr1 | 888659 | 888660 | t | c | c | 35 | NOC2L | CDS | Non-Synonymous | Non-Synonymous | rs3748597 |
| 5 | chr1 | 889638 | 889639 | g | c | c | 35 | NOC2L | Intron | | | rs13303206 |
| 6 | chr1 | 889713 | 889714 | c | a | a | 12.2483 | NOC2L | Intron | | | rs13303051 |
| 7 | chr1 | 897325 | 897326 | g | c | c | 35 | KLHL17 | CDS | Synonymous | Synonymous | rs4970441 |
| 8 | chr1 | 909238 | 909239 | g | c | g | 35 | PLEKHN1 | CDS | Non-Synonymous | | rs3829740 |
| 9 | chr1 | 909238 | 909239 | g | c | g | 35 | PLEKHN1 | CDS | Non-Synonymous | | rs3829740 |
| 10 | chr1 | 909419 | 909420 | c | t | c | 35 | PLEKHN1 | CDS | Synonymous | | rs28548431 |

**Fig. 1** Snapshot of the first ten entries in a typical OncoMiner input file for PROVEAN scoring. The file must contain the 12 required columns as described in Table 1 above, and may be in CSV or Excel format

OncoMiner computes the PROVEAN score for all the isoforms of each protein that are annotated in the GenBank build; therefore, a single input row frequently results in multiple output rows corresponding to different isoforms. Furthermore, a single input row describing a heterozygous variant may require two nucleotide variants to be processed. All output rows arising from a single input row appear contiguously in the output file.

After all PROVEAN scoring is complete, OncoMiner builds a summary output file by selecting, for each input row, the output row that yielded the lowest (implying most damaging) PROVEAN score. The summary output file thus contains exactly one row for each row in the input dataset. Both the fully scored output containing all protein isoforms and the summary file are included in the results sent to the user who submitted the job.

The PROVEAN scoring facility is implemented as a collection of Bash shell and Python scripts, driven by a single main Bash shell script *provean_pipeline.sh* that manages all processing for a submitted exome sequence variant dataset. When a dataset is submitted via the OncoMiner web interface, the web server saves the dataset as a temporary file and invokes *provean_pipeline.sh* via the Platform LSF scheduler in order to process the temporary file. All data related to a given input job are stored in a uniquely named job-specific disk directory created at the time of submission, which allows OncoMiner to handle multiple simultaneous submissions. A large dataset can take days to complete on the current hardware platform, if none of the sequence variants have previously been scored; therefore, user-submitted jobs must be accompanied by an email address for the output file to be sent when complete.

Each instance of *provean_pipeline.sh* will perform the following activities:

1. Analyze the structure of the input file, verifying that the necessary data columns exist. OncoMiner is able to accept any input file that contains the specific columns required to produce PROVEAN input from nucleotide variant data. Additional columns are preserved and will appear unchanged in the output, aside from possible duplication of rows due to the processing of multiple isoforms as described above.
2. Remove input rows with low read quality, if those data are available. Read quality is an indication of how confident the sequencer is that a particular region has been sequenced correctly.

3. Compute the collection of protein variants implied by each nucleotide variation. OncoMiner uses a custom Python library to perform gene translation locally by mapping human genome sequence files into RAM and using the GenBank CDS annotation to assemble and translate the appropriate exon sequences. It computes the native protein sequence of each isoform documented in the GenBank annotation, and the protein variant for each isoform as implied by the nucleotide variant under consideration.
4. For each protein variant sequence, compute the Human Genome Variation Society (HGVS) notation describing the differences between the native protein and the variant. This HGVS variant is the input required by PROVEAN. HGVS is a very compact notation which describes the variant protein sequence as an "edit" to the native protein sequence. OncoMiner computes the HGVS notation for a protein by using the Python standard library's *difflib* module to compute the differences between the protein sequences, and then using a table lookup to convert difflib output to an HGVS string.
5. Compute an intermediate dataset suitable for generating input to PROVEAN, using the following procedure:

   (a) Build a map that associates with each native protein isoform a list of all protein variants found for that isoform. This operation is performed globally, so that all protein variants in the input dataset that affect a specific isoform appear together. Because a single PROVEAN job can compute the scores of multiple variants of a single protein, this permits OncoMiner to run the smallest possible number of PROVEAN jobs.
   (b) Generate a text file describing all the necessary PROVEAN scoring computations. For each scoring job, this file contains the native protein sequence, each protein variant, and the input data row that generated each variant.

6. Submit each PROVEAN scoring job in the intermediate file to the IBM LSF cluster job scheduler. Each scoring job first consults a local database to see whether a score for each of its variants has already been computed, and uses any such scores found. If any previously unknown variants of the job's specific isoform require scores to be computed, a PROVEAN job is started. This is by far the most computationally intensive stage of processing. On the hardware available at the time of writing, it takes approximately 40 CPU-weeks to complete all PROVEAN scoring for a full-exome sequence dataset whose variants must all be scored rather than looked up in the database.
7. Collect all scores as they become available and assemble an output file that preserves the structure of the input, with columns indicating protein variants and PROVEAN scores appended, named "AA.variation" and "PROVEAN.score."
8. Interrogate the PubMed database for each output row and annotate the output file with the following additional information in new columns:

   (a) A URL linking to search results for the row's dbSNP ID, if supplied, and if that search returned any results.

(b) A URL linking to search results for the row's gene name and HGVS variant, if that search returned any results.

(9) Email the output file to the email address specified in the submitted job.

### *3.3 Visualization of PROVEAN Score Profiles*

Circos is a visualization tool, available at circos.ca [6], that produces circular genome and chromosome diagrams and allows data associated with chromosome loci to be visualized using heat-maps and inter-locus linkage.

OncoMiner can build Circos heat-map diagrams of multiple datasets containing PROVEAN score information, each dataset comprising the WES data for a single individual, with PROVEAN scores attached. In this manner, the PROVEAN-predicted impact of variants in different parts of the genome can be visually compared. This facility is available via OncoMiner's web-based user interface. While OncoMiner will accept as many datasets as uploaded, we recommend no more than ten datasets for a single visualization job, because visual crowding in the resulting diagrams would make interpretation difficult.

The user must supply a bin size (expressed in kilobases) for the heat-map resolution. OncoMiner extracts the PROVEAN scores for each bin and finds the minimum PROVEAN score in each bin. OncoMiner then generates Circos parameter files to produce separate Circos heat-map diagrams for each chromosome and the entire genome. The whole-genome heat-map is displayed in the browser when all diagrams have been generated.

In the displayed Circos diagram, lower scores are coded as warmer colors with the more damaging scores displayed in darker shades of red. The PROVEAN heat-map uses five colors to map the range of PROVEAN scores in all plotted datasets linearly. Each individual dataset is rendered as a separate concentric ring. A summary heat-map reflecting the most-damaging PROVEAN scores among all displayed datasets is plotted as a ring surrounding the individual dataset rings.

The outermost ring of the heat-map uses five colors from light blue to dark blue to plot the maximum number of PubMed references found among the variants in each bin, with darker colors reflecting more publications. Publication counts are capped at 20, so a variant with 150 PubMed results will be plotted as if it had 20 PubMed results and indicated as 20+. Each heat-map region is a link to the data rows for that region in the uploaded datasets. Each chromosome in the outer ring of the diagram is a link to the specific chromosome's heat-map. A sample genome heat-map for five individuals is shown in Fig. 2.

Circos diagrams can provide compact visualizations of the distribution of PROVEAN scores of multiple datasets. In addition, a comparison of the summary PROVEAN score ring (the outermost red ring) with the blue ring can help researchers identify chromosomal regions with possibly damaging exonic variants that have not been much reported in published literature. Such regions are found in chromosomes 1, 9, and X in Fig. 2.

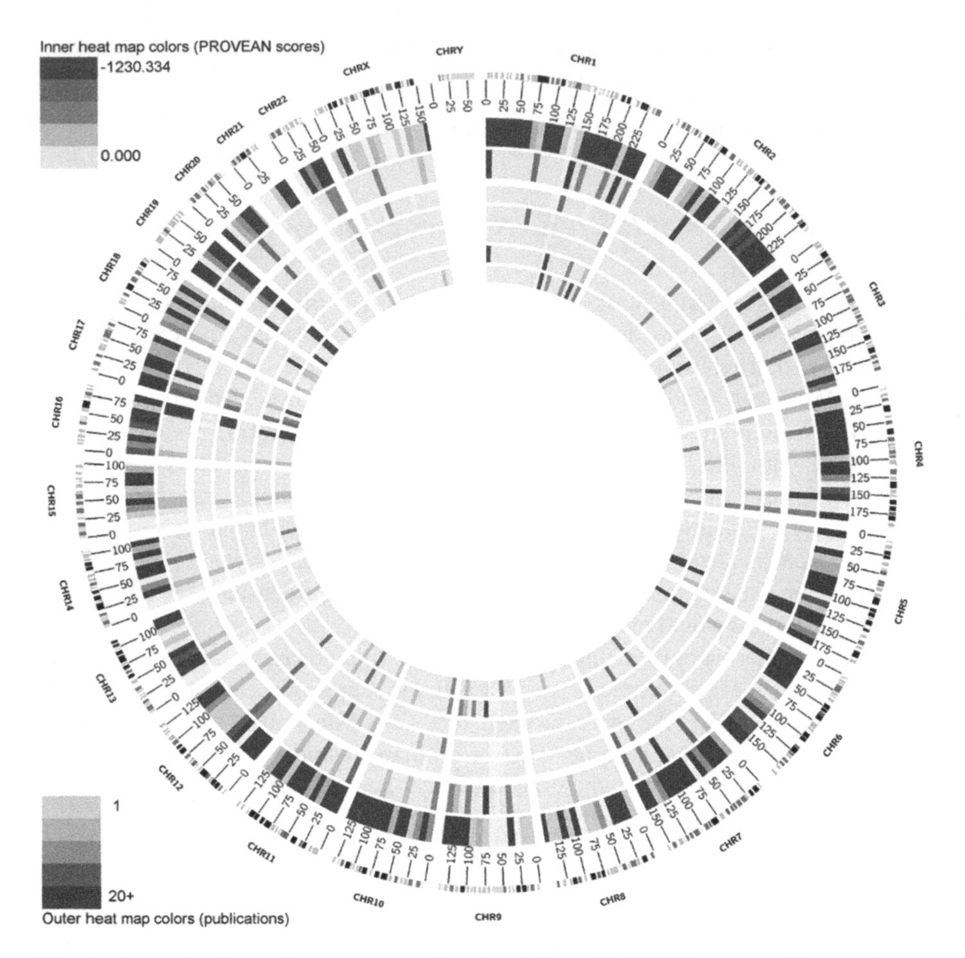


**Fig. 2** A sample Circos diagram generated by OncoMiner. PROVEAN score profiles are shown for five female individuals in the five inner rings. The outermost red ring is the summary plot reflecting the minimum PROVEAN scores of the individuals. OncoMiner allows users to choose the range of PROVEAN scores to be displayed. Here, only the negative scores, where the minimum score is −1230.334 in the datasets, are displayed. The blue ring indicates PubMed search result counts for the most-referenced variant in each bin. The outermost ring displays the ideograms representing the 24 human chromosomes (1–22, X, and Y), showing their relative sizes and banding patterns

## *3.4 Comparative Statistical Analysis of Exonic Variants*

To achieve the goal of identifying exonic variants that are likely to play critical roles in cancer so that experiments can be designed to test their biological relevance, the final step of OncoMiner is to summarize the occurrence frequencies of the combined collection of exonic variants in all the individual datasets grouped into two categories, cancer and control, compare them between the two groups, and then perform Fisher's exact tests to see if the differences are statistically significant. As these statistical comparisons are performed for a large number of variants, there are many hypotheses being tested simultaneously. Multiple-test corrections are therefore provided by OncoMiner to control familywise error and false discovery rates. The necessity of such error control, the conditions under which they should be applied, and the algorithms to implement them are explained in the next section.

## 4 Error Control for Multiple Tests

After the PROVEAN score computations are complete, users can then select subjects to produce a combined list of exonic sequence variants for comparing the proportions of occurrences of each variant between different groups of subjects. As such statistical comparisons would typically involve a large number of hypothesis test endpoints, adjustments for p-values to correct for multiple tests are often included to control familywise error or false discovery rates. These procedures are built into OncoMiner. To illustrate the necessity of multiplicity corrections, we present an example that will inform the subsequent discussion.

### *4.1 Motivating Example*

We first present a simple example to illustrate the importance of multiple-test corrections. To be clear, this toy example employs test endpoints based on simulated random variables and does not represent actual research results. Suppose p-values are reported for comparing the incidence of 40 specific variants between the patient and control groups. In this list, none of the reported p-values have been corrected for multiplicity and those statistically significant at 5 % are marked with asterisks:

| | | | | | | | | | |
|---|---|---|---|---|---|---|---|---|---|
| 0.815 | 0.466 | 0.405 | 0.298 | 0.096 | 0.681 | 0.664 | 0.401 | 0.834 | 0.415 |
| 0.005* | 0.208 | 0.675 | 0.723 | 0.685 | 0.995 | 0.826 | 0.782 | 0.387 | 0.142 |
| 0.467 | 0.391 | 0.896 | 0.101 | 0.692 | 0.094 | 0.319 | 0.517 | 0.035* | 0.982 |
| 0.029* | 0.870 | 0.012* | 0.637 | 0.674 | 0.881 | 0.518 | 0.251 | 0.086 | 0.709 |

If we consider a reported p-value smaller than 0.05 to be indicative of the particular variant being associated with the disease outcome, those p-values with asterisks associated with variants 11, 29, 31, and 33 would suggest that they are significantly associated with the disease outcome. However, note that there are 40 reported p-values in this example. This implies that for a pointwise 5 % significance level, on average 2 out of 40 of these endpoints will be spurious rejections of the null hypothesis of no association. Since the reported p-values were simulated as uniform [0,1] variables, the four seemingly significant endpoints, identified using a 0.05 pointwise significance level, actually occurred solely by chance. Indeed, when a multiplicity control like a simple p-value Bonferroni adjustment (as explained in Sect. 4.3 below) is applied, none of the test endpoints would be considered statistically significant, which reflects the chance model of null association used to generate the data. This example illustrates the "why" behind multiplicity controls. Without adjustments made for multiplicity to the data, pointwise significance often presents itself as compelling evidence, when in reality, it occurs solely by chance as in this example. This happens even more convincingly when the set of endpoints is larger than 40, as is often the case with WES studies.

After realizing the need for multiplicity correction, the next tasks are to define the family of inference and decide on the type of error control to employ. In general, an experimenter should first define the family of inference explicitly before considering the type of error control to employ. This is not always a clear task, but is probably one of the most important considerations. When faced with even a moderately sized set of p-values as in the motivating example, the first question asked of the investigator should be "what defines the family of inference?" In some cases, it may be reasonable to group test endpoints (e.g., by gender, age, or ethnicity) into smaller sets within which hypothesis tests are performed and multiplicity corrections are controlled.

Consideration of the family of inference is even more important for high-dimensional data. For example, a collection of individual WES datasets has a combined 25,000 variants for comparison between the cancer and control groups. It would not be prudent to carry out the analysis using a multiplicity control of any type with all 25,000 variants. Rather, experimental and exploratory analyses, along with relevant known biological information, can limit the family of inference without substantially inflating the overall type I error for the analysis. For instance, a researcher may be only interested in those variants found on a specific list of cancer-related genes, and the family of inferences should be defined only on the collection of relevant variants.

Following identification of the family (or families) of inference, a reasonable approach to error control should be decided upon. In general, there are two major approaches to limiting errors in large studies: familywise error rate (FWER) approaches that directly limit the incidence of one or more type I errors for the entire family of inferences and false discovery rate (FDR) approaches that limit the incidence of false positives among only the set of rejections. Deciding between which type of error control to employ is often difficult. For example, in the motivating example, the investigator may wish to make overall conclusions across all variants. Hence, there would be 40 association tests. If the investigator finds the claim of null association across all gene mutations plausible, then an FWER control procedure is most suitable. In contrast, if the investigator is cautious about making a type II error and finds the global null association claim implausible, then an FDR procedure is probably most suitable. When there are many test endpoints, as in WES studies, FDR control is an appropriate screening tool and FWER is usually only relevant once a smaller set of candidate variants is identified.

In the following description of FWER and FDR approaches to error control, the notations as shown in Table 2 will be used. Note that $V$ is the number of true

**Table 2** Notations representing the number of decisions when testing a set of null hypotheses

| Truth/decision | Non-significant | Significant | Total |
|---|---|---|---|
| True null hypothesis | U | V | I |
| False null hypothesis | T | S | M–I |
| Total | A | R | M |

null hypotheses that are rejected out of the total rejected null hypotheses ($R$). The total number of test endpoints is denoted by $M$. In all error-control problems, the magnitude of $V$, either relative to $M$ or $R$, is of primary interest.

## 4.2 Conditions for Error Control

Both FWER and FDR approaches to error control considered in this chapter are flexible methodologies that apply under a very general set of conditions. For $M$ tests, the set of null and alternative hypotheses may be denoted by $H_i$ and $H_i^{'}$, respectively, for $i = 1, \ldots, M$. For strong control of any set of null hypotheses $\{H_i\}$, there is almost always a subset of the $\{H_i\}$ that are true and FWER is controlled for any of these possible sets.

In WES settings, the number of inferences ($M$) is usually quite high and many subsets of the inferences can be highly correlated. Whenever particular subsets of the inferences are correlated, this should be recognized and, ideally, be capitalized on in order to lower the multiplicity corrected critical value and achieve more powerful error control. Resampling-based methodologies for making multiplicity adjustments are a powerful option for FWER or FDR error control for almost any setting with correlated inferences. See Westfall and Young [58] for an introduction. However, in very high-dimensional settings these are still very computationally demanding and, if many sets of analyses are being considered, the computational demand will be unwieldy. In this section, resampling-based methodologies will not be discussed, but other still powerful methods are available that do not require such extensive computational resources.

## 4.3 Control of Type I Errors

Conservative approaches of error control usually restrict attention to procedures that limit the probability of one or more false rejections of the null hypothesis. In WES analysis settings, this is usually far too conservative since researchers rarely find it is reasonable that all null hypotheses could simultaneously be true when testing a large number of test endpoints. Hence, this conservative approach, which focuses on the overall FWER of the family of inferences, is a classic methodology but limiting in many genomics scenarios since it directly controls $P(V \geq 1)$ for a set of inferences. At the same time, approaches that achieve FWER control provide a set of plausible significant endpoints that are likely to be reproducible in future related studies. Hence, genomics researchers should not completely disregard this type of error control, but instead attempt to address the conservative nature of FWER corrections while also employing other approaches to multiplicity control.

Utilizing the notations in Table 2, the following probability statements define the FWER:FWER $= P\,(\text{reject at least one } H_i : i \in \{1 \ldots M\})$

For any probability distribution, FWER has an upper bound provided by a pre-specified significance level $\alpha$. In order to employ multiplicity adjustments among any set of possibly correlated endpoints, we need to bound the probability of the union of all $M$ sets $A_i = \{Y_i > c\}$ for random variables $Y_i$ $(i = 1, \ldots, M)$ and a critical value $c$. FWER control requires that the union of all $A_i$ is bound by a pre-specified FWER $\alpha$, such that

$$P\left(\bigcup_{i=1}^{M} A_i\right) = \sum_{i=1}^{M} P\left(A_i\right) - \sum_{i<j} P\left(A_i \cap A_j\right) + \cdots + (-1)^M P\left(\bigcap_{i=1}^{M} A_i\right) \leq \sum_{i=1}^{M} P\left(A_i\right) = \alpha.$$

For error control, the objective is to bound the above inequality so that the probability of the union of all $A_i$ is as close to $\alpha$ as possible. The Bonferroni adjustment assumes $\sum_{i=1}^{M} P\left(A_i\right) = MP\left(A_i\right) = \alpha$, resulting in a per-comparison error of $P\left(A_i\right) = {}^{\alpha}\!/_{M}$. When the $Y_i$ endpoints are independent, the Bonferroni adjustment is exact. However, it can be conservative if the endpoints are dependent [59], meaning that the type I FWER will be lower than α, and the power for the set of inferences reduced. Hence, Bonferroni adjustments are very effective if the endpoints are not significantly correlated but lack power when endpoints significantly co-vary. Methods exist to correct for the conservative nature of the Bonferroni adjustment [59], and one such modification is incorporated into OncoMiner as an alternative to Bonferroni adjustments of FWER. A review of multiplicity corrections from this framework is available from Elsäßer et al. [60].

## *4.4 Control of False Discoveries*

Another approach to error control is to limit error control to the inferences that are rejections of the null hypothesis. As this class of procedures controls the number of false rejections ($V$) out of the total number of rejections ($R$), they are called FDR methods. With reference to the notations defined in Table 2, FDR is defined as $E(V/R)$, the proportion of times a null hypothesis is falsely rejected among the rejected hypotheses. For a specific $\gamma \in [0, 1]$, FDR is controlled by making $E\left(V/R\right) \leq \gamma$.

Recall that FWER control implies $P\left(V \geq 1\right)$ is bound by α for a set of inferences, and it protects against type I error in the strong sense. In practice, the proportion of times a null hypothesis is falsely rejected among the rejected hypotheses provides a more powerful method that still protects against "false discoveries" in the weak sense. The use of the proportion of type I errors among the significant tests ($R$) leads to a more powerful and "adaptive" procedure so that whenever a large number of tests are truly significant ($S$), the cutoff is lower using FDR than an FWER control. Thus, any procedure which controls FWER also controls FDR, but the reverse is not usually true. In fact, if there are a large number of false null hypotheses, which is usually the case with WES data analyses, then we would expect $S$ (in Table 2) to be large, resulting in

$$E(V/R) = E(V/(V+S)) \leq P(V \geq 1).$$

The FDR method was originally derived assuming independent endpoints [61]. Later, Benjamini and Yekutieli [62] demonstrate how the standard FDR methodology still controls $E(V/R)$ whenever the endpoints are positively dependent and Sabatti et al. [63] provide justification about why positive dependency is a reasonable assumption in genetic association studies. As a result, in large scale studies with multiple correlated endpoints, the FDR procedure is likely to result in greater power for the set of inferences. Moreover, the use of the FDR and related concepts has been recognized as providing a unified framework for these kinds of analyses [64].

A closely related alternative to FDR control is false discovery proportion (FDP) control that directly limits $P(V/R > \gamma)$ for a pre-specified threshold $\gamma \in [0, 1]$ [65, 66]. Both FDR and FDP control are useful in genomic settings for exploratory analysis where the investigator is primarily concerned with identifying subsets of the test endpoints that are statistically significant. In a hypothesis testing setting, if $R = 0$, then there is no probability of false discovery or a type I error (hence FDR = FWER = 0). Rather, whenever $R \geq 1$, the ratio $V/R$ is in the interval [0,1] and can be bounded to limit the proportion of the rejections of the null hypothesis that are false (note that $V$ denotes the number out of $R$ that are false rejections).

Note that the incidence of false discoveries may be controlled using a probabilistic formulation (FDP) or expectation (FDR). There are advantages and disadvantages to each approach. Since FDR control strictly limits the expectation or the proportion of false discoveries, the method is effective in limiting that expectation in the long run to a pre-specified significance level. However, in any particular application of the FDR, the empirical rate of false discoveries may exceed $\gamma$. Thus, in general, control of FDP seems more consistent with the classical framework of FWER control as it controls the probability of false rejections. However, due to the prevalence of FDR control, which is known to be adaptive to high-dimensional settings, in the genomics literature, we have selected to use FDR control in OncoMiner.

### *4.5 FDR Control Algorithm*

The FDR control algorithm used by OncoMiner is due to Benjamini and Hochberg [61]. For a set of test endpoints $\{Y_1, \ldots, Y_M\}$ there is a corresponding set of p-values $\{p_1, \ldots, p_M\}$ which are assumed to be independent and uniformly distributed. If these observed p-values are ordered $\{p_{(1)}, \ldots, p_{(M)}\}$, then the subset that are statistically significant with FDR controlled at some value $\alpha$ in [0,1] is partitioned using the threshold $\widehat{\lambda} = \max\left\{1 \le \lambda \le M : p_{(\lambda)} < \alpha\lambda/M\right\}$. Any test endpoints are significant for $(i) = (1), \ldots, \left(\widehat{\lambda}\right)$ whenever $p_{(i)} \le p_{\left(\widehat{\lambda}\right)}$. These test endpoints should be regarded as those that are not likely to be false positives out of the set of rejections of the null hypothesis and warrant further consideration. Follow-up analysis may be warranted to further investigate the practical and statistical significance of these endpoints using more stringent methods, such as FWER control.

Another approach for employing FDR control is to follow the two-stage strategy proposed by Benjamini and Yekutieli [62] and Reiner et al. [67]. This allows for a very liberal screening stage where FDR is controlled at a relatively large value ($q_1$). If a subset of genes is found to be significantly associated with the outcome at this stage, then this subset will be analyzed again where FDR is controlled at another rate ($q_2$). The result of this two-stage approach is that the FDR is conservatively controlled at $q_1q_2$. The values of $q_1$ and $q_2$ can be chosen at the discretion of the investigators such that $q_1q_2 = q$. This FDR correction is also being incorporated into the OncoMiner pipeline.

## 5 Using OncoMiner to Analyze Exonic Variants in Cancer

Genomic DNA was isolated from 14 tumor specimens and six human leukemia cell lines from the cancer bio-repository housed at UTEP using the Puregene Core Kit A (Qiagen) according to the manufacturer's instructions for genomic DNA purification. The ratio of absorbance at 260 and 280 nm was determined to establish purity of the DNA samples [68] prior to sending for WES analysis. The WES results, provided by Otogenetics [53], included CSV files containing the exonic variants of the individual subjects and all the required input information for OncoMiner.

Using the "Submit to PROVEAN" function at the OncoMiner site, the individual WES data files were submitted for PROVEAN score calculations. The output files for all the subjects in this study were then uploaded to the "Compute Statistics" page where they were merged into a summary file, in which the entire collection of over 70,000 variants in the subjects are listed, along with counts of occurrences for each variant in the cancer group and the control group.

Rather than analyzing this entire collection of variants, we focused on those occurring on a selected list of 290 cancer-related genes relevant to the investigations of our local researchers. This reduces the number of variants under consideration to

**Table 3** Variants in cancer-related genes with PROVEAN score < −2.5 that show statistically significant differences between the cancer and control groups with FDR controlled at 0.05

| Variant | Gene | Location | PROVEAN | AA.from | AA.to | Cancer | Control |
|---|---|---|---|---|---|---|---|
| chr13.80911525.g.a.g* | SPRY2 | 106 | −2.943 | P | S | 16 | 1 |
| chr17.37884037.c.g.c* | ERBB2 | 1140 | −2.848 | P | A | 14 | 0 |
| chr2.215645464.c.g.c | BARD1 | 281 | −2.672 | R | S | 13 | 1 |
| chr8.48805818..g.g | PRKDC | 1248 | −5.781 | S | Fr.Sh. | 16 | 2 |

Those marked with asterisks are also significant with FWER controlled at 0.05

1061. Among these, 483 variants had PROVEAN scores lower than −2.5 and hence would be considered potentially deleterious [5]. However these would still be too many to be practically investigated in the wet lab.

On the other hand, we could also select variants that show significant differences in their occurrence frequencies between the cancer group and the control group datasets. We first noted that if a variant occurred in the majority (or minority) of subjects in both the cancer and control groups, it would suggest there were no substantial differences between the groups. So we narrow down to those variants that occur in the majority (>50 %) of the cancer group but in the minority (<50 %) of the control group or vice versa. We refer to this as the majority/minority criterion, which was satisfied by 33 out of 1061 variants in the cancer-related genes. These 33 variants defined our family of inference. If the FWER was controlled at 0.05, 11 of these variants were found statistically significant using the Fisher's exact test [48]. If instead FDR was controlled at 0.05, 10 additional variants were found to be significant.

As our goal was to identify only a few variants that can be further investigated in the wet lab for their possible roles in cancer, we looked for those suggested by both the PROVEAN scores and the statistical comparison. Only two variants in the genes SPRY2 and ERBB2 with PROVEAN score < −2.5 showed significant differences between the cancer and control groups when FWER is controlled at 0.05. Two additional variants in the genes PRKDC and BARD1 satisfy the same criteria if, instead of FWER, we control FDR at 0.05 using the Benjamini and Hochberg [61] approach.

Table 3 displays the information about the four identified variants. The first three columns indicate the chromosomal location and the change in nucleotides, the name of the gene where the variant is located, and the position of the affected amino acid in the translated protein. The fourth column displays the PROVEAN scores, followed by the next two columns indicating the original amino acid at that position (AA.from) and what it is changed to (AA.to). The last two columns, respectively, show the frequencies of the variant found among the 20 samples in the cancer group and among the 10 samples in the control group. The first three variants in Table 3 are single-nucleotide substitutions, but the fourth represents an insertion of the nucleotide "*g*" on the PRKDC gene that causes a frame shift resulting in completely modifying the amino acid sequence starting from position 1248 of the protein.

It is interesting to note that although a large volume of biomedical literature (over 22,000 articles found by a PubMed search) already exists for the four cancer-related genes above, not much information about the specific variants have been reported. Would this imply that the variants may have certain roles in cancer that are still to be elucidated? Are possible deleterious biological effects of these significant variants not observable so far? Alternatively, it might be possible that the variants are more prevalent in patients with cancer in the El Paso region where over 70 % of the population is Hispanic. These call for further investigations with a larger collection of WES datasets and carefully designed experiments.

## 6 Future Work

We have implemented the OncoMiner pipeline as a web-server to which users can submit their own WES data to be analyzed and receive the results via email. It uses the resources of our local high-performance cluster to score, visualize, and make statistical comparisons of exonic variants for groups of individuals. We will continue to expand and fine-tune the pipeline's capabilities as its usage increases. For example, we are incorporating a version of JCVI-SIFT [69] and a customized scoring scheme that reflects biochemical properties of amino acids. Other features such as Circos diagram visualization of the statistical comparison results are also being added to OncoMiner.

Although requiring a good amount of time and wet-lab resources, we will proceed with experimental verification of selected exonic sequence variants from the OncoMiner's results in an effort to establish their biological relevance eventually. In the meantime, we would continue to assess the reliability and scalability of OncoMiner using different collections of datasets from public databases such as the International Cancer Genome Consortium (ICGC), The Cancer Genome Atlas (TCGA), and others as reviewed by Pavlopoulos et al. [41].

For the statistical analysis, there are situations where variants from more than two groups of individuals need to be compared. For example, an investigation may call for an overall comparison of three or more groups of patients with different cancer types. We are extending the current Fisher's exact test code on OncoMiner to handle multiple group comparisons.

It is well recognized that correlation structures among the variants will need to be taken into consideration in the statistical component of OncoMiner. The pairwise dependencies can be incorporated into the FWER multiplicity adjustments for large sets of association tests using methods adapted from Hunter [70] and Worsley [71]. Similarly, resampling methods could also be utilized for complex pairwise dependency structures. However, our preliminary analyses have revealed that the dependencies are not restricted to the usual linear correlations among pairs of variants. We have also detected higher-dimensional dependencies that may affect the resulting shortlist of variants. It would be an important task to capture the dependency structures and effectively incorporate them with suitable statistical error-control strategies in the variant selection process.

**Acknowledgments** This work was supported in part by grants from the Lizanell and Colbert Coldwell Foundation, the Edward N. and Margaret G. Marsh Foundation, as well as Grant 5G12MD007592 to the Border Biomedical Research Center from the National Institute on Minority Health and Health Disparities, a component of the National Institutes of Health, and an IBM Shared University Resources (SUR) grant to UTEP in 2012. The authors thank Jeremy Ross and Derrick Oaxaca for many helpful discussions in the early stages of this project, as well as Jaime Pena and Jonathon Mohl for their technical assistance in setting up the web interface.

## References

1. ENCODE Project Consortium (2012) An integrated encyclopedia of DNA elements in the human genome. *Nature* 489(7414):57–74.
2. Spinella J., Healy J, Saillour V, Richer C, Cassart P, Ouimet M, and Sinnett D (2015) Whole-exome sequencing of a rare case of familial childhood acute lymphoblastic leukemia reveals putative predisposing mutations in Fanconi anemia genes. *BMC Cancer* 15:539.
3. Li C, Gao Z, Li F et al (2015) Whole Exome Sequencing Identifies Frequent Somatic Mutations in Cell-Cell Adhesion Genes in Chinese Patients with Lung Squamous Cell Carcinoma. *Scientific Reports* 5:14237.
4. Robles AI, Traverso G, Zhang M et al (2016) Whole-exome Sequencing analyses of Inflammatory Bowel Disease-associated Colorectal Cancers. *Gastroenterology (in press)*.http://www.sciencedirect.com/science/article/pii/S0016508515018648. *Accessed 8 March 2016.*
5. Choi Y, Sims GE, Murphy S, Miller JR, and Chan AP (2012) Predicting the Functional Effect of Amino Acid Substitutions and Indels. *PLoS ONE* 7(10):e46688.
6. Krzywinski M, Schein J, Birol I et al (2009) Circos: An information aesthetic for comparative genomics. *Genome Research* 19(9):1639–1645.
7. Rhodes DR, Yu J, Shanker K et al (2004) ONCOMINE: A Cancer Microarray Database and Integrated Data-Mining Platform. *Neoplasia* 6(1):1–6.
8. McLaren W, Pritchard B, Rios D, Chen Y, Flicek P, and Cunningham F (2010) Deriving the consequences of genomic variants with the Ensembl API and SNP Effect Predictor. *Bioinformatics* 26(16):2069–2070.
9. Flicek P, Amode MR, Barrell D et al (2011) Ensembl 2011. *Nucleic Acids Research* 39:D800-D806.
10. De Baets G, Van Durme J, Reumers J et al (2012) SNPeffect 4.0: on-line prediction of molecular and structural effects of protein-coding variants. *Nucleic Acids Research* 40(D1):D935-D939.
11. Markarov V, O'Grady T, Cail G, Lihm J, Buxbaum JD, and Yoon S (2012) AnnTools: a comprehensive and versatile annotation toolkit for genomic variants. *Bioinformatics* 28:724–725.
12. Dees ND, Zhang Q, Kandoth C et al (2012) MuSiC: Identifying mutational significance in cancer genomes. *Genome Research* 22:1589–1598.
13. Cingolani P, Platts A, Wang LL et al (2012) A program for annotating and predicting the effects of single nucleotide polymorphisms, SnpEff: SNPs in the genome of Drosophila melanogaster strain w1118; iso-2; iso-3. *Fly* 6(2):80–92.
14. Medina I, De Maria A, Bleda M, Salavert F, Alonso R, Gonzalez CY, and Dopazo J (2012) VARIANT: Command Line, Web service and Web interface for fast and accurate functional characterization of variants found by Next-Generation Sequencing. *Nucleic Acids Research* 40:W54-W58.
15. Wang K, Li M, and Hakonarson H (2010) ANNOVAR: functional annotation of genetic variants from high-throughput sequencing data. *Nucleic Acids Research* 38(16):e164.
16. Yandell M, Huff C, Hu H et al (2011) A probabilistic disease-gene finder for personal genomes. *Genome Research* 21:1529–1542.

17. Ren J, Jiang C, Gao X et al (2010) PhosSNP for Systematic Analysis of Genetic Polymorphisms That Influence Protein Phosphorylation. *Molecular & Cellular Proteomics* 9:623–634.
18. Chang X and Wang K (2012) wANNOVAR: annotating genetic variants for personal genomes via the web. *Journal of Medical Genetics* 49:433–436.
19. Yang H and Wang K (2015) Genomic variant annotation and prioritization with ANNOVAR and wANNOVAR. *Nature Protocols* 10:1556–1566.
20. Dong C, Wei P, Jian X, Gibbs R, Boerwinkle E, Wang K, and Liu X (2015) Comparison and integration of deleteriousness prediction methods for nonsynonymous SNVs in whole exome sequencing studies. *Human Molecular Genetics* 24(8):2125–2137.
21. Vuong H, Che A, Ravichandran S, Luke BT, Collins JR, and Mudunuri US (2015) AVIA v2.0: annotation, visualization and impact analysis of genomic variants and genes. *Bioinformatics* 31(16):2748–2750.
22. Hu H, Huff CD, Moore B, Flygare S, Reese MG, and Yandell M (2013) VAAST 2.0: Improved Variant Classification and Disease-Gene Identification Using a Conservation-Controlled Amino Acid Substitution Matrix. *Genetic Epidemiology* 37:622–634.
23. Kennedy B, Kronenberg Z, Hu H et al (2014) Using VAAST to Identify Disease-Associated Variants in Next-Generation Sequencing Data. *Current Protocols in Human Genetics* 81:6.14.1–6.14.25.
24. Douville C, Carter H, Kim R et al (2013) CRAVAT: cancer-related analysis of variants toolkit. *Bioinformatics* 29(5):647–648.
25. Nadaf J, Majewski J, and Fahiminiya S (2015) ExomeAI: detection of recurrent allelic imbalance in tumors using whole-exome sequencing data. *Bioinformatics* 31(3):429–431.
26. Hansen MC, Nederby L, Roug A, Villesen P, Kjeldsen E, Nyvold CG, and Hokland P (2015) Novel scripts for improved annotation and selection of variants from whole exome sequencing in cancer research. *MethodsX* 2:145–153.
27. Wolfram Research, Inc (2015) Mathematica 10. http://www.wolfram.com/mathematica. *Accessed 8 March 2016.*
28. Gnad F, Baucom A, Mukhyala K, Manning G, and Zhang Z (2013) Assessment of computational methods for predicting the effects of missense mutations in human cancers. *BMC Genomics* 14(Supple. 3):S7.
29. Pabinger S, Dander A, Fischer M et al (2014) A survey of tools for variant analysis of next-generation genome sequencing data. *Briefings in Bioinformatics* 15(2):256–278.
30. Bao R, Huang L, Andrade J, Tan W, Kibbe W, Jiang H, and Feng G (2014) Review of Current Methods, Applications, and Data Management for the Bioinformatics Analysis of Whole Exome Sequencing. *Cancer Informatics* 13(S2):67–82.
31. Raphael BJ, Dobson JR, Oesper L, and Vandin F (2014) Identifying driver mutations in sequenced cancer genomes: computational approaches to enable precision medicine. *Genome Medicine* 6:5.
32. McCarthy DJ, Humburg P, Kanapin A et al (2014) Choice of transcripts and software has a large effect on variant annotation. *Genome Medicine* 6:26.
33. Granzow M, Paramasivam N, Hinderhofer K et al (2015) Loss of function of PGAP1 as a cause of severe encephalopathy identified by Whole Exome Sequencing: Lessons of the bioinformatics pipeline. *Molecular and Cellular Probes* 29:323–329.
34. Adzhubei I, Jordan DM, and Sunyaev S (2013) Predicting Functional Effect of Human Missense Mutations Using PolyPhen-2. *Current Protocols in Human Genetics* 76:7.20.1-7.20.41.
35. Ng PC and Henikoff S (2003) SIFT: predicting amino acid changes that affect protein function. *Nucleic Acids Research* 31(13):3812–3814.
36. Kircher M, Witten DM, Jain P, O'Roak BJ, Cooper GM, and Shendure J (2014) A general framework for estimating the relative pathogenicity of human genetic variants. *Nature Genetics* 46(3):310–315.
37. Liu X, Jian X, and Boerwinkle E (2013) dbNSFP v2.0: A Database of Human Nonsynonymous SNVs and Their Functional Predictions and Annotations. *Human Mutation Database in Brief* 34:E2393-E2402.

38. Choi Y and Chan AP (2015) PROVEAN web server: a tool to predict the functional effect of amino acid substitutions and indels. *Bioinformatics* 31(16):2745–2747.
39. National Center for Biotechnology Information (2015) RefSeq non-redundant proteins. http://www.ncbi.nlm.nih.gov/refseq/about/nonredundantproteins. *Accessed 8 March 2016.*
40. UniProt Consortium (2016) UniProt Knowledgebase. http://www.uniprot.org/uniprot. *Accessed 8 March 2016.*
41. Pavlopoulos GA, Malliarakis D, Papanikolaou N, Theodosiou T, Enright AJ, and Iliopoulos I (2015) Visualizing genome and systems biology: technologies, tools, implementation techniques and trends, past, present and future. *GigaScience* 4:38.
42. Teer JK, Green ED, Mullikin JC, and Biesecker LG (2012) VarSifter: visualizing and analyzing exome-scale sequence variation data on a desktop computer. *Bioinformatics* 28(4):599–600.
43. Niknafs N, Kim D, Kim R et al (2013) MuPIT interactive: webserver for mapping variant positions to annotated, interactive 3D structures. *Human Genetics* 132(11):1235–1243.
44. Berman HM, Westbrook J, Feng Z et al (2000) The Protein Data Bank. *Nucleic Acids Research* 28(1):235–242.
45. Cerami E, Gao J, Dogrusoz U et al (2012) The cBio Cancer Genomics Portal: An Open Platform for Exploring Multidimensional Cancer Genomics Data. *Cancer Discovery* 2(5):401–404.
46. Thorvaldsdottir H, Robinson JT, and Mesirov JP (2012) Integrative Genomics Viewer (IGV): high-performance genomics data visualization and exploration. *Briefings in Bioinformatics* 14(2):178–192.
47. Nielsen CB, Cantor M, Dubchak I, Gordon D, and Wang T (2010) Visualizing genomes: techniques and challenges. *Nature Methods* 7(3 Supple):S5-S15.
48. Agresti A (2013) *Categorical Data Analysis*. Wiley, New Jersey.
49. Sham PC and Purcell SM (2014) Statistical power and significance testing in large-scale genetic studies. *Nature Reviews Genetics* 15:335–346.
50. Wang GT, Peng B, and Leal SM (2014) Variant Association Tools for Quality Control and Analysis of Large-Scale Sequence and Genotyping Array Data. *American Journal of Human Genetics* 94:770–783.
51. IBM Platform Computing (2016) IBM Platform LSF (Load Sharing Facility). http://www-03.ibm.com/systems/platformcomputing/products/lsf. *Accessed 8 March 2016.*
52. HTCondor (2016) Center for High Throughput Computing, University of Wisconsin, Madison. https://research.cs.wisc.edu/htcondor. *Accessed 26 March 2016.*
53. Otogenetics Corporation (2016) Whole Exome and RNA Next Gen Sequencing Services. http://www.otogenetics.com/human-exome-ngs. *Accessed 8 March 2016.*
54. National Center for Biotechnology Information (2012) Human Annotation Release 104. http://www.ncbi.nlm.nih.gov/genome/guide/human/release_notes.html#b37. *Accessed 8 March 2016.*
55. WSGI (2016) Web Server Gateway Interface. http://wsgi.org. *Accessed 8 March 2016.*
56. Web.py (2016) A web framework for Python. http://webpy.org. *Accessed 8 March 2016.*
57. National Center for Biotechnology Information (2016) dbSNP Short Genetic Variations. http://www.ncbi.nlm.nih.gov/SNP. *Accessed 8 March 2016.*
58. Westfall PH and Young SS (1993) *Resampling-Based Multiple Testing: Examples and Methods for p-Value Adjustment*. Wiley, New York.
59. Hochberg Y (1988) A sharper Bonferroni procedure for multiple tests of significance. *Biometrika* 75(4):800–802.
60. Elsäßer A, Victor A, and Hommel G (2011) Multiple Testing in Candidate Gene Situations: A Comparison of Classical, Discrete, and Resampling-Based Procedures. *Statistical Applications in Genetics and Molecular Biology* 10(1).
61. Benjamini Y and Hochberg Y (1995) Controlling the False Discovery Rate: A Practical and Powerful Approach to Multiple Testing. *Journal of the Royal Statistical Society, Series B* 57(1):289–300.
62. Benjamini Y and Yekutieli D (2001) The control of the false discovery rate in multiple testing under dependency. *Annals of Statistics* 29(4):1165–1188.

63. Sabatti C, Service S, and Freimer N (2003) False Discovery Rate in Linkage and Association Genome Screens for Complex Disorders. *Genetics* 164(2):829–833.
64. Storey JD (2003) The positive false discovery rate: A Bayesian interpretation and the q-value. *Annals of Statistics* 31(6):2013–2035.
65. Genovese CR and Wasserman L (2002) Operating characteristics and extensions of the FDR procedure. *Journal of the Royal Statistical Society, Series B* 64(3):499–518.
66. Genovese CR and Wasserman L (2006) Exceedance Control for the False Discovery Proportion. *Journal of the American Statistical Association* 101(476):1408–1417.
67. Reiner A, Yekutieli D, and Benjamini Y (2003) Identifying differentially expressed genes using false discovery rate controlling procedures. *Bioinformatics* 19(3):368–375.
68. Sambrook J and Russell DW (2012) *Molecular Cloning: A laboratory manual.* Cold Spring Harbor Laboratory Press, New York.
69. JCVI-SIFT (2016) J. Craig Venter Institute, La Jolla. http://sift.jcvi.org. *Accessed 26 March 2016.*
70. Hunter D (1976) An upper bound for the probability of a union. *Journal of Applied Probability* 13(3):597–603.
71. Worsley KJ (1982) An Improved Bonferroni Inequality and Applications. *Biometrika* 69(2):297–302.

# A Bioinformatics Approach for Understanding Genotype–Phenotype Correlation in Breast Cancer

**Sohiya Yotsukura, Masayuki Karasuyama, Ichigaku Takigawa, and Hiroshi Mamitsuka**

**Abstract** Breast cancer (BC) patients can be clinically classified into three types, called ER+, PR+, and HER2+, indicating the name of biomarkers and linking treatments. The serious problem is that the patients, called "triple negative" (TN), who cannot be fallen into any of these three categories, have no clear treatment options. Thus linking TN patients to the main three phenotypes clinically is very important. Usually BC patients are profiled by gene expression, while their patient class sets (such as PAM50) are inconsistent with clinical phenotypes. On the other hand, location-specific sequence variants are expected to be more predictive to detect BC patient subgroups, since a variety of somatic, single mutations are well-demonstrated to be linked to the resultant tumors. However those mutations have not been necessarily evaluated well as patterns to predict BC phenotypes. We thus detect patterns, which can assign known phenotypes to BC TN patients, focusing more on paired or more complicated nucleotide/gene mutational patterns, by using three machine learning methods: limitless arity multiple procedure (LAMP), decision trees, and hierarchical disjoint clustering. Association rules obtained through LAMP reveal a patient classification scheme through combinatorial mutations in PIK3CA and TP53, consistent with the obtained decision tree and three major clusters (occupied 182/208 samples), revealing the validity of results from diverse approaches. The final clusters, containing TN patients, present sub-population features in the TN patient pool that assign clinical phenotypes to TN patients.

S. Yotsukura • H. Mamitsuka (✉)
Bioinformatics Center, Institute of Chemical Research, Kyoto University, Uji, Kyoto, Japan
e-mail: yotsus@kuicr.kyoto-u.ac.jp; mami@kuicr.kyoto-u.ac.jp

M. Karasuyama
Department of Computer Science, Nagoya Institute of Technology, Showa-ku, Nagoya, Japan
e-mail: karasuyama@nitech.ac.jp

I. Takigawa
Graduate School of Information Science and Technology, Hokkaido University, Sapporo, Hokkaido, Japan
e-mail: takigawa@ist.hokudai.ac.jp



K.-C. Wong (ed.), *Big Data Analytics in Genomics*,
DOI 10.1007/978-3-319-41279-5_13

This paper is an extended and detailed version on a pilot study conducted in Yotsukura et al. (Brief Bioinform, to appear).

**Keywords** Bioinformatics Approach • Genotypes • Phenotypes • Breast cancer • Correlation analysis

# 1 Background

Breast cancer (BC) statistics demonstrate that patients with estrogen receptors (ER+) or progesterone receptors (PR+) are 60 % likely to respond to endocrine therapy (such as tamoxifen, a preferred medication and endocrine-receptor blocker [2]) with positive outcomes [3]. Furthermore research has exhibited potential benefits to treating HER2+ enriched patients with specific anti-HER2 therapy (such as Herceptin) [4]. For the remaining patients who cannot fall into these three categories, i.e., negatives against ER, PR, and HER2, are known as triple negative (TN) patients, for which the treatment methods exhibit poor results [5, 6]. TN patients possess an aggressive histological phenotype with limited treatment options and very poor prognosis after extant treatments and standard chemotherapeutic regimens [7, 8]. Since TN is labeled by clinical markers, there is an urgent need to classify the similarities of TN cases consistently with clinical phenotypes to distinguish a novel treatment regiment for these patients.

Till now, the major methods for classifying breast cancers mostly incorporate gene expression data, which do not necessarily integrate its applicability to current treatments [6, 7, 9]. In gene expression profiling, PAM50 inherently demonstrates the various co-expression between 50 histologically and pathologically breast cancer related genes to classify the differences between breast cancer subgroups, while the PAM50 co-expression clusters cannot directly associate with the clinical markers, or the clinical treatments [10]. For instance, a Her2-enriched subtype detected by PAM50 can also include pools of TN patients, which is inconsistent, since a TN patient cannot be Her2+ by the clinical definition. Table 1 highlights the inconsistencies between biomarker subtypes and the corresponding clinical

**Table 1** Connection of molecular subtypes to clinical phenotypes, and prevalence rates for BC molecular subtypes

| Subtype | Clinical phenotype | Approximated prevalence (%) |
|---|---|---|
| Luminal A | ER+ and/or PR+, Her2−, low Ki67 | 40 |
| Luminal B | ER+ and/or PR+, Her2+, (or Her2− with high Ki67) | 20 |
| TN/basal-like | ER−, PR−, Her2− | 15–20 |
| Her2 type | ER−,PR−, Her2+ | 10–15 |

There are four molecular phenotypes currently used, whilst only two of them, Luminal A and Luminal B, are identified within ER+/PR+. This table is from the Susan G. Komen website [17]

phenotypes. In summary, the discrepancy of knowledge between clinical treatments and subtype classification has impeded the change in morbidity and mortality within cancer patients.

Sequence variants are believed to mark a region of the human genome which influences the risk of disease. More importantly, somatic mutations are specific to location and tumors cannot arise without them [11], which makes somatic variants useful to identify subgroups within a population. Several attempts have been already conducted to find mutation enrichment patterns linking to clinical phenotypes [9, 12, 13], while their analysis is at the level of only single nucleotides, single genes, or single gene families. The purpose of our work is to explore mutational combinations at numerous (at least three) levels of biomolecules, i.e., position-wise, gene-wise, and pathway-wise levels. We attempted to detect not only single but also complex reasonable patterns in exonic somatic variants from three levels, through a variety of viewpoints in machine learning. These patterns provide a strategic and schematic overview that outline the identified mutations and/or polymorphisms causing or predisposing to BC. In our work, we have employed limitless arity multiple procedure (LAMP), decision tree, and hierarchical disjoint clustering to investigate the linkage of the somatic variants to the clinical phenotypes, including TN cases.

## 2 Methods

We explain three machine learning methods, which we will use in our experiments: LAMP, decision tree, and hierarchical disjoint clustering. The data is a matrix with patients for rows and their features for columns. We first use mutation data, where features are mutations, taking binary values, where if a patient has some mutation at a particular location, the corresponding element is one; otherwise zero. We call this data *mutation-wise* features of patients. Next this data is transformed into a gene level, i.e., if a patient has some mutation at a particular gene, the corresponding gene is one; otherwise zero. We call this data *gene-wise* features of patients. We further set up *pathway-wise* features, which will be described later in this section.

### 2.1 *Limitless Arity Multiple Procedure*

Frequent itemset mining (FIM) is a technique to identify conjunctive combinations of items (i.e., mutations) that are frequently associated together in given data, being relevant to a particular label, such as TN. *LAMP* [14] is a tool, which efficiently scrutinizes the relevance of all possible conjunctive (AND) combinations of given features to the given class label. LAMP provides $p$-values further, which can be computed through multiple testing procedure of possible conjunctive combinations. The obtained combinations of features can be regarded as rules to explain the given class significantly.

## 2.2 Decision Tree

Decision tree is used to explain categorical response variables by given features. Methodologically, decision tree is to recursively split data based upon the predictors that best distinguish the response variable classes. We use the *rpart* package [15] to run the decision tree algorithm over our data, and the obtained results, such as the population demographics, are demonstrated by the *party* package [16].

## 2.3 Hierarchical Disjoint Clustering

Clustering is grouping instances (patients) using their features, based on some distance criterion. That is, two instances are likely to be in the same group (cluster) if their distance is closer. The distance can be measured by the similarity between features of given two instances. We use bottom-up hierarchical clustering, which starts with the closest (most similar) pairs of patients and repeats merging them until all patients are in one group, by which we can easily see the process of merging groups. So important points of clustering are two: generating features and computing some distance between instances by using features. In fact our clustering has three steps: (1) generating features, (2) computing the distance between patients, and (3) doing clustering with fixing the number of clusters by using gap statistic, which are all described below in more detail:

(1) Generating pathway-wise features

A patient can be first represented by a binary vector $x_i = (x_{i,1}, \ldots, x_{i,p})^\top \in \{0, 1\}^p$, where $p$ is the number of genes and $x_{i,j}$ is 1 if the $j$-th gene has a mutation; otherwise zero. We then focus on the pathway topology of the genes. For example, in our experiments, we use four genes ($p = 4$):FGFR2, PI3KCA, AKT1, and TP53, where their pathway is shown in Fig. 4, forming a forward, directed graph (Fig. 1a). We add, for one gene, one binary feature, which can be one if the corresponding gene or the upstream gene has a mutation; otherwise zero, indicating the mutation status of the two neighboring genes. This is *pathway-wise features*, more formally given as, $x_i^{path} = (x_{i,1}^{path}, \ldots, x_{i,p}^{path})^\top \in \{0, 1\}^p$:

$$x_{i,j}^{path} = \begin{cases} 1 & x_{i,j} = 1 \text{or} x_{i,\pi(j)} = 1, \\ 0 & \text{otherwise,} \end{cases}$$

where $\pi(j)$ is the index of the left-hand side gene of gene $j$ in Fig. 1a.

Figure 1b shows an example that $x^{path}$ can be computed from $x$.

(2) Cosine distance over gene-wise and pathway-wise features

We then use the cosine distance $D(x_i, x_{i'})$ of two patients $x_i$ and $x_{i'}$. In other words, we compute two $n \times n$ distance matrix over patients, $D^{gene}$ and

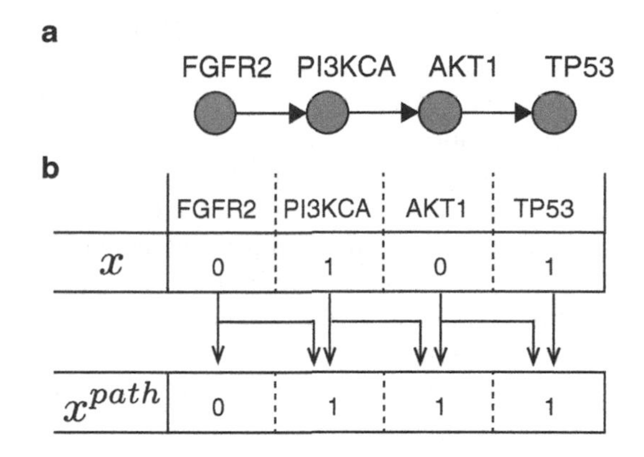


**Fig. 1** Generating pathway-wise features. (**a**) A possible (and real in this work) directed graph with four genes. (**b**) An example of generating pathway-wise feature $x^{path}$ from a gene-wise feature $x$

$D^{path}$, from the cosine distance over $\{x_i\}_{i=1}^n$ and $\{x_i^{path}\}_{i=1}^n$, respectively. The final distance matrix is then obtained by averaging over these two matrices: $D = \frac{1}{2}\left(D^{gene} + D^{path}\right)$.

The cosine distance $D(x_i, x_{i'})$ of two patients $x_i$ and $x_{i'}$ is

$$D(x_i, x_{i'}) = 1 - \frac{x_i^\top x_{i'}}{\|x_i\|_2 \, \|x_{i'}\|_2} = 1 - \frac{\mathscr{I}(x_i) \cap \mathscr{I}(x_{i'})}{\sqrt{|\mathscr{I}(x_i)| \, |\mathscr{I}(x_{i'})|}},$$

where $\mathscr{I}(x_i) = \{j \mid x_{ij} = 1\}$ is a set of genes with mutations. Note that the cosine distance is in between [0, 1], where zero (minimum) or one (maximum) is that two patients have totally consistent or inconsistent mutations, respectively.

(3) The optimal number of clusters with gap statistic for average linkage hierarchical clustering

We use the standard "hclust" function in R (in the "stat" library, an R default package) to run average linkage hierarchical clustering. The number of clusters is selected by a criterion, slightly modified from "gap statistic" [18], which examines the within-cluster dispersion by random sampling. This criterion is computed by modifying the `clusGap` function in the R "cluster" package. The number of bootstrap sampling is set to 100 (default), and the maximum of the gap statistic is chosen by the option `Tibs2001SEmax` which also follows the procedure to define the local maximum of the statistic [18].

We can compute the sum of within-cluster dispersion, $W_k$ for $k$ clusters:

$$W_k = \sum_{i=1}^{k} \frac{1}{n} \left( \sum_{j,j' \in \mathscr{C}_i} D_{jj'} \right),$$

where $\mathscr{C}_i$ is an index set of the $i$-th cluster. The basic idea of the gap statistic is to select $k$, which maximizes the difference of $W_k$ between that computed from the given data and the expected value under a null reference distribution:

$$\mathrm{Gap}(k) = E_n\{\log(W_k)\} - \log(W_k),$$

where $E_n$ is the expectation of a sample of size $n$ under a null reference distribution.

We follow [18] to generate reference features, which are further transformed into binary: Our $n \times p$ data matrix, $X$, is first scaled so that the mean of each column is zero. We then perform singular value decomposition of $X$: $UDV^\top$, where $D$ was computed by the same procedure as the distance matrix from the observed data. We computed the reference data $Z = Z'V^\top$, where $Z'$ is a uniform random matrix, for each column, with the same value range as $XV$. We then transform each $(i,j)$ element $Z_{ij}$ of $Z$ into a binary, as follows, after recovering the column scaling of $Z$:

$$\begin{cases} 1 & Z_{ij} \geq 0.5, \\ 0 & Z_{ij} < 0.5, \end{cases}$$

where if all values are zero for one row, the maximum value of this row was set at one. We finally generate $Z$ repeatedly to estimate $E_n\{\log(W_k)\}$.

# 3 Experiments

Again three levels of features were considered in this study: (1) *mutation-wise* defines the original SNP in respect to the location and nucleotide (A,C,T, or G), (2) *gene-wise* indicates whether the gene has any mutation or not, and lastly (3) *pathway-wise* indicates whether a specific pathway has a mutation or not.

## 3.1 Data

### 3.1.1 Main Data and Preparation

The somatic mutational data from *The Cancer Genome Atlas* Research Network (TCGA) (https://tcga-data.nci.nih.gov/tcgafiles/, IlluminaGA_DNASeq BRCA dataset from genome.wustl.edu Level_2.3.3.0.) was pre-processed patient data, consisting of 993 tumor samples and 1032 normal samples. Our rendition of the data analysis focuses on the intrinsic mutational codes, TCGA is sufficient to address our needs, due to its large size and previous publications for cohesive and methodological investigation of the breast cancer patients [19]. All SNV annotation was done through the GRCh37 human genome.

The original matrix of patients with somatic mutations was curated as follows (Table 2): we selected particular mutations by the annotated information of PAM50 subtypes, TN status, and somatic mutations, which appeared in three patients at minimum, because mutations appearing only once or twice would be unreliable (*Dataset A*). For statistical consistency, the matched normal samples were then

**Table 2** Molding of our working datasets

| Phenotype | Dataset A | Dataset B | Dataset C | Dataset D |
|---|---|---|---|---|
| TN | 77 | 77 | 77 | 30 |
| Her2 | 53 | 48 | 48 | 24 |
| ER+/PR+ | 341 | 341 | 341 | 154 |
| Matched normal | 547 | 466 | 0 | 0 |
| Total | 1013 | 932 | 466 | 208 |

This table shows the progression of how our datasets have been modified from the initially obtained Data from TCGA to the current working dataset. The initial dataset was (n=1013). We then decreased the matched normal number to equal to the total in other groups (n=932). Next, we removed all the matched normal (Dataset C) (n=466) to focus on the BC samples. Finally, Dataset D consists of only patients with exonic mutation (n=208). Note that all mutations appear at least three times in data

reduced to be equal in size to the cancer patient samples, resulting in a reduced number of patients in total (*Dataset B*). Out of only the cancer patients (*Dataset C*), the patients that exhibit at least one exonic mutation were selected (*Dataset D*).

Patients were then labeled by clinical phenotypes as follows: by using the absence of clinical molecular markers, the patients were divided into two groups (TN or non-TN), to label the TN population. Then the PAM50 classifier was applied to collocate the Her2-enriched patients in the non-TN patients, to divide the patients into the Her2+ labeled patients and those labeled by ER+/PR+ (Fig. 3b). This cascade manner was taken, because the original data contains samples labeled by both Her2 and TN, which, however, should be TN patients, since this is not consistent with clinical markers. By using the cascade manner, our population of Her2+ samples were clearly not TN.

### 3.1.2 Integrative Genomic Viewer

SNVs were associated with their genes through *Integrative Genomic Viewer (IGV)* [20], which allows to visually represent SNVs specific to chromosome numbers through the gene annotation of RefSeq (GRCh37/hg19). We obtained the RefSeq annotation of the genes for the SNV locations in our dataset and applied the mutation positions to obtain the relevant genes, through visualization of IGV. The result generated a unique gene-specific matrix, which consisted of all mutation positions combined into a binary vector for the corresponding gene. By using this matrix, our dataset was represented not only by mutation-wise features but also their corresponding genes, which we call *gene-wise* features.

#### 3.1.3 Database for Annotation, Visualization, and Integrated Discovery

To confirm the function of the gene, we used *Database for Annotation, Visualization and Integrated Discovery (DAVID)* [21], which associates the genes (identified from Refseq) to biochemical pathways (or diseases, etc.) and its significance in somatic breast cancer. For biochemical pathways and diseases, DAVID integrates KEGG pathways and OMIM, respectively:

**Kyoto Encyclopedia of Genomes and Genes (KEGG)** [22] is a manually curated gene network which visually displays biochemical pathways for query genes. We identified pathways (corresponding to genes) through KEGG, to be used as features, which we call *pathway-wise* features.
**Online Mendelian Inheritance in Man (OMIM)** [23] is a current comprehensive knowledgebase associating the relevant genes to all known diseases to enhance genomic analysis. Integrating all NCBI and Entrez sources, it allows for a global search of the query genes to related diseases.

The nine genes annotated by RefSeq were queried to check if they are connected to pathways in KEGG [22] through DAVID [21].

### *3.2 Evaluation Measures*

#### 3.2.1 Precision, Coverage, and F-Measure

The relevance of mutations (and their combinations) to clinical phenotypes was examined by *precision*, *coverage (recall)*, and *F-measure*, which were computed by using R [24] and Bioconductor.

Precision, coverage, and F-measure are given as follows:

$$\text{Precision} = \frac{N_{S,R}}{N_S},$$

$$\text{Coverage} = \frac{N_{S,R}}{N_R},$$

where $N_S$ be the number of patients with subtype $S$, and among those patients, let $N_{S,R}$ be the number of patients consistent with rule $R$, while $N_R$ be the number of all patients which are consistent with rule $R$. In general, a higher coverage will result in a lower precision value. Therefore, F-measure or F, the harmonic mean of precision and coverage, was used to evaluate rules, thinking about the balance between precision and coverage:

$$\frac{1}{\text{F}} = \frac{1}{2} \times \left( \frac{1}{\text{Precision}} + \frac{1}{\text{Coverage}} \right)$$

**Table 3** 30 exonic, somatic mutations selected

| ID | Chromosome | Position | Ref | Variant | rs id (dbSNP) | Class | Type | #patients |
|---|---|---|---|---|---|---|---|---|
| m1 | 1 | 16947064 | G | C | rs12041479 | SNV | RNA | 4 |
| m2 | 1 | 16950470 | C | T | rs12144467 | SNV | RNA | 4 |
| m3 | 2 | 198266834 | T | C | – | SNV | Missense | 5 |
| m4 | 3 | 178921553 | T | A | rs121913284 | SNV | Missense | 11 |
| m5 | 3 | 178936082 | G | A | rs121913273 | SNV | Missense | 19 |
| m6 | 3 | 178936091 | G | A | rs104886003 | SNV | Missense | 32 |
| m7 | 3 | 178938934 | G | A | – | SNV | Missense | 5 |
| m8 | 3 | 178952085 | A | G | rs121913279 | SNV | Missense | 70 |
| m9 | 3 | 178952085 | A | T | rs121913279 | SNV | Missense | 10 |
| m10 | 9 | 69501969 | C | A | rs7040086 | SNV | RNA | 4 |
| m11 | 10 | 123258034 | A | C | – | SNV | Missense | 1 |
| m12 | 10 | 123258034 | A | T | rs121913476 | SNV | Missense | 3 |
| m13 | 14 | 105246551 | C | T | rs121434592 | SNV | Missense | 11 |
| m14 | 15 | 23096921 | A | C | rs4778307 | SNV | RNA | 4 |
| m15 | 16 | 4432029 | A | C | rs3810818 | SNV | Missense | 4 |
| m16 | 17 | 7577120 | C | A | – | SNV | Missense | 1 |
| m17 | 17 | 7577120 | C | T | rs28934576 | SNV | Missense | 4 |
| m18 | 17 | 7577121 | G | A | rs121913343 | SNV | Missense | 4 |
| m19 | 17 | 7577539 | G | A | rs121912651 | SNV | Missense | 5 |
| m20 | 17 | 7578190 | T | C | – | SNV | Missense | 4 |
| m21 | 17 | 7578190 | T | G | rs121912666 | SNV | Missense | 1 |
| m22 | 17 | 7578190 | T | A | – | SNV | Missense | 1 |
| m23 | 17 | 7578212 | G | C | – | SNV | Nonsense | 1 |
| m24 | 17 | 7578212 | G | A | – | SNV | Nonsense | 3 |
| m25 | 17 | 7578263 | G | A | rs397516435 | SNV | Nonsense | 4 |
| m26 | 17 | 7578265 | A | G | – | NA | Missense | 4 |
| m27 | 17 | 7578271 | T | C | – | NA | Missense | 3 |
| m28 | 17 | 7578271 | T | G | – | NA | Missense | 2 |
| m29 | 17 | 7578271 | T | A | – | NA | Missense | 1 |
| m30 | 17 | 7578406 | C | T | rs28934578 | SNV | Missense | 10 |

From the most-left to the most-right columns, for each mutation, the ID, chromosome, position, original nucleotide, variant, id of dbSNP, class, type, and number of patients corresponding to the mutation are shown

### 3.3 *Results: 30 Single "Hotspot" Mutations*

The *D*ataset B with 466 positives and 466 negatives has 30 mutational positions (MP), which are shown in Table 3 and also we call *mutation-wise* features. 30 MP containing 23 missense, 3 nonsense, and 4 RNA variants and highlighted in Fig. 2. The majority of the SNPs were located in chromosomes 17 and 3, occupying 50 % and 20 %, respectively. Three hotspot locations were observed in chromosome 3 consisting of 70 (33.7 %), 32 (15.4 %), and 19 (9.1 %) patients.

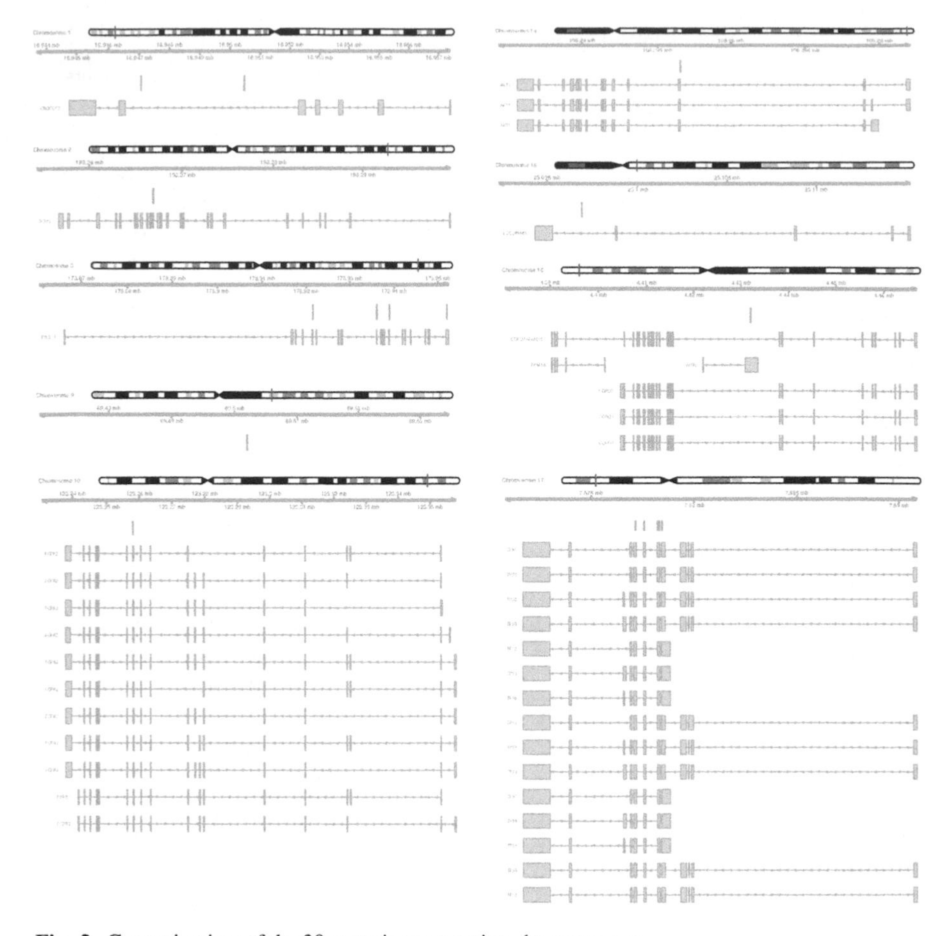

**Fig. 2** Genomic view of the 30 mutations over nine chromosomes

While in chromosome 17, the number of patients per mutation was all five or less except one with ten patients.

Interestingly, the four RNA variants appear only in relatively minor chromosomes 1, 9, and 15. *Dataset D* (Table 2), a mutation-only extraction set from *Dataset B*, contains only 208 samples comparatively less than the other relative studies [13], due to the minimum mutation restriction (n=3), where 182 (87.5 %), 25 (12 %), and one (0.5 %) patients exhibited one, two, and three mutations, respectively (Table 4). In this study, we focused mainly on *Dataset D*. All 208 samples are SNP-presenting BC patients, which allow for cancer patients to be distinguished from normal by nucleotide variants alone.

**Table 4** #patients (Dataset D)

| #mutations | ER+/PR+ | Her2+ | TN | Total |
|---|---|---|---|---|
| 1 | 133 | 21 | 28 | 182 |
| 2 | 20 | 3 | 2 | 25 |
| 3 | 1 | 0 | 0 | 1 |
| Total | 154 | 24 | 30 | 208 |

This table shows the number of patients within each subtype of the mutations-only (dataset D) Dataset. Please see the methodology for the explanation of the conditions for the subsetting n=208

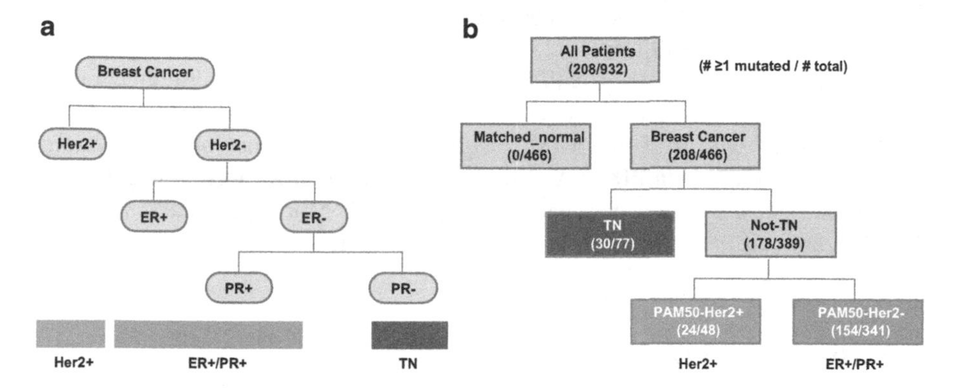


**Fig. 3** Classification cascade. (**a**:*left*) A cascade classification system on clinical breast cancer phenotypes, and (**b**:*right*) distribution of 208 patients over phenotypes

### 3.3.1 Phenotypes and Mutation Count Ratios

The patients in *Dataset D* were labeled in a cascade manner (Fig. 3), resulting in 154 (74 %) in ER+/PR+, 24 (11.5 %) in Her2+, and 30 (14.4 %) in TN. The BC patients (n=466) showed a similar ratio of 341 (73.2 %) in ER+/PR+, 48 (10.3 %) in Her2+, and 77 (16.5 %) in TN (Table 4). Thus, no correlation was observed between the clinical phenotypes and the number of mutations.

### 3.3.2 TP53 Status in TN Patients

The 30 MPs were located in the nine unique genes (*gene-wise* features), according to RefSeq annotation: CROCCP2, SF3B1, PIK3CA, LOC100132672, FGFR2, AKT1, LOC283683, VASN, and TP53 (Table 5). 143 (68.1 %) patients had mutations in PIK3CA, followed by 48 (23.1 %) in TP53 and 11 (5.3 %) in AKT1 (Table 12), where PIK3CA and TP53 are on chromosomes 3 and 17, respectively.

The ER+/PR+ patients occupied 154/208 (74 %) patients, by which the majority of patients were labeled by ER+/PR+. Also for each gene, the ER+/PR+ patients were 119/148 (83.2 %) patients with mutations in PIK3CA and 10/11 (90.9 %) in

AKT1, while they were only 22/48 (45.8 %) patients with mutations in TP53 and instead TN patients were 21/48 (43.8 %). The bulk of mutations of TP53 in TN amongst the nine genes (Table 5) establishes a clear but distinguishing link of TP53 gene to TN patients and different from other genes [13]. The existing percentage of genes with mutations appearing in patients of BC and also three clinical phenotypes for the five known genes was rather consistent with our data (Table 6).

#### 3.3.3 Mutations in Apoptosis Genes

Two statistically significant pathways were identified in KEGG: hsa05200 (pathways in cancer) with AKT1, FGFR2, TP53, and PIK3CA, and hsa04210 (apoptosis) with AKT1, TP53, and PIK3CA (Table 7). Figure 4 demonstrates the integration of the pathway-wise features to create a forward directed association in the cancer pathway with RAS and MIM2, highlighting its influence onto/by the four genes. Also, the SNPs in PIK3CA, AKT1, TP53, and FGFR2 were recognized as significant genes, within cancer pathways by OMIM. More specifically, SNPs in PIK3CA and AKT1 as pertinent for somatic breast cancer.

### *3.4 Results: LAMP*

#### 3.4.1 Patient Classification Tree

LAMP allows association analysis to detect hidden significant features, which can discriminate patients in one class from patients in other classes [14]. We considered three settings (Table 8): ($\alpha1$, $\alpha2$, and $\alpha3$) of class labels over BC patients: ($\alpha1$) the entire BC class, ($\alpha2$) two classes of non-TN or TN, and ($\alpha3$) three clinical phenotypes: ER+/PR+, Her2+, and TN. "non-BC" normal samples were added to each setting, meaning $\alpha1$, $\alpha2$, and $\alpha3$ had two, three, and four classes, respectively.

LAMP over mutation-only samples indicates that only $\alpha2$ and $\alpha3$ were tested with both mutation-wise and gene-wise features (Table 9). m6 and m8 mutations in PIK3CA were found to be significantly associated with the non-TN population in BC patients demonstrated by $\alpha2$ in both mutation-wise and gene-wise features. More importantly, m6 could further divide the non-TN patients to the respective phenotypic population ($\alpha3$). Overall, the mutation-wise and gene-wise analysis was relatively consistent, yet the gene-wise analysis was more informative, due to its less sparse feature matrix.

We further validated our mutation-wise results by analyzing the broad-scale analysis of using *Dataset B* under three LAMP settings (Table 10). $\alpha1$ detected seven significant combinations, through sequence variants mostly comprised of chromosome 3 (PIK3CA), followed by chromosomes 17 (TP53) and 14 (AKT1). $\alpha2$ detected three new mutations for the TN population. Finally, $\alpha3$ observed five more significant mutations compared to the only one found in *Dataset D*.

**Table 5** Nine genes from 30 mutations through NCBI annotation

| Gene | Mutation | Cytogenetic location | Description | Gene ID | Gene type | Pathway (KEGG) |
|---|---|---|---|---|---|---|
| CROCCP2 | m1–m2 | 1p36.13 | Ciliary rootlet coiled-coil, rootletin pseudogene 2 | 84809 | Pseudo | |
| SF3B1 | m3 | 2q33.1 | Splicing factor 3b, subunit 1, 155kDa | 23451 | Protein coding | |
| PIK3CA | m4–m9 | 3q26.32 | Phosphatidylinositol-4,5-bisphosphate 3-kinase, catalytic subunit $\alpha$ | 5290 | Protein coding | AKT signaling pathway |
| LOC100132672 | m10 | 9q21.11 | Glucoside xylosyltransferase 1 pseudogene 6 | | Pseudo | |
| FGFR2 | m11–m12 | 10q26.13 | Fibroblast growth factor receptor 2 | 2263 | Protein coding | FGF signaling pathway |
| AKT1 | m13 | 14q32.32 | V-akt murine thymoma viral oncogene homolog 1 | 207 | Protein coding | AKT signaling pathway |
| LOC283683 | m14 | 15q11.2 | Uncharacterized LOC283683 | | ncRNA | |
| VASN | m15 | 16p13.3 | Vasorin | 114990 | Protein coding | Inhibitor of TGF-$\beta$ signaling |
| TP53 | m16–m30 | 17p13.1 | Tumor protein p53 | 7157 | Protein coding | p53 pathway |

For each of the nine genes with 30 mutations, the corresponding mutations, location, description, ID, type, and corresponding pathways are shown

**Table 6** Mutational gene profiling of breast cancer on five genes

| Gene | Breast cancer | ER+ and/or PR+ | HER2+ | TN |
|---|---|---|---|---|
| PIK3CA | 26 % [25–27] | 34.5 % [26, 27] | 22–31 % [26, 27] | 8.3 % [27] |
| TP53 | 20–50 % [28, 29] | 62.1–72.6 % [30] | 47.6–48.3 % [30, 31] | 89 % [32] |
| AKT1 | 1.2–4 % [25, 33, 34] | 3.2 % [27] | >1 % [27] | >1 % [27] |
| FGFR2 | 1–2 % [35–38] | 4.8 % [35, 36] | 1 % [35, 36] | 2–4 % [35, 36, 38, 39] |
| SF3B1 | 1.8 % [40] | 2.1 % [40] | 1.8 % [40] | 0.5 % [40] |
| ATM | | | | 2 %[41] |

Frequency of somatic gene mutations in the apoptotic network in invasive breast cancer subtypes

**Table 7** Relevant KEGG pathways

| ID | Name | Relevant gene | *p*-value | Adjusted *p*-value | Fold enrichment |
|---|---|---|---|---|---|
| hsa05200 | Pathways in cancer | AKT1 | 2.66086E-4 | 0.012166 (Bonf.) | 15.503048 |
| | | FGFR2 | | 0.006102 (Benj.) | |
| | | TP53 | | 0.251217 (FDR) | |
| | | PIK3CA | | | |
| hsa04210 | Apoptosis | AKT1 | 8.58566E-4 | 0.03874 (Bonf.) | 43.83602 |
| | | TP53 | | 0.003943 (Benj.) | |
| | | PIK3CA | | 0.80856 (FDR) | |

Two pathways in KEGG are related with genes with 30 mutations. Their details, such as name, relevant genes, and *p*-value, are shown

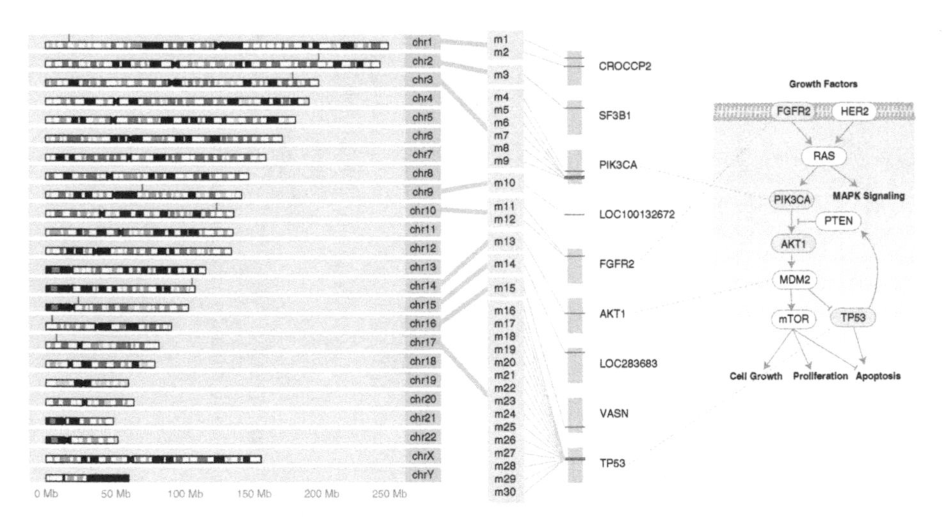


**Fig. 4** Genomic association from position to gene to pathway. The most *left-hand side* shows 24 chromosomes, where nine out of 24 have the 30 iconic mutations, which are considered in this work. A more detail view is given in Fig. 2. Also these 30 mutations are summarized into Table 3 with detailed information. The 30 mutations are further on nine genes, which are shown in the middle of the figure. Table 5 presents more detailed information on the relationships between the 30 mutations and nine genes. The *right-hand side* shows the apoptosis pathway obtained from the KEGG database, in which four (colored by *blue*) out of the nine genes are located. These four genes are PIK3CA, FGFR2, AKT1, and TP53

**Table 8** Data setting for LAMP

| Subtype groups | ($\alpha$1) | ($\alpha$2) | ($\alpha$3) | Subgroup name |
|---|---|---|---|---|
| BC | 466 | | | BC |
| BC_nonTN | | 389 | | BC |
| BC_TN | | 77 | | TN |
| matched_nonTN | 466 | 466 | | non-BC |
| BC_nonTN_Her2 | | | 48 | Her2+ |
| BC_nonTN_nonHer2 | | | 341 | ER+/PR+ |
| BC_TN_nonHer2 | | | 77 | TN |
| match_nonTN_nonHer2 | | | 466 | non-BC |
| Total | 932 | 932 | 932 | |

Summary of patient counts for Dataset B through each subgroup division (See Table 2). The counts for the various subtypes shown in Fig. 3

m6 and m8 ($\alpha$2) were consistent with those for non-TN (Table 9); however, four positions already established as significant features for BC-only setting in *Dataset D* failed to be detected in the non-TN sub-population of *Dataset B*. In summary, the broad-scale analysis confirmed the non-TN sequence variants and indicated that single mutations are not necessarily definitive and consistent in identifying clinical phenotypes.

Table 11 confirmed the gene-wise features for the previous positions for BC in $\alpha$1, yet also detected genes, such as CROCCP2, that were undetected through mutation-wise features alone. In contrast, $\alpha$2 and $\alpha$3 identified many significant mutation-wise features in the same genes. Our findings were biased to the ER+/PR+ phenotype in $\alpha$3, therefore we applied the PAM50 Her2-enriched label to detect Her2-enriched samples to have an association with PIK3CA even though it is less significant compared to its ER+/PR+ counterpart.

Significant gene-wise features (Table 11) were summarized into a classification scheme outlining BC patients (Algorithm 2). There were no single mutation markers, which could assign BC patients to either of the three clinical phenotypes directly, leading us to focus on the patients with more than one mutation. For instance, a combination of TP53 and PIK3CA was found as a significant feature (Table 11).

### 3.4.2 Examining the Combination of TP53 and PIK3CA

No novel decisive patterns could be found within the double mutations (Fig. 5a), since the distribution was rather broad. However, the double gene-wise mutation by TP53 and PIK3CA was an established pattern (Fig. 5b, c) [9, 12, 13, 19], observed in 9/23 patients. Overall, PIK3CA (P+) and TP53 (T+) exhibited high frequency of mutation (left of Fig. 6) over the three clinical phenotypes. The mutated/non-

**Table 9** Significant mutation-wise and gene-wise features from small-scale Dataset D by LAMP

| Setting | Label | Feature | #pat | Precision (Nopt/#pat) | Coverage (Nopt/Nos) | F-measure | Adjusted $p$-value |
|---|---|---|---|---|---|---|---|
| ($\alpha$2) | TN | TP53 | 48 | 0.4375 (21/48) | 0.7(21/30) | 0.538 | $2.9796^{-08}$ |
| ($\alpha$2) | Non_TN | PIK3CA | 143 | 0.958 (137/143) | 0.7697 (137/178) | 0.8536 | $6.2607^{-09}$ |
| ($\alpha$2) | Non_TN | m6 | 32 | 1.00 (32/32) | 0.180 (32/178) | 0.305 | $8.7111^{-03}$ |
| ($\alpha$2) | Non_TN | m8 | 70 | 0.929 (65/70) | 0.365 (65/178) | 0.524 | $4.7351^{-02}$ |
| ($\alpha$3) | ER+/PR+ | PIK3CA | 143 | 0.832 (119/143) | 0.7727 (119/154) | 0.8013 | $2.5418^{-05}$ |
| ($\alpha$3) | ER+/PR+ | m6 | 70 | 0.9063 (29/70) | 0.1883 (29/154) | 0.3118 | $3.8161^{-02}$ |

#pat is the counts of the patients with the feature. *Nopt* is the number of patients with the feature and in the class label. *Nos* is the number of patients in the class label

**Table 10** LAMP results of using mutation-wise features for Dataset B

| Setting | Label | Position | #pat | Precision (Nopt/#pat) | Coverage (Nopt/Nos) | F-measure | Adjust $p$-value |
|---|---|---|---|---|---|---|---|
| ($\alpha$1) | BC | m8 | 70 | 1.00 (70/70) | 0.150 (70/466) | 0.261 | $3.5946^{-22}$ |
| ($\alpha$1) | BC | m6 | 32 | 1.00 (32/32) | 0.0687 (32/466) | 0.129 | $9.3946^{-10}$ |
| ($\alpha$1) | BC | m5 | 19 | 1.00 (19/19) | 0.041 (19/466) | 0.0784 | $1.1072^{-05}$ |
| ($\alpha$1) | BC | m13 | 11 | 1.00 (11/11) | 0.0236 (11/466) | 0.0461 | $3.2199^{-03}$ |
| ($\alpha$1) | BC | m4 | 11 | 1.00 (11/11) | 0.0236 (11/466) | 0.0461 | $3.2199^{-03}$ |
| ($\alpha$1) | BC | m9 | 10 | 1.00 (10/10) | 0.0215 (10/466) | 0.042 | $6.5105^{-03}$ |
| ($\alpha$1) | BC | m30 | 10 | 1.00 (10/10) | 0.0215 (10/466) | 0.042 | $6.5105^{-03}$ |
| ($\alpha$2) | TN | m30 | 10 | 0.50 (5/10) | 0.0649 (5/77) | 0.115 | $1.4762^{-02}$ |
| ($\alpha$2) | TN | m25 | 4 | 0.75 (3/4) | 0.0390 (3/77) | 0.0741 | $4.9095^{-02}$ |
| ($\alpha$2) | TN | m26 | 4 | 0.75 (3/4) | 0.0390 (3/77) | 0.0741 | $4.9095^{-02}$ |
| ($\alpha$3) | ER+/PR+ | m8 | 70 | 0.8286 (58/70) | 0.1701 (58/341) | 0.282 | $1.7472^{-16}$ |
| ($\alpha$3) | ER+/PR+ | m6 | 32 | 0.9063 (29/32) | 0.0850 (29/341) | 0.156 | $1.4502^{-10}$ |
| ($\alpha$3) | ER+/PR+ | m13 | 11 | 0.9090 (10/11) | 0.0293 (10/341) | 0.0568 | $2.2927^{-04}$ |
| ($\alpha$3) | ER+/PR+ | m4 | 11 | 0.9090 (10/11) | 0.0293 (10/341) | 0.0568 | $2.2927^{-04}$ |
| ($\alpha$3) | ER+/PR+ | m5 | 19 | 0.7368 (14/19) | 0.0411 (10/341) | 0.0778 | $1.0063^{-03}$ |
| ($\alpha$3) | ER+/PR+ | m9 | 10 | 0.80 (8/10) | 0.0235 (8/341) | 0.0456 | $6.3485^{-03}$ |

#pat is the counts of the patients with the feature. *Nopt* is the number of patients with the feature and in the class label. *Nos* is the number of patients in the class label

**Table 11** LAMP results of using gene-wise features for Dataset B

| Setting | Label | Gene | #pat | Precision (Nopt/#pat) | Coverage (Nopt/Nos) | F-measure | Adjusted $p$-value |
|---|---|---|---|---|---|---|---|
| ($\alpha$1) | BC | PIK3CA | 143 | 1.00 (143/143) | 0.307 (143/466) | 0.4696 | $1.082^{-48}$ |
| ($\alpha$1) | BC | TP53 | 48 | 1.00 (48/48) | 0.103 (48/466) | 0.1868 | $4.9592^{-15}$ |
| ($\alpha$1) | BC | AKT1 | 11 | 1.00 (11/11) | 0.024 (11/466) | 0.0469 | $2.3^{-03}$ |
| ($\alpha$1) | BC | TP53&PIK3CA | 9 | 1.00 (9/9) | 0.0194 (9/466) | 0.0379 | $9.3923^{-03}$ |
| ($\alpha$1) | BC | CROCCP2 | 8 | 1.00 (8/8) | 0.017 (8/466) | 0.0334 | $1.8949^{-02}$ |
| ($\alpha$2) | TN | TP53 | 48 | 0.4375 (21/48) | 0.273 (21/77) | 0.5142 | $5.7652^{-11}$ |
| ($\alpha$3) | Her2 | PIK3CA | 53[a] | 0.375 (18/48) | 0.375 (18/48) | 0.1885 | $1.3187^{-03}$ |
| ($\alpha$3) | ER+/PR+ | PIK3CA | 143 | 0.832 (119/143) | 0.349 (119/341) | 0.5577 | $5.3698^{-35}$ |
| ($\alpha$3) | ER+/PR+ | AKT1 | 11 | 0.909 (10/11) | 0.0293 (10/341) | 0.0623 | $1.7561^{-03}$ |
| ($\alpha$3) | ER+/PR+ | SF3B1 | 5 | 1.00 (5/5) | 0.0147 (5/341) | 0.0289 | $3.8612^{-02}$ |

#pat is the counts of the patients with the feature. *Nopt* is the number of patients with the feature and in the class label. *Nos* is the number of patients in the class label.

[a]The five of 53 were dually typed as Her2 and TN by PAM50, and only 48 patients was used for analysis

**Algorithm 2:** Patient Classification Tree Generated by LAMP (Summary of Table 11)

| | |
|---|---|
| If mutation= PIK3CA $\vee$ TP53 $\vee$ AKT1 $\vee$ CROCCP2 $\vee$ SF3B1 $\vee$ FGFR2 then | $(\alpha 1)$ |
| If mutation =TP53 then | $(\alpha 2)$ |
| patient = TN (77/466) | |
| else | $(\alpha 2)$ |
| If mutation = PIK3CA then | (1st of $\alpha 3$) |
| patient = Her2+ (48/389) | |
| else | (1st of $\alpha 3$) |
| patient = ER+/PR+ (341/389) | |
| else | $(\alpha 1)$ |
| patient = matched_normal (466/932) | |

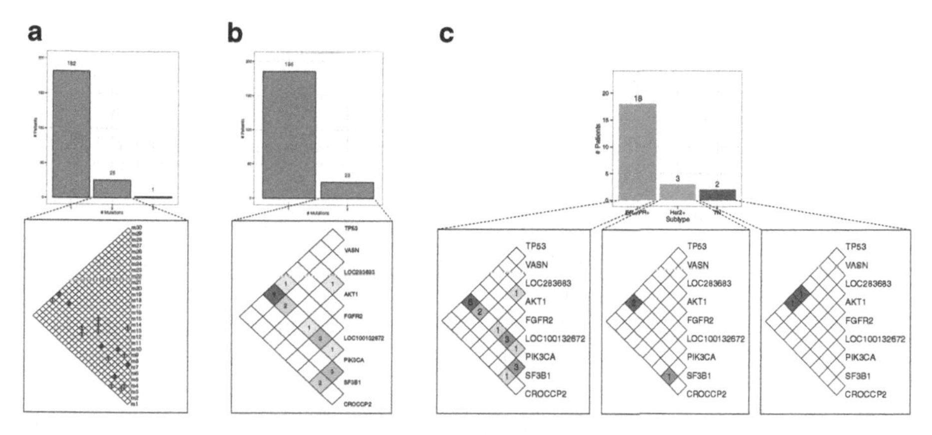


**Fig. 5** Co-occurrences. The distribution of patients over the number of mutations, focusing on double mutations, at the (**a**:*left*) mutation-wise, (**b**:*middle*) gene-wise, and (**c**:*right*) gene-wise over clinical phenotypes level.

mutated combination of the two genes: P+T+, P-T+, P+T-, and P-T-, subsequently contain high occurrences (right of Fig. 6).

A highly referenced driver gene, TP53, is responsible for 20–40 % of BC patients [42] and further 40–62 % TN patients [12, 13] (Table 12 and 21 (70 %) of 30 TN patients for T+ in out data). This suggests the possibility that TP53 can be a probable marker for TN. However, the distribution of P+T+ over the three phenotypes is similar to P+ or P+T-, contrasting from T+ or P-T+ (Fig. 6). Other minor pairwise mutations were mainly with PIK3CA and for ER+/PR+, while the pair of SF3B1 and CROCCP2 was also found for this phenotype (Fig. 5c). For Her2+, this trend was found but unclear, because of the small double-mutated Her2+ and TN populations, implying the subtleness of the trend.

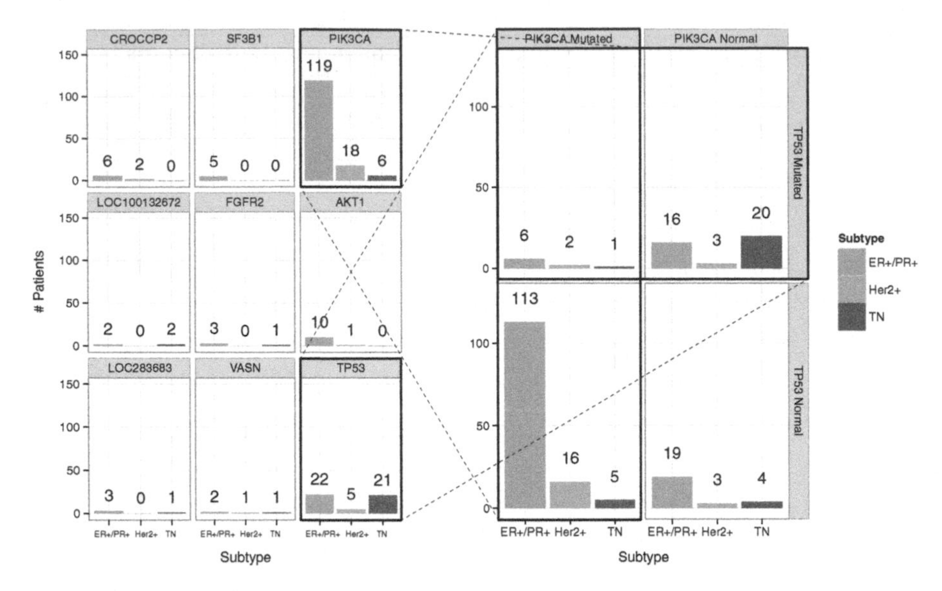


**Fig. 6** Double mutations. Patient distributions over clinical phenotypes, focusing on double mutations of PIK3CA and TP53

**Table 12** #patients with mutations in nine genes

| Gene | ER+/PR+ | Her2+ | TN | Total |
|---|---|---|---|---|
| PIK3CA | 119 | 18 | 6 | 143 (68.1 %) |
| TP53 | 22 | 5 | 21 | 48 (23.1 %) |
| AKT1 | 10 | 1 | 1 | 11 (5.3 %) |
| CROCCP2 | 6 | 1 | 1 | 8 (3.8 %) |
| SFSB1 | 5 | 0 | 0 | 5 (2.4 %) |
| FGFR2 | 3 | 0 | 1 | 4 (1.9 %) |
| VASN | 2 | 1 | 1 | 4 (1.9 %) |
| LOC100132672 | 2 | 0 | 2 | 4 (1.9 %) |
| LOC283683 | 3 | 0 | 1 | 4 (1.9 %) |
| All | 154 | 24 | 30 | 208 |

For each of the nine genes with 30 mutations, the number of patients in ER+/PR+, Her2+, TN, and the total number of patients are shown

## 3.5 Results: Decision Tree

We used *Dataset D* with gene-wise features to generate a decision tree [15] (Fig. 7), which separate *Dataset D* first by TP53, subsequently by PIK3CA for T+ pool. The tree model identified the T- and P+T+ groups as predominately containing patients with ER+/PR+ and Her2+, respectively, whereas P-T+ shows a comparatively equal distribution over ER+/PR+ and TN [16]. These results confirmed the importance of the significant gene features captured by LAMP and also the double mutations

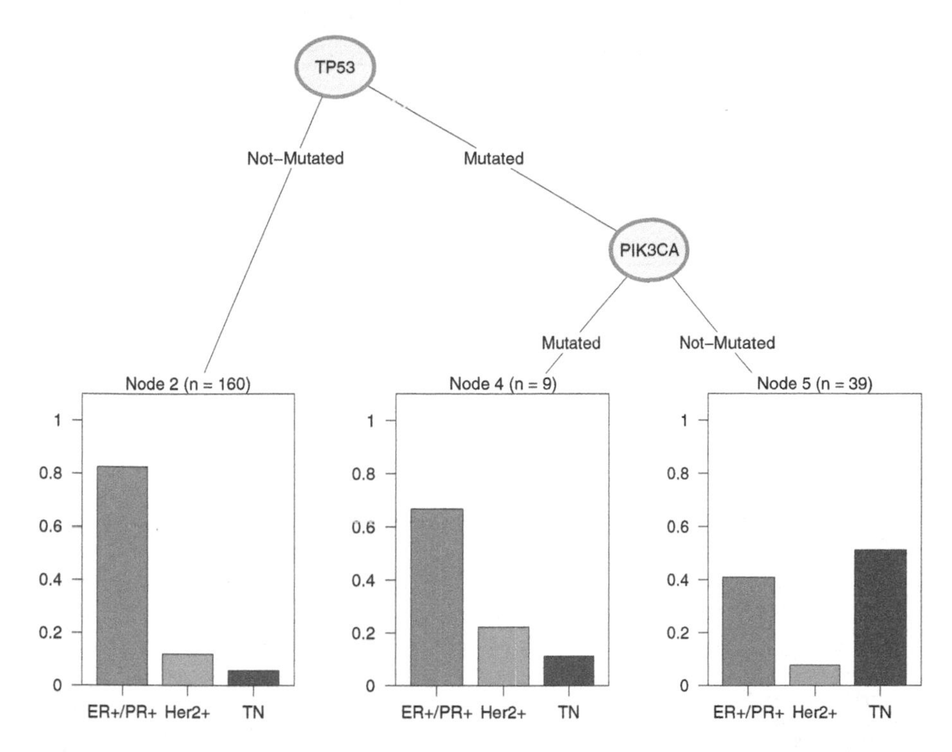


**Fig. 7** Resultant decision tree by (*rpart*). This decision tree shows that the first feature used to partition the data is the mutation in TP53, and then PIK3CA. That is, if there is a mutation in TP53, the instance is first classified as ER+/PR+; otherwise then if there is a mutation in PIK3CA further, the instance can be ER+/PR+ or Her2+. Here the number of instances without any mutation in TP53 is 160 (*the left panel*) and the number of instances with mutations both in PIK3CA and TP53 is 9 (*the middle panel*). The rest 39 instances are half likely to be TN (*the right panel*)

by TP53 and PIK3CA. It presented a similar pattern to the classification tree by LAMP (Algorithm 2) initially utilizing TP53, then PIK3CA as separation attributes. However, LAMP demonstrated that TP53 non-mutated patients, i.e., T-, were further separated, while through recursive partitioning, T+ patients were further divided. In other words, BC patients were classified into T+, P+T-, and P-T-, while T-, P+T+, and P-T+ in the decision tree. This is unexpected, since it was reported that TP53 mutations are strong characteristic of the TN phenotype [32, 42, 43], but the decision tree shows no direct correlation between TP53 and TN, implying the subtleness of their correlation. More importantly, the decision tree result presented the double mutation class of PIK3CA and TP53, confirming the importance of co-static interaction between these two genes.

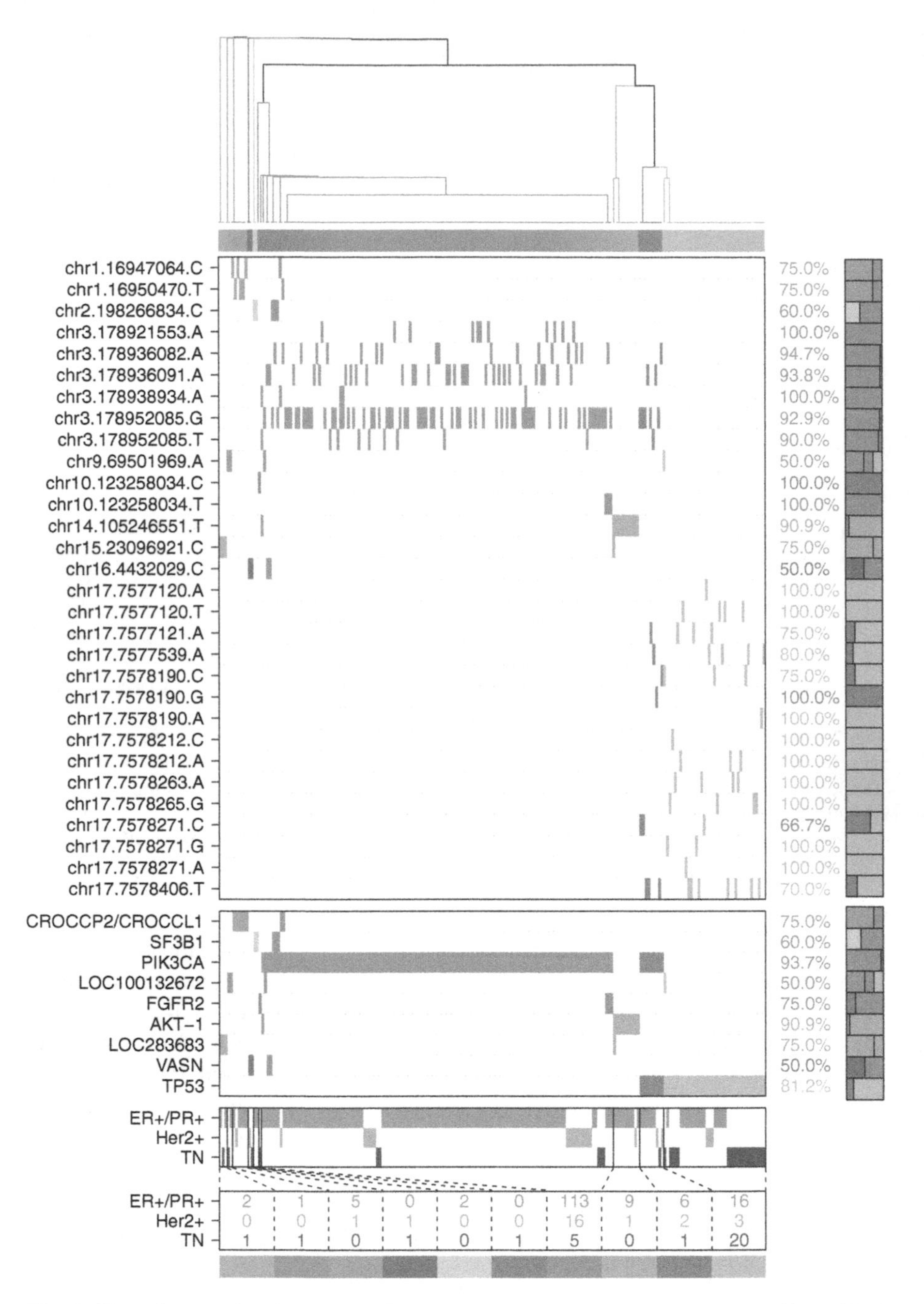


**Fig. 8** Clustering result. The *top dendrogram* is generated by average linkage hierarchical clustering, and below the distance matrix is shown as a heatmap. We selected $k = 10$, by using the gap statistic.The percentages in the *right side* indicate the ratio of patients who assigned to the largest cluster in the number of patients. The *color bars* in the *right side* represent the ratios of clusters sharing the same mutation. The *bottom table* represents the numbers of ER+/PR+, Her2+, and TN in each cluster

## *3.6 Results: Hierarchical Disjoint Clustering*

### 3.6.1 Minor Clusters Reveal the Complexity of Cancer

Clustering over *Dataset D* with gene-wise and pathway-wise features generated ten clusters by the dendrogram with PIK3CA as the primary split, followed by TP53 and AKT1 as the surrogate splits, resulting in the six minor and four major clusters (Fig. 8). Cluster 3 demonstrates a typical cluster emphasizing the complexity of cancer. Comprised of only one Her2+ and the rest as ER+/PR+, this cluster illustrates that Her2+ and ER+/PR+ cannot be perfectly distinguished from each other through gene-wise and pathway-wise information.

In contrast, clusters 1, 2, and 4 are comprised of both TN and one non-TN phenotype, implying some possible phenotype assignment to the TN patients, while the significance is questionable, due to the small size. In addition, clusters 1 and 2 are in chromosomes 15 and 9, respectively, are associated with unknown genes.

### 3.6.2 Mutational Landscapes Between PIK3CA and TP53

92.3 % of BC patients were located into the last four clusters. As indicated on the right panel of Fig. 6, clusters 7, 9, and 10 correspond to P+T-, P+T+, and P-T+ patients, respectively, which appeared in the LAMP, decision tree results. In particular, the interaction between PIK3CA and TP53 is iconically captured by cluster 9 ($p = 8.3087^{-15}$ for P+T+ by LAMP). Cluster 8 shows a typical cancer cluster, comprising both ER+/PR+ and Her2+ patients. Also this cluster comprised of the chromosomes 14 and 15 mutations, pertaining to AKT-1 and LOC283683, respectively, providing some insight into the interaction of the LOC283683 to the AKT1 gene, which warrants further investigation.

In conclusion, minor clusters implied rather insignificant patterns to assigning clinical phenotypes to TN patients, while major clusters revealed rather clear characteristics especially highlighting the interaction of two driver genes, PIK3CA and TP53. All ten clusters (except cluster 6) can be used to link TN patients to clinical phenotypes, if those patients have the mutation-wise or gene-wise features in the corresponding cluster (Tables 13 and 14). Finally, clinical phenotypes were proved to be more correlated to gene-wise and pathway-wise features than the mutation-wise feature alone.

# 4 Discussion

## *4.1 Importance of Somatic Mutations*

Key components of cells are gene elements or genetic codes of DNA, which form a basis of control and function through organized communication networks

**Table 13** Mutations appeared in TN patients

| Gene | Codon | Mutation | Cluster | Information |
|---|---|---|---|---|
| PIK3CA | 542 | m5 | 7, 9 | Exon 9 gain-of-function mutation requires interaction with RAS, hotspot mutation; helical domain-changes charge (− to +) [44] |
| | 1047 | m8 | 7, 9 | High levels of AKT; predictor to partial response to PI3K/AKT/mTOR inhibitors combinatorial therapy; hotspot mutation in kinase domain; enhanced lipid kinase activity; high apoptotic resistance; increased migration ability; independent of RAS binding; Exon 20 (kinase domain) mutations may indicate resistance to anti-EGFR therapy in wild type KRAS tumors [44] |
| LOC100132672 | NA | m10 | 2, 7 | Unknown (associated with lncRNA) |
| FGFR2 | 433 | m12 | 6 | Role in mammary development, maintenance of breast tumour initiating cells, altered activity of the ERa-associated transcriptional network, contributor to ER. disease, estradiol relevant stimulator of FGFR, IL8 increased secretion, enrichment and proliferation of PTTG1, regulator of SPDEF at protein level (interactor if p53), intron SNPS of FGFR2 mediate their effect through altering expression of the FGFR2 gene, regulator of the ERa network, FOXA1 and GATA3 are master regulators of FGFR2 response [45] |
| LOC283683 | NA | m14 | 1 | Unknown (ncRNA) |
| CASN | 384 | m15 | 4, 7 | Conservation in BRCA2 homologs, not located in a region of functional importance—not disease causing mutation [46] |
| TP53 | 273 | m17, m18 | 10 | Disordered growth in 3D cultures, induction of mevalonate pathway genes; induction of migration-related mutant p53 signature genes; inhibition of apoptosis; altered growth and cell polarity in 3D cultures, EMT induction; immortalization of normal mammary epithelial cells [47] |
| | 248 | m19 | 9, 10 | Altered growth and cell popularity in 3D cultures, EMT induction [47] |
| | 220 | m20, m22 | 9, 10 | Unknown |
| | 213 | m23, m24 | 10 | Loss of a TaqI site; polymorphism; suppress expression of ER |
| | 196 | m25, m26 | 10 | CpG dinucleotide; hotspot mutation; driving force for p53 inactivation |
| | 175 | m30 | 9, 10 | Increased growth rate, tumorigenic potential, chemoresistance; increased expression of pro-angiogenic genes, NF-Y and NF-$\kappa$B targets; inhibition of p73; inhibition of apoptosis mediated by the vit. D receptor (SK-BR-3); altered growth and cell polarity in 3D cultures, EMT induction [47] |

In order to explore the mutations appearing in TN patients, genes with those mutations appearing in TN patients, their codons, corresponding clusters, and also information on the codons are shown

**Table 14** Mutations and genes significantly relevant to clusters

| Cluster | Feature | #pat | Precision (Nopt/#pat) | Coverage (Nopt/Nos) | F-measure | Adj. *p*val |
|---|---|---|---|---|---|---|
| 1 | LOC283683 | 4 | 0.75 (3/4) | 1 (3/3) | 0.8572 | $3.7882^{-05}$ |
| 2 | LOC100132672 | 4 | 0.5 (2/4) | 1 (2/2) | 0.6667 | $5.0167^{-03}$ |
| 3 | CROCCP2 | 8 | 0.75 (6/8) | 1 (6/6) | 0.8571 | $3.75^{-09}$ |
| 4 | NOMO2 | 4 | 0.5 (2/4) | 1 (2/2) | 0.6667 | $5.0167^{-03}$ |
| 5 | SF3B1 | 5 | 0.4 (2/5) | 1 (2/2) | 0.5714 | $8.3612^{-03}$ |
| 7 | PIK3CA | 143 | 0.9371 (134/143) | 1 (134/134) | 0.9675 | $4.3716^{-44}$ |
| 8 | AKT1 | 11 | 0.9091 (10/11) | 1 (10/10) | 0.9524 | $4.5927^{-15}$ |
| 9 | TP53 & PIK3CA | 9 | 1 (9/9) | 1 (9/9) | 1 | $8.3087^{-15}$ |
| 9 | TP53 | 48 | 0.1875 (9/48) | 1 (9/9) | 0.3158 | $1.3935^{-05}$ |
| 10 | TP53 | 48 | 0.8125 (39/48) | 1 (39/39) | 0.8966 | $6.0571^{-33}$ |
| 1 | m14 | 4 | 0.5 (2/4) | 1 (2/2) | 0.6667 | $1.3378^{-02}$ |
| 3 | m1 | 4 | 0.75 (3/4) | 0.5 (3/6) | 0.6 | $1.6057^{-03}$ |
| 3 | m2 | 4 | 0.5 (2/4) | 1 (2/2) | 0.6667 | $1.3378^{-02}$ |
| 5 | m3 | 5 | 0.4 (2/5) | 1 (2/2) | 0.5714 | $2.2297^{-02}$ |
| 7 | m8 | 70 | 0.9286 (65/70) | 0.4851 (54/134) | 0.6373 | $4.2181^{-10}$ |
| 7 | m6 | 32 | 0.9375 (30/32) | 0.2239 (30/134) | 0.3615 | $2.3319^{-04}$ |
| 7 | m5 | 19 | 0.9474 (18/19) | 0.1343 (18/134) | 0.2353 | $9.445^{-03}$ |
| 7 | m4 | 11 | 1 (11/11) | 0.0821 (11/134) | 0.1517 | $3.4052^{-02}$ |
| 8 | m13 | 11 | 0.9091 (10/11) | 1 (10/10) | 0.9524 | $7.8733^{-15}$ |
| 10 | m30 | 10 | 0.7 (7/10) | 0.1795 (7/39) | 0.28572 | $7.7925^{-03}$ |
| 10 | m17 | 4 | 1 (4/4) | 0.1026 (4/39) | 0.1861 | $2.1713^{-02}$ |
| 10 | m25 | 4 | 1 (4/4) | 0.1026 (4/39) | 0.1861 | $2.1713^{-02}$ |
| 10 | m26 | 4 | 1 (4/4) | 0.1026 (4/39) | 0.1861 | $2.1713^{-02}$ |

There were ten clusters obtained from hierarchical clustering using the gene-wise and pair-wise features. #pat is the counts of the patients with the feature. *Nopt* is the number of patients with the feature and in the cluster. *Nos* is the number of patients in the cluster

known as biological pathways. Somatic mutations or "inherited errors" within the DNA, can influence the transcribed gene outputs, i.e., proteins, such as TP53 and PIK3CA in our work, which play key roles in the apoptotic and cancer pathways [22]. That is, the variant inherited by the patient is the cause of ceasing functions within the control pathway, leads to the disease. In the case of breast cancer identifying these aberrations with a varied clinical phenotype may be the key linkage for the development of novel treatments. For example, TN patients possessing TP53 mutations demonstrate complications in the p53 signaling pathway of the apoptotic biochemical framework, resulting in absence of "programmed cell death" [47, 48]. This leads to proliferation of cancer cells, resulting in a breast tumor. A ER+/PR+ patient possessing a PIK3CA or AKT1 mutation, subsequently affecting the PIK3CA-AKT1 signaling pathway of apoptosis, has a similar outcome. Therefore, the mutations of abnormal proteins may be potential targets for developing drugs. In the case of siRNA therapy, without a better understanding of the mutation-subtype link, viable targets cannot be designed.

The subset of patients, which express multiple somatic mutations in the same pathway, was expected to have a combined feature of the individual gene-wise mutation patterns. In fact, PIK3CA mutation co-occurrences are observed in other cancers, such as intestinal and endometrial cancer, with adenomatous polyposis coli gene and MAPK, respectively [49, 50]. In this study, we identified a minor population of TN patients that display correlations of mutation frequencies in both PIK3CA and TP53 [6, 12, 19]. That is, cluster 9 or P+T+, say m8 with m30, are the corresponding mutational combinations. We observed that the distribution of P+T+ was similar to P+T- rather than P-T+ or T+, implying the dominance of P+ over T+. Our dataset was (1) heavily biased to mutations in PIK3CA, and (2) with only nine patients of P+T+, by which P+T+ cannot be statistically examined, and so the similarity of P+T+ to P+T- might be just a result due to the limitation of our dataset. More importantly, these complex mutations reflect the complex nature of the biochemical control cross-talk in breast cancer. We believe double or combinatorial mutations like P+T+ would be useful, because P+T+ might be the combined-oncogenic driver that distinguishes a sub-population of the true-clinical TN phenotype [12, 27, 51], making P+T+ patients potential candidates for the extant ER+/PR+ hormone therapy regiment. This finding demonstrates that mutational combinations with additional pathways or molecular information might be an approach for finding a more precise prognostic biomarkers [47]. Thus our focus is on a mutational pair rather than a single SNP and detecting the sub-population of TN patients. We use machine learning methods to determine a better link of known clinical phenotypes with current treatments. These points make our work unique, even comparing with past and recent efforts on characterizing TN patients from somatic mutations mostly based on experiments and some statistical analysis [12, 13].

In general, the subtypes differ in genomic complexity. From the analysis, it became apparent that breast cancer is not one single entity but rather encompasses distinct characteristic biomolecular features within intrinsic subtypes. Gene-wise, there is a subtle difference from well-known drivers of TP53 and PIK3CA [12, 13]. Mutation-wise, however, these driver alterations can provide a modicum on treatment response and clinical prognosis. TN patients' treatment regimens are still mainly based on the application of chemotherapy. The molecular profiles of these patients may help interplay between optimal drug and the predictive value of molecular alterations. Drugs that selectively target molecular pathways correlated with the malignant phenotype will exhibit a maximum efficacy for the given patient. Due to the high clonality of TN patients, targeted therapy of the known aberration or mutation will be the best possible option to allow the body to fight the cancer cells in an efficient manner, while minimizing resistance [13]. That is, if a Her2 subtype patient exhibits a PIK3CA gene mutation, Trastuzumab, might not benefit from this drug, due to the drug being affect by the aberrations in PIK3CA [13]. Therefore, investing the molecular profiles can assist to identify the current optimal treatments for the non-TN patient.

### 4.2 Discriminating Her2+ Patients from ER+/PR+ Patients

LAMP analysis demonstrated that a PIK3CA mutation is a definitive genetic constraint for ER+/PR+, while gene-wise analysis demonstrated that PIK3CA can also produce a Her2+ phenotype (Algorithm 2). Similar to our analysis, past reports established that PIK3CA mutations can inhibit Her2+ signaling, i.e., expressing ER+/PR+ subtype [12], while 18.6–21.4 % of PIK3CA mutations expressed the Her2+ phenotype, being statistically insignificant [52, 53].

Our data further emphasized the pathway-dependent, domain-specific, yet non-nucleovariant specific nature of the PIK3CA mutations [54]. In other words, we observed that PIK3CA mutations are contributing factors in both the ER+/PR+ and Her2+ subgroups, but in different significance levels. As expected, most Her2+ patients fell into cluster 7, yet outliers were also split into the other groups. Due to the high sample ratio of ER+/PR+ (74 %), our analysis on the current data was limited to distinguish the ER+/PR+ and TN patients [52, 53].

### 4.3 Different Types of TN Patient Clusters

Currently, only the treatments for ER+/PR+ and Her2+ patients are promising, whereas TN patients, who usually possess an aggressive form of BC, are still in need of a probable treatment [5, 6]. Our clustering exhibits that TN patients can be further classified into two types: (1) a true TN or (2) characteristically similar ER+/PR+ phenotype. The first type is cluster 6 with m11, which has only one TN patient and might be a true TN, yet warrants further investigation. The latter is indicative of clusters 1 and 2, which have both TN and ER+/PR+ patients. In other words, these TN patients share same positional mutations as ER+/PR+ patients but with different expressed phenotype. These types of clusters are expressed in distinct clusters and therefore may be applicable to allocate the trivial distinct TN population of patients to extant treatable groups. Similar to the second but a different type is a mixture of the three phenotypes that are not simple in the interpretation. Cluster 7, mostly with mutations in PIK3CA, is typified by 13 mutational positions, where six out of 13 have mixed phenotypic distributions. An assumption would be to use the predominant subtype in that sub-population as a characteristic for that mutational position. More concretely, 81.8 % patients with m8 are with ER+/PR+ phenotype, whereas only 18.2 % express the other phenotypes, by which a patient with the expressional mutation of this position will be likely to be ER+/PR+. Another feature of the clustering specific to patients with a single mutation can be summarized, particularly for TN patients (Table 13: details of all mutations in TN patients). For instance, mutations in chromosome 17 were predominately grouped into cluster 10 with a mixture distribution of phenotypes (Fig. 8). Patients in cluster 10, mainly with only one mutation, were then linked to their mutation-wise features. In other words, TN patients can be allocated to the corresponding extant treatable phenotypes via positional mutations. For instance, cluster 10 has some positions, such as m22 and m23, which can be considered as the true TN. We believe that combining our analysis with the current methods would be helpful to identify the TN patients who can receive probable treatments.

## 5 Conclusion

Currently, combinatorial therapy is administered to prevent alternative growth, yet has also further conferred the patient to a faster development of resistance. As much as 50 % of ER+/PR+ patients administered endocrine therapy eventually acquire resistance, resulting in a relapse of the BC tumor [45]. For example, ER+/PR+ patients often demonstrate a slight up-regulation of the Her2 signaling [45]. It has also been reported that bidirectional cross-talk between the Her2, ER, and other signaling pathways has contributed to endocrine resistance. This can often cause misinterpretation of the BC subtype which is used for current treatments. In these cases, classification of mutational identification may be beneficial for these patients for therapy selection.

Due to the high costs of gene expression profiling, plus their poor ability to simultaneously compare the expression of related biological samples properly, we believe the "intrinsic code" is a more appropriate method to target the patients' BC clinical phenotype. That is, the classification of somatic mutations can contribute to identify viable targets for therapy selection. Our study showed that knowledge of the mutational patterns in the drivers, TP53 and PIK3CA, can give insight into the specific functional characteristics that lead to the biological selection of the breast cancer subtypes. We observed that the TP53 and PIK3CA combinational mutation pattern may influence a subset of TN phenotype. To access if the specific dual mutation pattern identified in our cluster directly influences the TN tumor phenotype, it warrants more investigation. The biomolecular approach, combining position-wise, gene-wise, and pathway-wise levels, instituted in our study has hinted into the associative complex nature and mechanisms involved in the clonality of the different subtypes. The biological context in which the mutations occur will help unravel new perspectives for novel therapeutic approaches, such as personalized targeted therapy and siRNA therapy, or even assist in current therapy selection. Furthermore, it may help to decipher the somatic based mechanisms that create the modifying effects that result in tumor development. Specific aberrations may provide a subtle link of the clinical impact that these drivers may play in gene-environment cross-talk to subtype differentiation. We believe our approach can be clinically applied to assist in proper treatment by specific targets to minimize the off-targets of conventional drugs and may help to delay the onset of antibiotic resistance for the patient.

## 6 Future Directives

The breast cancer tumor development undergoes numerous changes from the development to the progression of the various clinical stages. Our analysis used biomolecular features to characterize the various subtypes through clinical markers, but more investigation is needed to incorporate the cancer staging into the analysis, such that the somatic mutation architecture within stage progression can be characterized for the subtypes. Understanding the passenger mutations role in the progression may allow us to better understand molecular mechanisms of metastasis in a genomic level and improve the clinical management of the aggressive forms, such as TN. In our study [1], we have observed that the somatic mutations provide a "modicum in the intrinsic code" which can be used to classify the various subtypes, but a similar method can be used for the progression of metastasis of the disease. In essence, these transcriptional signatures can serve as prognostic markers to identify patients who are at the highest risk for developing metastases, which subsequently enable the development of tailored personalized treatment strategies.

**Acknowledgements** We acknowledge Dr. Ajit Bharti (Boston University, USA) on his innovative conception that a better apprehension of breast cancer subtypes is needed. We would like to thank the TCGA Data Access Committee (DAC) for providing us the opportunity to work with the data for this study.

S.Y. is supported by Grant-in-Aid for JSPS Fellows and JSPS KAKENHI #26-381. M.K. is supported by JSPS KAKENHI #26730120. I.T. is funded by Collaborative Research Program of Institute for Chemical Research, Kyoto University (Grant# 2014-27, #2015-33). H.M. is partially supported by JSPS KAKENHI #24300054.

## References

1. S. Yotsukura, I. Takigawa, M. Karasuyama, and H. Mamitsuka, "Exploring phenotype patterns of breast cancer within somatic mutations," *Briefings in Bioinformatics*. To appear. doi:10.1093/bib/bbw040
2. J. M. Rae, S. Drury, D. F. Hayes, V. Stearns, J. N. Thibert, B. P. Haynes, J. Salter, I. Sestak, J. Cuzick, and M. Dowsett, "CYP2D6 and UGT2B7 genotype and risk of recurrence in tamoxifen-treated breast cancer patients," *J. Natl. Cancer Inst.*, vol. 104, pp. 452–460, Mar 2012.
3. R. G. Margolese, G. N. Hortobagyi, and T. A. Buchholz, "Management of metastatic breast cancer," in *Holland-Frei Cancer Medicine* (D. W. Kufe, R. E. Pollock, R. R. Weichselbaum, *et al.*, eds.), Hamilton, ON: BC Decker, 6 ed., 2003.
4. L. R. Howe and P. H. Brown, "Targeting the HER/EGFR/ErbB family to prevent breast cancer," *Cancer Prev Res (Phila)*, vol. 4, pp. 1149–1157, Aug 2011.
5. K. R. Bauer, M. Brown, R. D. Cress, C. A. Parise, and V. Caggiano, "Descriptive analysis of estrogen receptor (ER)-negative, progesterone receptor (PR)-negative, and HER2-negative invasive breast cancer, the so-called triple-negative phenotype: a population-based study from the California cancer Registry," *Cancer*, vol. 109, pp. 1721–1728, May 2007.
6. A. Prat, C. Cruz, K. A. Hoadley, O. Diez, C. M. Perou, and J. Balmana, "Molecular features of the basal-like breast cancer subtype based on BRCA1 mutation status," *Breast Cancer Res. Treat.*, vol. 147, pp. 185–191, Aug 2014.

7. B. D. Lehmann, J. A. Bauer, X. Chen, M. E. Sanders, A. B. Chakravarthy, Y. Shyr, and J. A. Pietenpol, "Identification of human triple-negative breast cancer subtypes and preclinical models for selection of targeted therapies," *J. Clin. Invest.*, vol. 121, pp. 2750–2767, Jul 2011.
8. A. Prat, A. Lluch, J. Albanell, W. T. Barry, C. Fan, J. I. Chacon, J. S. Parker, L. Calvo, A. Plazaola, A. Arcusa, M. A. Segui-Palmer, O. Burgues, N. Ribelles, A. Rodriguez-Lescure, A. Guerrero, M. Ruiz-Borrego, B. Munarriz, J. A. Lopez, B. Adamo, M. C. Cheang, Y. Li, Z. Hu, M. L. Gulley, M. J. Vidal, B. N. Pitcher, M. C. Liu, M. L. Citron, M. J. Ellis, E. Mardis, T. Vickery, C. A. Hudis, E. P. Winer, L. A. Carey, R. Caballero, E. Carrasco, M. Martin, C. M. Perou, and E. Alba, "Predicting response and survival in chemotherapy-treated triple-negative breast cancer," *Br. J. Cancer*, vol. 111, pp. 1532–1541, Oct 2014.
9. D. C. Koboldt, R. S. Fulton, M. D. McLellan, H. Schmidt, J. Kalicki-Veizer, J. F. McMichael, et al., "Comprehensive molecular portraits of human breast tumours," *Nature*, vol. 490, pp. 61–70, Oct 2012.
10. J. S. Parker, M. Mullins, M. C. Cheang, S. Leung, D. Voduc, T. Vickery, S. Davies, C. Fauron, X. He, Z. Hu, J. F. Quackenbush, I. J. Stijleman, J. Palazzo, J. S. Marron, A. B. Nobel, E. Mardis, T. O. Nielsen, M. J. Ellis, C. M. Perou, and P. S. Bernard, "Supervised risk predictor of breast cancer based on intrinsic subtypes," *J. Clin. Oncol.*, vol. 27, pp. 1160–1167, Mar 2009.
11. I. R. Watson, K. Takahashi, P. A. Futreal, and L. Chin, "Emerging patterns of somatic mutations in cancer," *Nat. Rev. Genet.*, vol. 14, pp. 703–718, Oct 2013.
12. X. Bai, E. Zhang, H. Ye, V. Nandakumar, Z. Wang, L. Chen, C. Tang, J. Li, H. Li, W. Zhang, W. Han, F. Lou, D. Zhang, H. Sun, H. Dong, G. Zhang, Z. Liu, Z. Dong, B. Guo, H. Yan, C. Yan, L. Wang, Z. Su, Y. Li, L. Jones, X. F. Huang, S. Y. Chen, and J. Gao, "PIK3CA and TP53 gene mutations in human breast cancer tumors frequently detected by ion torrent DNA sequencing," *PLoS ONE*, vol. 9, no. 6, p. e99306, 2014.
13. S. P. Shah, A. Roth, R. Goya, A. Oloumi, G. Ha, Y. Zhao, G. Turashvili, J. Ding, K. Tse, G. Haffari, A. Bashashati, L. M. Prentice, J. Khattra, A. Burleigh, D. Yap, V. Bernard, A. McPherson, K. Shumansky, A. Crisan, R. Giuliany, A. Heravi-Moussavi, J. Rosner, D. Lai, I. Birol, R. Varhol, A. Tam, N. Dhalla, T. Zeng, K. Ma, S. K. Chan, M. Griffith, A. Moradian, S. W. Cheng, G. B. Morin, P. Watson, K. Gelmon, S. Chia, S. F. Chin, C. Curtis, O. M. Rueda, P. D. Pharoah, S. Damaraju, J. Mackey, K. Hoon, T. Harkins, V. Tadigotla, M. Sigaroudinia, P. Gascard, T. Tlsty, J. F. Costello, I. M. Meyer, C. J. Eaves, W. W. Wasserman, S. Jones, D. Huntsman, M. Hirst, C. Caldas, M. A. Marra, and S. Aparicio, "The clonal and mutational evolution spectrum of primary triple-negative breast cancers," *Nature*, vol. 486, pp. 395–399, Jun 2012.
14. A. Terada, M. Okada-Hatakeyama, K. Tsuda, and J. Sese, "Statistical significance of combinatorial regulations," *Proc. Natl. Acad. Sci. U.S.A.*, vol. 110, pp. 12996–13001, Aug 2013.
15. T. Therneau, B. Atkinson, and B. Ripley, *rpart: Recursive Partitioning and Regression Trees*, 2011.
16. T. Hothorn, K. Hornik, and A. Zeileis, "Unbiased recursive partitioning: A conditional inference framework," *Journal of Computational and Graphical Statistics*, vol. 15, no. 3, pp. 651–674, 2006.
17. 2014. http://ww5.komen.org/BreastCancer/SubtypesofBreastCancer.html.
18. R. Tibshirani, G. Walther, and T. Hastie, "Estimating the number of clusters in a dataset via the gap statistic," *Journal of the Royal Statistical Society: Series B (Statistical Methodology)*, vol. 63, no. 2, pp. 411–423, 2000.
19. C. Kandoth, M. D. McLellan, F. Vandin, K. Ye, B. Niu, C. Lu, M. Xie, Q. Zhang, J. F. McMichael, M. A. Wyczalkowski, M. D. Leiserson, C. A. Miller, J. S. Welch, M. J. Walter, M. C. Wendl, T. J. Ley, R. K. Wilson, B. J. Raphael, and L. Ding, "Mutational landscape and significance across 12 major cancer types," *Nature*, vol. 502, pp. 333–339, Oct 2013.
20. H. Thorvaldsdottir, J. T. Robinson, and J. P. Mesirov, "Integrative Genomics Viewer (IGV): high-performance genomics data visualization and exploration," *Brief. Bioinformatics*, vol. 14, pp. 178–192, Mar 2013.
21. d. a. W. Huang, B. T. Sherman, and R. A. Lempicki, "Systematic and integrative analysis of large gene lists using DAVID bioinformatics resources," *Nat. Protoc*, vol. 4, no. 1, pp. 44–57, 2009.

22. M. Kanehisa, S. Goto, Y. Sato, M. Kawashima, M. Furumichi, and M. Tanabe, "Data, information, knowledge and principle: back to metabolism in KEGG," *Nucleic Acids Res.*, vol. 42, pp. 199–205, Jan 2014.
23. 2014. Online Mendelian Inheritance in Man, OMIM. McKusick-Nathans Institute of Genetic Medicine, Johns Hopkins University (Baltimore, MD), World Wide Web URL: http://omim.org/.
24. R Core Team, *R: A Language and Environment for Statistical Computing*. R Foundation for Statistical Computing, Vienna, Austria, 2013.
25. C. O'Brien, J. J. Wallin, D. Sampath, D. GuhaThakurta, H. Savage, E. A. Punnoose, J. Guan, L. Berry, W. W. Prior, L. C. Amler, M. Belvin, L. S. Friedman, and M. R. Lackner, "Predictive biomarkers of sensitivity to the phosphatidylinositol 3' kinase inhibitor GDC-0941 in breast cancer preclinical models," *Clin. Cancer Res.*, vol. 16, pp. 3670–3683, Jul 2010.
26. L. H. Saal, K. Holm, M. Maurer, L. Memeo, T. Su, X. Wang, J. S. Yu, P. O. Malmstrom, M. Mansukhani, J. Enoksson, H. Hibshoosh, A. Borg, and R. Parsons, "PIK3CA mutations correlate with hormone receptors, node metastasis, and ERBB2, and are mutually exclusive with PTEN loss in human breast carcinoma," *Cancer Res.*, vol. 65, pp. 2554–2559, Apr 2005.
27. K. Stemke-Hale, A. M. Gonzalez-Angulo, A. Lluch, R. M. Neve, W. L. Kuo, M. Davies, M. Carey, Z. Hu, Y. Guan, A. Sahin, W. F. Symmans, L. Pusztai, L. K. Nolden, H. Horlings, K. Berns, M. C. Hung, M. J. van de Vijver, V. Valero, J. W. Gray, R. Bernards, G. B. Mills, and B. T. Hennessy, "An integrative genomic and proteomic analysis of PIK3CA, PTEN, and AKT mutations in breast cancer," *Cancer Res.*, vol. 68, pp. 6084–6091, Aug 2008.
28. H. G. Ahmed, M. A. Al-Adhraei, and I. M. Ashankyty, "Association between AgNORs and Immunohistochemical Expression of ER, PR, HER2/neu, and p53 in Breast Carcinoma," *Patholog Res Int*, vol. 2011, p. 237217, 2011.
29. P. de Cremoux, A. V. Salomon, S. Liva, R. Dendale, B. Bouchind'homme, E. Martin, X. Sastre-Garau, H. Magdelenat, A. Fourquet, and T. Soussi, "p53 mutation as a genetic trait of typical medullary breast carcinoma," *J. Natl. Cancer Inst.*, vol. 91, pp. 641–643, Apr 1999.
30. P. Yang, C. W. Du, M. Kwan, S. X. Liang, and G. J. Zhang, "The impact of p53 in predicting clinical outcome of breast cancer patients with visceral metastasis," *Sci Rep*, vol. 3, p. 2246, 2013.
31. H. Yamashita, M. Nishio, T. Toyama, H. Sugiura, Z. Zhang, S. Kobayashi, and H. Iwase, "Coexistence of HER2 over-expression and p53 protein accumulation is a strong prognostic molecular marker in breast cancer," *Breast Cancer Res.*, vol. 6, no. 1, pp. 24–30, 2004.
32. E. Biganzoli, D. Coradini, F. Ambrogi, J. M. Garibaldi, P. Lisboa, D. Soria, A. R. Green, M. Pedriali, M. Piantelli, P. Querzoli, R. Demicheli, P. Boracchi, I. Nenci, I. O. Ellis, and S. Alberti, "p53 status identifies two subgroups of triple-negative breast cancers with distinct biological features," *Jpn. J. Clin. Oncol.*, vol. 41, pp. 172–179, Feb 2011.
33. S. Banerji, K. Cibulskis, C. Rangel-Escareno, *et al.*, "Sequence analysis of mutations and translocations across breast cancer subtypes," *Nature*, vol. 486, pp. 405–409, Jun 2012.
34. C. X. Ma, T. Reinert, I. Chmielewska, *et al.*, "Mechanisms of aromatase inhibitor resistance," *Nat. Rev. Cancer*, vol. 15, pp. 261–275, May 2015.
35. E. Cerami, J. Gao, U. Dogrusoz, B. E. Gross, S. O. Sumer, B. A. Aksoy, A. Jacobsen, C. J. Byrne, M. L. Heuer, E. Larsson, Y. Antipin, B. Reva, A. P. Goldberg, C. Sander, and N. Schultz, "The cBio cancer genomics portal: an open platform for exploring multidimensional cancer genomics data," *Cancer Discov*, vol. 2, pp. 401–404, May 2012.
36. J. Gao, B. A. Aksoy, U. Dogrusoz, G. Dresdner, B. Gross, S. O. Sumer, Y. Sun, A. Jacobsen, R. Sinha, E. Larsson, E. Cerami, C. Sander, and N. Schultz, "Integrative analysis of complex cancer genomics and clinical profiles using the cBioPortal," *Sci Signal*, vol. 6, p. pl1, Apr 2013.
37. M. Heiskanen, J. Kononen, M. Barlund, J. Torhorst, G. Sauter, A. Kallioniemi, and O. Kallioniemi, "CGH, cDNA and tissue microarray analyses implicate FGFR2 amplification in a small subset of breast tumors," *Anal Cell Pathol*, vol. 22, no. 4, pp. 229–234, 2001.
38. V. K. Jain and N. C. Turner, "Challenges and opportunities in the targeting of fibroblast growth factor receptors in breast cancer," *Breast Cancer Res.*, vol. 14, no. 3, p. 208, 2012.

39. N. Turner, M. B. Lambros, H. M. Horlings, A. Pearson, R. Sharpe, R. Natrajan, F. C. Geyer, M. van Kouwenhove, B. Kreike, A. Mackay, A. Ashworth, M. J. van de Vijver, and J. S. Reis-Filho, "Integrative molecular profiling of triple negative breast cancers identifies amplicon drivers and potential therapeutic targets," *Oncogene*, vol. 29, pp. 2013–2023, Apr 2010.
40. S. L. Maguire, A. Leonidou, P. Wai, C. Marchio, C. K. Ng, A. Sapino, A. V. Salomon, J. S. Reis-Filho, B. Weigelt, and R. C. Natrajan, "SF3B1 mutations constitute a novel therapeutic target in breast cancer," *J. Pathol.*, vol. 235, pp. 571–580, Mar 2015.
41. A. C. Vargas, J. S. Reis-Filho, and S. R. Lakhani, "Phenotype-genotype correlation in familial breast cancer," *J Mammary Gland Biol Neoplasia*, vol. 16, pp. 27–40, Apr 2011.
42. A. Langerød, H. Zhao, Ø. Borgan, J. M. Nesland, I. R. Bukholm, T. Ikdahl, R. Kåresen, A. L. Børresen-Dale, and S. S. Jeffrey, "TP53 mutation status and gene expression profiles are powerful prognostic markers of breast cancer," *Breast Cancer Res.*, vol. 9, no. 3, p. R30, 2007.
43. J. Alsner, M. Yilmaz, P. Guldberg, L. L. Hansen, and J. Overgaard, "Heterogeneity in the clinical phenotype of TP53 mutations in breast cancer patients," *Clin. Cancer Res.*, vol. 6, pp. 3923–3931, Oct 2000.
44. G. Ligresti, L. Militello, L. S. Steelman, A. Cavallaro, F. Basile, F. Nicoletti, F. Stivala, J. A. McCubrey, and M. Libra, "PIK3CA mutations in human solid tumors: role in sensitivity to various therapeutic approaches," *Cell Cycle*, vol. 8, pp. 1352–1358, May 2009.
45. M. N. Fletcher, M. A. Castro, X. Wang, I. de Santiago, M. O'Reilly, S. F. Chin, O. M. Rueda, C. Caldas, B. A. Ponder, F. Markowetz, and K. B. Meyer, "Master regulators of FGFR2 signalling and breast cancer risk," *Nat. Commun.*, vol. 4, p. 2464, 2013.
46. B. Wappenschmidt, R. Fimmers, K. Rhiem, M. Brosig, E. Wardelmann, A. Meindl, N. Arnold, P. Mallmann, and R. K. Schmutzler, "Strong evidence that the common variant S384F in BRCA2 has no pathogenic relevance in hereditary breast cancer," *Breast Cancer Res.*, vol. 7, no. 5, pp. R775–779, 2005.
47. D. Walerych, M. Napoli, L. Collavin, and G. Del Sal, "The rebel angel: mutant p53 as the driving oncogene in breast cancer," *Carcinogenesis*, vol. 33, pp. 2007–2017, Nov 2012.
48. C. Coles, A. Condie, U. Chetty, C. M. Steel, H. J. Evans, and J. Prosser, "p53 mutations in breast cancer," *Cancer Res.*, vol. 52, pp. 5291–5298, Oct 1992.
49. D. A. Deming, A. A. Leystra, L. Nettekoven, C. Sievers, D. Miller, M. Middlebrooks, L. Clipson, D. Albrecht, J. Bacher, M. K. Washington, J. Weichert, and R. B. Halberg, "PIK3CA and APC mutations are synergistic in the development of intestinal cancers," *Oncogene*, vol. 33, pp. 2245–2254, Apr 2014.
50. B. Weigelt, P. H. Warne, M. B. Lambros, J. S. Reis-Filho, and J. Downward, "PI3K pathway dependencies in endometrioid endometrial cancer cell lines," *Clin. Cancer Res.*, vol. 19, pp. 3533–3544, Jul 2013.
51. B. D. Lehmann, J. A. Bauer, J. M. Schafer, C. S. Pendleton, L. Tang, K. C. Johnson, X. Chen, J. M. Balko, H. Gomez, C. L. Arteaga, G. B. Mills, M. E. Sanders, and J. A. Pietenpol, "PIK3CA mutations in androgen receptor-positive triple negative breast cancer confer sensitivity to the combination of PI3K and androgen receptor inhibitors," *Breast Cancer Res.*, vol. 16, no. 4, p. 406, 2014.
52. R. Arsenic, A. Lehmann, J. Budczies, I. Koch, J. Prinzler, A. Kleine-Tebbe, C. Schewe, S. Loibl, M. Dietel, and C. Denkert, "Analysis of PIK3CA mutations in breast cancer subtypes," *Appl. Immunohistochem. Mol. Morphol.*, vol. 22, pp. 50–56, Jan 2014.
53. S. Loibl, G. von Minckwitz, A. Schneeweiss, S. Paepke, A. Lehmann, M. Rezai, D. M. Zahm, P. Sinn, F. Khandan, H. Eidtmann, K. Dohnal, C. Heinrichs, J. Huober, B. Pfitzner, P. A. Fasching, F. Andre, J. L. Lindner, C. Sotiriou, A. Dykgers, S. Guo, S. Gade, V. Nekljudova, S. Loi, M. Untch, and C. Denkert, "PIK3CA mutations are associated with lower rates of pathologic complete response to anti-human epidermal growth factor receptor 2 (her2) therapy in primary HER2-overexpressing breast cancer," *J. Clin. Oncol.*, vol. 32, pp. 3212–3220, Oct 2014.
54. K. A. Hoadley, C. Yau, D. M. Wolf, A. D. Cherniack, D. Tamborero, S. Ng, et al., "Multiplatform analysis of 12 cancer types reveals molecular classification within and across tissues of origin," *Cell*, vol. 158, pp. 929–944, Aug 2014.

Printed in the United States
By Bookmasters